Proceedings of the Conference on

60 YEARS OF YANG–MILLS GAUGE FIELD THEORIES

C. N. Yang's Contributions to Physics

Proceedings of the Conference on

60 YEARS OF YANG–MILLS GAUGE FIELD THEORIES

C. N. Yang's Contributions to Physics

Nanyang Technological University, Singapore, 25–28 May 2015

Editors

L. Brink
Chalmers University of Technology, Sweden

K. K. Phua
Nanyang Technological University, Singapore

World Scientific

Published by

World Scientific Publishing Co. Pte. Ltd.
5 Toh Tuck Link, Singapore 596224
USA office: 27 Warren Street, Suite 401-402, Hackensack, NJ 07601
UK office: 57 Shelton Street, Covent Garden, London WC2H 9HE

Library of Congress Cataloging-in-Publication Data
Names: Conference on 60 Years of Yang-Mills Gauge Field Theories
 (2015 : Nanyang Technological University) | Brink, Lars, 1945– editor. |
 Phua, K. K., editor. | Yang, Chen Ning, 1922– honouree.
Title: 60 years of Yang-Mills gauge field theories : C.N. Yang's contributions to physics /
 editors, Lars Brink (Chalmers University of Technology, Sweden),
 Kok-Khoo Phua (Nanyang Technological University, Singapore).
Other titles: Sixty years of Yang-Mills gauge field theories | Yang-Mills gauge field theories
Description: Singapore ; Hackensack, NJ : World Scientific Publishing Co. Pte. Ltd., [2016] | 2016 |
 Proceedings of the Conference on 60 Years of Yang-Mills Gauge Field Theories:
 C.N. Yang's Contributions to Physics, Nanyang Technological University, Singapore, 25–28 May 2015.
Identifiers: LCCN 2016011546| ISBN 9789814725545 (hardcover) |
 ISBN 9814725544 (hardcover) | ISBN 9789814725552 (pbk.) | ISBN 9814725552 (pbk.)
Subjects: LCSH: Yang-Mills theory--History--Congresses. | Gauge fields (Physics)--Congresses. |
 Quantum theory--Congresses. | Yang, Chen Ning, 1922---Congresses.
Classification: LCC QC174.52.Y37 C66 2015 | DDC 530.14/35--dc23
LC record available at http://lccn.loc.gov/2016011546

British Library Cataloguing-in-Publication Data
A catalogue record for this book is available from the British Library.

Copyright © 2016 by World Scientific Publishing Co. Pte. Ltd.

All rights reserved. This book, or parts thereof, may not be reproduced in any form or by any means, electronic or mechanical, including photocopying, recording or any information storage and retrieval system now known or to be invented, without written permission from the publisher.

For photocopying of material in this volume, please pay a copying fee through the Copyright Clearance Center, Inc., 222 Rosewood Drive, Danvers, MA 01923, USA. In this case permission to photocopy is not required from the publisher.

Typeset by Stallion Press
Email: enquiries@stallionpress.com

Printed in Singapore

Preface

In 1954 Chen Ning Yang and Robert Mills met at Brookhaven National Laboratory. They started to discuss an idea that at least Yang had had for some time. What happens if you extend Quantum ElectroDynamics to involve a non-Abelian gauge symmetry? The symmetry they had in mind was the isotopic symmetry in the strong interactions. Could this model be a model for the strong interactions? The first one to ask this question at least in print was Oskar Klein who in 1938 had considered a five-dimensional QED model and reduced it to four dimensions. By some ingenious steps he managed to get an $SU(2)$ gauge symmetry albeit not getting the fully correct interactions. This was presented at a conference in Poland at the eve of the second world war and in these troubled times no one including Klein himself continued this idea. Yang–Mills Models were in the air in the beginning of the 1950s but they were beautifully exposed first in the Yang–Mills paper.

When Yang then gave a seminar at his home institution, the Institute for Advanced Study, he was quickly attacked by Wolfgang Pauli, who had certainly entertained the idea himself. Pauli argued that the model leads to long-range interactions and all knew that the strong interactions are short-range. Robert Oppenheimer intervened and Yang could conclude his talk. In the paper Yang and Mills addresses this issue and say prophetically that there might be a mechanism by which the interaction indeed becomes short-range.

The idea did not really catch on. This was the time when the big accelerators started to produce large numbers of new particles, and the ideas about the strong interactions were not to search for a theory within quantum field theory. Also the issues about the weak interactions were quite confusing and many ideas were put forward to get a grip on the weak interactions. It was clear that Fermi's old model from 1934 would not do the job. In 1956 Yang together with T. D. Lee put forward the suggestion that in the weak interactions parity is not conserved. This was a revolutionary idea that few believed but C. S. Wu and others set up experiments to test it. Her experiment tested ^{60}Co and indeed it was found that there is a maximal violation of parity in the decay. This was a world sensation and the paper was submitted on January 15, 1957. Later that year Yang and Lee were awarded the Nobel Prize in physics.

The idea about non-Abelian gauge theory as the theory for short-range interactions stayed on even though the main bulk of fundamental physicists worked on other problems. A fundamental new insight came with Yoichiro Nambu's discovery of spontaneous symmetry breaking and its consequences. In his remarkable explanation of superconductivity he used essentially QED coupled to phonons and showed that although the BCS vacuum does not conserve charge, gauge invariance is conserved although in a non-linear fashion. The Meissner effect looks just like QED with a mass term, i.e. as a theory with short-range forces. Nambu went on to use his knowledge about spontaneous symmetry breaking on pion physics understanding the low mass of the pion as an effect of the symmetry breaking, in a way explaining why nuclear physics is working.

One aspect of the spontaneous symmetry breaking was the insight that a spontaneously broken theory must involve a massless scalar particle, the Nambu–Goldstone boson. No such particle had been discovered and few people believed that the solution for a theory with short-range interactions could be harbored within gauge theories. In 1964 Robert Brout and François Englert and Peter Higgs found, however, that a gauge theory with a spontaneously broken symmetry can have a short-range interaction with no massless scalar particle.

Remarkably nothing happened for some three years until Steven Weinberg and Abdus Salam conjectured that the weak and the electromagnetic forces could be unified into a Yang–Mills theory with an $SU(2) \times U(1)$ symmetry spontaneously broken to a $U(1)$ symmetry. The gauge symmetry had previously been proposed by Sheldon Glashow and by Abdus Salam and John Ward, but it needed the spontaneous symmetry breaking to acquire short-range forces. Again nothing really happened. Few took notice and the authors themselves worked on other problems.

In 1971 a young Dutch student Gerhard 't Hooft shocked the world when he announced that the Yang–Mills theories are renormalizable even if the theory has a spontaneously broken symmetry. In the coming years he finished the proof with his advisor Martinus Veltman. This really started a revolution. Finally we had a consistent quantum field theory for the weak interactions. Over the years it also became proven that the particular gauge symmetry proposed was indeed the correct one.

What about the strong interactions? Murray Gell-Mann and George Zweig has introduced quarks as the ultimate constituents of the hadrons already back in 1964. One year later Nambu with Moo-Young Han suggested that the strong interactions should be a Yang–Mills theory with an $SU(3)$ gauge symmetry but they had quarks that had integral charges. However, it was not until David Gross and Frank Wilczek and David Politzer in 1973 found that non-Abelian Yang–Mills theories are asymptotically free, i.e. that the effective force goes logarithmically to zero at high energies making the quarks essentially free in a hadron, that a meaningful Yang–Mills theory could be suggested. Gross and Wilczek argued that a Yang–Mills theory with quarks in a triplet with the gauge group $SU(3)$ as Han and Nambu had suggested should be the correct theory for the strong interactions. This model had previously been suggested by Gell-Mann and Harald Fritzsch among several possibilities. Gell-Mann coined the word "QuantumChromoDynamics", QCD, for this theory. Over the years it has been amply proven that this is indeed the theory that governs the the interactions among quarks.

The Yang–Mills theories for the strong, weak and electromagnetic forces became the Standard Model of Particle Physics and over the years it has been proven to be the correct theory at the energies that we can measure with the present-day accelerators to a remarkable precision. There is no doubt that it is the correct theory at these energies. Over the years it has also been abundantly clear that the non-Abelian gauge theories, the Yang–Mills theories, are the only consistent relativistically invariant quantum field theory in four space–time dimensions. There are also strong connections to gravity and in some sense gravity can be seen as a certain kind of "square" of Yang–Mills theory. This is a remarkable journey from the first attempts by Yang and Mills that few believed in.

The Yang–Mills theories have also found other avenues to travel. In the hands of Edward Witten and many others it has been a tool to study many issues in mathematics. This is especially true for the supersymmetric Yang–Mills theories. Yang–Mills theories also play important roles in other fields of physics such as in condensed matter and in statistical physics.

Chen Ning Yang has been honored for many discoveries in physics. The crown jewel is the early Nobel Prize he got for the discovery of parity violations. However, when history is written about last century's greatest discoveries, the discovery of the Yang–Mills theory must be counted as his greatest discovery.

On May 25–28, 2015 a conference was held at the Institute of Advanced Study at Nanyang Technological University in Singapore to commemorate "60 Years of Yang–Mills Gauge Theories". Speakers included Chen Ning Yang, David Gross, Lars Brink, Michael Fisher, Ludwig Faddeev, Sau Lan Wu, Tai Tsun Wu, Tony Zee, Kazuo Fujikawa, George Sterman, Michael Creutz, George Savvidy, Henry Tye, Burt Ovrut, Tohru Eguchi, Yong-Shi Wu, Hong-Mo Chan, Sheung Tsun Tsou and others.

The conference also included interdisciplinary talks by Paul Chu on the search for higher T_c, Robert Crease on Yang–Mills for historians and Antti Niemi on protein folding. There were also reprises of C. N. Yang's contributions to physics by Yu Shi and Zhong-Qi Ma.

The conference was organized and chaired by K. K. Phua and Lars Brink. Over 189 participants attended from Europe, USA, India, China, Japan, Korea, Australia and Southeast Asia.

Conference Photos ix

x   Conference Photos

Chen Ning Yang

David Gross

Conference Photos xi

Michael Fisher

Ludwig Faddeev

Bruce McKellar

Paul Chu

Conference Photos xiii

Anthony Zee

George Sterman

Speakers at the Roundtable Discussion (from left): Henry Tye, Hishamuddin Zainuddin, Auttakit Chatrabhuti, Yifang Wang, Lars Brink, David Gross, Kok Khoo Phua and Ching Ray Chang.

At the Public Lecture (from left): Michael Fisher, David Gross, Guaning Su, Chen Ning Yang, Kok Khoo Phua and Lars Brink.

Contents

Preface		v
Conference Photos		ix
1.	The Future of Physics — Revisited *C. N. Yang*	1
2.	Quantum Chromodynamics — The Perfect Yang–Mills Gauge Field Theory *David Gross*	11
3.	Maximally Supersymmetric Yang–Mills Theory: The Story of $N=4$ Yang–Mills Theory *Lars Brink*	25
4.	The Lattice and Quantized Yang–Mills Theory *Michael Creutz*	41
5.	Yang–Mills Theories at High Energy Accelerators *George Sterman*	53
6.	Yang–Mills Theory at 60: Milestones, Landmarks and Interesting Questions *Ling-Lie Chau*	79
7.	Discovery of the First Yang–Mills Gauge Particle — The Gluon *Sau Lan Wu*	111
8.	Yang–Mills Gauge Theory and Higgs Particle *Tai Tsun Wu & Sau Lan Wu*	139

9. Scenario for the Renormalization in the 4D Yang–Mills Theory   161
L. D. Faddeev

10. Statistical Physics in the Oeuvre of Chen Ning Yang   167
Michael E. Fisher

11. Quantum Vorticity in Nature   199
Kerson Huang

12. Yang–Mills Theory and Fermionic Path Integrals   217
Kazuo Fujikawa

13. Yang–Mills Gauge Theory and the Higgs Boson Family   233
Ngee-Pong Chang

14. On the Physics of the Minimal Length: The Questions of Gauge Invariance   245
Lay Nam Chang, Djordje Minic, Ahmed Roman, Chen Sun & Tatsu Takeuchi

15. Generalization of the Yang–Mills Theory   269
G. Savvidy

16. Some Thoughts about Yang–Mills Theory   301
A. Zee

17. Gauging Quantum Groups: Yang–Baxter Joining Yang–Mills   315
Yong-Shi Wu

18. The Framed Standard Model (I) — A Physics Case for Framing the Yang–Mills Theory?   327
Chan Hong-Mo & Tsou Sheung Tsun

19. The Framed Standard Model (II) — A First Test Against Experiment   339
Chan Hong-Mo & Tsou Sheung Tsun

20. On the Study of the Higgs Properties at a Muon Collider   351
Mario Greco

21.	Aharonov–Bohm Types of Phases in Maxwell and Yang–Mills Field Theories *Bruce H. J. McKellar*	355
22.	Yang–Mills for Historians and Philosophers *R. P. Crease*	377
23.	Gauge Concepts in Theoretical Applied Physics *Seng Ghee Tan & Mansoor B. A. Jalil*	387
24.	Yang–Yang Equilibrium Statistical Mechanics: A Brilliant Method *Xi-Wen Guan & Yang-Yang Chen*	399
25.	Chern–Simons Theory, Vassiliev Invariants, Loop Quantum Gravity and Functional Integration Without Integration *Louis H. Kauffman*	415
26.	The Scattering Equations and Their Off-Shell Extension *York-Peng Yao*	443
27.	Feynman Geometries *Sen Hu & Andrey Losev*	453
28.	Particle Accelerator Development: Selected Examples *Jie Wei*	473
29.	A New Storage-Ring Light Source *Alex Chao*	487
30.	New Contributions to Physics by Prof. C. N. Yang: 2009–2011 *Zhong-Qi Ma*	499
31.	Brief Overview of C. N. Yang's 13 Important Contributions to Physics *Yu Shi*	505

The Future of Physics — Revisited*

C. N. Yang
Tsinghua University, China

I

In April 1961 there was a big Centennial Celebration at MIT. That was when science and technology were making unprecedented progress for mankind and when the United States had just inaugurated her young and ambitious new President. Naturally it was a proud, joyous, even intoxicating Celebration. At the week-long event there was a panel discussion on "The Future of Physics", chaired by Francis Low, with four speakers in the following order: Cockcroft, Peierls, Yang and Feynman. It was originally understood that the talks were to be published by MIT, but somehow that never happened. Much later the talks by Cockcroft and Peierls were summarized by Schweber in his 2008 book "Einstein and Oppenheimer". My talk and Feynman's were published, respectively in 1983[1] and in 2005.[2]

In my talk I said,

> "But since there seems to be too ready a tendency to have boundless faith in a "future fundamental theory", I shall sound some pessimistic notes. And in this Centennial celebration, in an atmosphere charged with excitement, with pride for past achievements and an expansive outlook for the future, it is perhaps not entirely inappropriate to interject these somewhat discordant notes."

I then argued that to reach the present level of understanding of field theory, according to Wigner's counting, one must penetrate four levels of physical concepts formed out of experiments. Furthermore to reach deeper levels of penetration will become more and more difficult.

> "Here physicists are handicapped by the fact that physical theories have their justification in reality. Unlike the mathematicians, or the artists, physicists cannot create new concepts and construct new theories by free imagination."

---

*Based on talk given at the Conference on 60 Years of Yang–Mills Gauge Field Theories, 25–28 May 2015, NTU, Singapore.

Editorial comment: We think this is a very interesting article. For the convenience of the readers, we reprint below the talks by Yang and by Feynman at that 1961 panel (App. A and App. B).

I was followed by Feynman, who began as follows:

> "As I listened to Professor Cockcroft, Professor Peierls and Professor Yang, I found that I agreed with almost everything they said. (But) I don't agree with Professor Yang's idea that the thing is getting too difficult for us. I still have courage. I think that it always looked difficult at any stage."
>
> "One possibility is that a final solution will be obtained. I disagree with Professor Yang that it's self-evident that this is impossible."
>
> "What I mean by a final solution is that a set of fundamental laws will be found, such that each new experiment only results in checking laws already known."

Feynman was one great intuitive theoretical physicist of my generation. Reading these passages today, I wonder

(1) what type of "final solution" he had in mind in 1961, and
(2) whether he still held such very optimistic views later in his life.

II

What have we learned in the fifty odd years since that 1961 panel?
A lot. Through very intensive collaborative efforts between theorists and experimentalists important conceptual developments were proposed and verified:

- One specific model of Symmetry Breaking
- Electroweak Theory
- Renormalizability of non-Abelian Gauge Theory
- Asymptotic Freedom and QCD

Capping these developments was the dramatic experimental discovery in 2012 of the Higgs particle.

We have now a workable

standard model, a **SU(3) × SU(2) × U(1) gauge theory**.

Thus in these fifty odd years since 1961, we have reached *one more layer* of physical concepts, which is based on *all* the previous layers *plus* a large number of very large experiments.

III

Are there additional layers of physical concepts deeper than those that we have reached so far? *I believe yes, many more.*

When can we expect to reach the next level? *I believe in the distance future, if ever.*

Why are you so pessimistic? *I am not pessimistic, I am just realistic.*

References

1. C. N. Yang, *Selected Papers* (Freeman, 1983), p. 319; E. P. Wigner, *Proc. Amer. Phil. Soc.* **94**, 422 (1950) [Reprinted below as Appendix B].
2. M. Feynman (ed.), *Perfectly Reasonable Deviations from the Beaten Track*: *The Letters of Richard P. Feynman* (Basic Books, 2005), Appendix III [Reprinted below as Appendix A].

Appendix A

The Future of Physics[‡]

Richard P. Feynman

As I listened to Professor Cockcroft, Professor Peierls and Professor Yang, I found that l agreed with almost everything they said. I don't agree with Professor Yang's idea that the thing is getting too difficult for us. I still have courage. I think that it always looked difficult at any stage. On the other hand, I agree, as you will see, with something about this pessimism, and I don't think that I can add anything sensible to anything that the other speakers said. So in order to proceed, I have to add something that is not sensible, and if you will excuse me therefore, I am going to try to say something quite different than what they said.

First of all, to make the subject not infinite, I am going to limit myself very much and discuss only the problem of the discovery of the fundamental laws in physics — the very front line. If I were talking about that which is behind the front line, things like solid-state physics, and other applications of physics and so on, I would say very different things. So please appreciate this limitation of the discussion.

I do not think that you can read history without wondering what is the future of your own field, in a wider sense. I do not think you can predict the future of physics alone with the context of the political and social world in which it lies. If you want to predict, as Professor Peierls does, the physics a quarter of a century in the future, you must remember that you are trying to predict the physics of 1984.

[‡]As published in The Technology Review, 1961–1962. Extracted from "Perfectly Reasonable Derivations from the Beaten Track: The Letters of Feynman" (Basic Books, 2005), Appendix III.

The other speakers seem to want to be safe in their predictions, to they predict for 10, perhaps 25, years ahead. They are not so safe because you will catch up with them and see that they were wrong. So, I am going to be really safe by predicting 1,000 years ahead.

What we must do according to the method of prediction used by the other speakers is to look at the physics of 961 and compare it to the present, 1961. We must compare the physics even a century before the age when Omar Khayyam could come out the same door as in he went, to the physics of today as we open one door after the other and see rooms with treasures in them, and in the backs of the rooms five or six doors obviously going into other rooms with even greater treasures. This is a heroic age. This is an exciting time of very vital development in the fundamental physics and the study of the fundamental laws. It is not fair to compare it to 961, but to find another heroic age in the development of physics, the age of, perhaps, Archimedes, of Aristarchus; say, the third century B.C. Add 1,000 years and you find the future of their physics: the physics of the year 750! The future of physics depends on the circumstances of the rest of the world and is not merely a question of extrapolation of the present rate of progress into the future. If I go a thousand years, I have a difficult problem. Is it possible that it's all going to be over?

One of the most likely things, from a political and social point of view, is that we have soon a terrific war and a collapse. What will happen to physics after such a collapse? Will it recover? I would like to suggest that the physics, fundamental physics, may possibly not recover, and to give some arguments as to why not.

In the first place it is very likely that if there were sufficient destruction in the Northern Hemisphere the high-energy machines, which seem to be necessary for further research, would become inoperative. The machines themselves may be destroyed, the electrical power to operate them may be unavailable, and the industrial technology needed to repair or maintain them may no longer exist, at least for a while. Experimental physics techniques are the quintessence of our technological and industrial abilities, and so they must suffer some temporary setback.

Can physics slide back temporarily and then recover? I don't think so. Because, in order to have this heroic age an exciting one, one must have a series of successes. If you look at the grand ages of different civilizations, you see that people have an enormous confidence in success, that they have some new thing that is different, and that they are developing it by themselves. If one were to slide back, you would find for a while, no great successes. You would be doing experiments that were done before. You would be working on theories that "the ancients" knew very well. What could result would be a lot of mouthing and philosophizing; a great effort to do the physics in the sense that one should do it to be civilized again, but not really to do it. To write, instead, commentaries, that disease of the intellect, which appears in so many fields. Physics is technically too hard to recover immediately. There would be practical problems at that time that would occupy the attention of intelligent people.

The difficulty is that there would be no fun in it. The new discoveries wouldn't come for a while. The other feature is that it would not be useful. No one has yet thought of a use of the results of the experiences we have with the high-energy particles. And finally, it is possible that antagonism is produced by the terrible calamity; there might be a universal antagonism toward physics and physicists as a result of the destruction which people might blame on the scientists who made it possible. Another thing to remember is that the spirit of research may not build itself up again because this spirit is concentrated in the Northern and advanced industrial countries, and this spirit does not exist fully in the other countries.

Well, I said 1,000 years. Maybe in that much time there will be another renaissance. What kind of machinery could there be for a recovery from this thing? (I said I wouldn't talk about anything sensible. I can't.) Some success somewhere must be the cause of a new renaissance. Where will this success lie? Perhaps in other fields. Perhaps in some other field than physics one would find a new age developing a success above "the ancients" and then getting a new confidence and growing. When this grows, it can pass its enthusiasm to physics. Or perhaps there would be a new aspect to physics, some other point of view, or some other completely different thing. That I cannot tell.

Another interesting possibility is that the renaissance may lie in some nation or people discovering a success by using a scientific attitude as a kind of morality, in society, government and business. You know what I mean — when someone says something, looking for what it is that they are saying, not why they are saying it. Propaganda would be a dishonest thing, and no one would pay attention to someone who would say something not for the content of the idea they want to get across, but because they want to show that they are big, or good, or some such reason. It is possible that if any success results from using such a scientific attitude, a country would be encouraged to go on from this success in its society and develop a re-interest in the scientific problems.

Well, now let's take the opposite view. Suppose there is no collapse. How, I don't know, but suppose there is no collapse. Then what? Suppose we can imagine a society somewhat like our own continuing for a thousand years. (Ridiculous!) What would happen to fundamental physics, the fundamental problems, the study of the laws of physics?

One possibility is that a final solution will be obtained. I disagree with Professor Yang that it's self-evident that this is impossible. It has not been found yet, but if you were walking through a building to get from one side to the other, and you hadn't yet reached the door, you could always argue, "Look, we've been walking through this building, we haven't reached the door, therefore there is no door at the other end." So it seems to me that we are walking through a building and we do not know whether it is a long infinite building, or a finite building, so a possibility is that there is a final solution.

What I mean by a final solution is that a set of fundamental laws will be found, such that each new experiment only results in checking laws already known, and it gets relatively more and more boring as we find that time after time nothing new is discovered that disagrees with the fundamental principles already obtained. Of course, attention would then go to the second line about which I am not speaking. But the fundamental problem will have been solved.

I would say that one thing that would happen if such a final solution were found, would be the deterioration of a vigorous philosophy of science. It seems to me that the reason that we are so successful against the encroachment of professional philosophers and fools on the subject of knowledge, and the way of obtaining new knowledge, is that we are not completely successful in physics. We can always say to such people, "That was very clever of you to have explained why the world just has to be the way we have found it to be so far. But what is it going to look like to us tomorrow?" Since they are absolutely unable to make any predictions, we see that their philosophy does not have real understanding of the situation. But if the solution is all present, how many people are going to prove it had to be four dimensions, it just had to be this way because of such and such and so forth. And so the vigor of our philosophy, which is a vigor which conies from the fact that we are still struggling, I think that may fail.

What other possibilities are there? Suppose that the building we are walking through is infinite, as Professor Yang feels. Then there will be a continual exciting unfolding. We will rush through this house, one door after another, one treasure after another. A thousand years! Three unfoldings in sixty years is fifty unfoldings in a thousand years. Is the world going to have fifty exciting revolutions of our basic physical ideas? Is there that much treasure in fundamental physics? If there is, it will become somewhat boring. It will be boring to have to repeat it twenty times, this fact that things change always when you look deeper. I do not believe that it can last 1,000 years of active investigation. Well, if it doesn't stop (I mean if you can't get the final solution), and if I can't believe that it will keep on being excitingly developed for fifty revolutions, what else is there?

There is another possibility, and that is, that it will slow up. The questions will become more difficult, How will it look then? The strong couplings are analyzed, the weak couplings partially analyzed, but there are still weaker couplings that are harder to analyze. To obtain useful experimental information has become extremely difficult because the cross-sections are so tiny. Data comes in slower and slower. The discoveries are made more and more slowly, the questions get harder and harder. More and more people find it a relatively uninteresting subject. So it is left in an incomplete state with a few working very slowly at the edge on the question of what is this third-order tensor field that has a coupling constant 10 to the power of -30 times smaller than gravity?

It is possible, of course, that what we call physics will expand to include other things. I believe, for example, that physics will expand, just as Professor Peierls says, into the studies of astronomical history and cosmology. The laws of physics, as we presently know them, are of this kind. Given the present condition, what is the future? The laws are given by differential equations in time. But there must be another problem: what determines the present condition? That is, what is the whole history of the development of the universe? One way to see that this may someday be a part of physics and will not always be called astronomical history is to note there is at least a possibility that the laws of physics change with time. If the laws of physics change with absolute time, then there will be no way to separate the problems of formulating the laws and of finding the history I think that it is very likely that cosmological problems will be enmeshed in physics.

Finally, I must remind you that I limited myself to talking about the future in fundamental physics. I'd say that there will be an important return from the front line into the applications and the development of the consequences of the laws. This will be a very exciting thing, and I would say quite different things about its future than I would say about the future of the fundamental laws.

We live in a heroic, a unique and wonderful age of excitement. It's going to be looked at with great jealousy in the ages to come. How would it have been to live in the time when they were discovering the fundamental laws? You can't discover America twice, and we can be jealous of Columbus. You say, yes, but if not America, then there are other planets to explore. That is true. And if not fundamental physics, then there are other questions to investigate.

I would summarize by saying that I believe that fundamental physics has a finite lifetime. It has a while to go. At the present moment it is going with terrific excitement, and I do not want to retire from the field at all. I take advantage of the fact that I live in the right age, but I do not think it will go on for a thousand years.

Now, to finish, I would summarize two points. First, I did not talk about applied physics or other fields, about which I would give a considerably different talk. And second, in these modern times of high-speed change, what I am forecasting for a thousand years will probably occur in a hundred.

Thank you.

Appendix B

The Future of Physics[†]

C. N. Yang

In the last four or five years much effort and attention have been devoted by theoretical physicists to the analytic continuation from physically observable experience into unphysical regions. In particular, it has been tried by extrapolation to study properties of the singularities in the unobserved region. Such attempts have been beset from the beginning with great difficulties. But interest in them maintained. In a similar spirit, this morning we try to pursue a parallel approach: By extrapolation

---

[†]Panel discussion by C. N. Yang at the MIT Centennial Celebration, April 8, 1961.

we want to look beyond our past experience and learn something about the so far unobserved future development of physics. We cannot hope for any real success in this pursuit. But I believe we all agree the attempt is highly interesting.

By all standards the achievements in physics in the twentieth century so far have been spectacular. Whereas at the turn of the century the atomic aspect of matter was just emerging as a new subject of study, we have today progressed dimension-wise by a refinement of a factor of a million: from atomic dimensions to subnuclear dimensions. Energywise the progress is even more impressive: from a few electron volts to multibillion volts. And the power and ingenuities of the experimental techniques have fully kept pace with the increased depth of the physicists' inquiries. The influence and impact brought about by the advances in physics upon other sciences — upon chemistry, astronomy, and even upon biology — have been important beyond description. Similar influence upon technology, upon human affairs, has been so preoccupying in the postwar years as to need no further emphasis here.

But it is not in these influences that the glory of physics and the heart of the physicist lie. It is not even in the continued enlargement of the domain of physical experimentation, important as it is, that the physicists take the greatest pride and satisfaction. What makes physics so unique as an intellectual endeavor lies in the possibility of the formulation of concepts out of which, in the words of Einstein,[1] a "comprehensive workable system of theoretical physics" can be constructed. Such a system embodies universal elementary laws, "from which the cosmos can be built up by pure deduction."

Judged by such a lofty and rigorous standard, the sixty years of the twentieth century stand out as nothing short of heroic; for besides the numerous important discoveries that *widened* our knowledge of the physical world, this period has witnessed not one, not two, but three revolutionary changes in physical concepts: the special relativity, the general relativity and the quantum theory. Out of these conceptual revolutions were constructed *deepened*, comprehensive and unified systems of theoretical physics.

Endowed with such a distinguished heritage of the recent past, what are the prospects for its future?

Surely the rapid widening of knowledge will continue, both in what Professor Peierls refers to as the foundation of physics and in what he calls behind the front line.

In the former our present knowledge is sufficient to enable us to say with some certainty that great clarification will come in the field of weak interactions in the next few years. With luck on our side we might even hope to see some integration of the various manifestations of the weak interactions.

Beyond that we are on very uncertain grounds. To be sure we can already formulate a number of questions, the answers to which seem to us at this moment to be crucial: How should one treat a system with an infinite degree of freedom? Is the continuum concept of space time extrapolatable to regions of space 10^{-14} cm to 10^{-17} cm, and to regions smaller than 10^{-17} cm? What are the bases of the invariance under charge conjugation, and the invariance under isotopic spin rotation, both of which, unlike space–time symmetries, are known to be violated? What is the unifying basis of the strong, the electromagnetic and the weak interactions? What is the role of the gravitational field relative to all these? The list can go on, but we are not even sure that these questions are meaningful as we here phrased them: in fact much of the progress in physics had developed from the very recognition of the meaninglessness of some previously asked questions.

Of one thing, however, we can be sure. The accumulation of knowledge will proceed at a very rapid pace. We need only remind ourselves that not so very long ago, the time scale of discoveries in physics was measured in years, if not in decades. The Michelson–Morley experiment, for example, was first performed in 1881, then repeated with greater precision in 1887. To explain the negative result of the experiment Fitzgerald invented a contraction hypothesis in 1892, and then Lorentz invented the Lorentz transformation in 1902, culminating in Einstein's special theory of relativity in 1905. Imagine that Michelson's first experiment were done today!

The general awakening of mankind to the importance of science and the amazing ingenuity of the human mind at technological creativity virtually ensure for us further quickening of the pace of the experimental sciences.

But where shall we stand with respect to the "comprehensive system of theoretical physics" that was referred to a few minutes ago? Can we reasonably expect further success in the glorious tradition of the first sixty years of the twentieth century?

If it is difficult to locate singularities of functions by extrapolation, it is as difficult to predict revolutionary changes in physical concepts by forecasting. But since there seems to be too ready a tendency to have boundless faith in a "future fundamental theory," I shall sound some pessimistic notes. And in this Centennial celebration, in an atmosphere charged with excitement, with pride for past achievements and with an expansive outlook for the future, it is perhaps not entirely inappropriate to interject these somewhat discordant notes.

First, let us emphasize again that the mere accumulation of knowledge, while interesting, and beneficial to mankind, is nonetheless quite different from the aim of fundamental physics.

Second, the subject matter of subnuclear physics is already very remote from the direct sensory experience of mankind, and this remoteness is certainly going to increase as we delve into even smaller dimensions. A direct and graphic proof of this can be easily found in the growing physical size of the laboratories, the accelerators, the detectors and the computers.

What we call an experiment today consists of elaborate operations with elaborate equipments. To make any sense at all of the results of an experiment, concepts have to be formulated on all levels between our direct sensory experience and the actual experimental arrangement. The difficulty inherent in this state of affairs is as follows. Each level of concepts is connected with and in fact built upon the previous levels. When inadequacies manifest themselves, one must reach for greater depth by examining the whole complex of previous concepts. The difficulty of this task rapidly diverges with the depth of the considerations, much like in chess playing, it becomes increasingly difficult to practice always examining one more move ahead as one's skill improves.

According to Wigner's counting,[2] to reach our present investigations in field theory one must penetrate at least four levels of concepts. The details of the counting may be a subject of discussion, but there is no denying that what we envisage as a construction of a deeper and more comprehensive theoretical system represents at least one more level of penetration. Here physicists are handicapped by the fact that physical theories have their final justification in reality. Unlike the mathematicians, or the artists, physicists cannot create new concepts and construct new theories by free imagination.

Third, Eddington[3] has once given the example of a marine biologist who uses a net with 6 inch holes and who after careful and long studies formulates the law that all fish are larger than 6 inches. If this imaginary example sounds ridiculous, we can easily find examples in modern physics where because of the complexity and indirectness of the experiments, it has happened that one does not realize the selective nature of one's experiments, a selection that is based on concepts which may be inadequate.

Fourth, in the day-to-day work of a physicist it is very natural to implicitly believe that the power of the human intellect is limitless and the depth of natural phenomena finite. It is of course useful, or as is sometimes said, healthy, to have faith so as to have courage. But the belief that the depth of natural phenomena is finite is inconsistent and the faith that the power of the human intellect is limitless is false. Furthermore, important considerations have also to be given to the fact that psychological and social limitations on the development of the creative ability of each individual may be effectively even more stringent than natural limitations.

Having voiced these cautionary remarks, we must ask are they relevant to the development of physics, say, in the remaining forty years of this century? We cannot know the answer to this question now, but let us hope that it is in the negative.

References

1. A. Einstein, *Essays in Science* (Philosophical Library, New York, 1934).
2. E. P. Wigner, *Proc. Amer. Phil. Soc.* **94**, 422 (1950).
3. A. S. Eddington, *The Philosophy of Science* (MacMillan, New York, 1939).

Quantum Chromodynamics — The Perfect Yang–Mills Gauge Field Theory

David Gross

University of California, Santa Barbara, CA 93106, USA

David Gross: My talk today is about the most beautiful of all Yang–Mills Theories (non-Abelian gauge theories), the theory of the strong nuclear interactions, Quantum Chromodynamics, QCD. We are celebrating 60 years of the publication of a remarkable paper which introduced the concept of non-Abelian local gauge symmetries, now called the Yang–Mills theory, to physics. In the introduction to this paper it is noted that the usual principle of isotopic spin symmetry is not consistent with the concept of localized fields. This sentence has drawn attention over the years because the usual principle of isotopic spin symmetry is consistent, it is just not satisfactory. The authors, Yang and Mills, introduced a more satisfactory notion of local symmetry which did not require one to rotate (in isotopic spin space) the whole universe at once to achieve the symmetry transformation. Global symmetries are thus are similar to 'action at a distance', whereas Yang–Mills theory is manifestly local.

Symmetry has been one of the main themes of 20th century science, and especially one of the main themes of Professor Yang's research. Symmetries have played such an important role because they organize the laws of nature, much as physical laws organize our observations of nature. The laws of nature summarize the regularities of events that are independent of initial conditions. Symmetries organize the laws of nature themselves. They summarize the regularities of the laws of nature that are independent of specific dynamics. Symmetry principles provide a structure and coherence to the laws of nature just as the laws of nature provide a structure and coherence to the set of events we use to explore nature and test our ideas. By now we recognize that we have two general types of symmetries: global symmetries and local symmetries.

Global symmetries, the oldest to be recognized and used in physics, are regularities of the laws of nature and are formulated in terms of physical events. The application of a symmetry transformation yields a different physical situation, but the laws of physics are invariant under that transformation. That is what we mean by global symmetry. We can rotate our laboratory globally and the results of all experiments will remain unchanged, because the laws of nature are invariant under rotations. Rotating the laboratory does not change the experimental results and consequently we do not have to note in our papers at which angle our laboratory was situated. Similarly, we need not note where and when the experiment was carried out, as the laws of nature are unchanged by spatial and time translations. Associated with such continuous global symmetries are conservation laws. In the case of rotational invariance the ensuing conservation law is that of angular momentum; whereas, space and time translation invariance lead to momentum and energy conservation.

Local or gauge symmetries are more abstract, more subtle. They are formulated only in terms of the laws of nature themselves. The application of a gauge symmetry transformation merely changes our description of the same physical situation. It does not lead to a different physical situation. Local (gauge) symmetry first appeared, unbeknown to the inventors, in Maxwell's theory of electrodynamics. Maxwell's electrodynamics can be summarized as:

> The electromagnetic field, $\mathbf{F}_{\mu\nu}$ is defined in terms of an auxiliary vector potential, $\mathbf{A}_\mu$: $\mathbf{F}_{\mu\nu} = \partial_\mu \mathbf{A}_\nu - \partial_\nu \mathbf{A}_\mu$; the dynamical law is Maxwell's equation: $\partial^\mu \mathbf{F}_{\mu\nu} = \mathbf{J}_\nu$, where $\mathbf{J}_\nu$ is the conserved electric current; and the field strength obeys the Bianchi condition: $\varepsilon^{\alpha\beta\gamma\delta}\partial_\beta \mathbf{F}_{\gamma\delta} = 0$.

These are Maxwell's equations in modern form. There was a great debate in the latter part of the 19th century as to what the meaning of the auxiliary vector potential, $\mathbf{A}_\mu$, was, since it does not really enter into the equations of motion and is not an observable. At least it was not thought to be an observable in any sense. The equations of motion, and the field strength, are invariant under the local gauge symmetry transformation: $\mathbf{A}_\mu \to \mathbf{A}_\mu + \partial_\mu \lambda(x)$. Under this transformation, the equations of motion as well as the observables, such as $\mathbf{F}_{\mu\nu}$ or $\mathbf{J}_\nu$, are invariant. But associated with this symmetry are no new charges. Emmy Noether proved that once you have any continuous symmetry of the Lagrangian, whose variation yields the equations of motion, there will be associated a conserved current. So given a local symmetry of motion: $\mathbf{A}_\mu \to \mathbf{A}_\mu + \partial_\mu \lambda(x)$, there should exist a conserved current, labeled by the field $\lambda(x)$. There should exist an infinite number of conserved charges, and indeed there are. Following Noether's prescription, you can construct a conserved current, $\mathbf{J}_{(\lambda)\mu} = \mathbf{F}_{\mu\nu}\partial^\nu \lambda(x)$, one for each function $\lambda(x)$. Thus we have an infinite number of conserved currents and supposedly an infinite number of conserved charges; but, if you actually construct the charges, they are integrals of the total divergence of the electric field times $\lambda(x)$. Therefore, these charges, for all but λ's that approach a constant at infinity, vanish. There are no new charges in addition to the electric charge. In Quantum Mechanics, a local gauge symmetry produces no new operators that would generate new symmetries of the Hamiltonian or transformations of the states in the Hilbert space. So local gauge symmetries are symmetries of the Lagrangian, symmetries of the action and of the equations of motion; but, they do not produce new transformations of the physical Hilbert space, they do not produce new transformations of physical objects. They describe a redundancy in our description of our laws of nature. Gauge Symmetry in a sense equals Redundancy.

For many years, before quantum mechanics, people asked: Is gauge symmetry a real symmetry? Is the vector potential a real physical object? Wigner famously wrote:

"This gauge symmetry is, of course, an artificial one, similar to that which we could obtain by introducing into our equations the location of a ghost. The equations must then be invariant with respect to changes of coordinates of that ghost. One does not see, in fact, what good the introduction of the coordinate of the ghost does."

Wigner was not convinced that local gauge symmetries expressed some real feature of the underlying physics. It was Einstein who first realized that there was something wrong about insisting on global transformations in the case of space–time transformations of special relativity, as Yang and Mills later pointed out in the case of isotopic spin. Special relativity was based on global Lorentz-invariance wherein you transform observables from one inertial frame to another, but the theory of general relativity was based on the local reparametrization invariance of space–time. This year we celebrate the 100th anniversary of Einstein's equations which embody that local diffeomorphism invariance and give rise to general relativity. Around the time that quantum mechanics appeared, Weyl introduced what he called, for the first time, a "gauge symmetry," a local scaling symmetry of space–time, in the hope of obtaining a geometrical derivation of electricity and magnetism. This proposal did not exactly work, but London explained that you could give a gauge interpretation of electromagnetism as a complex phase symmetry, which was quite natural in quantum mechanics which had introduced complex wave functions to describe charged particles. Yang and Mills then extended this gauge symmetry to non-Abelian gauge groups, where even to write down the equations of motion in a local form one needs to introduce the gauge potential satisfying a non-Abelian gauge symmetry. The fact that you need this redundant field and the consequent local gauge symmetry to express the laws of physics in a manifestly local fashion actually goes all the way back to Maxwell. He realized that in order to understand the local nature of electromagnetism, it was necessary to introduce the gauge potential. Then in quantum mechanics it turns out that in order to describe the Hamiltonian equation of motion of charged particles, in a local fashion, you also need to use the gauge potential. So this redundancy, which is one way of regarding local gauge invariance, is a necessary condition needed to formulate the laws of physics in a manifestly local fashion.

But what are the implications of this gauge symmetry? The gauge transformation do not transform the laboratory or the physical observables. The symmetry does not lead to new conservation laws. It is a redundant way of describing local physics. But, as Yang has emphasized, gauge symmetry dictates the nature of the interactions. The fundamental laws of Nature (with the exception of the "Higgs sector") are dictated by gauge symmetry. In addition to Einstein's gauge theory of space–time dynamics, local gauge symmetries are the basis of the standard model, in which we have three separate local gauge theories that describe the three sub-atomic and sub-nuclear forces, and govern the structure of atoms and of nuclei and of sub-nuclei and of just about everything. This standard model or theory has been tested with extraordinary precision. It seems to apply to just about everything that we observe and measure, with the exception of one force of nature, whose simple form has only recently been confirmed, the "Higgs sector." Unless you believe in supersymmetry, the Higgs scalar does not fit naturally into the standard model and has to be added in with much fine tuning. The standard model gauge groups can be labeled: one, two, three; with the labels being the rank of the three gauge groups. Electromagnetism is based on an Abelian gauge group, with rank 1. Yang and Mills extended this Abelian symmetry to the non-Abelian case. These appear in the theory of the weak and the strong interactions, with not just the one electric charge of Maxwell, but two charges in the case of the weak and three charges in the case of the strong interactions.

Yang–Mills theory appeared in 1954. Why did it take so long for it to find its way into successful physics? It is because of the many problems that it posed, some of which were recognized by the authors of the famous paper and by its first critics. Indeed, as Lars Brink already noted, it was quite brave for Yang and Mills, in the face of such obvious problems, to proceed with such a bold hypothesis. The main problems in trying to apply Yang–Mills theory to the weak interactions were two-fold: first, the gauge mesons, the carriers of the force, the analogues of the quanta of light (photons), were according to the theory, massless. If the underlying gauge symmetry was a local version of isotopic spin, there must exist charged, massless vector particles. Massless long ranged forces, except for gravity and electromagnetism, have not been observed. But it was evident to Yang and Mills, and to their critics (such as Pauli), that given the underlying symmetry that was the basis of the theory, the masslessness of these particles was protected, and that there was no way to easily alter this property. Thus it seemed that nature clearly did not like this physical idea. The other problem was that the symmetries that one was promoting from global to local symmetries, such as isotopic spin or any of the other flavor symmetries, as we now call them, were not exact symmetries. Again, it makes absolutely no sense to explicitly break gauge invariance. In the 1950s there was little understanding of the phenomenon of spontaneously broken symmetries, symmetries that are broken by the nature of the vacuum state, not by the equations of motion. Broken symmetries are commonplace in physics. It is often the case that the energy landscape of a many body or relativistic quantum mechanical theory is such that symmetric state is an unstable state of higher energy than the many possible states which represent vacua with broken symmetry. The ground state in such case breaks the global symmetry. This happens in crystals, which break translational symmetry, or in magnetism, which breaks rotational symmetry. But in all these cases, where a continuous symmetry is broken, a severe problem arises; namely, for every broken global symmetry there must exist zero energy, long wavelength (Goldstone) modes, which in relativistic quantum mechanics correspond to massless particles. So attempts to apply Yang–Mills theory to the weak interactions gave rise to the problem of massless gluons or, if you try to break the underlying symmetry, massless Goldstone bosons. As you know both problems were solved in the early 1960s by Brout and Englert and Higgs (of course the BCS explanation of superconductivity was an earlier example of the spontaneous breaking of an Abelian gauge symmetry). This mechanism not only explains that when you consistently gauge a spontaneously broken symmetry, you both produce a mass for the carriers of the force (the W and the Z in the case of the weak interactions) as well as eliminate the massless Goldstone bosons. This simple perturbative mechanism has now been confirmed brilliantly at the LHC, about three years ago, when the Higgs particle was discovered, as hypothesized in the simplest version of standard model.

In the case of the strong interactions the problems were even more severe since it was not evident at all what the nuclear charges were, or what non-Abelian symmetry could be gauged. Originally again, it was thought to be flavor symmetry (SU(2) of isospin, or SU(3)). All such attempts to use flavor symmetry as a basis for a gauge theory of strong interactions failed. Carriers of the force would have been ρ mesons, which were certainly not massless; the flavor symmetries were not exact, and so on and so on. Also, in the case of the strong interactions, one could not do much anyway even if one could write down a consistent theory, since the forces were strong, one did not know how to handle them. Most important, the nuclear charges, which we now know are centered on the color quantum number of quarks and gluons, were totally hidden. It is not that the symmetry is broken, the color symmetry of the strong force is not broken. Rather the charges were dynamically confined or hidden inside nucleons, which are color singlets. Even more bizarre, no matter how hard one smashed hadrons together, one could never produce quarks. So, although it looked to some extent as if hadrons were made out of quark-like constituents — fractionally charged, half integer particles which had three color labels assigned to them — there was no evidence of physical quarks in Nature.

For me, the breakthrough in understanding the strong interactions came from the deep inelastic scattering experiments at SLAC, which discovered that the proton did look like it was made out of quarks. Quarks that were freely moving around, almost non-interacting when viewed at very short distances and at very short times. The scaling behavior of the cross sections that was observed was only understandable if the quarks interacted very weakly. Furthermore, the cross sections obeyed quark model sum rules and many of us were convinced that this was evidence that the proton is indeed made out of quarks. But then: how come these quarks which are bound together so strongly inside protons, are moving freely as seen by these experiments? The SLAC experiment was appropriately awarded the Nobel Prize for the discovery of quarks.

The freely moving behavior of quarks inside the proton, when viewed at short distances, is rather strange. By then we had learned enough about quantum field theory to understand that forces could look different at different scales, due to the physical effects that give rise to the renormalization of the charge. But all theories that had been studied up to then, such as the very successful quantum electrodynamics, exhibited dynamics familiar from the phenomenon of the screening of electric charges. It is natural that if you have a medium such as a solid, or for that matter the quantum mechanical vacuum, and you put a charge in it, the various dipoles in the material, or the virtual dipoles that make up the vacuum, will rearrange and screen the charge. Consequently, the electric force that an observer observes will be larger as you get closer and closer to the bare charge (less screening) and weaker at large distances (more screening). This is exactly the opposite of what was seen in the SLAC experiment where the force between the quarks decreases at short distances. In fact, this screening behavior of many-body systems in general, including quantum field theories, is just about true of all theories in four space–time dimensions. This is why there are so few consistent four-dimensional quantum field theories, as the only ones that seem to make physical sense are asymptotically free theories. This was the reason that Landau in 1960 reached the conclusion that QED is not consistent. If you try to calculate the bare electron charge in QED (defined at zero distance) it becomes infinite. Those infinities actually occur at a finite distance (the so-called Landau poles) and Landau concluded that local quantum field theory is inconsistent and must be abandoned, especially in the case of the strong force.

We discovered that non-Abelian gauge theories, Yang–Mills theories, have a very remarkable property, unique among renormalizable theories in four dimensions, of asymptotically approaching free field theory at short distance — Asymptotic Freedom. This phenomenon, could explain the scaling observed at SLAC, with calculable corrections, and that is why we proposed that Yang–Mills theory be used to construct a theory of the strong interactions — QCD. Anti-screening in QCD can actually be understood using a simple physical argument. Think, in magnetic terms, about the quantum vacuum as a medium containing virtual quarks and gluons. Note that the once you make the gauge symmetries, as Yang and Mills did, non-Abelian, the gluons, the carriers of the force, are themselves charged. In the case of QCD the gluons transform in the eight-dimensional adjoint representation of SU(3). So gluons are charged vector bosons with spin 1, and behave as paramagnets, permanent magnets. The vacuum is full of these and if you insert a test magnetic charge its magnetic field will be enhanced by these virtual magnets, leading to an increase of the force at large distances and a decrease at shorter distances, exactly the opposite of QED. Consequently, that the strong force, which is mediated by the chromodynamic non-Abelian gauge field set up between colored quarks, color being the charge, does not obey Coulomb's law as it would classically (which would of course lead to a finite ionization energy). In the complicated QCD vacuum with fluctuating fields, the flux of the field lines is squeezed to form flux tubes, leading to a force that remains constant as you pull the particles apart, and decreases as you push them together; producing both asymptotic freedom at short distances and confinement at large distances. Asymptotic freedom solved the problems with the quarks at both short and large distances. Asymptotic freedom explains why quarks are a good perturbative description of hadrons at short distances and the strong force that ensues at large distances explains why they are permanently confined inside large hadrons. So the key dynamical property of non-Abelian gauge theories is asymptotic freedom.

Detailed confirmation of the predictions of QCD took many, many years to obtain. Nowadays, the standard model as a whole, including QCD and its asymptotic freedom, has been tested with exquisite precision. To use such a theory to explain the properties of hadrons, the protons and the neutrons and the strange particles, is not easy as the strong force gets stronger and stronger at large distances and one must control the large distance structure of the vacuum. Lattice QCD remains, so far, the only tool for precise, controllable, first principle calculations of large scale hadronic properties. Lattice QCD has been enormously successful, reaching a new level of maturity due to Moore's Law and to theoretical ingenuity and I recommend to you the talk by Michael Creutz, later in this conference. But I do want to just point to a comparison of the light hadron spectrum of QCD as calculated by lattice gauge theory with measured values. We can now calculate, from first principles and with controllable errors, to one percent or better, the mass spectrum of all of the particles that were being discovered when I was a graduate student. To a theorist this is enormously pleasing.

Let me now justify the title of my talk and explain why QCD is a perfect theory. It is the first quantum field theory that possesses three remarkable features:

It has no infinities.
It has no adjustable parameters.
It has no new physics at short distances or at large energies.

This is somewhat of an exaggeration. First of all, it has no infinities. Infinite ultraviolet divergences plagued quantum field theory since its birth in the 1920s; so much so that the inventors of quantum field theory, Wigner, Dirac, etc. gave up on quantum field theory. They expected that it was fundamentally flawed and that some new revolution was needed to make sense of it. QCD has no ultraviolet divergences at all, there are no infinities. The local (bare) coupling, the quantum dynamical charge of the quark, vanishes. That is the essence of asymptotic freedom. The only way infinities ever appear is when we calculate Feynman diagrams, using perturbation theory the way we all have learned, and are simply due to the fact that one expresses physical observables, which always refer to some finite distance, in terms of the coupling that is defined at zero distance in a local theory. Now if you divide one meter by zero meter, you obviously get infinity, but this infinity has nothing to do with physics. These infinities never appear if you use the most rigorous, non-perturbative, definition of the theory, namely put the theory on a space–time lattice of lattice spacing **a**, and reduce the calculation of the partition function, say to a multidimensional (but finite) integral. The coupling only appears as the weight in this integral: $\exp[-S_{\rm YM}/g^2(\mathbf{a})]$; where $S_{\rm YM}$ is the Yang–Mills action, and the coupling constant, $g^2(\mathbf{a})$, is defined at the scale of the lattice spacing $\mathbf{a}$. No infinite quantities appear at all. One of course must take the continuum limit, $\mathbf{a} \to 0$. In this limit, in a fixed volume, the number of integrals gets large and larger, and one must take $g^2(\mathbf{a})$ to approach zero as calculated using asymptotic freedom: $g^2(\mathbf{a}) \to c/\ln(\mathbf{a}\Lambda)$, c is a (known) number and Λ the (arbitrary) QCD scale. After taking this limit one gets a finite answer. This is what people doing lattice QCD do to calculate the masses of the hadrons. Infinite quantities never appear, there are simply no divergences.

Second, there are no adjustable parameters. This is a bit of an exaggeration, as quarks have masses which have nothing to do with QCD, so we cannot calculate the quark masses but must take them as experimentally measured parameters of the theory. But they do not play a big or essential role in hadronic physics. The light quarks are very light, the heavy quarks are very heavy. To a very good approximation, and we can set the mass of the up and down (and even the strange) quarks to zero and set the mass of the heavy quarks to infinity (at which point they decouple) and we get a very good approximation to the real world. All masses are calculable then as pure numbers, in units of the QCD scale Λ. This is because in the original theory, as Yang and Mills pointed out in their original paper, no dimensional parameter appears. The classical theory is scale invariant. The coupling that defines this strength of the theory will then define what a physical scale of length is, because it varies with length. We need some parameter, some physical mass to define physics, but having chosen a particular mass or distance scale, then the coupling is fixed at that scale, and all mass ratios are calculable.

I love ask young students, in public lectures, what the origin of the proton mass is, and to point out to that this mass, which they have no idea where it comes from, mostly consists of confined energy. And this mass is calculable in terms of the one mass scale you need to introduce, say the scale where say the Yang–Mills coupling is equal to 1. All mass ratios are then pure numbers, which in principle, up to quark mass corrections, can be calculated with arbitrary precision. For the first time we have had a theory with no adjustable parameters.

Finally, every theory we have ever had in physics has contained in it, the seeds of its own destruction, an indication of some place where it will break down. This breakdown usually occurs (say for classical or quantum electrodynamics) at short distances. In QCD there is no new physics at short distances or arbitrary high energies. In fact, to the contrary, at higher and higher energies, QCD becomes simpler and perturbation theory more exact. We can in fact even control the divergent asymptotic expansion, which is what perturbation theory in any quantum field theory is, by going to shorter and shorter distances.

This ability to extrapolate QCD to arbitrary high energies has important and conceptual implications. It enables us to calculate the physics of the universe back to close to its beginning. It enables us to extrapolate the strong force to the point where it joins with the other forces. Together with the fact that gravity becomes strong at about the same Planck distance, this extrapolation provides the strongest clue that the standard model forces should be unified with gravity. And finally, it solves one of Dirac's famous large number problems, namely why is the ratio of the mass of a proton divided by the Planck mass, which is about equal to 10^{-19}, so small. QCD provides an explanation as to why this number is so small. It is simply that at the unification scale, the nuclear coupling, g^2_{Planck}, is a small number, say 1/25, similar to the electromagnetic coupling at that scale and then as you go to larger distances the force strengthens, and finally you get to a distance where the force between the quarks is strong enough to hold them together. That distance scale defines the size of the proton and thereby the mass of the proton, which is mostly the kinetic energy of the quarks and the gluons in this confining region of space. So you can roughly calculate the mass of the proton compared to the Planck mass and one gets: $M_{\text{Proton}}/M_{\text{Planck}} \sim \text{Exp}[-c/g^2_{\text{Planck}}]$, which is exactly of the right order of magnitude.

QCD is so perfect because it is the first example of a complete theory, where if you ignore all other forces in the world, has no adjustable parameters and no indication, within the theory, of a distance scale where it might break down. It has infinite bandwidth. Of course, in the real world, we would like to understand why is the underlying symmetry group given by SU(3), why are there quarks, and why do they have these masses. And, of course, we know that there are other forces of Nature including gravity. Thus we must go on.

Over the many years after the completion of the standard model, there have been extraordinary impressive tests of quantum chromodynamics as well as impressive calculations. The calculation of the running of the coupling has now been carried out to many loop orders and George Sterman, who will talk about the application of perturbative QCD to LHC physics, correct George?

George Sterman: Something like that.

David Gross: Something like that. These calculations and their comparison with experiment are absolutely essential. At the LHC we look for new physics, but new physics is bound to be rare — typically one new interesting event per billion old and boring events. In order to pick out one new event out of a billion events, you must understand those one billion events with enormous accuracy. The technology of calculating within QCD is just totally amazing. In another experimental development there is the fact that, in quantum chromodynamics, quarks and gluons can exist in diffcrent phases that we never dreamed of; not just the hadronic gas of confined quarks and gluons that we see at low temperature and low density, but a quark–gluon plasma whose properties are being explored at RHIC and at LHC, and perhaps fascinating new phases that might play a role in neutron stars or quark stars or high density nuclear matter.

Perhaps the most interesting new aspect of non-Abelian gauge theories, especially QCD, that has been revealed in the last 30 years is its connection to string theory. When you pull quarks apart, because of the properties of the vacuum, the chromodynamic flux between the quarks is squeezed into flux-tubes. Mesons look somewhat like fat strings and this is why string theory emerged from an attempt to understand the nuclear force and the properties of hadrons. The focus originally indeed was on open strings, relativistic open strings with labels at the ends, so-called Chan–Paton labels. In gauge theory these would be the charges or the labels of a gauge meson in the adjoint representation of the gauge group. Indeed, this connection between gauge theory and string theory is now gone from wild hypothesis to something that everyone believes. What is truly exciting is that one discovered that once you have open strings you must close them and that closed strings describe gravity. Thus string theory led to a unification between gauge theories, theories of open strings with fundamental quarks or labels at their ends, with close strings that had no labels, and describe fluctuations of space–time. By now that understanding has grown and has proven to be so fruitful that I think of the framework of theoretical physics not as quantum field theory, the basis of the standard model, plus this separate bizarre theory of extended objects called string theory, but as an enlarged framework that contains both. Neither quantum field theory nor string theory are really theories, rather they are broad frameworks, and now we believe that these frameworks are one and the same thing. We have many indications that this is the case. In some special cases (AdS/CFT) the evidence is beyond question. This combined field/string framework is the arena for which exploratory research in theoretical fundamental physics nowadays, with the hope of eventually of understanding how the standard model, the standard theory eventually fits in or is picked out.

The best example of the duality between Quantum Field Theory and string theory goes back to the study of large N_c QCD, namely the structure of Yang–Mills theory when you take N_c (the rank of the gauge group or the number of colors of the quarks) to infinity. The remarkable thing is that there is incredible simplification in this limit. First, the theory becomes a theory of an infinite number of stable, non-interacting, glueballs. In fact, the limit N_c goes to infinity, is a classical limit, and all quantum corrections and interactions between the glueballs are governed by, powers of $1/N_c$. Large N_c, is a classical limit of QCD, a limit that we believe could be described by a classical string theory with a coupling proportional to $1/N_c$. There is a much evidence for this duality, from the structure of Feynman diagrams in non-Abelian gauge theories in the $1/N_c$ expansion (which can be mapped onto the genus expansion of string world sheets). There is also experimental evidence that the $1/N_c$ expansion is in fact (even though $1/N_c = 1/3$ is in the real world) a good approximation for hadronic phenomenology. Most importantly, within string theory, is the precise mapping of ten-dimensional superstring theory, in the background space–time $AdS_5 \times S^5$, to maximally supersymmetric ($N = 4$) gauge theory in four dimensions. In fact, since Lars is going to talk about the maximally supersymmetric gauge theory, I want to say a word about it as well. I regard it as the hydrogen atom of quantum field theory. The hydrogen atom is an integrable model which is a good approximation to the real world and the fact that it is integrable is of great importance. Such integrable models, ones that are good approximations to the real world, have had an enormous impact on physics over the years, in fact they have guided the historical development of modern physics. The simplest example of an integrable model that is a good approximation to the real world is the Earth's orbit around the Sun, which is governed to a good approximation by just a central two-body force law between the Earth and the Sun, proportional to $1/r^2$. This inverse square law, $1/r^2$, in four dimensions is nice because it is integrable. It is lucky that it is the case (and the effect of the other planets can be neglected to a good approximation) because otherwise it would have taken Newton (or somebody else) much longer to discover classical mechanics and Newton's law of gravitation. The same feature played an absolutely important role in quantum mechanics, wherein the Bohr model, which essentially assumes simple electronic orbits with a $1/r^2$ force law, is such a good approximation to hydrogen atom. This is because the hydrogen atom is integrable and the Bohr's model was a semi-classical quantization of a quantum theory which didn't exist. So if one hadn't had before in one's eyes the simple integrable model, it would have been very difficult for Bohr to have guessed the semi-classical approximation to something that didn't exist yet. The success (and subsequent problems) of the Bohr model led to the development of non-relativistic quantum mechanics. The Dirac equation is also integrable and that helped enormously in the development of relativistic quantum mechanics. Historically we have made much progress by starting with an integrable model that is a very good approximations to the real world and then exploring the corrections to it. This path led to classical mechanics, non-relativistic quantum mechanics, and relativistic quantum field theory. To some extent asymptotically free QCD is already an example of this approach, which is one reason why it has been so successful. For very large momenta, because of asymptotic freedom, we have an trivially integrable system of free non-interacting quarks and gluons, about which we can perturb and which becomes a better and better approximation to the real world as one goes to higher and higher energies. But currently, the most interesting example of such a situation is taking the maximally supersymmetric gauge theory, with N (the number of supersymmetric extensions of Yang–Mills theory) $= 4$. When the number of colors becomes infinite, $1/N_c \to \infty$, the theory is believed to be integrable. We are very close to deriving the exact analytic solution of this limit of non-Abelian gauge theories. In order to make contact with QCD, we would have, of course, to break supersymmetry. This is not impossibly difficult. We would have to add in quarks, which is not difficult. But then the calculations then become more difficult as one perturbs away from the integrable limit. Thus, $N = 4$ SUSY-YM theory is the closest thing to the hydrogen atom in relativistic quantum field theory that we have ever had.

The AdS/CFT duality between string theory and gauge theory has been enormously productive. It has given us new insights into gauge theory by producing a useful string theory dual of this supersymmetric cousin of QCD that allows us to tackle problems in QCD. One famous example is the calculation of the transport coefficients of the quark–gluon fluid produced at RHIC, something which is beyond our capability to calculate in QCD, but easy to calculate in string theory by solving Einstein's equations, with surprising success. Similarly, one can use this duality to learn about difficult issues in strongly correlated many-body systems. The supersymmetric gauge theory is conformally invariant and there is evidence that any conformally invariant quantum field theory has a geometric dual which a string theory and in certain cases can be described by the low energy limit of string theory, which is simply Einstein's theory of gravity. So condensed matter physicists are now solving Einstein's equations in order to learn about quantum critical points, critical points that occur at zero temperature as some physical parameter (doping, . . .) is varied. Finally, most importantly, from the point of view of fundamental physics these dualities give us new tools to study quantum gravity. Starting with the gauge theory, which is a standard quantum field theory that we can put on a computer, one can study quantum gravity and string theory. It is not surprising therefore that the dualities have been used to try to resolve some of the puzzles of quantum gravity, such as the fate of black holes and the nature of quantum cosmology.

Finally, these dualities relate theories in different space–time dimensions and hint that space–time itself is an emergent concept. Currently there is a large research program devoted to understanding which properties of a quantum field theory (entanglement, . . .) are responsible for the emergence and the dynamics of space–time.

But at the end of this talk I go back to my original point that local gauge symmetry is a redundancy of our description of nature. Is the beautiful non-Abelian gauge symmetry, discovered by Yang and Mills, just a redundant description of physical phenomenon? Does the AdS/CFT duality shed light on this issue? Yes, in this new world of string theory/field theory correspondence, the meaning of gauge symmetry becomes a more even interesting question. In string theory, you do not see the gauge degrees of freedom of the dual theory at all. When you, to make the correspondence you can associate those gauge degrees of freedom as dynamical modes of branes. Branes might be sheets or three-dimensional volumes, points where open strings end, which is where the gauge charges lie. So in string theory, to construct gauge particles, you have to put in collective vibrations of string theory, of space–time if you want, which are branes, and open strings can end on those branes and the strings that connect them are the gauge degrees of freedom and the places where they end, are where the gauge charges are. So in a sense, from the string theory you can construct these redundant gauge degrees of freedom if you want, but they are clearly redundant and are not visible directly from the string theory. Conversely in gauge theory, you can regard the space–time of the corresponding string description, and its dynamics, as a way of describing the dynamics of the gauge field. Dozens of theorists are now engaged in trying to develop a precise understanding of how space–time emerges, or could be constructed out of the dynamics of a gauge theory on a rigid four-dimensional space. So this makes the question of what is a redundant description more interesting. Are gauge degrees of freedom emergent from modes of dynamical space–time or is space–time emergent from gauge dynamics or both? I will leave you with that question, it is at the center of research nowadays at the frontiers of fundamental physics. How it will help us answer the question of how does our standard model fit in to this much wider and richer framework that we are now studying — I don't know. But one thing I am absolutely sure of, and that is that Yang–Mills gauge theories will be at the forefront of fundamental physics for centuries to come. Thank you.

Lars Brink: Thank you, let me before we start asking questions, just tell you a little story about what happened in 2004 when we called David to tell him that he had got a certain prize and this was 2 am in California. But anyhow he can tell himself whatever but we had the press conference afterwards, we had David on the phone so I was sitting there and had a speaker next to me, that was David and so then one of the, you know, hundreds of journalists from all over the world, asked him, so when did you start to believe that your theory really was the correct one? And David was very polite and he said, "Yes, well you know during the 80s when you know we saw the results come in and they, you know, agreed so well, then I started to believe in...," but I clapped the speaker and said that is wrong, you knew it one microsecond after you had discovered it, which I still think is true.

David Gross: You could ask Frank Yang: "when did you know that Yang–Mills theory was part of the real world." Now he might very well answer: "Immediately, it is so beautiful that it must be part of the real world." But then, there is the second knowing, which requires the agreement of the ultimate judge, namely nature and that comes much later.

Kwek Leong Chuan: What do you mean when you talk about lattice gauge theory when you mentioned about Moore's Law?

David Gross: Moore's Law, as you know, is the fact that computers become twice as powerful every 18 months or so. All I meant was that this rapid development has helped lattice gauge theory enormously. Lattice gauge theory began in the middle 70s. As a tool for calculating hadrons it originated with Ken Wilson. But after some years, Ken gave up. Mike Creutz was one of the few to continue, but the computers they were working with, by today's standards, were extraordinary primitive. You really needed modern computers to tackle this very difficult calculation.

Lars Brink: Okay, now to all you young people, feel free to ask David and he will be so glad to answer your questions outside here, so I think we take a break but we first have to take a group picture and then we have to get coffee and we have to be back here by 11, so we do it quickly.

Acknowledgments

The published version of the transcribed talk was reviewed and edited by Lars Brink and David Gross.

Maximally Supersymmetric Yang–Mills Theory: The Story of $N=4$ Yang–Mills Theory

Lars Brink

Department of Fundamental Physics, Chalmers University of Technology,
S-412 96 Göteborg, Sweden
lars.brink@chalmers.se

1. History

All roads lead to Rome. This historic saying is also true in fundamental physics where the Milliarium Aureum is the Yang–Mills Theory. This was not at all obvious fifty years ago. The particle physics of the 1960's was dominated by the efforts to find the theory for the strong interactions. The domineering figure was Murray Gell-Mann who also got his Nobel Prize in 1969 for "for his contributions and discoveries concerning the classification of elementary particles and their interactions". He introduced the $SU(3)$ flavor symmetry in 1961 to classify the strongly interacting particles[1] and and a few years later the quarks.[2]

He was also one of the driving forces to understand the weak interactions among all these new particles and introduced the current algebras to study these relations.[3] All these works were precursors to the Standard Model that grew out of the attempts during the 1960's and early 1970's. His first comments in this direction was in his famous $V-A$ paper from 1957.[4] These attempts were however overshadowed by other attempts to find viable models in particle physics. That history is described in the scientific background text to the Nobel Prizes in physics in 2004, 2008 and 2013.[5]

The ideas to build a field theory model for the strong interactions had failed rather miserably, and many argued that quantum field theory was dead and that we needed something different. One major effort that was very popular in order to find a model for the strong interactions was the *S-Matrix approach*. It is described in two classical text books.[6] It used many consequences from quantum field theories such as crossing symmetry, analyticity, Poincaré invariance intertwined with new concepts such Regge asymptotic behavior.[7] By introducing the elementary particles as poles in the scattering matrix it was hoped that by self-consistency one should be led to a unique S-matrix. In this approach gauge invariance played no rôle and the key ingredient was analyticity rather than gauge invariance. Another one was the (seemingly) linear Regge-trajectories of the strongly interacting particles, when the spin was plotted against the m^2 of the resonance particles with the same quantum numbers as the lowest lying state such as the π or the ρ.

In the next sections I will give a brief and by no means complete history of *Dual Models* and *String Theory* and how eventually all these attempts to construct an S-Matrix Theory led us back to the center of fundamental physics, the Yang–Mills Theories. There were more than 1000 research papers on Dual Models in the period 1968–72[a] which would justify a full review of only that field. Here I will have gauge invariance as a red thread.

2. Dual Models

In 1967 a new concept was introduced, "Finite Energy Sum Rules" (FESR),[8] where the authors found that a description of πN scattering in terms of all the resonances in the direct s-channel was dual to an asymptotic description in the exchange t-channel in the sense that one should use either description but not the sum of the two. This started a major effort to find models that reproduced this *duality* and in 1968 Gabriele Veneziano[9] found a formula for the scattering amplitude for the process $\pi\pi \to \pi\eta$ as

$$A(s,t,u) = [B(1-\alpha(t), 1-\alpha(s)) \\ + B(1-\alpha(t), 1-\alpha(u)) \\ + B(1-\alpha(s), 1-\alpha(u))], \quad (1)$$

where $\alpha(s) = \frac{1}{2} + \alpha' s$, the ρ-trajectory, and α' is the slope of the trajectory, the strength of the strong interactions. $s = (p_1+p_2)^2$, $t = (p_2+p_3)^2$ and $u = (p_1+p_3)^2$ are the Mandelstam variables with p_i the momenta of the four particles in the amplitude and B is the Jacobi β-function.

This amplitude was an explicit solution for an S-matrix element for the process involving four particles. Could it be extended to an arbitrary number of external particles? A race all over the particle physics community was started and within less than half a year a solution was found first by Chan and Tsou.[10] Slightly after another way to formulate the amplitudes was found by Koba and Nielsen.[11] This form was very useful for the future developments that I will describe and here I give their solution

$$B_N = \int_{-\infty}^{\infty} \frac{\prod_1^N dz_i\, \theta(z_i - z_{i+1})}{dV_{abc}} \prod_{i=1}^N (z_i - z_{i+1})^{\alpha_0 - 1} \prod_{j>i} (z_i - z_j)^{2\alpha' p_i \cdot p_j}. \quad (2)$$

This is a remarkable formula for the scattering of N scalar particles. It has the correct resonance particles in every channel and the correct asymptotic behavior in terms of Regge behavior. One could also introduce isospin with the help of so-called Chan–Paton factors.[12] In a field theoretic language one would say that this was the

---

[a] All the papers were collected by Paul Frampton.

Born-term, the lowest order in a perturbation series but instead it was referred to as the "narrow resonance approximation". In fact there were leading scientists that claimed that one should not unitarize like in a field theory and that there must be another way of doing it. We were very far from Yang–Mills Theories!

A first problem to solve was whether the amplitudes could factorize. Indeed they could as Nambu[13] and Fubini, Gordon and Veneziano[14] discovered. They found that the amplitudes could be written as

$$B_N = \langle 0|V\,D\,V\,......\,D\,V|0\rangle \tag{3}$$

with V being a three-point vertex and D a propagator. The state space was now an infinite Hilbert space constructed from an infinite set of harmonic oscillators and Nambu considered that in his paper and wrote "Eq (17) suggests that the internal energy of a meson is analogous to that of a quantized string of finite length." This was the first understanding that the "Dual Model" amplitudes were indeed scattering of states of a relativistic string. Similar ideas was also put forward by Susskind[15] and Nielsen[16] but it took quite some time before the Dual Models became String Theory.

Even though some of the senior physicists involved in Dual Models refused to consider them in a field theoretic framework the form of (3) opened for questions about the norms of the ingoing states and about the possible symmetries of the amplitudes. The harmonic oscillators carry a spacetime index and thus there is an infinity of negative-norm states in the spectrum. Could it be that they decouple? In order for that to happen one needs an infinite symmetry in the amplitudes. Such a one was found by Virasoro.[17] However, there was a price to pay. It was only by shifting the intercept of the leading trajectory such that the lowest lying particle, the alleged ρ-particle is massless. A massless vector particle should ring a bell but most people involved was so impressed by the phenomenological success of the Veneziano Model that the immediate reaction was to try to move the intercept back to $1/2$. I was certainly one of those. The Virasoro algebra with the c-number discovered slightly later by Weis is

$$[L_n, L_m] = (n-m)L_{n+m} + \frac{d}{12}(n^3-n)\delta_{m+n,0}, \tag{4}$$

where d is the dimension of spacetime which was taken for granted to be 4, and the physical state conditions could be seen to be

$$L_n|\text{phys}\rangle = 0,\ n > 0$$
$$(L_0 - 1)|\text{phys}\rangle = 0. \tag{5}$$

The big issue now was if the conditions (5) were enough to warrant a positive-norm physical spectrum. This was solved by Brower[18] and Goddard and Thorn[19] in 1972. They found indeed that for $d \leq 26$ the spectrum is positive definite and for the case $d = 26$ the spectrum consists of only transverse states.

The first one to find that $d = 26$ was special was Lovelace a year earlier.[20] He had studied possible loop diagram by just sewing tree diagrams together, which did not allow him to get the correct measure, but the correct pole structure (although with negative-norm states propagating in the loop), and found that for a certain type of diagrams new non-unitary cuts appeared. In a brave analysis he found that if $d = 26$ these discontinuities became a new set of poles with a leading trajectory with an intercept double the original one. The result was not taken too seriously then but in the light of the no-ghost theorem a year later people started to realize the importance of the "critical dimension".

The Veneziano Model only contained bosons and to have a real physical model one also needed fermions. The problem was solved by Pierre Ramond[21] in a remarkable paper from the Christmas time of 1970. In the bosonic case one writes the spacetime coordinate as a function of a Koba–Nielsen variable, see (2), and expand it in an infinite set of harmonic oscillators. Ramond argued that the Dirac γ-matrices in a similar way should be given a dependence on a Koba–Nielsen variable and be expanded in an infinite set of anticommuting harmonic oscillators extending the Virasoro algebra into the algebra

$$[L_n, L_m] = (n-m)L_{n+m} + \frac{d}{8}(n^3 - n)\delta_{m+n,0},$$

$$[L_n, F_m] = \left(\frac{n}{2} - m\right) F_{n+m},$$

$$\{F_n, F_m\} = 2L_{n+m} + \frac{d}{2}(n^2 - 1)\delta_{m+n,0}. \qquad (6)$$

This is the first supersymmetry algebra, indeed it is a superconformal algebra in two dimensions in modern language. This was the key step on the road to the Superstring Theory.

Soon after André Neveu and John Schwarz[22] constructed a new bosonic model including also anticommuting harmonic oscillators and Charles Thorn[23] constructed a new model with two Ramond fermions and N Neveu–Schwarz bosons. Also these models were found to be ghost-free and the critical dimension turned out to be 10.

We now had a ghost-free model with both bosons and fermions but with massless vector particles and also a sector with massless particles with spin-2. Later they were called the open and the closed string. This was a great achievement but it was really useless for the purpose it had been invented, namely to describe strong interaction amplitudes and the interest faded.

At this stage David Olive and myself started a program to find out precisely if the models are unitarizable and if all amplitudes indeed are ghost-free. In order to construct correct one-loop graphs we had to follow the way that Feynman once did for Yang–Mills Theories.[24] He used the so called "tree theorem" in which a loop graph can be constructed by sewing tree-amplitudes together introducing a physical-state projection operator in one propagator to ensure that no unphysical state is propagating in the loop. We constructed those physical state projection operators in all sectors[25] and showed that they provided another proof of the no-ghost theorem and then used these to construct all the relevant one-loop amplitudes.[26] They all exhibited the correct discontinuities. We also used the projection operators to show that that the couplings between open and closed strings and between bosons and fermions maintained unitarity. Even though the models still contained tachyons the models exhibited the same unitarity structure as gauge field theories.

Now it should have been the time to take the Yang–Mills aspect seriously but it was still not the time. In fact Joël Scherk had already in 1971 asked the question what happens when the slope of the Regge trajectories is taken to zero. In his first paper[27] he argued that one would get a $\lambda\phi^3$ theory, but shortly after he found with Neveu[28] that the correct theory should be a Yang–Mills Theory. A similar question was what happens in the zero-slope limit for the closed string amplitudes? We have seen that it involves a massless spin-2 particle. The first one in print to connect to quantum gravity was Yoneya.[29] However slightly later Scherk and Schwarz[30] put forward the idea that the open and closed strings were really extensions of a quantum theory involving gravity and Yang–Mills particles and that the intercepts were indeed correct. The idea did not really catch immediately. Still there was a hope that a new dual model would indeed bring in the strong interaction trajectories. In 1975 the next dual model was finally constructed by myself and a full football team of Italians.[31] The model was even further away from strong interaction amplitudes and we interpreted the critical dimension to be 2 hence overshooting the wanted dimension of 4.

3. Supersymmetric Field Theories and Supergravity

Ramond had found a superconformal algebra to be behind the Ramond–Neveu–Schwarz Model. A realization of that algebra in terms of two-dimensional fields on a world-sheet was rapidly found by Gervais and Sakita.[32] In 1973 Wess and Zumino asked the question if there exists a four-dimensional version of such an algebra[33] and found one. They also realized that one can relax the conformal invariance and only demand a Poincaré invariance and constructed then supersymmetric field theories. Such ones had previously been constructed in the Soviet Union by Golfand and Likhtman[34] in an attempt to describe neutrinos, but that paper had not been noticed in the West. However after the papers by Wess and Zumino the field of supersymmetric four-dimensional field theory exploded. Rather quickly a supersymmetric version of Yang–Mills Theory was constructed by Ferrara and Zumino.[35] A year later Fayet[36] constructed an extended $N=2$ Yang–Mills Theory.

Also Gell-Mann got interested in the problem and very quickly he classified all the superPoincaré algebras with Ne'eman.[37] They particularly pointed out the CPT-invariant representations $N=4$ and $N=8$.

These developments were quite independent from developments in string theory. This changed in 1976 when supergravity was constructed by Freedman, van Nieuwenhuzen and Ferrara[38] and Deser and Zumino.[39] For me it was a revelation and we could use this technique to solve a problem I had worked on for quite some time, to get a formulation of the Lagrangian for strings and superstrings as a σ-model.[40,41] Quickly supergravites with extended supersymmetry were constructed and it became clear that the zero-slope limit of the supersymmetric string theory as it was called then should be an extended supergravity coupled to an extended Yang–Mills Theory. However, most people working in supergravity were either relativists or field theorists deeply anchored in four spacetime dimensions while only a few of us came into the field from string theory. We did have an advantage since we were used to work in higher dimensions. Higher dimensions had been discussed in the early 1900's first by Nordström and by Weyl and then by Kaluza and Klein but those attempts were completely forgotten in the 1970's. (Not even Gell-Mann knew that Kaluza was German and forced us to pronounce his name in Polish.) We string theorists regarded supergravity and super Yang–Mills Theory as zero-slope limits of the ten-dimensional Superstring Theory, while all others thought of them as extensions of Einstein gravity and Yang–Mills Theory. Both groups were of course in retrospect right. However, the ability to work in higher dimensions was advantageous as I will now describe.

4. Maximally Supersymmetric Yang–Mills Theories

In the fall of 1976 I came to Caltech to work with John Schwarz. Knowing that the zero-slope limit of the open superstring theory in $d = 10$ is the maximally supersymmetric Yang–Mills theory, we realized that this theory should be the mother of all supersymmetric Yang–Mills theories either by dimensional reduction or by that together with a truncation. Hence we set out to construct the ten-dimensional theory.[42] It is easy to write the action for it since it is really unique

$$S = \int d^{10}x \left\{ -\frac{1}{4} F^a_{\mu\nu} F^{a\mu\nu} + i\bar{\lambda}^a \gamma \cdot D\lambda^a \right\} \tag{7}$$

with the usual definitions for the field strength and the covariant derivative.

In order for this to be supersymmetric the spinor must be both Majorana and Weyl which is possible in $d = 10$. In fact this form works in three, four, six and ten dimensions with the spinors properly chosen. The supersymmetry transformations are easy to construct and also looks the same in all these dimensions

$$\begin{aligned} \delta A^a_\mu &= i\bar{\epsilon}\gamma_\mu \lambda^a - i\bar{\lambda}^a \gamma_\mu \epsilon, \\ \delta \lambda^a &= \sigma^{\mu\nu} F^a_{\mu\nu} \epsilon, \\ \delta \bar{\lambda}^a &= -\bar{\epsilon}\sigma^{\mu\nu} F^a_{\mu\nu}. \end{aligned} \tag{8}$$

The checking of the closure of these transformations is quite tricky though and one has to use a famous Fierz identity that we discovered

$$f^{abc} \bar{\lambda}^a \gamma_\mu \lambda^b (\gamma^\mu \lambda^c)_\alpha = 0, \tag{9}$$

which is valid again in $d = 3, 4, 6, 10$.

Now the road was open to construct all possible supersymmetric Yang–Mills theories which we did. Here I only describe the maximally supersymmetric theory in $d = 4$ which is obtained by a straight dimensional reduction. We put $x^4, x^5, \ldots, x^9 = 0$. The 16-dimensional spinor is divided up into four Weyl spinors and some γ-gymnastics has to be performed leading to the following action (I write it as in the original paper.) The final action is then

$$S = \int d^4x \left\{ -\frac{1}{4} F^a_{\mu\nu} F^{a\mu\nu} + i \bar{\chi}^a \gamma \cdot DL\chi^a + \frac{1}{2} D_\mu \Phi^a_{ij} \, D_\mu \Phi^a_{ij} \right.$$

$$- \frac{i}{2} g f^{abc} (\bar{\tilde{\chi}}^{ai} L \chi^{jb} \Phi^c_{ij} - \bar{\chi}^a_j R \tilde{\chi}^b_j \Phi^{ijc})$$

$$\left. - \frac{1}{4} f^2 f^{abc} g^{ade} \Phi^b_{ij} \Phi^c_{kl} \Phi^{ijd} \Phi^{kle} \right\}. \tag{10}$$

The resulting supersymmetry transformations are then

$$\delta A^a_\mu = i(\bar{\epsilon}_i \gamma_\mu L \chi^{ia} - \bar{\chi}^a_i \gamma_\mu L \epsilon^i),$$

$$\delta \Phi^a_{ij} = i(\bar{\epsilon}_j R \tilde{\chi}^a_i - \bar{\epsilon}_i R \tilde{\chi}^a_j + \epsilon_{ijkl} \bar{\tilde{\epsilon}}^k L \chi^{al}),$$

$$\delta \chi^{ia} = \sigma_{\mu\nu} F^{\mu\nu a} L \epsilon^i - \gamma \cdot D\Phi^{ija} R \tilde{\epsilon}_j + \frac{1}{2} g f^{abc} \Phi^{bik} \Phi^c_{kj} L \epsilon^j,$$

$$\delta R \tilde{\chi}^a_i = \sigma_{\mu\nu} F^{\mu\nu a} R \tilde{\epsilon}_i + \gamma \cdot D\Phi^a_{ij} L \epsilon^j + \frac{1}{2} g f^{abc} \Phi^b_{ik} \Phi^{ckj} R \tilde{\epsilon}_j. \tag{11}$$

R and L stands for the right-handed and left-handed projections respectively and for the rest of the notation please see the original paper.

This is the "$N = 4$ theory" with an $SU(4)$ symmetry. If we had used Majorana spinors it would have been an $SO(4)$ symmetry. Note that it is quite hard to find the four-dimensional action from scratch. We were very much helped by our knowledge of higher dimensions. However, in most circles at this time it was not appropriate to talk about higher dimensions, so we were very careful to say that this was just a means to find the four-dimensional action.

Similar ideas were pursued at the same time by Gliozzi, Scherk and Olive[43] who concentrated on the ten-dimensional string to make it consistent and as a by-product they also got the supersymmetric Yang–Mills theories of above. When comparing notes with Joël Scherk we realized that we had the same models and hence we wrote our paper together with him.

When supersymmetric theories came around it was noticed that the quantum properties got improved. The issue arose if such a theory could be perturbatively finite but it looked implausible.

After having constructed the $N = 4$ theory we turned to other problems but in the summer of 1977 Murray Gell-Mann heard that the β-function for this theory is zero at the one-loop level. He then prophetically declared that it is probably zero to all orders. He did not commit himself to write it in any report keeping up his promise to himself never to print anything that could be wrong. However, his comments were taken ad notam and a few months later Poggio and Pendleton [44] could report that the two-loop contribution to the β-function is indeed zero. Now there was a race to compute the three-loop contribution and three years later Caswell and Zanon [45] and Grisaru, Rocek and Siegel [46] could indeed confirm that it is zero. These were formidable calculations in Feynman diagrams. In the first case they had to consider some 600 diagrams while in the second case they worked with super-Feynman diagrams and had to consider 53 such diagrams. It became then clear that other techniques were needed for a general proof.

5. The Light-Cone Gauge Formulation of $N = 4$ Yang–Mills Theory

Around 1980 and the years after I was busy with my colleagues Michael Green and John Schwarz to set up the Superstring Theory and to check its physical properties. We did it mostly in a light-cone formulation, i.e. we were only using the dynamical degrees of freedom in a non-covariant way. In 1981 we asked what happens when we take the zero-slope limit of the one-loop graphs that we had constructed. Both for the closed string that leads to maximal supergravity and for the open string where we could isolate the maximally supersymmetric Yang–Mills Theory we could check the one-loop graph in various dimensions and see how finiteness could come about. The results were remarkably simple. Both in the supergravity case and the Yang–Mills case the complete one-loop graph for a four-particle scattering is just the box diagram of a ϕ^3 theory with kinematical factors taking care of the spin of the particles appearing in the loop graph.

This gave me the idea to check the light-cone gauge formulation of the $N = 4$ theory more explicitly and I did it with my collaborators Olof Lindgren and Bengt Nilsson.[47] Since the formalism we used is still not too well known I will describe the paper rather carefully. We did it by starting with the action (7). We then chose the gauge $A^+ = \frac{1}{\sqrt{2}}(A^0 + A^3) = 0$ and solved for the kinematical field $A^- = \frac{1}{\sqrt{2}}(A^0 - A^3)$ leaving us with only the the transverse degrees of freedom. (We used x^+, see below, as the evolution coordinate.) Similarly for the spinor field we used the decomposition

$$\lambda = \frac{1}{2}(\gamma_+\gamma_- + \gamma_-\gamma_+)\lambda = \lambda_+ + \lambda_- \qquad (12)$$

where again we could solve for λ_- since again it satisfies a kinematical equation of motion. (We did it in a path integral formulation and integrated out the non-dynamical fields). We could then rewrite the action and dimensionally reduce it to $d = 4$. Finally we could introduce a superspace and rewrite the action in that space. An alternative way which we have used in many cases[48] later is to simply look for a representation of the superPoincaré algebra. The final result is as follows which I now describe.

With the space-time metric $(-,+,+,\ldots,+)$, the light-cone coordinates and their derivatives are

$$x^{\pm} = \frac{1}{\sqrt{2}}(x^0 \pm x^3); \quad \partial^{\pm} = \frac{1}{\sqrt{2}}(-\partial_0 \pm \partial_3); \tag{13}$$

$$x = \frac{1}{\sqrt{2}}(x_1 + ix_2); \quad \bar{\partial} = \frac{1}{\sqrt{2}}(\partial_1 - i\partial_2); \tag{14}$$

$$\bar{x} = \frac{1}{\sqrt{2}}(x_1 - ix_2); \quad \partial = \frac{1}{\sqrt{2}}(\partial_1 + i\partial_2), \tag{15}$$

so that

$$\partial^+ x^- = \partial^- x^+ = -1; \quad \bar{\partial}x = \partial\bar{x} = +1. \tag{16}$$

In four dimensions, any massless particle can be described by a complex field, and its complex conjugate of opposite helicity, the $SO(2)$ coming from the little group decomposition

$$SO(8) \supset SO(2) \times SO(6). \tag{17}$$

Particles with no helicity are described by real fields. The eight vector fields in ten dimension reduce to

$$\mathbf{8}_v = \mathbf{6}_0 + \mathbf{1}_1 + \mathbf{1}_{-1}, \tag{18}$$

and the eight spinors to

$$\mathbf{8}_s = \mathbf{4}_{1/2} + \bar{\mathbf{4}}_{-1/2}. \tag{19}$$

The representations on the right-hand side belong to $SO(6) \sim SU(4)$, with subscripts denoting the helicity: there are six scalar fields, two vector fields, four spinor fields and their conjugates. To describe them in a compact notation, we introduce anticommuting Grassmann variables θ^m and $\bar{\theta}_m$,

$$\{\theta^m, \theta^n\} = \{\bar{\theta}_m, \bar{\theta}_n\} = \{\bar{\theta}_m, \theta^n\} = 0, \tag{20}$$

which transform as the spinor representations of $SO(6) \sim SU(4)$,

$$\theta^m \sim \mathbf{4}_{1/2}; \quad \bar{\theta}^m \sim \bar{\mathbf{4}}_{-1/2}, \tag{21}$$

where $m, n, p, q, \ldots = 1, 2, 3, 4$, denote $SU(4)$ spinor indices. Their derivatives are written as

All the physical degrees of freedom can be captured in one complex superfield

$$\phi(y) = \frac{1}{\partial^+} A(y) + \frac{i}{\sqrt{2}} \theta^m \theta^n \overline{C}_{mn}(y) + \frac{1}{12} \theta^m \theta^n \theta^p \theta^q \epsilon_{mnpq} \partial^+ \bar{A}(y)$$
$$+ \frac{i}{\partial^+} \theta^m \bar{\chi}_m(y) + \frac{\sqrt{2}}{6} \theta^m \theta^n \theta^p \epsilon_{mnpq} \chi^q(y). \tag{22}$$

In this notation, the eight original gauge fields A_i, $i = 1, \ldots, 8$ appear as

$$A = \frac{1}{\sqrt{2}}(A_1 + iA_2), \quad \bar{A} = \frac{1}{\sqrt{2}}(A_1 - iA_2), \tag{23}$$

while the six scalar fields are written as antisymmetric $SU(4)$ bi-spinors

$$C^{m\,4} = \frac{1}{\sqrt{2}}(A_{m+3} + iA_{m+6}), \quad \overline{C}^{m4} = \frac{1}{\sqrt{2}}(A_{m+3} - iA_{m+6}), \tag{24}$$

for $m \neq 4$; complex conjugation is akin to duality,

$$\overline{C}_{mn} = \frac{1}{2} \epsilon_{mnpq} C^{pq}. \tag{25}$$

The fermion fields are denoted by χ^m and $\bar{\chi}_m$. All have adjoint indices (not shown here), and are local fields in the modified light-cone coordinates

$$y = \left(x, \bar{x}, x^+, y^- \equiv x^- - \frac{i}{\sqrt{2}} \theta^m \bar{\theta}_m \right). \tag{26}$$

This particular light-cone formulation we call LC_2 since all the unphysical degrees of freedom have been integrated out, leaving only the physical ones.

Introduce the chiral derivatives,

$$d^m = -\partial^m - \frac{i}{\sqrt{2}} \theta^m \partial^+; \quad \bar{d}_n = \bar{\partial}_n + \frac{i}{\sqrt{2}} \bar{\theta}_n \partial^+, \tag{27}$$

which satisfy the anticommutation relations

$$\{d^m, \bar{d}_n\} = -i\sqrt{2} \delta^m{}_n \partial^+. \tag{28}$$

One verifies that ϕ and its complex conjugate $\bar{\phi}$ satisfy the chiral constraints

$$d^m \phi = 0; \quad \bar{d}_m \bar{\phi} = 0, \tag{29}$$

as well as the "inside-out" constraints

$$\bar{d}_m \bar{d}_n \phi = \frac{1}{2} \epsilon_{mnpq} d^p d^q \bar{\phi}, \tag{30}$$

$$d^m d^n \bar{\phi} = \frac{1}{2} \epsilon^{mnpq} \bar{d}_p \bar{d}_q \phi. \tag{31}$$

The Yang–Mills action is then simply

$$\int d^4x \int d^4\theta d^4\bar{\theta}\mathcal{L}, \tag{32}$$

where

$$\mathcal{L} = -\bar{\phi}\frac{\Box}{\partial^{+2}}\phi + \frac{4g}{3}f^{abc}\left(\frac{1}{\partial^+}\bar{\phi}^a\phi^b\bar{\partial}\phi^c + \text{complex conjugate}\right)$$

$$- g^2 f^{abc}f^{ade}\left(\frac{1}{\partial^+}(\phi^b\partial^+\phi^c)\frac{1}{\partial^+}(\bar{\phi}^d\partial^+\bar{\phi}^e) + \frac{1}{2}\phi^b\bar{\phi}^c\phi^d\bar{\phi}^e\right). \tag{33}$$

Grassmann integration is normalized so that $\int d^4\theta\theta^1\theta^2\theta^3\theta^4 = 1$, and f^{abc} are the structure functions of the Lie algebra.

6. The Perturbative Finiteness $N = 4$ Yang–Mills Theory

After having obtained the light-cone formulation of the this theory we set out to check its perturbation expansion to see if it could be UV finite.[49] There are well-defined techniques to find superFeynman rules and to construct supergraphs. One direct difficulty though is that the superfield satisfies different constraints (29), (30) and (31). This means that the functional derivatives used when computing the Feynman rules become a bit intricate.

$$\frac{\delta\phi^a(y,\theta)}{\delta\phi^b(y',\theta')} = \frac{1}{4!2}d^4\delta^4(x-x')\delta^4(\theta-\theta')\delta^4(\bar{\theta}-\bar{\theta}')\delta^a_b, \tag{34}$$

$$\frac{\delta\bar{\phi}^a(y,\theta)}{\delta\phi^b(y',\theta')} = 12\frac{1}{4!4}\frac{\bar{d}^4 d^4}{\partial^{+4}}\delta^4(x-x')\delta^4(\theta-\theta')\delta^4(\bar{\theta}-\bar{\theta}')\delta^a_b. \tag{35}$$

With this knowledge one can derive the expressions for the propagator, the three-point and four-point vertices and build up superFeynman diagrams. The explicit form can be seen in the paper. To estimate the naive dimension of a diagram one is helped by the fact that the δ-functions and the θ-integrals appearing in the propagator and in the vertex functions are not to be taken into account in computing the dimensionality. This fact is due to the property of supergraphs that they can always be reduced to a local expression in θ. One can now check that the naive dimension of any diagram is zero. To prove finiteness one has to show that for any diagram one can integrate out some momenta. In the paper we consider a general diagram and extract either a three-point vertex or a four-point one and show that for all contributions from them one can perform this extraction. Again I refer to the paper for the details.

The final key point is to show that one can make a Wick rotation to implement Weinberg's theorem.[50] The obstacles here are the poles in ∂^+. When we derived the formalism above we integrated out a determinant in ∂^+. This means the remaining freedom we have is in the choice of the exact form of the poles in ∂^+. By making the choice that we interpret the pole as $(p^+ + i\epsilon p^-)^{-1}$ we indeed can make the Wick rotation and use Weinberg's theorem to complete the proof that the perturbation expansion is finite.

At the same time as we were doing this analysis Mandelstam[51] gave a similar proof using a slightly different light-cone formulation.

There were also other arguments put forward for finiteness around this time. Sohnius and West[52] considered anomaly multiplets, and concluded that the $N = 4$ Yang–Mills theory is conformally invariant, if one can assume that the theory is supersymmetric and $O(4)$-invariant and that the structure of anomalies is given by the breakdown of conformal invariance in its coupling to supergravity. These have later been confirmed to be correct assumptions.

In the Soviet Union the group at ITEP[53] attacked the problem by studying instanton calculus and could also argue that the β-function should be zero.

The proofs by us and by Mandelstam depended on a non-linear realization of supersymmetry. It was also very important to find a proof within the covariant formulation of supersymmetry. This was found by Howe, Stelle and Townsend.[54]

It became now an established fact that the $N = 4$ theory is indeed a perturbatively finite quantum field theory. For quite some time there was a discussion if this meant that the theory is trivial. For us with a superstring background this was obviously not true since it is the zero-slope limit of the open string theory and as such is an integral part of a theory construction that was getting more and more established as the correct framework for a unified theory of all the interactions. It was amply shown by Sen[55] how beautiful and intricate the structure of the $N = 4$ theory is when he showed the full structure of the dyons and monopoles in the theory.

When we had proved the finiteness I went into Murray Gell-Mann's office and told him that we now had proven his conjecture. He then replied that you cannot have a field theory without a scale. This was indeed an objection to be taken seriously. In the string theory there is a scale, the slope α', and the Yang–Mills particles couple to other massive particles. However, the crucial observation was made by Maldacena[56] when he suggested that the $N = 4$ Yang–Mills theory is dual to the superstring theory in the sense that a strong-coupling limit of one of them corresponds to the weak-coupling limit of the other. This has been very carefully studied since then and verified in all attempts. We should not think about the $N = 4$ Yang–Mills theory as just an ordinary quantum field theory with no dimensionful coupling but as a string theory in disguise and where conformal dimensions instead play an important role. Finally the $N = 4$ Yang–Mills theory had found its place in fundamental physics. A journey that started in S-Matrix theory in opposition to Yang–Mills theories gave us string theory and the Superstring Theory and generalized Yang–Mills theories to end back into the new Milliarium Aureum of fundamental physics, the $N = 4$ Yang–Mills theory.

References

1. M. Gell-Mann, "The Eightfold Way: A Theory of strong interaction symmetry," CTSL-20, TID-12608,
   Y. Ne'eman, "Derivation of strong interactions from a gauge invariance," *Nucl. Phys.* **26**, 222 (1961).
2. M. Gell-Mann, "A Schematic Model of Baryons and Mesons," *Phys. Lett.* **8**, 214 (1964).
   G. Zweig, "An $SU(3)$ model for strong interaction symmetry and its breaking", CERN preprint 8182/TH.401, 1964.
3. M. Gell-Mann and Y. Ne'eman, "Current-generated algebras," *Ann. of Phys.* **30**, 360 (1964).
4. R. P. Feynman and M. Gell-Mann, "Theory of Fermi interaction," *Phys. Rev.* **109**, 193 (1958).
5. See "The Nobel Prize in Physics 2004, 2008 and 2013 — Advanced Information". Nobelprize.org.
6. R. J. Eden, P. V. Landshoff, D. I. Olive, J. C. Polkinghorne, "The Analytic S-Matrix", (Cambridge University Press, 1966),
   Geoffrey F. Chew, "The Analytic S Matrix", (W. A. Benjamin, 1966).
7. T. Regge, "Introduction to complex orbital momenta," *Nuovo Cim.* **14**, 951 (1959).
8. R. Dolen, D. Horn and C. Schmid, "Finite energy sum rules and their application to π N charge exchange," *Phys. Rev.* **166**, 1768 (1968).
9. G. Veneziano, "Construction of a crossing — symmetric, Regge behaved amplitude for linearly rising trajectories," *Nuovo Cim. A* **57**, 190 (1968).
10. H. M. Chan and S. T. Tsou, "Explicit construction of the n-point function in the generalized Veneziano model," *Phys. Lett. B* **28**, 485 (1969).
11. Z. Koba and H. B. Nielsen, "Reaction amplitude for n mesons: A Generalization of the Veneziano-Bardakci-Ruegg-Virasora model," *Nucl. Phys. B* **10**, 633 (1969).
12. J. E. Paton and H. M. Chan, "Generalized veneziano model with isospin," *Nucl. Phys. B* **10**, 516 (1969).
13. Y. Nambu, "Quark model and the factorization of the Veneziano amplitude," In *Detroit 1969, Proceedings, Conference On Symmetries.
14. S. Fubini, D. Gordon and G. Veneziano, "A general treatment of factorization in dual resonance models," *Phys. Lett. B* **29**, 679 (1969).
15. L. Susskind, "Dual symmetric theory of hadrons. 1.," *Nuovo Cim. A* **69**, 457 (1970).
16. H. B. Nielsen, XV Int. Conf.on High Energy Physics, Kiev (1970).
17. M. S. Virasoro, "Subsidiary conditions and ghosts in dual resonance models," *Phys. Rev. D* **1**, 2933 (1970).
18. R. C. Brower, "Spectrum generating algebra and no ghost theorem for the dual model," *Phys. Rev. D* **6**, 1655 (1972).

19. P. Goddard and C. B. Thorn, "Compatibility of the Dual Pomeron with Unitarity and the Absence of Ghosts in the Dual Resonance Model," *Phys. Lett. B* **40**, 235 (1972).
20. C. Lovelace, "Pomeron form-factors and dual Regge cuts," *Phys. Lett. B* **34**, 500 (1971).
21. P. Ramond, "Dual Theory for Free Fermions," *Phys. Rev. D* **3**, 2415 (1971).
22. A. Neveu and J. H. Schwarz, "Factorizable dual model of pions," *Nucl. Phys. B* **31**, 86 (1971).
23. C. B. Thorn, "Embryonic Dual Model for Pions and Fermions," *Phys. Rev. D* **4**, 1112 (1971).
24. R. P. Feynman, "Quantum theory of gravitation," *Acta Phys. Polon.* **24**, 697 (1963).
25. L. Brink and D. I. Olive, "The physical state projection operator in dual resonance models for the critical dimension of space-time," *Nucl. Phys. B* **56**, 253 (1973).
26. L. Brink and D. I. Olive, "Recalculation of the the unitary single planar dual loop in the critical dimension of space time," *Nucl. Phys. B* **58**, 237 (1973).
27. J. Scherk, "Zero-slope limit of the dual resonance model," *Nucl. Phys. B* **31**, 222 (1971).
28. A. Neveu and J. Scherk, "Connection between Yang-Mills fields and dual models," *Nucl. Phys. B* **36**, 155 (1972).
29. T. Yoneya, "Quantum gravity and the zero slope limit of the generalized Virasoro model," *Lett. Nuovo Cim.* **8**, 951 (1973).
30. J. Scherk and J. H. Schwarz, "Dual Models for Nonhadrons," *Nucl. Phys. B* **81**, 118 (1974).
31. M. Ademollo *et al.*, "Supersymmetric Strings and Color Confinement," *Phys. Lett. B* **62**, 105 (1976).
32. J. L. Gervais and B. Sakita, "Field Theory Interpretation of Supergauges in Dual Models," *Nucl. Phys. B* **34**, 632 (1971).
33. J. Wess and B. Zumino, "Supergauge Transformations in Four-Dimensions," *Nucl. Phys. B* **70**, 39 (1974).
34. Y. A. Golfand and E. P. Likhtman, "Extension of the Algebra of Poincare Group Generators and Violation of p Invariance," *JETP Lett.* **13**, 323 (1971) [*Pisma Zh. Eksp. Teor. Fiz.* **13**, 452 (1971)].
35. S. Ferrara and B. Zumino, "Supergauge Invariant Yang–Mills Theories," *Nucl. Phys. B* **79**, 413 (1974).
36. P. Fayet, "Fermi-Bose Hypersymmetry," *Nucl. Phys. B* **113**, 135 (1976).
37. Unpublished but there were extensive discussion in Aspen in 1974.
38. D. Z. Freedman, P. van Nieuwenhuizen and S. Ferrara, "Progress Toward a Theory of Supergravity," *Phys. Rev. D* **13**, 3214 (1976).
39. S. Deser and B. Zumino, "Consistent Supergravity," *Phys. Lett. B* **62**, 335 (1976).

40. L. Brink, P. Di Vecchia and P. S. Howe, "A Locally Supersymmetric and Reparametrization Invariant Action for the Spinning String," *Phys. Lett. B* **65**, 471 (1976).
41. S. Deser and B. Zumino, "A Complete Action for the Spinning String," *Phys. Lett. B* **65**, 369 (1976).
42. L. Brink, J. H. Schwarz and J. Scherk, "Supersymmetric Yang–Mills Theories," *Nucl. Phys. B* **121** (1977) 77.
43. F. Gliozzi, J. Scherk and D. I. Olive, "Supersymmetry, Supergravity Theories and the Dual Spinor Model," *Nucl. Phys. B* **122**, 253 (1977).
44. E. C. Poggio and H. N. Pendleton, "Vanishing of Charge Renormalization and Anomalies in a Supersymmetric Gauge Theory," *Phys. Lett. B* **72**, 200 (1977).
45. W. E. Caswell and D. Zanon, "Zero Three Loop Beta Function in the $N=4$ Supersymmetric Yang–Mills Theory," *Nucl. Phys. B* **182**, 125 (1981).
46. M. T. Grisaru, M. Rocek and W. Siegel, "Zero Three Loop beta Function in N=4 Superyang-Mills Theory," *Phys. Rev. Lett.* **45**, 1063 (1980).
47. L. Brink, O. Lindgren and B. E. W. Nilsson, "N=4 Yang–Mills Theory on the Light Cone," *Nucl. Phys. B* **212**, 401 (1983).
48. A. K. H. Bengtsson, I. Bengtsson and L. Brink, "Cubic Interaction Terms for Arbitrarily Extended Supermultiplets," *Nucl. Phys. B* **227**, 41 (1983).
49. L. Brink, O. Lindgren and B. E. W. Nilsson, "The Ultraviolet Finiteness of the N=4 Yang–Mills Theory," *Phys. Lett. B* **123**, 323 (1983).
50. S. Weinberg, "High-energy behavior in quantum field theory," *Phys. Rev.* **118**, 838 (1960).
51. S. Mandelstam, "Light Cone Superspace and the Ultraviolet Finiteness of the N=4 Model," *Nucl. Phys. B* **213**, 149 (1983).
52. M. F. Sohnius and P. C. West, "Conformal Invariance in N=4 Supersymmetric Yang–Mills Theory," *Phys. Lett. B* **100**, 245 (1981).
53. V. A. Novikov, M. A. Shifman, A. I. Vainshtein and V. I. Zakharov, "Exact Gell-Mann-Low Function of Supersymmetric Yang–Mills Theories from Instanton Calculus," *Nucl. Phys. B* **229**, 381 (1983).
54. P. S. Howe, K. S. Stelle and P. K. Townsend, "Miraculous Ultraviolet Cancellations in Supersymmetry Made Manifest," *Nucl. Phys. B* **236**, 125 (1984).
55. A. Sen, "Dyon — monopole bound states, selfdual harmonic forms on the multi — monopole moduli space, and $SL(2,Z)$ invariance in string theory," *Phys. Lett. B* **329**, 217 (1994) [hep-th/9402032].
56. J. M. Maldacena, "The Large N limit of superconformal field theories and supergravity," *Int. J. Theor. Phys.* **38**, 1113 (1999) [*Adv. Theor. Math. Phys.* **2**, 231 (1998)].

The Lattice and Quantized Yang–Mills Theory

Michael Creutz

*Physics Department, Brookhaven National Laboratory,
Upton, NY 11973, USA*
creutz@bnl.gov

Quantized Yang–Mills fields lie at the heart of our understanding of the strong nuclear force. To understand the theory at low energies, we must work in the strong coupling regime. The primary technique for this is the lattice. While basically an ultraviolet regulator, the lattice avoids the use of a perturbative expansion. I discuss the historical circumstances that drove us to this approach, which has had immense success, convincingly demonstrating quark confinement and obtaining crucial properties of the strong interactions from first principles.

1. Introduction

As we have been hearing throughout this meeting,[a] the Yang–Mills theory[1] was developed in an attempt to generalize the gauge symmetry of electromagnetism to the non-Abelian SU(2) symmetry of isospin. Remarkably, this simple idea has developed into a core ingredient of all modern theories of elementary particles. With the particular application to the strong interactions, quarks interact by exchanging non-Abelian gauge gluons. This gives rise to some rather unique issues. In particular, asymptotic freedom and dimensional transmutation imply that low energy physics is controlled by large effective coupling constants. Long distance phenomena, such as chiral symmetry breaking and quark confinement, lie outside the realm of accessibility to the traditional Feynman diagram approach. This has driven theorists to new approaches, amongst which the lattice has proven the most successful.

Here I will present an overview of what motivated the lattice approach and how it grew to become the dominant technique to study non-perturbative effects in quantum field theory. Much of this presentation is adapted from a previous review.[2] Along the way we will see that there are both practical and fundamental issues with the lattice method. On the practical side, quantitative computer calculations are now routine for non-perturbative effects in the strong interactions. On the more conceptual side, the lattice gives deep insights into the workings of relativistic field theory.

---

[a]Conference on *60 Years of Yang–Mills Gauge Field Theories* (25–28 May 2015, Nanyang Technological University, Singapore).

2. Before the Lattice

I begin with the situation in particle physics in the late '60s, when I was a graduate student. Quantum-electrodynamics was the model field theory, with immense success. While hard calculations remained, and indeed still remain, the feeling was that this theory was understood. Some subtle conceptual issues do still remain, such as the likely breakdown of the perturbative expansion at ultra high energies.

These were the years when the "eight-fold way" for describing multiplets of particles had gained widespread acceptance.[3] The idea of "quarks" was around, but with considerable caution about assigning them any physical reality; were they nothing but a useful mathematical construct? A few insightful theorists were working on the weak interactions, and the basic electroweak unification was emerging.[4,5] The SLAC experiments were observing substantial inelastic electron–proton scattering at large angles. This was quickly interpreted as evidence for substructure, and the idea of "partons" became popular. While there were speculations on connections between quarks and partons, people tended to be rather cautious about pushing this too hard.

A crucial feature of the time was the failure of extension of quantum electrodynamics to a meson–nucleon field theory. The pion–nucleon analog of the electromagnetic coupling had a value about 15, in comparison with the 1/137 of QED. This meant that higher order corrections to perturbative processes were substantially larger than the initial calculations. There was no known small parameter in which to expand.

In frustration over this situation, much of the particle theory community abandoned traditional quantum field theoretical methods and explored the possibility that particle interactions might be completely determined by fundamental postulates such as analyticity and unitarity. This "S-matrix" approach raised the deep question of just "what is elementary?" A delta baryon might be regarded as a combination of a proton and a pion, but it would be just as correct to regard the proton as a bound state of a pion with a delta. All particles were to be thought of as bound together by exchanging themselves.[6] These "dual" views of the basic objects of the theory have evolved into many of the ideas of string theory.

3. The Birth of QCD

In the early '70s, partons were increasingly identified with quarks. This shift was pushed by two dramatic theoretical accomplishments. First was the proof of renormalizability for Yang–Mills theories,[7] giving confidence that these elegant mathematical structures[1] might indeed have something to do with reality. Second was the discovery of asymptotic freedom, the fact that interactions in Yang–Mills theories become weaker at short distances.[8,9] Indeed, this was quickly connected with the point-like structures hinted at in the SLAC experiments. Out of these ideas evolved QCD, the theory of quark confining dynamics.

The viability of this picture depends upon the concept of "confinement." While there exists strong evidence for quark substructure, no free quarks have ever been observed. This is particularly puzzling given the nearly free nature of their interactions inside the nucleon. Indeed, the question of "what is elementary?" reappears. Are the fundamental objects the physical particles we see in the laboratory or are these postulated quarks and gluons?

Struggling with this paradox led to the now standard flux-tube picture of confinement. The eight gluons are analogues of photons except that they carry "charge" with respect to each other. Gluons would presumably be massless like the photon were it not for confinement. But a massless charged particle would be a rather peculiar object. Indeed, what happens to the self energy in the electric fields around a gluon? Such questions naturally lead to a conjectured instability of the æther that removes zero mass gluons from the physical spectrum. This should be done in a way that does not violate Gauss's law. Note that a Coulombic $1/r^2$ field is a solution of the equations of a massless field, not a massive one. Without massless particles in the spectrum, such a spreading of the gluonic flux is not allowed since it cannot satisfy the appropriate equations in the weak field limit. According to Gauss's law, the field lines emanating from a quark cannot simply end. Instead of spreading in an inverse square manner, the gluo-electric flux lines cluster together, forming a tube emanating from the quark and ultimately ending on an anti-quark as sketched in Fig. 1. This structure should be regarded as a real physical object, which grows in length as the quark and anti-quark are pulled apart. The resulting force is constant at long distance, and is measured via the spectrum of high angular-momentum states, organized into the famous "Regge trajectories." In physical units, the flux tube pulls with a tension of about 14 tons.

In essence, the reason a quark cannot be isolated is similar to the fact that a piece of string cannot have just one end. Of course a piece of string cannot have three ends either. This is resolved by the underlying SU(3) group theory, wherein three fundamental charges can form a neutral object. It is important to emphasize that the confinement phenomenon cannot be seen in perturbation theory; when the coupling is turned off, the spectrum becomes free quarks and gluons, dramatically different from the pions and protons of the interacting theory.

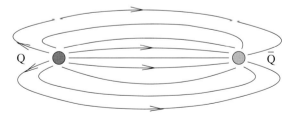

Fig. 1. A tube of gluonic flux connects quarks and anti-quarks. The strength of this string is 14 tons.

4. The '70s Revolution

The discoveries related to the Yang–Mills theory were just the beginning of a revolutionary period in particle physics. Perhaps the most dramatic event was the discovery of the J/ψ particle.[10] The interpretation of this object and its partners as bound states of heavy quarks provided what might be regarded as the hydrogen atom of QCD. The idea of quarks became inescapable; field theory was reborn. The SU(3) non-Abelian gauge theory of the strong interactions was combined with the recently developed electroweak theory to become the durable "Standard Model."

This same period also witnessed several remarkable events on a more theoretical front. Nonlinear effects in classical field theories were shown to have deep consequences for their quantum counterparts. Classical "lumps" represented a new way to get particles in a quantum field theory.[11] Much of the progress here was in two dimensions, where techniques such as "bosonization" showed equivalences between theories of drastically different appearance. A boson in one approach might appear as a bound state of fermions in another, but in terms of the respective Lagrangian approaches, they were equally fundamental. Again, we were faced with the question of "what is elementary?"

The interest in classical solutions quickly led to the discovery[12] of "pseudo-particles" or "instantons," solutions of the four-dimensional Yang–Mills theory in Euclidean spacetime. These turned out to be intimately related to the famous anomalies in current algebra, and gave a simple mechanism to generate the masses of such particles as the η'.[13] These effects are inherently nonperturbative, having an explicit exponential dependence in the inverse coupling.

This slew of discoveries had deep implications: field theory had many aspects that could not be seen via the traditional analysis of Feynman diagrams. This has crucial consequences for practical calculations. Field theory is notorious for divergences requiring regularization. The bare mass and charge are divergent quantities. They are not physical observables. For practical calculations, a "regulator" is required to tame the divergences, and when physical quantities are related to each other, any regulator dependence should drop out.

The need for controlling infinities had been known since the early days of QED. But all regulators in common use were based on Feynman diagrams; one would calculate until a divergent diagram appeared, and that diagram was then cutoff. Numerous schemes were devised for this purpose, ranging from the Pauli–Villars approach[14] to the forest formulae[15] to dimensional regularization.[16] But with the increasing realization that non-perturbative phenomena were crucial, it was becoming clear that we needed a "non-perturbative" regulator, independent of Feynman diagrams.

5. The Lattice

The necessary tool appeared with Wilson's lattice theory. He originally presented this as an example of a model exhibiting confinement. The strong coupling expansion has a nonzero radius of convergence, allowing a rigorous demonstration of confinement, albeit in an unphysical limit. The resulting spectrum has exactly the desired properties; only gauge singlet bound states of quarks and gluons can propagate.

This was not the first time that the basic structure of lattice gauge theory had been written down. A few years earlier, Wegner[17] presented a Z_2 lattice gauge model as an example of a system possessing a phase transition but not exhibiting any local order parameter. In his thesis, Jan Smit[18] described using a lattice to formulate gauge theories outside of perturbation theory. Very quickly after Wilson's suggestion, Balian, Drouffe, and Itzykson[19-21] explored an amazingly wide variety of aspects of these models.

To reiterate, the primary role of the lattice is to provide a non-perturbative regulator. Spacetime is not really meant to be a crystal; the lattice is a mathematical trick. It provides a minimum wavelength through the lattice spacing a, i.e. a maximum momentum of π/a. Path summations become well-defined ordinary integrals. By avoiding the convergence difficulties of perturbation theory, the lattice provides a route towards a rigorous definition of a quantum field theory as a limiting process.

This approach had a marvelous side effect. After discreetly making the system discrete, the lattice system becomes sufficiently well-defined to be placed on a computer.[22] This was fairly straightforward, and came at the same time that computers were growing rapidly in power. Indeed, numerical simulations and computer capabilities have continued to grow together, making these efforts the mainstay of modern lattice gauge theory.

6. Gauge Fields and Phases

As formulated by Wilson, the lattice cutoff is remarkable in remaining true to many of the underlying concepts of a gauge theory. At the most simplistic level, a Yang–Mills theory is simply electrodynamics embellished with isospin symmetry. By working directly with elements of the gauge group, this is inherent in lattice gauge theory from the start.

At another level, a gauge theory is a theory of phases acquired by a particle as it passes through spacetime. Using group elements on links directly gives this connection, with the phase associated with some world-line being the product of these elements along the path in question. For the Yang–Mills theory, the concept of "phase" becomes a rotation in the internal symmetry group.

At a still deeper level, a gauge theory is a theory with a local symmetry. With the Wilson action being formulated in terms of products of group elements around closed loops, this symmetry remains exact even with the regulator in place.

In perturbative discussions, the local symmetry forces one to adopt a gauge-fixing prescription to remove a formal infinity on integrating over gauges. The lattice formulation, in contrast, uses a compact representation for the group elements, making the integration over all gauges finite. For gauge invariant observables, no gauge fixing is required. While gauge fixing can still be done, it only needs to be introduced to study such conventional gauge-variant quantities such as gluon or quark propagators.

One property of continuum gauge theory that the lattice approach violates involves transformations under Lorentz transformations. In a continuum theory the basic vector potential can change under a gauge transformation when transforming between frames.[23-25] The lattice, of course, breaks Lorentz invariance, and thus this concept loses meaning until the continuum limit is taken.

7. The Wilson Action

The concept of gauge fields representing path dependent phases leads directly to the conventional lattice formulation. We approximate a general quark world-line by a set of hoppings lying along lattice bonds, as sketched in Fig. 2. We then introduce the gauge field as group valued matrices on these bonds. Thus the gauge fields form a set of SU(3) matrices, one such associated with every nearest neighbor bond on our four-dimensional hyper-cubic lattice.

In terms of these matrices, the gauge field dynamics takes a simple natural form. In analogy with regarding electromagnetic flux as the generalized curl of the vector potential, we are led to identify the flux through an elementary square, or "plaquette," on the lattice with the phase factor obtained on running around that plaquette; see Fig. 3. Spatial plaquettes represent the "magnetic" effects and plaquettes with one time-like direction give the "electric" fields. This motivates the conventional "action" used for the gauge fields as a sum over all the elementary squares of the lattice. Around each square we multiply the phases and to get a real

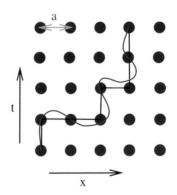

Fig. 2. In lattice gauge theory the world-line describing the motion of a quark through spacetime is approximated by a sequence of discrete hops. On each of these hops the quark wave function picks up a "phase" described by the gauge fields. For the strong interactions, this phase is a unitary matrix in the group SU(3).

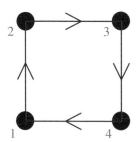

Fig. 3. Analogous to Stoke's law, the flux through an elementary square of the lattice is found from the product of gauge matrices around that square. The dynamics is determined by adding the real part of the trace of this product over all elementary squares. This "action" is inserted into a "path integral." The resulting construction is formally equivalent to a partition function for a system of "spins" existing in the group SU(3).

number we take the real part of the trace

$$S_g = \sum_p \text{Re Tr} \prod_{l \in p} U_l \sim \int d^4x\, E^2 + B^2 .\tag{1}$$

Here the fundamental squares are denoted p and the links l. As we are dealing with non-commuting matrices, the product around the square is meant to be ordered, while because of the trace, the starting point of this ordering drops out.

To formulate the quantum theory of this system one usually uses the Feynman path integral. To construct this, exponentiate the action and integrate over all dynamical variables

$$Z = \int (dU) e^{-\beta S} ,\tag{2}$$

where the parameter β controls the bare coupling. This converts the three space dimensional quantum field theory of gluons into a classical statistical mechanical system in four spacetime dimensions. Such a many-degree-of-freedom statistical system cries out for Monte Carlo simulation, which now dominates the field of lattice QCD. Note the close analogy with a magnetic system; we could think of our matrices as "spins" interacting through a four-spin coupling expressed in terms of the plaquettes.

The formulation is conventionally taken in Euclidean four-dimensional space. In effect this replaces the time evolution operator e^{-iHt} by e^{-Ht}. Despite involving the same Hamiltonian H, excited states are inherently suppressed and information on high energy scattering is particularly hard to extract. However, low energy states and matrix elements are the natural physical quantities to explore numerically. This is the bread and butter of the lattice theorist. Indeed, the simulations reproduce the qualitative spectrum of stable hadrons quite well.[26] Matrix elements currently under intense study are playing a crucial role in ongoing tests of the standard model of particle physics.

8. A Paucity of Parameters

It is important to emphasize one of the most remarkable aspects of QCD: the small number of adjustable parameters. To begin with, the lattice spacing itself is not an observable. We are using the lattice to define the theory, and thus for physics we must take the continuum limit $a \to 0$. Then there is the coupling constant, which is also not a physical parameter due to the phenomenon of asymptotic freedom. The lattice works directly with a bare coupling, and in the continuum limit this should vanish as predicted by asymptotic freedom

$$g_0^2 \sim \frac{1}{\log(1/\Lambda a)} \to 0. \tag{3}$$

In the process, the coupling is replaced by an overall scale Λ, which can be regarded as an integration constant for the renormalization group equation. Coleman and Weinberg[27] gave this phenomenon of replacing a dimensionless coupling with a scale the marvelous name "dimensional transmutation." An overall scale is not really something we should expect to calculate from first principles. Its numerical value would depend on the units chosen, be they furlongs or light-fortnights.

Next consider the quark masses. These also renormalize to zero as a power of the coupling in the continuum limit. Factoring out this divergence, we can define a renormalized quark mass, a second integration constant of the renormalization group equations. One such constant M_i is needed for each quark "flavor" or species i. Up to an irrelevant overall scale, the physical theory is then a function only of the dimensionless ratios M_i/Λ. These are the only free parameters in the strong interactions. The origin of the underlying masses remains one of the outstanding mysteries of particle physics.

With multiple flavors, the massless quark limit gives a rather remarkable theory, one with no undetermined dimensionless parameters. This limit is not terribly far from reality; chiral symmetry breaking should give massless pions, and experimentally the pions are considerably lighter than the next hadron, the rho. A theory of two massless quarks is a fair approximation to the strong interactions at intermediate energies. In this limit all dimensionless ratios should be calculable from first principles, including quantities such as the rho-to-nucleon mass ratio.

Since it is absorbed into an overall scale, the strong coupling constant at any physical scale is not an input parameter, but should be determined from first principles. Such calculations have placed lattice gauge theory into the famous particle data group tables.[28]

9. Numerical Simulation

While other techniques exist, such as strong coupling expansions, large scale numerical simulations currently dominate the practice of lattice gauge theory. They are based on evaluating the path integral

$$Z = \int (dU) e^{-\beta S} \qquad (4)$$

with β proportional to the inverse bare coupling squared. A direct evaluation of such an integral has pitfalls. At first sight, the basic size of the calculation is overwhelming. Considering a 10^4 lattice, small by today's standards, there are 40,000 links. On each is an SU(3) matrix, parametrized by 8 numbers. Thus we have a $10^4 \times 4 \times 8 = 320{,}000$ dimensional integral. One might try to replace this with a discrete sum over values of the integrand. If we make the extreme approximation of using only two points per dimension, this gives a sum with

$$2^{320{,}000} = 3.8 \times 10^{96{,}329} \qquad (5)$$

terms! Of course, computers are getting pretty fast, but one should remember that the age of universe is only $\sim 10^{27}$ nanoseconds.

These huge numbers suggest a statistical treatment. The above integral is formally a partition function. Consider a more familiar statistical system, such as a glass of beer. There are a huge number of ways of arranging the atoms of carbon, hydrogen, oxygen, etc. that still leave us with a glass of beer. We do not need to know all those arrangements, we only need a dozen or so "typical" glasses to know all the important properties.

This is the basis of the Monte Carlo approach. The analogy with a partition function and the role of $\frac{1}{\beta}$ as a temperature enables the use of standard techniques to obtain "typical" equilibrium configurations, where the probability of any given configuration is given by the Boltzmann weight

$$P(C) \sim e^{-\beta S(C)}. \qquad (6)$$

For this we use a Markov process, making changes in the current configuration

$$C \to C' \to \cdots \qquad (7)$$

biased by the desired weight.

The idea is easily demonstrated with the example of Z_2 lattice gauge theory.[29] For this toy model, the links are allowed to take only two values, either plus or minus unity. One sets up a loop over the lattice variables. When looking at a particular link, calculate the probability for it to have value 1

$$P(1) = \frac{e^{-\beta S(1)}}{e^{-\beta S(1)} + e^{-\beta S(-1)}}. \qquad (8)$$

Then pull out a roulette wheel and select either 1 or -1 biased by this weight. Lattice gauge Monte Carlo programs are by nature quite simple. They are basically a set of nested loops surrounding a random change of the fundamental variables.

Extending this to fields in larger manifolds, such as the SU(3) matrices representing the gluon fields, is straightforward. The algorithms are usually based on a detailed balance condition for a local change of fields taking configuration C to configuration C'. If probabilities for making these changes in one step satisfy

$$\frac{P(C \to C')}{P(C' \to C)} = \frac{e^{-\beta S(C')}}{e^{-\beta S(C)}}, \qquad (9)$$

a straightforward argument shows that any ensemble of configurations will approach the equilibrium ensemble.

The results of these simulations have been fantastic, giving first principles calculations for interacting quantum field theories. I will just mention a few examples. The early result that bolstered the lattice into mainstream particle physics was the convincing demonstration of the confinement phenomenon. The force between two quark sources indeed remains constant at large distances.

A major goal of lattice simulations is to understand the hadronic spectrum. This is done by studying the long distance behavior of correlation functions. Let $\phi(t)$ be an operator that can create a specific particle at time t. Then as t becomes large the correlator

$$\langle \phi(t)\phi(0) \rangle \to e^{-mt} \qquad (10)$$

where m is the mass of the lightest hadron that can be created by ϕ. In these calculations the bare quark masses are parameters that can be determined by fitting a few of the light mesons. Chiral symmetry is useful here, with the pion mass squared predicted to be proportional to the light quark mass. Using the pion mass to fix the light quark mass and the kaon mass to fix the strange quark, all other particle masses should be determined. In this way, recent simulations with physical mass pions have successfully mapped out much of the low energy hadron spectrum.[26]

Another accomplishment for which the lattice excels over all other methods has been the study (using an approximation to QCD) of the deconfinement of quarks and gluons into a plasma at a temperature of about 170–190 MeV.[30] Indeed, the lattice is a unique quantitative tool capable of making precise predictions for the value of this temperature. The method is based on the fact that the Euclidean path integral in a finite temporal box directly gives the physical finite temperature partition function, where the size of the box is proportional to the inverse temperature. This transition represents the confining flux tubes becoming lost in a background plasma of virtual flux lines.

10. Concluding Remarks

In summary, lattice gauge theory provides the dominant framework for investigating non-perturbative phenomena in quantum field theory. The approach is currently dominated by numerical simulations, although the basic framework is potentially considerably more general. With the recent developments towards implementing chiral symmetry on the lattice, including domain-wall fermions, the overlap formula, and variants on the Ginsparg–Wilson relation, parity conserving theories, such as the strong interactions, are fundamentally in quite good shape.

A particularly fascinating unsolved issue is the chiral gauge problem. Without a proper lattice formulation of a chiral gauge theory, it is unclear whether such models make any sense as a fundamental field theories. This is important for understanding how neutrinos can couple in only one helicity state. A marvelous goal would be a fully finite, gauge invariant, and local lattice formulation of the Standard Model. The problems encountered with chiral gauge theory are closely related to similar issues with supersymmetry, another area that does not naturally fit on the lattice. Understanding these issues will be necessary to make ties with the explosive activity in string theory and a possible regularization of gravity.

The other major unsolved problems in lattice gauge theory are algorithmic. Current fermion algorithms are extremely awkward and computer intensive. It is unclear why this has to be so, and may only be a consequence of our working directly with fermion determinants. One could to this for bosons too, but that would clearly be terribly inefficient. At present, the fermion problem seems completely intractable when the fermion determinant is not positive. This is of more than academic interest since interesting superconducting phases are predicted at high quark density. Similar sign problems appear in other fields of physics, such as doped strongly coupled electron systems.

Finally, throughout history the question of "what is elementary?" continues to arise. This is almost certainly an ill-posed question, with variant approaches being simpler in distinct contexts. At a more mundane level, for low energy chiral dynamics we lose nothing by considering the pion as an elementary pseudo-goldstone field, while at extremely short distances string structures may become more fundamental. Quarks and their confinement may only be a useful temporary construct along the way.

References

1. C.-N. Yang and R. L. Mills, *Phys. Rev.* **96**, 191 (1954).
2. M. Creutz, Yang–Mills fields and the lattice, in *50 Years of Yang–Mills Theory*, ed. G. 't Hooft (World Scientific, 2005), pp. 357–374.
3. M. Gell-Mann and Y. Neemam, *The Eightfold Way: A Review with a Collection of Reprints* (1964).
4. S. Weinberg, *Phys. Rev. Lett.* **19**, 1264 (1967).
5. A. Salam, *Conf. Proc.* **C680519**, 367 (1968).
6. G. F. Chew, The analytic S-matrix: A theory for strong interactions, in *Physique des Hautes Energies* (1965).
7. G. 't Hooft and M. J. G. Veltman, *Nucl. Phys. B* **44**, 189 (1972).
8. H. D. Politzer, *Phys. Rev. Lett.* **30**, 1346 (1973).
9. D. Gross and F. Wilczek, *Phys. Rev. Lett.* **30**, 1343 (1973).

10. E. Eichten, K. Gottfried, T. Kinoshita, K. D. Lane and T.-M. Yan, *Phys. Rev. D* **21**, 203 (1980).
11. S. R. Coleman, *Subnucl. Ser.* **13**, 297 (1977).
12. A. A. Belavin, A. M. Polyakov, A. S. Schwartz and Yu. S. Tyupkin, *Phys. Lett. B* **59**, 85 (1975).
13. G. 't Hooft, *Phys. Rev. D* **14**, 3432 (1976).
14. W. Pauli and F. Villars, *Rev. Mod. Phys.* **21**, 434 (1949).
15. M. Gomes, J. H. Lowenstein and W. Zimmermann, *Commun. Math. Phys.* **39**, 81 (1974).
16. G. 't Hooft, *Nucl. Phys. B* **61**, 455 (1973).
17. F. J. Wegner, *J. Math. Phys.* **12**, 2259 (1971).
18. J. Smit, UCLA Thesis, 24 (1974).
19. R. Balian, J. M. Drouffe and C. Itzykson, *Phys. Rev. D* **10**, 3376 (1974).
20. R. Balian, J. M. Drouffe and C. Itzykson, *Phys. Rev. D* **11**, 2098 (1975).
21. R. Balian, J. M. Drouffe and C. Itzykson, *Phys. Rev. D* **11**, 2104 (1975) [Erratum-*ibid* **19**, 2514 (1979)].
22. M. Creutz, *Phys. Rev. D* **21**, 2308 (1980).
23. S. Weinberg, *Phys. Rev. B* **133**, 1318 (1964).
24. S. Weinberg, *Phys. Rev. B* **134**, 882 (1964).
25. S. Weinberg, *Phys. Rev.* **181**, 1893 (1969).
26. S. Durr *et al.*, *Phys. Lett. B* **701**, 265 (2011).
27. S. R. Coleman and E. J. Weinberg, *Phys. Rev. D* **7**, 1888 (1973).
28. K. A. Olive *et al.*, *Chin. Phys. C* **38**, 090001 (2014).
29. M. Creutz, L. Jacobs and C. Rebbi, *Phys. Rev. Lett.* **42**, 1390 (1979).
30. Y. Aoki, Z. Fodor, S. Katz and K. Szabo, *Phys. Lett. B* **643**, 46 (2006).

Yang–Mills Theories at High Energy Accelerators

George Sterman

C. N. Yang Institute for Theoretical Physics and
Department of Physics and Astronomy
Stony Brook University, Stony Brook, NY 11794-3840, USA
george.sterman@stonybrook.edu

I will begin with a brief review of the triumph of Yang–Mills theory at particle accelerators, a development that began some years after their historic paper. This story reached a culmination, or at least local extremum, with the discovery at the Large Hadron Collider of a Higgs-like scalar boson in 2012. The talk then proceeds to a slightly more technical level, discussing how we derive predictions from the gauge field theories of the Standard Model and its extensions for use at high energy accelerators.

Keywords: Gauge theory; colliders; quantum chromodynamics; factorization.

1. On the Triumph of Yang–Mills Theory at Accelerators

High energy accelerators offer the most direct window to short-lived quantum processes. The strategy of probing matter at short distances has resulted in the identification/discovery of the gauge and matter fields of the Standard Model. Accelerator programs, however complex and costly, remain experiments that follow the scientific canon. They are capable of design, replication and variation in response to the demands of nature and the imagination. The series of accelerator-based experiments of the past fifty years led ineluctably to the triumph of the gauge concept.[1]

I will review a little of how quantum field theory is applied in accelerator experiments, but we can sum it up with Fig. 1, a picture worth a thousand words.

On the left, the figure shows a quantum-mechanical history that includes all the essential elements of the spontaneously-broken gauge theories[2–4] of the Standard Model.[5–9] Once it became clear that such theories lend themselves to quantum-mechanical renormalization,[10] and with the discovery of asymptotic freedom,[11,12] it was possible to create systematic methods to propose, predict and test for experimental signatures that reflect directly their fundamental structure. The left of Fig. 1 shows the merger of $SU(3)$-colored gluons through a top quark loop, followed by the production of a Higgs boson,[13,14] linking the top quarks to the $SU(2)_L \times U(1)$ electroweak sector of the Standard Model, with the transient appearance of Z bosons, and their subsequent decay into lepton pairs.

In quantum field theory, every observed final state is the result of a quantum-mechanical set of stories of this type. So far, the stories supplied by the Standard

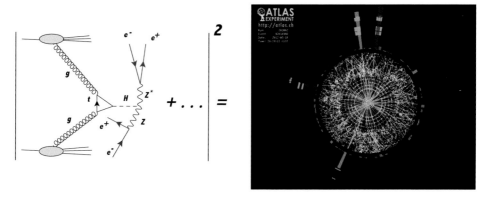

Fig. 1. The schematic equivalence between the squared microscopic amplitude and the ATLAS experiment detector signal for Higgs production followed by decay to four leptons through Z bosons.

Model, built on an unbroken $SU(3)$ color gauge theory (very much like the original Yang–Mills Lagrangian) with gluons the gauge bosons, and a spontaneously-broken $SU(2)_L \times U(1)$, with $W^\pm$, Z and photons, account for and explain essentially all observations at accelerators. The gluons and electroweak bosons in this process are themselves gauge bosons, whose identities and self-interactions disclose the underlying group structure[1] of the Standard Model. All other particles observed at colliders, with a variety and range of masses that remain mysterious, are seen precisely because of their gauge-theory interactions. Indeed, without these interactions we would have no way to produce them at all.

The signature feature of the Yang–Mills extension of gauge invariance to non-Abelian groups is the interaction of gauge fields among themselves. This, of course, is a direct consequence of the form of the gauge field strength, given in the notation of Ref. 1 as

$$F_{\mu\nu} = \frac{\partial B_\mu}{\partial x^\nu} - \frac{\partial B_\nu}{\partial x^\mu} + i\epsilon(B_\mu B_\nu - B_\nu B_\mu), \tag{1}$$

in terms of gauge fields expressed as matrices. In "pure-Yang–Mills" the Lagrange density is just $\mathcal{L} = -(1/4)F^{\mu\nu}F_{\mu\nu}$, invariant under local group (gauge) transformations. The quadratic terms of the field strength are necessary for this invariance, so that the interactions among the gauge bosons, and with other fields "are essentially determined by the requirement of gauge invariance."[1]

The self-interactions of the $SU(3)$ component of the Standard Model, quantum chromodynamics (QCD) are the origin of asymptotic freedom, and also contribute directly to nearly every multijet cross-section, as described below. The self-couplings of the electroweak gauge bosons also hew closely to the Standard Model (for a review, see Chap. 10 of Ref. 15), and their study at high energies continues at the Large Hadron Collider (LHC).[16,17] At LHC energies the mutual scattering of

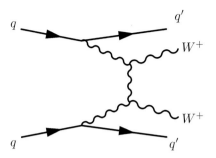

Fig. 2. An example of electroweak vector boson scattering at a hadron collider, from Ref. 18.

electroweak bosons, through processes like those shown in Fig. 2, is for the first time coming into focus.[18,19]

All this could be the "end of the story," except that: (1) Cosmological observations strongly suggest that there are other sources of gravitation in the universe: dark matter, dark energy. Dark matter is by definition bereft of at least electromagnetic interactions. (2) The mass of the Higgs particle in the Standard Model in isolation is unstable to overwhelming quantum corrections, a conundrum often referred to as the "hierarchy" problem. If the existence of as-yet unobserved particles resolves this problem, they may or may not participate in the gauge interactions of the Standard Model.

Contemporary distress with the hierarchy problem of the Standard Model may be compared to 17th Century objections to action at a distance in Newtonian gravity. The objection comes from profound intuition, but does not immediately suggest a resolution. It is attractive to suggest that dark matter plays a role, although this is just a guess. Putting all this aside, as the progress of science put gravitational action at a distance aside until 1915, the success of the gauge-theory based Standard Model is extraordinary. And resolutions of the Standard Model's puzzles, and even of dark matter and energy, may in the fullness of time come from theories with many or most of the Standard Model's properties, or generalizations inspired by it, like supersymmetry. Hopefully, we will not have to wait as long! For the remainder of this talk, I will try to explain how accelerator studies helped us get to this stage, how we learned to recount and recognize the stories like those of Fig. 1 that led to the Standard Model's successes.

2. Techniques from Quantum Field Theory

High energy collisions make possible large momentum transfers, and correspondingly processes that develop over very short distances and times. As we shall see, for short distances accessible to accelerators, we can expand around the free field theory. Starting with an initial state, the system evolves via transitions through one or more intermediate states, finally ending up in an observed final state. The list of possible transitions between states *are* the stories that provide predictions through the calculation of quantum-mechanical amplitudes. The systematic computation of amplitudes in this way is perturbation theory. A related discussion of the following can be found in Ref. 20.

2.1. Perturbation theory

Perturbation theory really just follows from the Schrödinger equation, describing the mixing of free particle states (more on this later),

$$i\hbar \frac{\partial}{\partial t}|\psi(t)\rangle = \left(H^{(0)} + V\right)|\psi(t)\rangle, \qquad (2)$$

with $H^{(0)}$ the "free Hamiltonian," and V some potential. The form of the free Hamiltonian determines the list of possible free states. Usually we start with free-state "in" boundary conditions,

$$|\psi(t=-\infty)\rangle = |m_0\rangle = |p_1^{\text{in}}, p_2^{\text{in}}\rangle, \qquad (3)$$

corresponding to two particles that approach each other from the distant past, as prepared in an accelerator.

Theories differ in their lists of particles and their (hermitian) potentials, sets of operators represented by V. The expansions of perturbation theory are given in terms of matrix elements of V between free states, for which we adopt the notation,

$$V_{j \leftarrow i} = \langle m_j | V | m_i \rangle. \qquad (4)$$

These matrix elements are represented graphically by vertices in diagrams like those on the left of Fig. 1. The states $|m_i\rangle$ and $|m_j\rangle$ differ by the absorption or emission of a particle, the annihilation (creation) of a pair of particles into (from) a heavier particle, etc. In gauge theories, the difference between the numbers of particles is limited to three. Many particles can be created, but only by repeated actions of operator V.

Solutions to the Schrödinger equation (2) are sums of ordered time integrals over the matrix elements in (4). The result is often referred to as "old-fashioned perturbation theory." The scattering amplitudes computed this way are precisely equivalent to the result of computing with the more familiar Feynman diagrams. Scattering experiments measure the quantum-mechanical overlap between a state prepared far in the past (an "in" state, as in Eq. (3)) and a state observed far in the future (an "out" state). In states are what high energy accelerators like the LHC provide; out states are what detectors like ATLAS and CMS detect.

It is not difficult to verify[20] that, in the notation of Eq. (4) the overlap between a solution to Eq. (2) with initial condition $|m_0\rangle$ at $t = -\infty$ and a "final state"

$|m_{\text{out}}\rangle$ can be written in as

$$\langle m_{\text{out}}(\infty)|m_0\rangle = \sum_{n=0}^{\infty} \sum_{m_1\ldots m_{n-1}} \prod_{a=0}^{n} \left(\frac{-i}{\hbar} V_{a \leftarrow a-1}\right) \int_{-\infty}^{\infty} d\tau_n \int_{-\infty}^{\tau_n} d\tau_{n-1}$$

$$\times \cdots \times \int_{-\infty}^{\tau_2} d\tau_1 \, \exp\left[-\frac{i}{\hbar} \sum_{\text{states } b=0}^{n-1} \left(\sum_{j \text{ in } b} E(\vec{p}_j)\right) (\tau_{b+1} - \tau_b)\right], \quad (5)$$

where the ath factor $V_{a \leftarrow a-1}$ is the matrix element of Eq. (4) that takes state $|m_{a-1}\rangle$ into the state $|m_a\rangle$, with $|m_n\rangle \equiv |m_{\text{out}}\rangle$. In fact, this expression is given in the interaction picture, where we remove the phases associated with $|m_0(-\infty)\rangle$ and $|m_{\text{out}}(\infty)\rangle$, thus dropping two of the terms in the phase proportional to τ_0 and τ_{n+1}. The sums over states are suitably-normalized integrals over free-particle phase space, which we may reinterpret as integrals over the loop momenta of a sum of perturbation theory (Feynman) diagrams.

Generically, the sums over states in Eq. (5) are divergent whenever subsets of τ's coincide, $\tau_i \to \tau_j$, and also when some set of τ_i go to infinity. The former, "UV" problem is handled by renormalization, and the solution is summarized by scaling each term in V by an appropriate coupling constant $g(\mu)$, with $(\tau_i - \tau_j)_{\min} = 1/\mu$. In four dimensions, only Yang–Mills theories have the property of asymptotic freedom, $g(\mu) \sim 1/\ln(\mu)$.[11,12] The couplings of the Standard Model are either asymptotically free, or are small enough to not change much over experimentally-accessible energies. This makes an expansion in powers of $\alpha(\mu) = g^2(\mu)/4\pi$ plausible, at least in principle.

Once we do the expansion to calculate the amplitude for a process, the form of an "ideal cross-section," the square of the amplitude, would be one with only a single kinematic scale, to which we can set renormalization scale μ,

$$Q^2 \hat{\sigma}_{\text{SD}}(Q^2, \mu^2, \alpha_s(\mu)) = \sum_n c_n(Q^2/\mu^2)\alpha_s{}^n(\mu) + \mathcal{O}\left(\frac{1}{Q^p}\right)$$

$$= \sum_n c_n(1)\alpha_s{}^n(Q) + \mathcal{O}\left(\frac{1}{Q^p}\right), \quad (6)$$

up to corrections that vanish as Q^{-p}, for some positive power p, typically an integer. The key is to find quantities that are observable, and for which the coefficients are well-behaved, and do not depend on scales for which the coupling is too large. Such quantities are sometimes called "infrared safe." For proton accelerators or hadronic final states, the problem is that there are essentially no cross-sections that qualify as infrared safe without further analysis. What is reason for this problem?

2.2. Mass-shell enhancements in perturbation theory

As we have seen, the Schrödinger equation gives transition matrix elements as sums of ordered time integrals. These time integrals extend to infinity, but usually oscillations damp them and they provide finite answers. Long-time, "infrared" divergences (logs) come about only when phases vanish so that the time integrals diverge.

When does this happen? We can tell by reorganizing the phase in Eq. (5),

$$\exp\left[-\frac{i}{\hbar}\sum_{\text{states } b}\left(\sum_{j \text{ in } b} E(\vec{p}_j)\right)(\tau_{b+1} - \tau_b)\right]$$

$$= \exp\left[-\frac{i}{\hbar}\sum_b \left(\sum_{j \text{ in } b-1} E(\vec{p}_j) - \sum_{j \text{ in } b} E(\vec{p}_j)\right)\tau_b\right] \quad (7)$$

Divergences can occur in the integrals of Eq. (5) for $\tau_i \to \infty$ if two requirements are met.

(1) As is shown by the right-hand side of (7), the phase must vanish, corresponding to sequences of degenerate states,

$$\sum_{j \text{ in } b-1} E(\vec{p}_j) = \sum_{j \text{ in } b} E(\vec{p}_j). \quad (8)$$

(2) From the left-hand side, even if it vanishes, the phase must also be stationary with respect to the momentum integrals that are implicit in the sums over states in Eq. (5). Imposing momentum conservation, this translates into the vanishing of their derivatives with respect to loop momenta ℓ_i^μ,

$$\frac{\partial}{\partial \ell_{i\mu}}[\text{phase}] = \sum_{\text{states } b}\sum_{j \text{ in } b}(\eta_{ij}\beta_j^\mu)(\tau_{b+1} - \tau_b) = 0, \quad (9)$$

where the β_j's are 4-velocities,

$$\frac{\partial E(\vec{p}_j)}{\partial \ell_{i\mu}} = \frac{\partial p_{j\mu}}{\partial \ell_{i\mu}}\frac{\partial E(\vec{p}_j)}{\partial p_{j\mu}} = \eta_{ji}\beta_j^\mu, \quad (10)$$

with $\eta_{ji} = \pm 1, 0$, depending on whether loop i flows on line j, and if so whether momentum ℓ_i flows in the same or opposite direction as p_j.

Now, any vector of the form $\beta_j^\mu \Delta\tau = \Delta x^\mu$ is a translation that is consistent with free-particle classical equations of motion. Equation (9) implies that around each loop, every sequence of free propagations is consistent. Infrared divergences, then, arise from regions in the sums over states where particles describe free, classical propagation that extends to $t \to \infty$, even as the numbers of particles may change through the actions of vertices. This is easy to satisfy in subdiagrams where all the β_j's are equal, but is otherwise quite restrictive. Thus, whenever fast partons emerge from the same point in space–time, amplitudes are enhanced by corrections

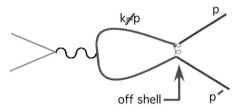

Fig. 3. Example of a degenerate intermediate state that cannot give long-time divergences.

that describe the rescattering, creation and absorption of collinear partons at large times.

A simple example, in Fig. 3, illustrates the surprising power of the requirement of classical propagation. The intermediate state involving momentum k may be degenerate with the final state, but if the two lines emerging from the decay are not parallel, they can never meet again though free-particle propagation. Hence these degenerate states are not associated with a stationary point in the integral, and there is no possibility of an infrared divergence. This kind of reasoning makes identifying infrared enhancements a lot simpler. For particles emerging from a local scattering, (only) collinear or soft (infinite wavelength) lines can give long-time behavior and enhancement. The most straightforward examples of configurations that do give divergences are in the hadronic decay of electroweak bosons, or e^+e^- annihilation in the single-electroweak boson approximation, illustrated in Fig. 4. This pattern generalizes to any order, and any field theory, but gauge theories are the only renormalizable theories with soft $(k \to 0)$ divergences.

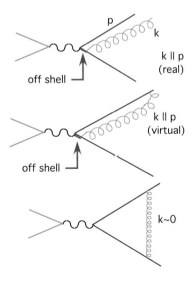

Fig. 4. Low order infrared configurations for electroweak boson decay.

For e^+e^- annihilation, this implies by the optical theorem that the total cross-section is infrared safe.[21] To lowest order in electroweak interactions, but to all orders in QCD, the total cross-section is proportional to the imaginary part of the electroweak boson self-energy. The same reasoning that applies to Fig. 3 shows that there are no stationary points with finite-energy on-shell lines at any loop order. Because the self-energy is finite, so is its imaginary part, and thus so is the total cross-section. The same reasoning, based on the optical theorem, also applies to jet cross-sections,[22,23] all of whose singularities can be derived from a rotationally non-invariant, but still hermitian, truncation of the quantum field theory Lagrangian.

2.3. *Jets: Rare but highly structured events*

At energies much above the mass of the nucleon, certain events include subsets of particles $\{q_j\}$ with anomalously small invariant mass, $(\sum_i q_i)^2 \ll (\sum_i E_i)^2$, and not embedded among other particles of similar energy. Such sets are "jets," and, as depicted in Fig. 5 from the CMS experiment at the LHC, a single scattering event may produce a number of jets. Events containing jets, in which the flow of momentum is radically changed between initial and final states, are a signature of large momentum transfers through local interactions, and as such direct evidence of processes taking place on distances of the order of 1/(momentum transfer).

The history of the term "jets" applied to final states in particle collisions goes back to the 1950's, with sprays of particles observed in collisions of cosmic rays

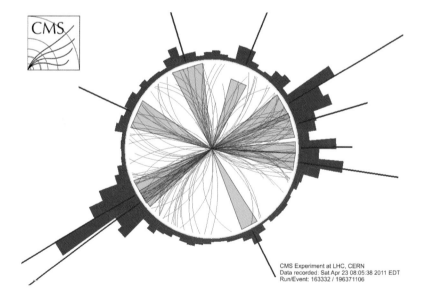

Fig. 5. A multi-jet event as observed by the CMS experiment at the LHC.

with detector materials. In Ref. 24, we read "The average transverse momentum resulting from our measurements is $p_T = 0.5$ BeV/c for pions ... [a table] gives a summary of jet events observed to date ...". These jets emerged from cosmic rays, with energies far above what could then be achieved in the laboratory. The jets of these events are by now interpreted as fragments of the projectile nucleus, whose collider analogs are sometimes referred to as "beam jets." They are not, for the most part, a signal of very large momentum transfer processes, or of the decay of newly-created heavy particles. The observation of jets of the latter types had to wait for the era of high energy physics and the discovery of the Standard Model.

This modern story begins with the parton model for inclusive deep inelastic electron–nucleon scattering (DIS).[25] In the parton model, this *inclusive* process is approximated by the *exclusive* scattering of an electron by a charged constituent of the nucleon, multiplied by a function $F(x)$ that depends only on a "scaling variable," $x \equiv Q^2/2p \cdot q$, with q the 4-momentum transfer and $Q^2 \equiv -q^2 > 0$,

$$\sigma^{\text{incl}}_{e\,\text{proton}}\left(Q, x = \frac{Q^2}{2p \cdot q}\right) \to \sigma^{\text{excl}}_{e\,\text{parton}}(Q, xp) \times F_{\text{proton}}(x). \tag{11}$$

Variable x is interpreted as the fractional momentum of the nucleon carried by the parton, when all masses are neglected. The elastic electron–parton scattering is calculated to lowest order in quantum electrodynamics,

$$e(k) + a(xp) \to e(k - q) + a(xp + q), \tag{12}$$

where a represents the parton, a quark or antiquark in QCD. The value of x is determined simply by requiring the scattered parton to remain massless, $(xp + q)^2 = 0$. In the parton model, the function $F(x)$ has the interpretation of a probability distribution of momentum fractions for parton a in the nucleon. Its independence of the momentum transfer is known as "scaling." Scaling turned out to be a striking and successful description of early DIS data. Its explanation in terms of asymptotic freedom[11,12] was an electrifying development in the discovery of the Standard Model. For an asymptotically free theory, the coupling becomes weak at short distances, even as it increases at longer distances. This made sense for the inclusive cross-section, but the question arises as to what happens to partons in the final state? Does confinement forbid a direct phenomenological expression for quarks?

The unequivocal answer, "no," came from SLAC in 1975: In electron–positron annihilation to hadrons, the angular distribution for energy flow follows the Born expression for the creation of spin-1/2 pairs of fermions: the quarks and antiquarks.[26] Jets are "rare" because the high momentum transfer scattering of partons is rare, but once a hard scattering has occurred they are inevitable, and the rates of their appearances are calculable.

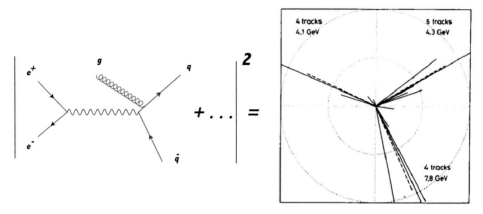

Fig. 6. The partonic process for a three jet event in lowest order QCD and the corresponding TASSO experiment signal at the PETRA accelerator.[28]

After the quark jets of SLAC, came hints of jets associated with gluons in Upsilon decay,[27] and by 1979 clear gluon jets at PETRA,[28–30] illustrated in Fig. 6.

To compute the probabilities of such events we compute jet cross-sections directly in perturbative QCD as though the final state consisted of quarks and gluons.[22,23] Such a prescription, purely in terms of partonic language, seemed strange at first, knowing that quarks and gluons are confined. Nevertheless, the theory gives a prediction, and the theory can tell us when this prediction is not self-consistent. That is, we assume that infrared safe perturbation theory provides an asymptotic expansion.

A good illustration of the kind of cross-section we can calculate is the "thrust,"[31] in e⁺e⁻ annihilation, defined by

$$T \equiv \max_{\hat{n}} \frac{1}{Q} \sum_{\text{particles } i} \vec{p}_i \cdot \hat{n}, \qquad (13)$$

where the maximum is taken over all unit directions $\hat{n}$ in the sphere of the center-of-mass frame. The thrust equals unity for perfectly pencil-like, back-to-back jets of massless particles. By now, the coefficients of α_s/π (lowest order, or LO), $(\alpha_s/\pi)^2$ (next-to-lowest, NLO) and $(\alpha_s/\pi)^3$ (next-to-next-to-lowest, NNLO) are known,[32] and agreement with experiment extends over several orders of magnitude, as illustrated in Fig. 7.

2.4. Factorization

Machines for hadron–hadron scattering allowed for the scattering of quarks and gluons from pre-existing hadrons, whose internal interactions long predate their entrance into a collider beam — indeed, they generally extend back to baryosynthesis in the early universe! Such interactions, involving confinement, are clearly not calculable in perturbation theory. In this case, we apply the method of factorization, in which perturbatively calculable short-distance effects are separated

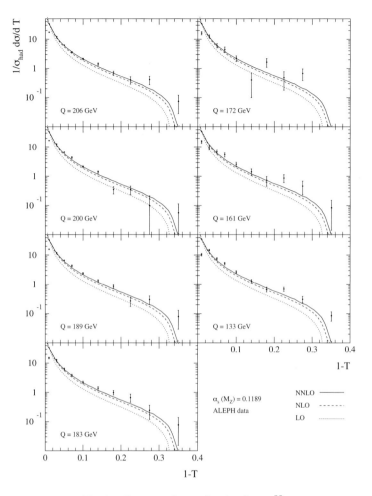

Fig. 7. Data vs. theory for the thrust.[32]

from long-distance nonperturbative dynamics, which enter as parton distributions. Parton distributions are the pre-existing probability densities for individual quarks and gluons to carry the fractional momenta of the colliding nucleons. Factorization is the key to predictions for proposed and established theories, and in general takes the form

$$Q^2 \sigma_{\text{phys}}(Q,m) = \hat{\sigma}(Q/\mu, \alpha_s(\mu)) \otimes f_{\text{LD}}(\mu, m) + \mathcal{O}\left(\frac{1}{Q^p}\right), \qquad (14)$$

which can be thought of as a generalization of both Eqs. (6) and (11). In expressions like this, μ is referred to as the factorization scale, which is the lowest momentum scale on which the "short-distance" function $\hat{\sigma}$ depends, and m represents the infrared scales whose dependence is factorized into the "long-distance" function f_{LD}. "New physics" is in $\hat{\sigma}$, while f_{LD} is "universal." Factorizaton of this sort is required for almost all collider applications, and the determination of parton distributions requires a synthesis of measurements of many factorized cross-sections.[33] The lowest-order process of (11) is now generalized beyond electron–quark scattering, and dressed in $\hat{\sigma}$ by QCD quantum corrections, as in (6). In place of the simple product in (11), the product $\otimes$ represents a convolution in terms of partonic fractional momenta, or other kinematic degrees of freedom that are not observed directly.

What we actually do is to compute the "physical cross-section" σ on the left-hand side of (14) and $f_{\rm LD}$ on the right-hand side, in an infrared-regulated variant of QCD, where we can prove the factorization explicitly. We then extract the perturbative quantity $\hat{\sigma}$, assuming it is the same for true QCD as for its IR-regulated cousin. Factorization for a given cross-section generally requires that it be sufficiently inclusive that no small parameters are introduced in the selection of final states. The form of factorization may also depend on the measurement. Calculations of the short distance functions generally become very complex beyond the lower order in α_s. An enormous amount of work has been put into the calculation of the perturbative amplitudes[34,35] upon which these cross-sections are based, which has led to fruitful interplay between QCD phenomenology, abstract quantum field theory and pure mathematics. The steps from amplitudes to cross-sections with prescribed phase space presents further challenges. Although great progress has been made over the past few years, much remains to be done to exploit the full potential of accelerator data.[36] A recent milestone is the completion of the order α_s^3 corrections to $\hat{\sigma}$ for the inclusive production of the Higgs boson in the Standard Model.[37]

3. Evolution/Resummation

The full power of asymptotic freedom and factorization requires evolution, by which we can control dependence on the factorization scale μ in factorized cross-sections like (14). This enables us to compute the short distance scattering $\hat{\sigma}$ in terms of a single scale, with $\mu \sim Q$, and hence to derive the most accurate expansions in $\alpha_s(Q)$ as Q increases. Whenever there is a factorized physical quantity, we can derive an evolution equation, starting with the independence of any such quantity from the factorization scale,

$$0 = \mu \frac{d}{d\mu} \ln \sigma_{\rm phys}(Q, m). \tag{15}$$

For simplicity, suppose $\sigma_{\rm phys}$ is a simple product of $\hat{\sigma}$ and f. Then by an elementary separation of variables we find that

$$\mu \frac{d \ln f}{d\mu} = -P(\alpha_s(\mu)) = -\mu \frac{d \ln \hat{\sigma}}{d\mu}, \tag{16}$$

where the function $P(\alpha_s)$ depends on only the (dimensionless) variables held in common by the short- and long-distance functions. We can calculate $P(\alpha_s)$, which may be referred to as a splitting function or an anomalous dimension depending on the context, because it is the derivative of $\hat{\sigma}$, which we can calculate. Equations of this sort are said to describe evolution or resummation. Applied to Eq. (14), for example, Eq. (16) implies that

$$\sigma_{\text{phys}}(Q,m) = \sigma_{\text{phys}}(q,m) \left[\frac{\hat{\sigma}(Q,\alpha_s(Q))}{\hat{\sigma}(q,\alpha_s(q))}\right] \exp\left\{\int_q^Q \frac{d\mu'}{\mu'} P(\alpha_s(\mu'))\right\}, \quad (17)$$

in which the second and third factors on the right-hand side, which contain all Q-dependence, are computable in perturbation theory so long as $\alpha_s(Q)$ and $\alpha_s(q)$ are both small. This means that in a theory with asymptotic freedom, observations at moderate scales, q say, lead to predictions for all larger scales, Q.

Such factorization, and hence evolution applies, for example, to the dimensionless structure function $F_2(Q^2, x)$ in DIS, which can be thought of as a cross-section with dimensionful kinematic factors removed. Specifically, we consider its Mellin moments with respect to the scaling variable, x, $\tilde{F}_2(Q^2, N) = \int_0^1 dx\, x^{N-1} F_2(Q^2, x)$. These moments resolve the convolution in partonic momentum fraction into a product of the form

$$\tilde{F}_2(Q^2, N) = \sum_{\text{partons } a} C_{2a}\left(N, \frac{Q}{\mu}, \alpha_s(\mu)\right) f_a(N, \mu). \quad (18)$$

Here, the sum is over parton flavors: quarks, antiquark, gluons. On the right, the C_{2a} are short-distance "coefficient functions", one for each parton flavor, and the $f_a(N, \mu)$ are Mellin moments of parton distributions, $f_a(\xi, \mu)$, in the spirit of Eq. (14), with respect to momentum fraction, ξ. In this case, the separation constants $P_{ab}(N, \alpha_s)$ depend on the moment variable N and are denoted $\gamma_{ab}(N, \alpha_s)$. This is a matrix, as variations with the factorization scale reflect the effects of quark pair creation and gluon radiation, which can change the flavor of the parton that initiates the hard scattering. The evolution equation for the long-distance parton distributions can then be written as

$$\mu \frac{\partial}{\partial \mu} f_a(N, \mu) = \sum_b \gamma_{ab}(N, \alpha_s(\mu)) f_b(N, \alpha_s(\mu)), \quad (19)$$

which is one of the forms of the celebrated Dokshitzer–Gribov–Lipatov–Altarelli–Parisi (DGLAP) equation,[38–40] the bedrock of high energy phenomenology at colliders.

If for simplicity we suppress the matrix structure, and neglect the analog of the perturbative prefactor in Eq. (17), we find for our structure functions the Q-dependence,

$$\tilde{F}_2(Q^2, N) = \tilde{F}_2(Q_0^2, N) \exp\left[-\frac{1}{2} \int_{Q_0^2}^{Q^2} \frac{d\mu'^2}{\mu'^2} \gamma(N, \alpha_s(\mu'))\right], \quad (20)$$

again enabling the use of lower-energy observations to give higher energy predictions. Expanding $\gamma(N, \alpha_s) = \gamma_N^{(1)}(\alpha_s/\pi) + \cdots$, and using the lowest-order version of the running coupling, $\alpha_s(\mu) = 4\pi/b_0 \ln(\mu^2/\Lambda_{QCD}^2)$, we find the approximation

$$\tilde{F}_2(Q^2, N) = \tilde{F}_2(Q_0^2, N) \left(\frac{\ln(Q^2/\Lambda_{QCD}^2)}{\ln(Q_0^2/\Lambda_{QCD}^2)} \right)^{-2\gamma_N^{(1)}/b_0} \tag{21}$$

In its full form, which depends on all of the particle content and dynamics of quantum chromodynamics, this procedure works really well. It implies the approximate scaling seen in early DIS experiments at moderate x and pronounced evolution for smaller x, as shown in Fig. 8.

Hard hadron–hadron scattering, as at the LHC, involves partons from both colliding hadrons, which involves only a modest extension of the factorization formalism. For an inclusive process in hadron–hadron scattering that requires a

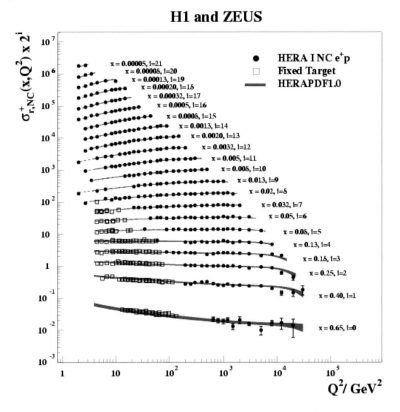

Fig. 8. DIS cross-sections from ZEUS and H1 experiments at the HERA accelerator.[41]

large momentum transfer M to produce final state $F(M) + X$, the corresponding factorized cross-section can be written as

$$d\sigma_{H_1 H_2}(p_1, p_2, M) = \sum_{a,b} \int_0^1 dx_a\, dx_b\, d\hat{\sigma}_{ab \to F+X}(x_a p_1, x_b p_2, M, \mu)$$

$$\times f_{a/H_1}(x_a, \mu)\, f_{b/H_2}(x_b, \mu), \quad (22)$$

in terms of a perturbative short distance (differential) cross-section $d\hat{\sigma}$, combined with the same parton distribution functions as above, but evolved to higher scales in general. This form is a straightforward variant of the generic factorization (14), but of course, it requires a proof. Factorization proofs, which have been the subject of considerable effort, justify the "universality" of the parton distributions, that is, that parton distributions measured in deep-inelastic scattering are the same functions that appear in hadron–hadron scattering. If this is indeed the case, then data from one class of experiments can be used to make predictions for another class. Thus, for example, the measurements of parton distributions at HERA in the 1990s can be used to predict cross-sections for hypothetical super-partners at the LHC in Run II.

We will come back to the how and why of factorization, which is fundamental to applications of gauge theories at accelerators, in the closing pages. We can use jet cross-sections to illustrate the success of the factorization paradigm.

Through the early 1980's, there were strong suggestions of scattered parton jets at CERN and Fermilab.[42,43] Fuller clarification, however, awaited experiments at the SPS collider, whose large angular coverage made possible plots that exhibited energy flow over the whole detector. This led to the observation of high-p_T jet pairs that unequivocally represent the scattering of partons at short distances. The underlying equivalence of the quantum-mechanical picture of gluon exchange between quarks and detector signals is represented by Fig. 9, in which the towers on the "unrolled" detector surface reveal the flow of energy.[44,45]

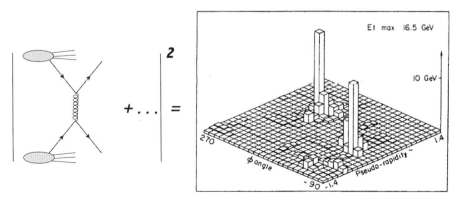

Fig. 9. Equivalence of QCD quark–quark scattering and the observation of jets by the flow of energy in the UA1 detector.[45]

The nineteen nineties ushered in what we may call the era of the great "Standard Model Machines." HERA for DIS, LEP I and II for e^+e^- annihilation, and the Tevatron Run I for $p\bar{p}$ collisions together provided jet cross-sections over multiple orders of magnitude, as they established and confirmed the electroweak structure and flavor content of the Standard Model. The current decade has brought a new era of jets at the limits of the Standard Model, initiated by Tevatron Run II, and realized with the ongoing LHC program, at 7 TeV, then 8 TeV, and now 13 TeV in the center of mass. Events transpiring at the scale $\delta x \sim \frac{\hbar c}{1\,\text{TeV}} \sim 2 \times 10^{-19}$ meters are now routinely observed about 10 meters away, an observational bridge of twenty orders of magnitude. The impressive success of theory predictions based on factorized cross-sections like Eq. (22) is shown in Fig. 10.

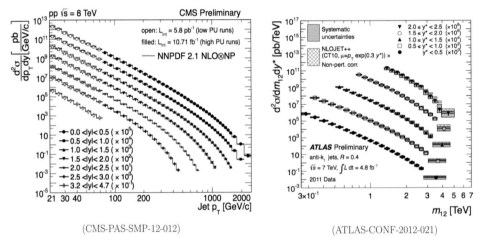

Fig. 10. Jet p_T distributions measured by the CMS experiment, and pair mass distributions measured by the ATLAS experiment, showing agreement with Standard Model predictions over many orders of magnitude.

At this point, we may mention another life for jets, "shining from the inside" as a probe for new phases of strongly-interacting matter in nuclear collisions[46,47] at the Relativistic Heavy Ion Collider (RHIC) and the LHC, and, prospectively in "cold nuclei" at a electron–ion collider.[48]

With this motivation, we turn to the physics behind, and the arguments for, the factorization properties on which the calculation of jet and related cross-sections are based.

4. Gauge Theory Factorization

Understanding factorization is about learning how to calculate with a theory that acts differently on different scales. For the purposes of jet cross-sections, the underlying factorization takes a form first recognized in the late 1970s,[50–55]

$$d\sigma(A + B \to \{p_i\}) = \sum_{a,b} \int_0^1 dx_a dx_b \, f_{a/A}(x_a, \mu_F) \, f_{b/B}(x_b, \mu_F)$$

$$\times dC \left(x_a p_A, x_b p_B, \frac{Q}{\mu_F}, \frac{p_i \cdot p_j}{p_k \cdot p_l} \right)_{ab \to c_1 \ldots c_{N_{\text{jets}}} + X}$$

$$\times \left[\prod_{i=1}^{N_{\text{jets}}} J_{c_i}(p_i, \mu_F) \right]. \tag{23}$$

Like Eqs. (14) and (22), this formula includes parton distributions and short-distance "coefficient" functions, as well as a further factorization into functions that represent the jets. Closely-related factorizations apply to processes involving exclusive final states,[56–58] and have found renewed development in the language of effective field theory.[59–61]

Expressions of this form tell a story of nearly on-shell propagations in the initial and final states, punctuated by a single short-distance interaction. As mentioned above, the definitions of the jets must be sufficiently inclusive that no small scales are introduced.

The jets themselves are characterized by correlated internal dynamics, which is "autonomous" relative to the remainder of the process. We have already seen that enhancement of correlations between collinear particles is built into quantum field theory. Where does the automonomy come from?

To distill the essence of this argument, we will think of classical fields seen by scattered charges.[65] Even though we are working in quantum field theory, a classical picture is not so far-fetched, because the correspondence principle is the key to infrared divergences in perturbation theory. An accelerated charge must produce classical radiation, and infinite numbers of soft gauge vectors are required to make a classical field.

We imagine a "jet-parton," whose coordinate space position we label as x' in its own rest frame, moving with velocity v in the 3-direction toward or away from a "recoiling parton" of charge q, at rest in a system with coordinates x. We can imagine both of these particles as products of the hard scattering. They may interact with particles in their immediate neighborhood, but here we concentrate on the effect of the recoil parton on the jet parton. The relevant Lorentz transformation between the two frames is

$$x_3 = \gamma(x'_3 + vt') = \gamma \Delta, \tag{24}$$

which serves to define Δ. We can think of x'_3 as a small, fixed scale, while x_3 changes rapidly as the particle recedes, as in Fig. 11.

We imagine that the "collision" was at $\Delta = 0$, i.e. $t' = -\frac{1}{v}x'_3$, at which time the particles were separated by only a small transverse distance, $x_T = x'_T$, when a large momentum transfer could have taken place between the two. An estimate

Fig. 11. Illustration of separating charges.

of the gauge theory physical effects at later times is found from expressions for the electric field in an abelian theory due to a charge q at rest in the x frame and as seen in the x' frame:

$$E'_3(x') = \frac{q\gamma\Delta}{(x_T^2 + \gamma^2\Delta^2)^{3/2}} \sim \frac{1}{\gamma^2}, \frac{q}{\Delta}. \tag{25}$$

The force in the 3-direction felt by the parton of charge q' traveling with the jet due to the recoiling charge is just q' times this field. The electric, **E** field, however, seen by the receding (or approaching) particle is highly contracted, falling off as $1/\gamma^2$ at all times except during an interval whose width decreases as $1/\gamma^2$, and hence as a power of the momentum of the jet. This suggests that the time development of scattered charges is indeed independent of the hard scattering.

Even in this classical, abelian example, however, the richness of the gauge theory description can be seen. In contrast to the field strengths, the vector potential, A^μ is uncontracted, but is mostly a total derivative as seen in the x' frame:

$$A^\mu(t', x'_3) = q\frac{\partial}{\partial x'_\mu} \ln(\Delta(t', x'_3)) + \mathcal{O}(1-\beta). \tag{26}$$

This "large" part of A^μ can be removed by a gauge transformation in principle. The need to implement this freedom makes proofs of factorization challenging in gauge theories.[62–64] Nevertheless, apparent nonfactorization cancels for inclusive cross-sections, and corrections to Eq. (23) are of the same order as the residual "drag" forces remaining from the vector potential left over from the total derivative. We can estimate the energy-dependence of these corrections by the relation

$$1 - \beta \sim \frac{1}{2}\left[\sqrt{1-\beta^2}\right]^2 \sim \frac{1}{2}\left[\frac{m}{\sqrt{\hat{s}}}\right]^2, \tag{27}$$

with $\sqrt{\hat{s}}$ the invariant mass of the system made of the jet parton and the recoil parton, assumed to have some mass m. Corrections to the autonomous, i.e. factorized, description of high energy processes in this model are thus power suppressed in momentum transfer, suggesting the size of corrections to factorization.

In QCD, these same features are embedded in Feynman diagrams.[65,66] When a gluon's momentum, k becomes collinear to the momentum p of a particle that has emitted it, diagrams that contribute to the amplitude are singular. This is the quantum field theory analog of the "uncontracted" classical gauge potential in Eq. (26). In covariant gauges, these singularities appear in cross-sections through interference with emission from other lines in the collision process. In general, all diagrams contribute, but in the limit that the gluon is parallel to the particle that emits it, the sum of all diagrams is independent of the other momentum directions, and hence is the same for all hard scattering kinematics (see Fig. 12). This is the mechanism that makes possible the universality of long-distance factors (parton distributions) in hadron–hadron scattering in QCD, and in general it only emerges after a sum over many diagrams.

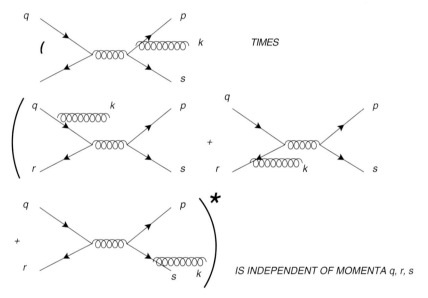

Fig. 12. In the limit that k is parallel to p the interference between the emission of gluon k from the p line and its emission from lines q, r and s is independent of momenta q, r and s.

On a still more technical level, the singular, "collinear" gluons emitted by a charge carry polarizations that are proportional to their own momentum ($\epsilon^\mu(k) \propto k^\mu$), and can give non-factoring contributions that grow with energy in individual diagrams. In QCD and other gauge theories, the gauge invariance of physical quantities ensures that such lines with unphysical polarizations organize themselves into gauge rotations on physical particles.[59,61,64] Such gauge rotations are generalizations of phase factors associated with the gauge potential of quantum electrodynamics,

$$\Phi_\beta(0) = P \, \exp\left[\int_0^\infty d\lambda \, \beta \cdot A(\lambda\beta^\mu)\right], \tag{28}$$

with P path ordering in the group space of the non-Abelian field A, along the semi-infinite lightlike path in the direction β. Ambiguities in the choice of β are in many ways analogous to the arbitrariness of the precise factorization scale, described above.

In the manner of a classical charge moving near the speed of light, the jet only knows the rest of the world as a source of unphysically-polarized gluons. These non-Abelian phases[67] or Wilson lines[69] are themselves related to the geometric structure underlying gauge theories (see the commentary on Ref. 67 in Ref. 68), and appear in numerous contexts beyond the perturbative description on which we've focused in this talk.[70–73] The future will surely see new applications, especially in the transition from infrared safe, perturbative observables to infrared-sensitive hadronic observables. In this context, many results are known to all orders in perturbation theory. Such results typically point at how perturbation theory transcends itself,[74,75] making room for the true long-time behavior of the system, emergent from its gauge interactions.

5. Conclusions

Accelerators have confirmed the fundamental degrees of freedom in the gauge theories of the Standard Model directly, complementing and motivating the great advances in technology that were necessary to probe nature at distance scales down to twenty orders of magnitude below the size of the apparatus. For the most part, contemporary observations are designed for identifying partonic states, in an effort to detect and reject QCD backgrounds in the search for physics beyond the Standard Model.

Time will tell whether the gauge theories accessible at accelerators will offer a resolution to dark matter identity, the hierarchy problem, and related mysteries, as they did for once-exotic manifestations of the Standard Model. I hope so. Reflecting on the extraordinary range of success for the theories inspired by the gauge concept, sixty years after its extension to non-Abelian groups, we may have cause for optimism.

Acknowledgments

I thank the organizers, especially K. K. Phua and L. Brink, for the invitation and support to participate in the *Conference on 60 Years of Yang–Mills Gauge Theories*, and the Institute of Advanced Studies, Nanyang Technological University for its hospitality. This work was supported in part by the National Science Foundation under award PHY-1316617.

References

1. C. N. Yang and R. L. Mills, Conservation of isotopic spin and isotopic gauge invariance, *Phys. Rev.* **96**, 191 (1954).

2. P. W. Higgs, Broken symmetries and the masses of gauge bosons, *Phys. Rev. Lett.* **13**, 508 (1964).
3. F. Englert and R. Brout, Broken symmetry and the mass of gauge vector mesons, *Phys. Rev. Lett.* **13**, 321 (1964).
4. G. S. Guralnik, C. R. Hagen and T. W. B. Kibble, Global conservation laws and massless particles, *Phys. Rev. Lett.* **13**, 585 (1964).
5. S. L. Glashow, Partial symmetries of weak interactions, *Nucl. Phys.* **22**, 579 (1961).
6. S. Weinberg, A model of leptons, *Phys. Rev. Lett.* **19**, 1264 (1967).
7. A. Salam, Weak and electromagnetic interactions, in *Elementary Particle Theory: Relativistic Groups and Analyticity* (Ed. Nils Svartholm), Almqvist & Wiksell, Stockholm, *Conf. Proc. C* **680519**, 367 (1968).
8. H. Fritzsch, M. Gell-Mann and H. Leutwyler, Advantages of the color octet gluon picture, *Phys. Lett. B* **47**, 365 (1973).
9. S. Weinberg, Nonabelian gauge theories of the strong interactions, *Phys. Rev. Lett.* **31**, 494 (1973).
10. G. 't Hooft and M. J. G. Veltman, Regularization and renormalization of gauge fields, *Nucl. Phys. B* **44**, 189 (1972).
11. D. J. Gross and F. Wilczek, Ultraviolet behavior of nonabelian gauge theories, *Phys. Rev. Lett.* **30**, 1343 (1973).
12. H. D. Politzer, Reliable perturbative results for strong interactions?, *Phys. Rev. Lett.* **30**, 1346 (1973).
13. G. Aad *et al.* [ATLAS Collaboration], Observation of a new particle in the search for the Standard Model Higgs boson with the ATLAS detector at the LHC, *Phys. Lett. B* **716**, 1 (2012) [arXiv:1207.7214 [hep-ex]].
14. S. Chatrchyan *et al.* [CMS Collaboration], Observation of a new boson at a mass of 125 GeV with the CMS experiment at the LHC, *Phys. Lett. B* **716**, 30 (2012) [arXiv:1207.7235 [hep-ex]].
15. K. A. Olive *et al.* [Particle Data Group Collaboration], Review of particle physics, *Chin. Phys. C* **38**, 090001 (2014).
16. G. Aad *et al.* [ATLAS Collaboration], Measurement of the $WW + WZ$ cross-section and limits on anomalous triple gauge couplings using final states with one lepton, missing transverse momentum, and two jets with the ATLAS detector at $\sqrt{s} = 7$ TeV, *JHEP* **1501**, 049 (2015) [arXiv:1410.7238 [hep-ex]].
17. V. Khachatryan *et al.* [CMS Collaboration], Measurement of the W^+W^- cross-section in pp collisions at $\sqrt{s} = 8$ TeV and limits on anomalous gauge couplings, CMS-SMP-14-016, CERN-PH-EP-2015-122, arXiv:1507.03268 [hep-ex].
18. G. Aad *et al.* [ATLAS Collaboration], Evidence for electroweak production of $W^{\pm}W^{\pm}jj$ in *pp* collisions at $\sqrt{s} = 8$ TeV with the ATLAS detector, *Phys. Rev. Lett.* **113**, 141803 (2014) [arXiv:1405.6241 [hep-ex]].
19. V. Khachatryan *et al.* [CMS Collaboration], Study of vector boson scattering and search for new physics in events with two same-sign leptons and two jets, *Phys. Rev. Lett.* **114**, 051801 (2015) [arXiv:1410.6315 [hep-ex]].

20. G. F. Sterman, Perturbative quantum field theory, *Int. J. Mod. Phys. A* **16**, 3041 (2001) [hep-ph/0012237].
21. T. Appelquist and H. Georgi, e^+e^- annihilation in gauge theories of strong interactions, *Phys. Rev. D* **8**, 4000 (1973).
22. G. F. Sterman and S. Weinberg, Jets from quantum chromodynamics, *Phys. Rev. Lett.* **39** (1977) 1436;
23. G. F. Sterman, Mass divergences in annihilation processes. 1. Origin and nature of divergences in cut vacuum polarization diagrams, *Phys. Rev. D* **17**, 2773 (1978); Mass divergences in annihilation processes. 2. Cancellation of divergences in cut vacuum polarization diagrams, *Phys. Rev. D* **17** 2789 (1978).
24. B. Edwards, J. Losty, D. H. Perkins, K. Pinkau and J. Reynolds, Analysis of nuclear interactions of energies between 1,000 and 100,000 BeV, *Phil. Mag.* (Ser. 7) 3, 237 (1958).
25. J. D. Bjorken and E. A. Paschos, Inelastic electron proton and gamma proton scattering, and the structure of the nucleon, *Phys. Rev.* **185**, 1975 (1969).
    R. P. Feynman, Very high-energy collisions of hadrons, *Phys. Rev. Lett.* **23**, 1415 (1969).
26. G. Hanson *et al.*, Evidence for jet structure in hadron production by e^+e^- annihilation, *Phys. Rev. Lett.* **35**, 1609 (1975).
27. C. Berger *et al.* [PLUTO Collaboration], Jet analysis of the Υ (9.46) decay into charged hadrons, *Phys. Lett. B* **82**, 449 (1979); C. Berger *et al.* [PLUTO Collaboration], Topology of the Υ decay, *Z. Phys. C* **8**, 101 (1981).
28. S. L. Wu, e^+e^- Physics at PETRA: The First 5-Years, *Phys. Rept.* **107**, 59 (1984).
29. J. R. Ellis, M. K. Gaillard and G. G. Ross, Search for gluons in e^+e^- annihilation, *Nucl. Phys. B* **111**, 253 (1976) [Erratum-*ibid.* **130**, 516 (1977)].
30. J. R. Ellis and I. Karliner, Measuring the spin of the gluon in e^+e^- annihilation, *Nucl. Phys. B* **148**, 141 (1979).
31. E. Farhi, A QCD test for jets, *Phys. Rev. Lett.* **39**, 1587 (1977).
32. A. Gehrmann-De Ridder, T. Gehrmann, E. W. N. Glover and G. Heinrich, NNLO corrections to event shapes in e^+e^- annihilation, *JHEP* **0712** (2007) 094 [arXiv:0711.4711 [hep-ph]].
33. J. Rojo *et al.*, The PDF4LHC report on PDFs and LHC data: Results from Run I and preparation for Run II, *J. Phys. G* **42**, 103103 (2015) [arXiv:1507.00556 [hep-ph]].
34. C. F. Berger and D. Forde, Multi-parton scattering amplitudes via on-shell methods, *Ann. Rev. Nucl. Part. Sci.* **60**, 181 (2010) [arXiv:0912.3534 [hep-ph]].
35. H. Elvang and Y. T. Huang, Scattering amplitudes, arXiv:1308.1697 [hep-th].
36. J. M. Campbell *et al.*, Working group report: Quantum chromodynamics, arXiv:1310.5189 [hep-ph].
37. C. Anastasiou, C. Duhr, F. Dulat, F. Herzog and B. Mistlberger, Real-virtual contributions to the inclusive Higgs cross-section at N^3LO, *JHEP* **1312**, 088 (2013) [arXiv:1311.1425 [hep-ph]].

38. V. N. Gribov and L. N. Lipatov, Deep inelastic e p scattering in perturbation theory, *Sov. J. Nucl. Phys.* **15**, 438 (1972) [*Yad. Fiz.* **15**, 781 (1972)].
39. G. Altarelli and G. Parisi, Asymptotic freedom in parton language, *Nucl. Phys.* B **126**, 298 (1977).
40. Y. L. Dokshitzer, Calculation of the structure functions for deep inelastic scattering and e^+e^- annihilation by perturbation theory in quantum chromodynamics., *Sov. Phys. JETP* **46**, 641 (1977) [*Zh. Eksp. Teor. Fiz.* **73**, 1216 (1977)].
41. F. D. Aaron et al. [H1 and ZEUS Collaborations], Combined measurement and QCD analysis of the inclusive e^+- p scattering cross sections at HERA, *JHEP* **1001**, 109 (2010) [arXiv:0911.0884 [hep-ex]].
42. M. Della Negra et al. [CERN-College de France-Heidelberg-Karlsruhe Collaboration], Observation of jet structure in high p_T events at the ISR and the importance of parton transverse momentum, *Nucl. Phys.* B **127**, 1 (1977); P. Darriulat, *Ann. Rev. Nucl. Part. Sci.* **30**, 159 (1980).
43. M. D. Corcoran et al., Evidence that high P(t) jet pairs give direct information on parton parton scattering, *Phys. Rev. Lett.* **44**, 514 (1980).
44. M. Banner et al. [UA2 Collaboration], Observation of very large transverse momentum jets at the CERN anti-p p collider, *Phys. Lett.* B **118**, 203 (1982).
45. G. Arnison et al. [UA1 Collaboration], Observation of jets in high transverse energy events at the CERN proton — anti-proton collider, *Phys. Lett.* B **123**, 115 (1983).
46. J. Adams et al. [STAR Collaboration], Experimental and theoretical challenges in the search for the quark gluon plasma: The STAR Collaboration's critical assessment of the evidence from RHIC collisions, *Nucl. Phys.* A **757**, 102 (2005) [nucl-ex/0501009].
47. K. Adcox et al. [PHENIX Collaboration], Formation of dense partonic matter in relativistic nucleus-nucleus collisions at RHIC: Experimental evaluation by the PHENIX collaboration, *Nucl. Phys.* A **757**, 184 (2005) [nucl-ex/0410003].
48. A. Accardi et al., Electron Ion Collider: The next QCD frontier — Understanding the glue that binds us all, arXiv:1212.1701 [nucl-ex].
49. G. Aad et al. [ATLAS Collaboration], Observation of a centrality-dependent dijet asymmetry in lead-lead collisions at $\sqrt{s_{NN}} = 2.77\,\text{TeV}$ with the ATLAS detector at the LHC, *Phys. Rev. Lett.* **105** (2010) 252303 [arXiv:1011.6182 [hep-ex]].
50. S. B. Libby and G. F. Sterman, Jet and lepton pair production in high-energy lepton-hadron and hadron-hadron scattering, *Phys. Rev.* D **18**, 3252 (1978).
51. R. K. Ellis, H. Georgi, M. Machacek, H. D. Politzer and G. G. Ross, Perturbation theory and the parton model in QCD, *Nucl. Phys.* B **152**, 285 (1979).
52. A. V. Efremov and A. V. Radyushkin, Field theoretic treatment of high momentum transfer processes. 3. Gauge theories, *Theor. Math. Phys.* **44**, 774 (1981) [*Teor. Mat. Fiz.* **44**, 327 (1980)].

53. C. T. Sachrajda, Inclusive production at large transverse momentum in QCD, *Phys. Lett. B* **76**, 100 (1978).
54. A. H. Mueller, Cut vertices and their renormalization: A generalization of the Wilson expansion, *Phys. Rev. D* **18**, 3705 (1978).
55. A. H. Mueller, Perturbative QCD at high-energies, *Phys. Rept.* **73**, 237 (1981).
56. A. V. Efremov and A. V. Radyushkin, Factorization and asymptotical behavior of pion form-factor in QCD, *Phys. Lett. B* **94**, 245 (1980).
57. G. P. Lepage and S. J. Brodsky, Exclusive processes in perturbative quantum chromodynamics, *Phys. Rev. D* **22** (1980) 2157.
58. X. D. Ji, Deeply virtual Compton scattering, *Phys. Rev. D* **55**, 7114 (1997) [hep-ph/9609381].
59. C. W. Bauer, S. Fleming, D. Pirjol and I. W. Stewart, An effective field theory for collinear and soft gluons: Heavy to light decays, *Phys. Rev. D* **63**, 114020 (2001) [hep-ph/0011336].
60. C. W. Bauer, D. Pirjol and I. W. Stewart, Soft collinear factorization in effective field theory, *Phys. Rev. D* **65**, 054022 (2002) [hep-ph/0109045].
61. T. Becher, A. Broggio and A. Ferroglia, Introduction to soft-collinear effective theory, arXiv:1410.1892 [hep-ph].
62. G. T. Bodwin, Factorization of the Drell-Yan cross-section in perturbation theory, *Phys. Rev. D* **31**, 2616 (1985) [Erratum *ibid.* **34**, 3932 (1986)].
63. J. C. Collins, D. E. Soper and G. Sterman, Factorization for short distance hadron-hadron scattering, *Nucl. Phys. B* **261**, 104 (1985);
J. C. Collins, D. E. Soper and G. Sterman, Soft gluons and factorization, *Nucl. Phys. B* **308**, 833 (1988).
64. J. C. Collins, D. E. Soper and G. F. Sterman, Factorization of hard processes in QCD, *Adv. Ser. Direct. High Energy Phys.* **5**, 1 (1989) [hep-ph/0409313].
65. R. Basu, A. J. Ramalho and G. F. Sterman, Factorization at higher twist in hadron-hadron scattering, *Nucl. Phys. B* **244**, 221 (1984).
66. J. W. Qiu and G. F. Sterman, Power corrections to hadronic scattering. 2. Factorization, *Nucl. Phys. B* **353**, 137 (1991).
67. C. N. Yang, Integral formalism for gauge fields, *Phys. Rev. Lett.* **33**, 445 (1974).
68. C. N. Yang, Selected papers (1945–1980) with commentary, World Scientific Series in 20th Century Physics Vol. 36, (World Scientific, 2005).
69. K. G. Wilson, Confinement of quarks, *Phys. Rev. D* **10**, 2445 (1974).
70. I. Bialynicki-Birula, Gauge invariant variables in the Yang–Mills theory, *Bull. Acad. Polon. Sci.* **11**, 135 (1963).
71. S. Mandelstam, Feynman rules for electromagnetic and Yang–Mills fields from the gauge independent field theoretic formalism, *Phys. Rev.* **175**, 1580 (1968).
72. A. M. Polyakov, Thermal properties of gauge fields and quark liberation, *Phys. Lett. B* **72**, 477 (1978).
73. L. Susskind, Lattice models of quark confinement at high temperature, *Phys. Rev. D* **20**, 2610 (1979).

74. A. H. Mueller, "On the structure of infrared renormalons in physical processes at high-energies," *Nucl. Phys. B* **250**, 327 (1985).
75. G. F. Sterman, Approaching the final state in perturbative QCD, *Int. J. Mod. Phys. A* **18**, 4329 (2003) [*Annales Henri Poincare* **4**, S259 (2003)] [hep-ph/0301243].

Yang–Mills Theory at 60:
Milestones, Landmarks and Interesting Questions[*]

Ling-Lie Chau

Department of Physics, University of California, Davis, CA 95616, USA
chau@physics.ucdavis.edu
www.physics.ucdavis.edu

On the auspicious occasion of celebrating the 60th anniversary of the Yang–Mills theory, and Professor Yang's many other important contributions to physics and mathematics, I will highlight the impressive milestones and landmarks that have been established in the last 60 years, as well as some interesting questions that are worthy of answers from future researches. The paper is written (without equations) for the interest of non-scientists as well as of scientists.

Keywords: Yang–Mills theory at 60.

Contents

1	Overview	79
2	Yang–Mills Theory: Milestones and Interesting Questions	81
3	The (Anti-)Self-dual Yang–Mills Fields: Landmarks and Interesting Questions	83
4	Concluding Remarks	87
	Acknowledgments	88
	Attachment A	89
	Attachment B	94
	Attachment C	100
	Attachment D	101
	References	102

1. Overview[a]

The 1954 Yang–Mills (YM) paper,[1] that set forth the YM theory with YM equations, made at least the following three major advances, if not revolutions, in theoretical and mathematical physics.

---

[*]Invited contribution to the Proceedings of the Conference on 60 Years of Yang–Mills Gauge Theories, IAS, NTU, Singapore, 25–28 May 2015. Also by invitation an earlier version of the paper (with slight differences) was published in the Dec. 30, 2016 issue of International Journal of Modern Physics A, *Int. J. Mod. Phys. A* **30**, 1530068 (2015).

[a]The author would recommend that readers first read Secs. 1 (Overview) and 4 (Concluding Remarks) before reading Secs. 2 and 3, which cover a vast amount of information in highlight fashion, supplemented with references and ample sources for references for those who would like to dive into details.

First, the YM equations generalized the Maxwell equations[2] from the Abelian gauge field equations, in which the Maxwell fields do not interact with themselves, to the more general non-Abelian gauge field equations, in which the YM fields do interact with themselves. It is akin to Einstein's general relativity (GR) equations[3] generalizing Newton's equations for gravity,[4] though in different ways. The Maxwell equations are exact by themselves and the YM equations are generalizations, leaping conceptually from the simpler special case of the Maxwell equations. In contrast, Newton's equations for gravity are weak-field approximations of Einstein's GR equations, so they still work for our daily lives and for the motions of the planets, except for minor but measurable GR improvements to agree with Nature, e.g., in the workings of the global positioning system (GPS) and in the precession of the perihelion of Mercury.

Second, the YM equations (including the anti-self-dual and the self-dual YM equations) are beautiful nonlinear partial differential equations not only for physics but also for mathematics, which have inspired the development of fundamental physics as well as beautiful mathematics, as has been the case with the GR equations.

Third, the YM theory manifested the local gauge covariance, which was first established in Maxwell's equations for electromagnetic interactions, as a principle for more general interactions (later found to be the weak and strong interactions, see below). In contrast, the GR theory manifests the local coordinate covariance as the principle for gravitational interactions.

After decades of further theoretical development and experimental work, amazingly, Nature is found to make use of these theoretical advances for interactions among elementary matter fields.[b] The SU(3) Yang–Mills fields are the mediating force fields for strong interactions. The SU(2) Yang–Mills fields are the mediating force fields for weak interactions, unifying in an intricate way with the U(1) Maxwell fields to become the U(1) × SU(2) Maxwell–Yang–Mills (MYM) mediating force fields for the unified electroweak interactions. Nature surprisingly chooses to use the simplest types of groups,[c] U(1) × SU(2) × SU(3), for electroweak and strong interactions, three of the four major interactions known in Nature.

For the last thirty years of his life, Einstein had searched for, unsuccessfully, the answer to the question[5]: Does Nature have a unified theory for all interactions? As the end of 2015 is approaching, sixty years after his passing in 1955, the answer to his question is still yet to be found. However, he might be pleased to know the progress mankind has made toward answering his question. At the time of his passing, when Yang–Mills theory was less than one year old, among the four

---

[b]The experimentally established elementary matter fields, as of 2015, are the fields of quarks, leptons, and the Higgs boson of mass 125 GeV/c^2.

[c]A group is a set of mathematical elements, together with an operation called group-multiplication, under which the group has an identity and every group element has its inverse. Groups are the second simplest mathematical structure, next to sets, yet are so prominently used by Nature in formulating laws of interactions.

experimentally established major interactions: electromagnetic, weak, strong, and gravity, only for electromagnetism had a quantum field theory been established: quantum electrodynamics (QED), with the quantum Maxwell fields being the U(1) "glues". Now a quantum field theory has been established for electroweak interactions, quantum flavor dynamics (QFD), which unifies electromagnetic and weak interactions in an intricate way — the quantum Maxwell fields remain to be the U(1) "glues" as in QED and the quantum Yang–Mills fields are the SU(2) "glues". Also, a quantum field theory has been established for strong interactions, quantum chromodynamics (QCD), in which the quantum Yang–Mills fields are the SU(3) "glues". So Einstein's question can be asked in more specific terms: Will we find that QFD and QCD (the combination of the two is called the Standard Model of particle physics) are parts of a grand unified theory (GUT)? Will we find the ultimate quantum description of Einstein's general relativity (QGR)? Will we find the ultimate unified theory, The Unified Theory (TUT) or The Theory of Everything (TOE), of GUT and QGR? To know how these questions are answered will make life richer.[d]

2. Yang–Mills Theory: Milestones and Interesting Questions

The milestones to be highlighted in this section, except for Milestone 4, have all been theoretically developed and experimentally established, and set forth in modern graduate textbooks on quantum field theories. See the authoritative volumes with comprehensive and exhaustive references by S. Weinberg,[6] and the celebration volume for the 50th anniversary of the Yang–Mills theory edited by 't Hooft,[7] for which Attachment A provides a copy of its Title page, Preface and Contents. Therefore, l will be very brief in highlighting the milestones in this section without giving references (except a few) to original papers which Refs. 6 and 7 comprehensively and exhaustively provide.

It is interesting to note that among the sixty-one Nobel Prizes in Physics[8] from 1954 to 2014, twenty-six are related to the making of the Standard Model, while the 2015 one honors the experiment that gave evidence for the need of going beyond the Standard Model (see Milestone 5).

Milestone 1. Quantum flavor dynamics QFD has been theoretically developed and experimentally established for electroweak interactions,[e] unifying electromagnetic and weak interactions. In QFD the quantum SU(2) Yang–Mills fields are the quantum mediating force fields for weak interactions, i.e., the SU(2) "weak glues", among elementary particles with SU(2) "weak charges", while the quantum U(1)

---

[d]Nature has been kind to humans in allowing their brains to understand the complexity of Nature bit by bit over time and then bring the pieces of understanding together into larger structures of meaning. How long will this process continue? Will the human brain ever discover the totality of Nature (finite or infinite)?
[e]The letter F in QFD is to indicate the fact that quarks and leptons exist in three different types called "flavors".

Maxwell fields remain, as in QED, to be the quantum U(1) "electromagnetic glues" among particles with U(1) "electromagnetic charges". In this intricate quantum unification the Maxwell field quanta, photons or light, are massless (as in QED) and can travel large distance in space void of matter with electromagnetic properties. Thus we can see each other, the moon, the sun, stars and galaxies. On the other hand, the SU(2) Yang–Mills quanta gain masses, through the Higgs mechanism (see Milestone 4. below), to become the massive W^+, W^- and Z^0 particles that can travel only very short distances.

Interestingly, among the four major interactions, **only** weak interactions, mediated by the SU(2) YM quanta, W^+, W^- and Z^0, violate the discrete symmetries P (parity) and CP (parity and charge conjugation), as well as T (time reversal), while CPT symmetry stays true in all interactions.

Milestone 2. Quantum chromodynamics QCD has been theoretically developed and experimentally established for strong interactions.[f]

In QCD the quantum SU(3) Yang–Mills fields are the quantum mediating forces for strong interactions, the SU(3) "color glues" of quanta called gluons. They interact with quarks according to their strong-interaction charge, the SU(3) "color charge". (The leptons and the Higgs particle are SU(3) "colorless", and therefore do not participate in strong interactions).

The QFD quanta with the U(1) and the SU(2) electroweak charges and "flavors" can exist freely. In contrast, the QCD quanta with non-zero SU(3) "color charges" cannot exist freely. Or, they are "color confined" or "color enslaved". They have to combine to form SU(3) "colorless" particles to exist freely. For example, protons and neutrons are SU(3) "colorless" particles formed from "colorful" quarks and gluons. The SU(3) "colorless" particles formed from "colorful" gluons are called glueballs, which are still under theoretical study[9] and experimental search.[10]

Also unusual is that the binding forces from the gluons decrease as distances decrease (corresponding to increasing energies) between the quarks — so QCD is an asymptotically free theory, quite opposite to that of QED.[11] These make QCD manageable to calculate in high energies and in high temperatures.[12] This has implications for cosmological studies on the early universe and for high energy experiments.[13]

Milestone 3. Local gauge covariance has been established to be the underlying principle of how Maxwell and Yang–Mills fields act as "glues" among the basic matter fields[14,15] — the Maxwell fields as the "electromagnetic glues" and the Yang–Mills fields as the "weak glues" as well as the "strong (color) glues". This principle has been used to develop new theories, e.g., the supersymmetric Yang–Mills theories, SYM,[16] and even for supersymmetric gravity, SGR.[17]

---

[f]The letter C in QCD is to indicate the fact that the "SU(3) gluons" have different SU(3) quantum identities called SU(3) "colors", conjured up by the color of photons, the "U(1) gluons".

Milestone 4. A Higgs particle (possibly the first of many) of the Higgs mechanism (the ubiquitous "molasses"),[18] which produces the masses of quarks and the masses of W^+, W^- and Z^0 (the SU(2) Yang–Mills quanta in weak interactions), was finally observed in 2012,[19,20] almost fifty years after the original theoretical proposals. It was an admirable triumph of the collaborative spirit in scientific research!

Milestone 5. Toward the end of the twentieth century, experiments established that the mass differences between any two of the three-flavored neutrinos are not zero, in contradiction to the all-zero-mass neutrinos in QFD of the Standard Model. This fact forces consideration of going beyond the Standard Model.[21,22,23]

Interesting questions

- As highlighted above, Nature has made versatile uses of the Yang–Mills fields, but not for uses on the macroscopic scales like the Maxwell fields and the Newton–Einstein fields.[24] Can a precise mathematical physics explanation for this be found?[25]
- Is there a deeper guiding principle for the Higgs mechanism? How many Higgs particles there are? Will the mass matrices of quarks and of leptons (which are generated by the presence of the Higgs fields) be derived, so hopefully a deeper understanding of the origin of CP violations in weak interactions will also follow?
- What will be established to be the Beyond Standard Model? Will the supersymmetric extensions[26] become experimentally observed realities?
- Will a grand unification theory, GUT, be found in which the electroweak and the strong interactions are unified?
- Will GR be established theoretically and experimentally as a quantum theory?

For a succinct and insightful perspective on the last three questions, see Witten.[27]

3. The (Anti-)Self-dual Yang–Mills Fields: Landmarks and Interesting Questions

The Yang–Mills fields in even dimensions of spacetime can be expressed as a sum of two terms: the self-dual term and the anti-self-dual term. Fields that are self-dual (SDYM) or anti-self-dual (ASDYM) are special because they automatically satisfy the Yang–Mills equations, and in addition, by definition, satisfy nonlinear differential equations, one-order lower than the Yang–Mills equations, which are called the (A)SDYM equations. (A)SDYM fields are simpler to study, while still remaining interesting and important for researches in physics and mathematics.

Landmark 1. The (A)SDYM fields have interesting solutions called instantons, which possess specific topological numbers (the Pontryagin or the second-Chern number) and minimize the Yang–Mills action.

In 1975, Belavin, Polyakov, Schwartz and Tyupkin constructed explicitly the first such solutions,[28] which stimulated much interest in mathematics as well as in physics.

In 1977 Ward showed[29] that (A)SDYM fields are naturally described by the twistor formulation that Penrose developed[30] to describe massless fields and the self-dual Einstein fields, and derived what are now called the Penrose–Ward transformations. Not long afterwards, this breakthrough by Ward led to the full description of the space of instantons in 1979 by Atiyah, Drinfeld, Hitchin and Mannin (now called the ADHM construction).[31] See the comprehensive review lecture by Atiyah and references to original papers.[32]

The instanton solutions prompted intensive research on their non-perturbative effects in QCD and QFD,[33] and led to experimentally testable predictions, for example the axions[34] and the related sphaleron phenomena.[35]

Landmark 2. For (A)SDYM fields in four-dimensions Yang showed[36] that (A)SDYM fields can be written as second-order nonlinear partial differential equations, now called the Yang equations, in terms of group-value local fields.[g]

Later the Yang equations were found to have many characteristics of the classical integrable systems in lower dimensions: Backlund transformations,[37] non-local conservation laws,[38] Riemann–Hilbert transformation properties,[39] Painleve properties,[40] generalized Riccarti equations,[41] and, most importantly, the linear systems (the Lax systems),[42] and Kac-Moody algebras.[43,44,45,46] For an overview of the work and references during this period of discoveries, see "*Integrable Systems*",[47] the lectures therein[48,49] and Attachment B below, which gives the memorable front pages of the book. For a glimpse of the landscape of mathematical physics in the late 1980s, see Chau and Nahm.[50]

Additionally, many integrable systems in lower dimensions, including the famed Nahm equations,[51] can be directly reduced from the Yang equations. See the in-depth reviews[52,53] and the extensive references therein. See the overview[54] of (A)SDYM as a classical integrable system in four dimensions and its relations to those in lower and higher than four dimensions. The review Ref. 49 includes discussions on extensions to supersymmetric theories, SYM and SGR, which will be discussed in Landmark 4.

Therefore, (A)SDYM equations have been shown to be classical integrable systems in **four** dimensions that interestingly also serve as conceptual pathways between those integrable systems in dimensions lower and higher than four dimensions, as well as in supersymmetric dimensions.

The next important step was to develop (A)SDYM equations into a quantum integrable field theory in four dimensions. In the 1990s, the Yang equations were put to quantization, using the action constructed by Nair and Schiff[55] and by Hou

---

[g]Group-valued local fields are special. Usually group-valued fields are non-local, defined on a loop, not at points as local fields are.

and Song,[56] in terms of the Lie-valued fields[55,57] and in terms of the group-valued local fields.[58] In Ref. 58 Yamanaka and I found that the quantized group-valued local fields are bi-module fields which satisfy intricate exchange algebras with structure coefficients satisfying generalized Yang–Baxter equations. So here Yang–Mills met Yang–Baxter for the first time with the quantum group properties,[59,60] originated from the Yang–Baxter equations in low dimensions, appearing in a four-dimensional quantum theory for the first time. Further, the algebraic relations satisfied by the Lie-valued fields derived in Refs. 55 and 57 can be derived from the exchange algebras of the group-value local fields. For further developments see Popov–Preitschopf[61] and Popov,[62] in 1996 and in 1999 respectively.

However, the quantum contents of the Yang equations have not been fully developed in detail, and the full quantum versions of its classical integrability properties are yet to be revealed.

Landmark 3. Using (A)SDYM fields as tools, Donaldson[63,64] made major discoveries about the four-manifolds.[h]

Here was what Atiyah said on the occasion of the awarding of the 1986 Fields Medal to Donaldson,[65] referring to Donaldson's paper Ref. 63 as [1]:

"In 1982, when he was a second-year graduate student, Simon Donaldson proved a result [1] that stunned the mathematical world. Together with the important work of Michael Freedman (described by John Milnor), Donaldson's result implied that there are "exotic" 4-spaces, i.e., 4-dimensional differentiable manifolds which are topologically but not differentiably equivalent to the standard Euclidean 4-space R4. What makes this result so surprising is that $n = 4$ is the only value for which such exotic n-spaces exist. These exotic 4-spaces have the remarkable property that (unlike R4) they contain compact sets which cannot be contained inside any differentiably embedded 3-sphere!

To put this into historical perspective, let me remind you that in 1958 Milnor discovered exotic 7-spheres, and that in the 1960s the structure of differentiable manifolds of dimension > 5 was actively developed by Milnor, Smale (both Fields Medalists), and others, to give a very satisfactory theory. Dimension 2 (Riemann surfaces) was classical, so this left dimensions 3 and 4 to be explored. At the last Congress, in Warsaw, Thurston received a Fields Medal for his remarkable results on 3-manifolds, and now at this Congress we reach 4-manifolds. I should emphasize that the stories in dimensions 3, 4, and $n > 5$ are totally different, with the low-dimensional cases being much more subtle and intricate.

The surprise produced by Donaldson's result was accentuated by the fact that his methods were completely new and were borrowed from theoretical physics, in the form

---

[h]A manifold is a space that can be covered by local patches that are like the Euclidean space, or the Minkowski space.

of the Yang–Mills equations. These equations are essentially a nonlinear generalization of Maxwell's equations for electro-magnetism, and they are the variational equations associated with a natural geometric functional. Differential geometers study connections and curvature in fibre bundles, and the Yang–Mills functional is just the L2-norm of the curvature. If the group of the fibre bundle is the circle, we get back the linear Maxwell theory, but for nonabelian Lie groups, we get a nonlinear theory. Donaldson uses only the simplest non-Abelian group, namely $SU(2)$, although in principle other groups can and will perhaps be used.

Physicists are interested in these equations over Minkowski space–time, where they are hyperbolic, and also over Euclidean 4-space, where they are elliptic. In the Euclidean case, solutions giving the absolute minimum (for given boundary conditions at ∞) are of special interest, and they are called instantons.

Several mathematicians (including myself) worked on instantons and felt very pleased that they were able to assist physics in this way. Donaldson, on the other hand, conceived the daring idea of reversing this process and of using instantons on a general 4-manifold as a new geometrical tool. In this he has been brilliantly successful: he has unearthed totally new phenomena and simultaneously demonstrated that the Yang–Mills equations are beautifully adapted to studying and exploring this new field.

Donaldson's works continue to be highly honored.[66]

Then in 1994 Witten[67] showed that the Seiberg–Witten equations[68] of $N = 2$ SYM also provide a powerful tool for analyzing four-manifolds, confirming those by Donaldson and discovering new properties, which generated much excitement among mathematicians.[69]

Landmark 4. The landmarks above resulting from studying the (A)SDYM fields have inspired extensions into supersymmetric theories.

In 1978, Witten generalized the concept advanced by Ward in Landmark 1,[70] that (A)SDYM equations are the results of certain integrability conditions,[29] to the full extent of YM and SYM. Of particular interest is that the integrability along light-like lines precisely gives the $N = 3$ SYM equations.[16] This naturally led to the discovery of many classical integrability properties of the theory and also to the construction of group-valued local fields and the derivation of their equations, giving the generalized Yang equations of SYM.[71]

In 1986 Lim and I showed that $N = 5, 6, 7, 8$ SGR, in the linearized approximation, had a similar interpretation.[72]

In 1992, I derived the linear systems for high-N SGR,[73] which signal the classical integrability properties, and constructed group-valued local fields and derived the equations they satisfy, giving the generalized Yang equations of SGR.

Recently, quantum $N = 4$ SYM and its quantum integrability properties have been studied very actively,[74] giving a renaissance to studies on S-matrices.[75,76] For the current status of such studies, see the paper by Brink in this *Proceedings*.[77]

So the landmarks of this section have revealed that (A)SDYM fields in four dimensions are important to study, both for physics and for mathematics. They also serve importantly as conceptual pathways between integrable systems in lower and higher dimensions, as well as to those in supersymmetric spaces.[i]

Interesting questions

- What will be the experimental observations of the effects related to instantons?
- What will be the quantum manifestations of the classical integrality properties of the Yang equations? Do quantum bi-module fields have physical reality?
- Will experiments find that Nature uses supersymmetric theories, SYM and SGR?

If yes, such a huge discovery will inevitably lead us to ask whether SYM or SGR theories have quantum integrability properties?

- What are the implications to physics of the four-manifolds discovered by Donaldson?

4. Concluding Remarks

In the 60 years since the Yang–Mills paper, we have seen the establishment of quantum flavor dynamics, QFD, unifying electromagnetic and weak interactions into electroweak interactions, with Maxwell fields acting as the "U(1) glues" and Yang–Mills fields acting as the "SU(2) glues". The quanta of the U(1) Maxwell fields, photons, are massless. They can travel far and are our light (in its full spectrum). In contrast, the quanta of SU(2) Yang–Mills fields for weak interactions gain masses from the Higgs mechanism and become massive W^+, W^- and Z^0 particles. They can travel only for very short distances.

We have also seen the development of quantum chromodynamics QCD for strong interactions, with the SU(3) Yang–Mills fields acting as the "SU(3) color glues" and its quanta, the massless gluons, interacting among themselves and mediating interactions among quarks that have "SU(3) color charges". Unlike particles with U(1) electric charges and/or SU(2) weak-interaction charges, the "SU(3) colored" gluons and quarks cannot exist freely. To be free they must form "SU(3) colorless" particles: glueballs made of gluons, hadrons (protons, neutrons, etc.) made of quarks and gluons, or be at super high energies or super high temperatures.

Nature chooses these versatile ways to use the Yang–Mills fields; however, it does not allow us to experience the Yang–Mills fields in our daily lives as we do the Maxwell fields and the Newton–Einstein fields. Why this is so is an interesting mathematical physics question worthy of an answer.

After the celebration of 60 years of the Yang–Mills Theory in 2014–2015 comes the celebration of 100 years of Einstein's theory of general relativity in 2015–2016.

---

[i]In Attachment C, I propose to add the topic of (A)SDYM fields to the list of the other thirteen topics that Professor Yang considered important and to which he made major contributions.

We are now being treated to progress reports about the advances made in our understanding of it,[78,79] in particular its quantum theoretical description and experimental attempts to observe its possible quantum phenomena. As of now, near the end of 2015, no definitive quantum theoretical description for general relativity has been established and no quantum phenomena of general relativity have yet been observed — no gravitational waves (happily this statement needs to be revised on February 11, 2016),[80] not to mention the more illusive gravitons, their quanta.

The ultimate question is whether Nature makes use of a unified theory for all interactions.

With all these interesting questions, mankind will be kept busy intellectually for generations to come. We look forward to seeing what new discoveries will be celebrated when Yang–Mills theory turns 70 in 2024 and Einstein's general relativity turns 110 in 2025.[81]

Acknowledgments

I thank Prof. Phua, Kok Khoo, the Founding Director of the Institute of Advanced Studies (IAS), Nanyang Technological University (NTU). Singapore, for his gracious invitation to me to participate in the momentous event of celebrating the 60th anniversary of the Yang-Milly theory at NTU in 2015 (see poster in Attachment D), and to contribute a paper for the Proceedings, and for his further invitation to publish the paper also in the International Journal of Modern Physics A. I appreciate very much the excellent work performed by the editorial and production team of World Scientific Publishing Co Pte Ltd, of which Prof. Phua is the Chairman and the Editor-in-Chief.

Also, I would like to give my heartfelt thanks to many friends, especially Ms. Karen Andrews, Prof. Nicolai (Kolya) Reshetikhin, Ms. Gloria Rogers, Prof. William (Bill) Rogers, and Mr. Weiben Wang, for reading the drafts and making helpful comments which improved the presentation of the paper, and for their warm friendship and encouragement that made the task of writing this paper more enjoyable; and offer my deep gratitude to Kolya, Professor of Mathematics who has served two multi-year terms as the editor of Communications in Mathematical Physics (Springer), and Bill, Professor of English, for their expert critiques that sharpened the paper. Of course, any errors in the paper are mine alone.

Attachment A

The Title Page, Preface and Contents of *50 Years of Yang–Mills Theory*, ed. G. 't Hooft (World Scientific, 2004).

edited by **Gerardus 't Hooft**
Utrecht University, The Netherlands

 **World Scientific**

NEW JERSEY · LONDON · SINGAPORE · BEIJING · SHANGHAI · HONG KONG · TAIPEI · CHENNAI

Attachment A (Continued)

PREFACE

At the time of the publication of this volume, fifty years have passed since the appearance of an article in *The Physical Review* by Chen Ning Yang and Robert L. Mills, entitled "Conservation of Isotopic Spin and Isotopic Gauge Invariance". This book on the one hand serves as a tribute to that monumental piece of work, and on the other intends to show how its subject has evolved since that time, highlighting the landmarks that followed after the original paper emerged, and allowing its authors to indulge in new ideas and concepts. Gauge Theory has indeed grown into a pivotal concept in the Theory of Elementary Particles, and it is expected to play an equally essential role in even more basic theoretical constructions that are speculated upon today, with the aim of providing an all-embracing picture of the universal Laws of Physics.

Some of the chapters in this book are contributions that have appeared elsewhere; most of the contributions are original pieces of work. All are accompanied by brief comments by the Editor. Needless to state that this volume is far from complete. There are numerous well-known landmarks that we could not cover. Furthermore, like most developments in Science, progress not only comes from the relatively small set of papers by famous authors that enjoy enormous scores on citation indices, but it predominantly comes from the large crowds of scientists who confirm and reproduce the original research while adding inconspicuous but essential bits of understanding, not only by writing papers, but also by lecturing to students, by performing experiments and doing calculations. Without them, this book could not have been written.

Attachment A (Continued)

CONTENTS

Preface   v

Introduction   *1*

Chapter 1
Gauge Invariance and Interactions;   7
Remembering Robert Mills
   C. N. Yang

Quantizing Gauge Field Theories   *13*

Chapter 2
The Space of Gauge Fields: Its Structure and Geometry   15
   Bryce DeWitt

Ghosts for Physicists   *33*

Chapter 3
Perturbation Theory for Gauge-Invariant Fields   39
   V. N. Popov and L. D. Faddeev

Breaking the Symmetry   *61*

Chapter 4
Broken Symmetry and Yang–Mills Theory   65
   François Englert

Towards the Standard Model   *97*

Chapter 5
The Making of the Standard Model   99
   Steven Weinberg

Attachment A (*Continued*)

Renormalization 119

Chapter 6
Renormalization 121
    Gerardus 't Hooft

Chapter 7
From Koszul Complexes to Gauge Fixing 137
    R. Stora

Chapter 8
The Renormalization Content of Slavnov–Taylor Identities 168
    Carlo Becchi

Anomalies 183

Chapter 9
Anomalies to All Orders 187
    Stephen L. Adler

Chapter 10
Fifty Years of Yang–Mills Theory and Our Moments of Triumph 229
    Roman Jackiw

Asymptotic Freedom 253

Chapter 11
Yang–Mills Theory In, Beyond, and Behind Observed Reality 255
    Frank Wilczek

Magnetic Monopoles 269

Chapter 12
To Be Or Not to Be? 271
    F. Alexander Bais

Quark Confinement and Strings 309

Chapter 13
Confinement and Liberation 311
    A. M. Polyakov

Attachment A (Continued)

Contents	ix

Fixing in Gauge Condition Non-Perturbatively 331

Chapter 14
Non-Perturbative Aspects of Gauge Fixing 333
    Pierre van Baal

The Lattice 355

Chapter 15
Yang–Mills Fields and the Lattice 357
    Michael Creutz

Fermions on the Lattice 375

Chapter 16
Chiral Symmetry on the Lattice 377
    P. Hasenfratz

Confrontation with Experiment 399

Chapter 17
Fifty Years of Yang–Mills Theories: A Phenomenological Point of View 401
    Alvaro De Rújula

Supersymmetry and Supergravity 431

Chapter 18
Supergravity as a Yang–Mills Theory 433
    Peter van Nieuwenhuizen

Physics of the 21st Century 457

Chapter 19
Gauge/String Duality for Weak Coupling 459
    Edward Witten

Attachment B

The Title Page, Forewords by S. S. Chern and C. N. Yang, and Contents of *Integrable Systems*, Nankai University, Tianjin, China; ed. X. C. Song, (World Scientific, 1987).

Nankai Lectures on Mathematical Physics

INTEGRABLE SYSTEMS

Nankai Institute of Mathematics, China
August 1987

Editor:
Song Xing-Chang
Peking University

Singapore • New Jersey • London • Hong Kong

Attachment B (Continued)

数學物理

永結一体

陳省身

一九八八年九月

Mathematics and Physics
One Body Forever

S.S.Chern

Attachment B (Continued)

数学物理是很古老的学科．近年来发展尤其迅速．南开数学所理论物理研究室将出版一系列关於此学科的书籍，这是很有意义的事．

杨振宁

```
    Mathematical physics is a subject of
study which dated far back to ancient times,
and has found speedy development in recent
years.
    It is most significant that the Theore-
tical Physics Division of Nankai Institute
of Mathematics will publish a series of books
on the subject.
                              C.N.Yang
```

Attachment B (Continued)

CONTENTS

INSCRIPTIONS — IV

PREFACE — IX

SPEECHES

On the Relation between Mathematics and Physics — 3
  L.D. Faddeev

Journey through Statistical Mechanics — 11
  C.N. Yang

LECTURES

Lectures on Quantum Inverse Scattering Method — 23
  L.D. Faddeev

  Lecture I — 24
  Lecture II — 34
  Lecture III — 48
  Lecture IV — 59

Geometrical Integrability and Equation of Motion in Physics: A Unifying View — 71
  L.L. Chau

CONTRIBUTIONS

Generalization of Sturm-Liouville Theory — 154
  C.N. Yang

Generalized Levinson's Theorem — 155
  Z.Q. Ma

Attachment B (Continued)

xii

On the Darboux Form of Bäcklund Transformations *C.H.Gu*	162
A Trace Identity and Its Applications in Hamiltonian Theory of Integrable Systems *G.Z.Tu*	169
On the Hierarchies of Integrable Systems in 2+1 Dimensions *Y.Cheng*	177
The Relationships between Inverse Scattering Techniques and Conformal Symmetry Transformations in Certain Nonlinear Systems *B.Y.Hou and W.Li*	183
Hamiltonian Structures of the Two Dimensional Classical Integrable System, Loop Algebras and Basic Poisson Brackets *M.L.Ge and J.F.Lü*	192
Nonlinear Realization of Hidden Symmetry U(1) and the Kac-Moody Algebra of Chiral Field in Curved Space *X.C.Song*	206
The Bosonization of (1+1) Dimensional Fermion Fields by Path Integration Method *G.J.Ni, R.T.Wang and S.Q.Chen*	220
Bethe Ansatz Equations for Multicomponent Nonlinear Schrödinger Model *F.C.Pu and B.H.Zhao*	226
LIST OF PARTICIPANTS	245

Attachment B (Continued)

WORKSHOP ON GEOMETRICAL INTEGRABILITY
AND EQUATIONS OF MOTION IN PHYSICS

	Aug. 18	Aug. 19	Aug. 20	Aug. 21	Aug. 22
8:30-9:30	Chau	Chau	Chau	Chau	Chau
10:00-11:00	"	"	"	. "	"
Lunch					
2:30-3:30	Song	Yan	Ni	Contribu-	Panel
4:00-5:00	Hou	Ge		tions	Discussion
				"	

Topics:

L.-L. Chau (乔玲丽), University of California, Davis
    I. Generic properties of geometrically integrable systems:
       Flow Chart, and two dimensional systems;
    II. Self-dual Yang-Mills Equations;
    III. Supersymmetric Yang-Mills systems;
    IV. Supergravity;
    V. Summary and outlook.

M.-L. Ge (葛墨林), Lanchou University and Nankai University, Tianjing
    Integrability of Belinski-Zakharov Gravity and Solitons in
    Gravitational Wave.

B.-Y. Hou (侯伯宇), Xibei University, Xian
    Virasoro and Kac-Moody of Ernst equation.

G.-J. Ni (倪光炯), Fudan University, Shanghai
    A general method of calculation of anomalies.

X.-C. Song (宋行長), Beijing University, Beijing
    Integrability properties of Ernst equation.

M.-L. Yan (问沐霖), Science and Technology University, Hofei
    Potts Model, Yang-Baxter Equation's Solutions; Surfaces with Genus > 1.

Attachment C

C. N. Yang's Commandments in Physics

When celebrating Professor Yang's 90th birthday, 2012, Tsinghua University presented him a black rock cube with his thirteen important contributions to physics listed and carved on the vertical faces, grouped into four major categories in physics (one on each vertical face of the cube), which I list in the table below (the first thirteen in black letters).

(The two photos of the cube below are copies of [Ph 72] and [Ph 73] in C. N. Yang, *Selected Papers II with Commentaries* (World Scientific, 2013). In the table below, I adopt the same numbering system and English translation as given by Yu Shi, "Beauty and physics: 13 important contributions of Chen Ning Yang," *Int. J. Mod. Phys. A* **29**, 1475001 (2014).)

Face-A: Statistical Mechanics:		Face-B. Condensed Matter Physics:	
A.1.	1952 Phase Transition	B.1.	1961 Flux Quantization
A.2.	1957 Bosons	B.2.	1962 ODLRO
A.3.	1967 Yang–Baxter Equation		
A.4.	1969 Finite Temperature		
Face-C. Particle Physics:		Face-D. Field Theory:	
C.1.	1956 Parity Nonconservation	D.1.	1954 Gauge Theory
C.2.	1957 T, C and P	D.2.	1974 Integral Formalism
C.3.	1960 Neutrino Experiment	D.3.	1975 Fiber Bundle
C.4.	1964 CP Nonconservation	D.4.	1977 (Anti-)Self-dual Gauge Fields

Here I would propose to call the listed important contributions

C. N. Yang's Commandments in Physics

and to add "1977 (Anti-) Self-dual Gauge Fields" to the list on Face-D (as D.4., shown in red) so to become the Fourteenth Commandment.

Attachment D

The poster of the Conference on 60 Years of Yang–Mills Gauge Theories, IAS, NTU, Singapore, 25–28 May 2015.

References

1. C. N. Yang and R. Mills, Isotopic spin conservation and a generalized gauge invariance, *Phys. Rev.* **95**, 631 (1954).
2. J. M. Maxwell, A dynamical theory of the electromagnetic field, *Philos. Trans. R. Soc. London* **155**, 459 (1865), doi:10.1098/rstl.1865.000 (This article followed a December 8, 1864 presentation by Maxwell to the Royal Society.)
3. A. Einstein, Feldgleichungen der gravitation (English translation: The field equations of gravitation) *Preussische Akademie der Wissenschaften, Sitzungsberichte*, 844–847, part 2 (1915); according to A. Pais of reference below (Chap. 14, Ref. E1; Chap. 15, Ref. E15), this is the defining paper of general relativity (GR) after several earlier publications on GR in 1915, which was then followed by many more in the following years, see https://en.wikipedia.org/wiki/List_of_scientific_publications_by_Albert_Einstein#Journal_articles.
4. I. Newton, *Philosophiæ Naturalis Principia Mathematica*, three books in Latin, first published 5 July 1687, often referred to as *The Principia*; English translation: *The Mathematical Principles of Natural Philosophy* (1729).
5. A. Pais, *Subtle Is the Lord: The Science and the Life of Albert Einstein* (Oxford University Press, 1982).
6. S. Weinberg, *The Quantum Theory of Fields Vol. I: Foundations* (Cambridge University Press, 1995); S. Weinberg, *The Quantum Theory of Fields Vol. II: Modern Applications* (Cambridge University Press, 1996); S. Weinberg, *The Quantum Theory of Fields Vol. III: Supersymmetry* (Cambridge University Press, 2000).
7. G. 't Hooft (ed.), *50 Years of Yang–Mills Theory* (World Scientific, Singapore, 2005).
8. For list of all Nobel Prizes in Physics, see http://www.nobelprize.org/nobel_prizes/physics/laureates/.
9. M. Creutz, The lattice and quantized Yang–Mills theory, in this *Proceedings*.
10. W. Ochs, The status of glueballs, *J. Phys. G: Nucl. Part. Phys.* **40**, 67 (2013), arXiv:1301.5183v3; F. Close and P. R. Page, Glueballs, *Sci. Amer.* **279**, 80 (1998).
11. D. J. Gross, Quantum chromodynamics: The perfect Yang–Mills gauge field theories, in this *Proceedings*; F. Wilczek, Yang–Mills theory in, beyond, and behind observed reality, in Ref. 7.
12. G. Sterman, Yang–Mills theories at high energy accelerators, in this *Proceedings*; R. Hofmann, *The Thermodynamics of Quantum Yang–Mills Theory: Theory and Applications* (World Scientific, 2012); $SU(2)$ Yang–Mills thermodynamics, talk at the Conference.
13. CMS Collab. (V. Khachatryan *et al.*), Evidence for collective multiparticle correlations in p-Pb collisions, *Phys. Rev. Lett.* **115**, 012301 (2015); and news http://earthsky.org/human-world/lhc-creates-liquid-from-big-bang, LHC creates liquid from Big Bang. earthsky.org/human-world, Sep 15, 2015.
14. C. N. Yang, The conceptual origins of Maxwell's equations and gauge theory, *Phys. Today* **67**, 45 (2014).

15. C. N. Yang, Gauge invariance and interactions, in Ref. 7. It was a copy of his notes hand-written in 1947 when he was a graduate at University of Chicago — impressive that a graduate student would think about such a big question! A typed transcript of the notes is given in Attachment-1 of L. L. Chau's "Almost everything Yang touched eventually turned into gold", in *Proceedings of the Conference in Honor of C. N. Yang's 85th Birthday*, eds. C. H. Oh, K. K. Phua, N. P. Chang and L. C. Kwek (World Scientific, 2008), http://chau.physics.ucdavis.edu/Yang85-byChau-fromWS081104f.pdf, Also Attachment-2 of this Chau's paper gives an anecdote about Wigner's attempt to understand gauge potential.
16. L. Brink, J. H. Schwarz and J. Scherk, Supersymmetric Yang–Mills theories, *Nucl. Phys. B* **121**, 77 (1977); G. Gliozzi, J. Scherk and D. Olive, Supersymmetry, supergravity theories and the dual spinor model, *Nucl. Phys. B* **122**, 253 (1977); and references therein for earlier papers.
17. P. van Nieuwenhuizen, Supergravity as a Yang–Mills theory, in Ref. 7; D. Z. Freedman and A. Van Proeyen, *Supergravity* (Cambridge University Press, 2012).
18. P. W. Higgs, Spontaneous symmetry breakdown without massless bosons, *Phys. Rev.* **145**, 1156 (1966); F. Englert, R. Brout and M. F. Thiry, Vector mesons in presence of broken symmetry, *Nuovo Cimento A* **43**(2), 244 (1966); For an overview and more references, see http://www.nobelprize.org/nobel_prizes/physics/laureates/2013/advanced-physicsprize2013.pdf.
19. ATLAS Collab., Observation of a new particle in the search for the Standard Model Higgs boson with the ATLAS detector at the LHC, *Phys. Lett. B* **716**, 1 (2012), arXiv:1207.7214.
20. CMS Collab., Observation of a new boson at a mass of 125 GeV with the CMS experiment at the LHC, *Phys. Lett. B* **716**, 30 (2012).
21. Particle Data Group (S. Raby), Grand unified theories (2011), http://pdg.lbl.gov/2014/reviews/rpp2014-rev-guts.pdf; See also Wikipidia, *Physics Beyond the Standard Model*, https://en.wikipedia.org/wiki/Physics_beyond_the_Standard_Model.
22. The leaders of the experiments are honored by the 2015 Nobel Prize in Physics, http://www.nobelprize.org/nobel_prizes/physics/laureates/2015/ which gives an overview and references about neutrino oscillation at https://www.nobelprize.org/nobel_prizes/physics/laureates/2015/advanced-physicsprize2015.pdf; They are also honored by the Breakthrough Prize, https://breakthroughprize.org/, in the category of Fundamental Physics 2016, https://en.wikipedia.org/wiki/Fundamental_Physics_Prize.
23. Both Advanced Information for the 2015 and for the 2008 Nobel Prizes in Physics, https://www.nobelprize.org/nobel_prizes/physics/laureates/2015/advanced-physicsprize2015.pdf and http://www.nobelprize.org/nobel_prizes/physics/laureates/2008/advanced-physicsprize2008.pdf, have used the parameterization for the flavor mixing matrix originally given by L. L. Chau and W. Y. Keung, Comments on the parameterization of the Kobayashi-Maskawa matrix, *Phys. Rev. Lett.* **53**, 1802 (1984), http://chau.physics.ucdavis.edu/Chau-Keung-paraKM-PRL53-p1802-1984.pdf, which has also been adopted by the Particle Data Group as the "standard choice",

http://pdg.lbl.gov/2015/reviews/rpp2014-rev-ckm-matrix.pdf, Ref. 3, after an interesting evolution of choices. More importantly, the parameters of the parameterization which is in product form and in terms of exact variables (not approximate) have physical meaning in experiments (such as those honored by the 2015 Nobel Prize in Physics). They are precisely used by Nature to characterize flavor oscillations among quarks and among neutrinos, as well as to characterize CP violations. While CP violation in weak interactions of quark flavor changing has been established (after a long interesting history of theoretical and experimental work), CP violation or no CP violation in neutrino flavor changing is yet to be settled by experiments. It will surely be a milestone in the history of physics when that is settled, and then the parameterization will once again be highlighted. [Right now the data are consistent with three flavors of quarks and three flavors of leptons. However, in the event we are surprised in the future and find more than three flavors, the Chau-Keung type of parameterizations (in product form and in terms of exact variables) for any N flavors have already been derived and given in L. L. Chau, Phase direct CP violations and general mixing matrices, *Phys. Lett. B* **651**, 293–297 (2007), http://chau.physics.ucdavis.edu/Chau-Direct-CP-PLb651-p293.pdf].

Another close encounter with Nobel-Prize-honored discoveries was through the paper by R. F. Peierls, T. L. Trueman and L. L. (Chau) Wang, Estimates of production cross sections and distributions for W bosons and hadrons jets in high energy pp and $p\bar{p}$ collisions, *Phys. Rev. D* **16**, 1397 (1977). The results in this paper for the quantities given in the title agreed with those of the 1983 experiments in which the observation of W^+, W^- and Z^0 earned the 1984 Nobel Prize in Physics for C. Rubbia and S. van der Meer, http://www.nobelprize.org/nobel_prizes/physics/laureates/1984/. After the suspension of several months between the observation of the W's and the observation of Z^0, amazingly (or boringly, depending on one's point of view) the masses of W^+, W^- and Z^0 were observed at the predicted values based upon the earlier 1979 Nobel-Prize-honored theoretical works by S. L. Glashow, by A. Salam and by S. Weinberg, http://www.nobelprize.org/nobel_prizes/physics/laureates/1979/. This success story concurs once more with Einstein's famous statement, *"Subtle is The Lord, but malicious He is not!"* (The English translation from German is that of A. Pais, Ref. 5.)

The reason that Peierls, Trueman and (Chau) Wang did the calculations reported in that paper was because Brookhaven National Laboratory (BNL or Brookhaven) was building the then highest energy pp (proton-proton) collider, ISABELLE, in search of W^+, W^- and Z^0. Unfortunately Brookhaven failed to produce magnets with the required properties. This resulted in the cancellation of ISABELLE, after already spending $200M, https://en.wikipedia.org/wiki/ISABELLE (so goes the joke, "IS-A-BELLE became WAS-A-BELLE"). As a consequence, leading experimental physicists left BNL and went to CERN, Geneva, Switzerland to build the proton-antiproton $p\bar{p}$ collider and then made the observation of W^+, W^- and Z^0. (As the authors of the paper we were told by the experimentalists that the results of substantial $p\bar{p}$ production cross sections for W^+, W^- and Z^0 in our paper gave them confidence.) The momentous event also planted the seed that eventually led to the building of the now world-famed Large Hadron Collider LHC at Geneva, where the discovery of the Higgs particle was made in 2012, Refs. 18, 19, 20. Brookhaven had lost such a golden chance. One can only imagine what it would be like for Brookhaven and for US high energy physics had Brookhaven been successful in building ISABELLE.

24. This is related to the problem of "Yang–Mills and mass gap", which is the first of the seven Millennium Problems posted by the Clay Mathematics Institute http://www.claymath.org/millennium-problems, with official problem description at http://www.claymath.org/sites/default/files/yangmills.pdf, and the status as of 2004 at http://www.claymath.org/sites/default/files/ym2.pdf.
25. E. Witten, The problem of gauge theory, http://arxiv.org/pdf/0812.4512v3.pdf, where he gave the following concluding statements: "*The mass gap is the reason, if you will, that we do not see classical nonlinear Yang–Mills waves. They are a good approximation only under inaccessible conditions. I have spent most of my career wishing that we had a really good way to quantitatively understand the mass gap in four-dimensional gauge theory. I hope that this problem will be solved one day.*" See also his, *What We Can Hope to Prove About 3d Yang–Mills Theory*, http://media.scgp.stonybrook.edu/presentations/20120117_3_Witten.pdf Simons Center, January 17, 2012.
26. S. Weinberg, *Supersymmetry*, Vol. III, of Ref. 6; J. Terning, *Modern Supersymmetry*: *Dynamics and Duality* (Clarendon Press, Oxford, 2006).
27. E. Witten, Quest for unification, http://arxiv.org/pdf/hep-ph/0207124.pdf, based on his Heinrich Hertz lecture at SUSY 2002 at DESY, June, 2002.
28. A. A. Belavin, A. M. Polyakov, A. S. Schwartz and Yu. S. Tyupkin, Pseudoparticle solutions of the Yang–Mills equations, *Phys. Lett. B* **59**, 85 (1975).
29. R. S. Ward, On self-dual gauge fields, *Phys. Lett. A* **61**, 81 (1977).
30. R. Penrose, Zero rest-mass fields including gravitation: Asymptotic behaviour, *Proc. R. Soc. A* **284**, 159 (1965); The twistor programme, *Rep. Math. Phys.* **12**, 65 (1977).
31. M. F. Atiyah, V. G. Drinfeld, N. J. Hitchin and Yu. I. Manin, Construction of instantons, *Phys. Lett. A* **65**, 185 (1978).
32. M. F. Atiyah, *Geometry of Yang–Mills Fields* (Publications of the Scuola Normale Superiore, Pisa, 1979).
33. See the discussions and references on instantons in Chapter 23, Vol. II. of Ref. 6.
34. M Kuster, G. Raffelt and B. Beltrán (eds.), *Axions*: *Theory, Cosmology, and Experimental Searches*, Lecture Notes in Physics, Vol. 741 (Springer-Verlag, Berlin, Heidelberg, 2008).
35. H. Tye, Sphaleron physics in Yang–Mills theory, talk at the Conference.
36. C. N. Yang, Conditions of self-duality for SU(2) gauge fields on Euclidean four-dimensional space, *Phys. Rev. Lett.* **38**, 1377 (1977).
37. M. K. Prasad, A. Sinha and L. L. Chau Wang, Parametric Backlund transformation for self-dual SU(N) Yang–Mills fields, *Phys. Rev. Lett.* **43**, 750 (1979).
38. M. K. Prasad, A. Sinha and L. L. Chau Wang, Non-local continuity equations for self-dual SU(N) Yang–Mills fields, *Phys. Lett. B* **87**, 237 (1979).

39. K. Ueno and Y. Nakamura, Transformation theory for anti-(self)-dual equations and the Riemann–Hilbert problem, *Phys. Lett. B* **109**, 273 (1982).
40. R. S. Ward, The Painleve properties for the self-dual gauge-field equations, *Phys. Lett. A* **102**, 279 (1984).
41. L. L. Chau and H. C. Yen, Generalized Riccati equations for self-dual Yang–Mills fields, *J. Math. Phys.* **28**, 1167 (1987).
42. L. L. Chau, M. K. Prasad and A. Sinha, Some aspects of the linear systems for self-dual Yang–Mills fields, *Phys. Rev. D* **24**, 1574 (1981).
43. L. L. Chau, M. L. Ge and Y. S. Wu, Kac-Moody algebra in the self-dual Yang–Mills equations, *Phys. Rev. D* **25**, 1086 (1982); L. L. Chau, M. L. Ge, A. Sinha and Y. S. Wu, Hidden-symmetry algebra for the self-dual Yang–Mills equation, *Phys. Letts. B* **121**, 391 (1983).
44. L. Dolan, Kac-Moody symmetry group of real self-dual Yang–Mills, *Phys. Lett. B* **113**, 378 (1982).
45. L. L. Chau and Y. S. Wu, More about hidden-symmetry algebra for the self-dual Yang–Mills system, *Phys. Rev. D* **26**, 3581 (1982), see the discussion of comparing results with Ref. 44.
46. L. Dolan, Kac-Moody algebras and exact solvability in hadronic physics, *Phys. Rep.* **109**, 1 (1984).
47. X. C. Song (ed.), *Integrable Systems* (World Scientific, 1987), (Nankai University, Tianjin, China).
48. L. D. Faddeev, Lectures on quantum inverse scattering method, in Ref. 47.
49. L. L. Chau, A unifying description of equations of motion in physics from geometrical integrability, in Ref. 47.
50. L. L. Chau and W. Nahm (eds.), *Differential Geometric Methods in Physics: Physics and Geometry, Proceedings of the NATO Advanced Research Workshop and the 18th International Conference*, University of California, Davis, 1988 (Plenum Press, 1990); and see its Title Pages and Contents at http://chau.physics.ucdavis.edu/Text/pdfs/NATO-Intl-Conf-Diff-Geo-Th-Phys-UCD-1988-Jul2-8.pdf.
51. W. Nahm, The construction of all self-dual multimonopoles by the ADHM method, in *Monopoles in Quantum Field Theory* (World Scientific, Singapore, 1982), pp. 87–94, See also F. Alexander Bais, Magnetic monopoles in non-Abelian gauge theories, in Ref. 7.
52. L. J. Mason and N. M. J. Woodhouse, *Integrability, Self-Duality, and Twistor Theory*, LMS Monograph, New Series, Vol. 15 (Oxford University Press, Oxford, 1996).
53. M. J. Ablowitz, S. Chakravarty and R. G. Halburd, Integrable systems and reductions of the self-dual Yang–Mills equations, *J. Math. Phys.* **44**, 3147 (2003).
54. R. S. Ward, Integrable and solvable systems, and relations among them, *Phil. Trans. R. Soc. London A* **315**, 451 (1985).
55. V. E. Nair and J. Schiff, A Kähler-Chern-Simons theory and quantization of instanton moduli, *Phys. Lett. B* **246**, 423 (1990); Kahler Chern–Simons theory and symmetries of anti-self-dual gauge fields, *Nucl. Phys. B* **371**, 329 (1992).

56. B. Y. Hou and X. C. Song, Stony Brook ITP- SB-83-66, (1983), unpublished; Lagrangian form of the self-dual equations for SU(N) gauge fields on four-dimension Euclidean space, *Commun. Theor. Phys.* **29**, 443 (1998).
57. A. Losev, G. Moore, N. Nekrasov and S. Shatashvili, Four-dimensional avatars of two-dimensional RCFT, *Nucl. Phys. B (Proc. Suppl.)* **46**, 130 (1996).
58. L. L. Chau and I. Yamanaka, Canonical formulation of the self-dual Yang–Mills system: Algebras and hierarchy, *Phys. Rev. Lett.* **68**, 1807 (19932); L. L. Chau and I. Yamanaka, Quantization of the self-dual Yang–Mills system: Exchange algebras, and local quantum group in four-dimensional quantum field theories, *Phys. Rev. Lett.* **70**, 1916 (1993).
59. V. G. Drinfeld, Hopf algebras and the quantum Yang–Baxter equation, *Dokl. Akad. Nauk, SSSR* **283**, 1060 (1985); Quantum groups, in *Proc. Int. Congr. Math. (Berkeley)*, Vol. I (Academic Press, New York, 1986), pp. 798–820; M. Jimbo, A q-analog of U(gl($N+1$)) Hecke algebras and the Yang–Baxter equation, *Lett. Math. Phys.* **11**, 247 (1986).
60. N. Yu. Reshetikin, L. A. Takhtadzhyan and L. D. Faddeev, Quantization of Lie groups and Lie algebras, *Leningrad Math. J.* **1**(1) (1990); M. Jumbo (ed.), *Yang–Baxter Equation in Integrable Systems*, Advanced Series in Mathematical Physics, Vol. 10 (World Scientific, 1990).
61. A. D. Popov and C. R. Preitschopf, Extended conforrnal symmetries of the self-dual Yang–Mills equations, *Phys. Lett. B* **374**, 71 (1996).
62. A. D. Popov, Holomorphic Chern–Simons–Witten theory: From 2D to 4D conformal field theories, *Nucl. Phys. B* **550**, 585 (1999).
63. S. K. Donaldson, Self-dual connections and the topology of smooth 4-manifolds, *Bull. Amer. Math. Soc.* **8**, 81 (1983).
64. S. K. Donaldson's web page, http://wwwf.imperial.ac.uk/ skdona/, gives web links to all his publications; complete at http://www.ams.org/mathscinet/and, since 2004 at http://arxiv.org/.
65. M. Atiyah, On the work of Simon Donaldson, in *Proceedings of the International Congress of Mathematicians*, Berkeley, California, USA, 1986.
66. See the profile of S. K. Donaldson at https://en.wikipedia.org/wiki/Simon_Donaldson, The citation of his work for his 1986 Fields Medal at https://en.wikipedia. org/wiki/Fields_Medal, The citation of his work for the 2009 Shaw Prizes in Mathematical Sciences at http://www.shawprize.org/en/shaw.php?tmp= 3&twoid=12&threeid=41&fourid=20, and see him as one of the seven recipients of the inaugural Breakthrough Prizes in Mathematics, 2014, https://en.wikipedia.org/wiki/Breakthrough_Prize_in_Mathematics, And the citation of his work, https://breakthroughprize.org/?controller=Page&action=laureates&p=3&laureate_id=55.
67. E. Witten, Monopoles and four-manifolds, *Math. Res. Lett.* **1**, 769 (1994).
68. N. Seiberg and E. Witten, Electric-magnetic duality, monopole condensation, and confinement in $N=2$ supersymmetric Yang–Mills theory, *Nucl. Phys. B* **426**, 19 (1994) [Erratum: *ibid.* **430**, 485 (1994)].
69. A. Jackson, A revolution in mathematics, http://web.archive.org/web/20100426172959/http:/www.ams.org/samplings/feature-column/mathnews-revolution; D. Kotschick, Gauge theory is dead! — Long live gauge theory!,

http://www.ams.org/notices/199503/kotschick.pdf; A. Scorpan, *The Wild World of 4-Manifolds* (American Mathematical Society, 2005).
70. E. Witten, An interpretation of classicall Yang–Mills theory, *Phys. Lett. B* **77**, 394 (1978).
71. V. Volovich, Supersymmetric Yang–Mills theories and twistors, *Phys. Lett. B* **129**, 429 (1983), *Theor. Math. Phys.* **57**, 39 (1983); C. Devchand, An infinite number of continuity equations and hidden symmetries in supersymmetric gauge theories, *Nucl. Phys. B* **238**, 333 (1984); L. L. Chau, M-L. Ge and Z. Popowicz, Riemann–Hilbert transforms and Bianchi–Backlund transformations for the supersymmetric Yang–Mills fields, *Phys. Rev. Lett.* **52**, 1940 (1984); L. L. Chau, M-L. Ge and C. S. Lim, Constrains and equitions of motion in supersymmetric Yang-Mills Theories, *Phys. Rev. D* **33**, 1056 (1986); Review by L.-L Chau, Supersymmetric Yang–Mills fields as an integrable system and connections with other linear systems, in *Proceedings of the Workshop on Vertex Operators in Mathematical Physics*, Berkeley, 1983, eds. J. Lepowsky, S. Mandelstam and I. M. Singer (Springer, New York, 1984).
72. L. L. Chau and C. S. Lim, Geometrical constraints and equations of motion in extended supergravity, *Phys. Rev. Lett.* **56**, 294 (1986).
73. L. L. Chau, Linear systems and conservation laws of gravitational fields in four plus extended superspace, *Phys. Lett. B* **202**, 238 (1992).
74. N. Beisert *et al.*, Review of AdS/CFT integrability: An overview, *Lett. Matt. Phys.* **99**, 3 (2012), and the references therein for all the review papers in the same volume.
75. L. D. Landau and E. M. Liesfiftz, *Quantum Mechanics, Non-relativistic Theory*, Chapters XIV & XV, Vol. 3, *Theoretical Physics* (translated by J. B. Sykes and J. S. Bell, 1958).
76. G. F. Chew, *S-Matrix Theory of Strong Interactions*, Lecture Note & Reprint Series, (1961); *The Analytic S Matrix: A Basis for Nuclear Democracy*, (1966); G. F. Chew (author) and D. Pines (editor), *S-Matrix Theory Of Strong Interactions*, (2011).
77. L. Brink, Maximally supersymmetric Yang–Mills theory: The story of N=4 Yang–Mills theory, in this *Proceedings*.
78. C. M. Will (Guest ed.), Focus issue: Milestones of general relativity, *Class. Quantum Grav.* **32**, 120201 (2015), http://iopscience.iop.org/0264-9381/page/Focus%20issue%20on%20Milestones%20of%20general%20relativity.
79. W. T. Ni (ed.), One hundred years of general relativity: From genesis and empirical foundations to gravitational waves, cosmology and quantum gravity, *Int. J. Mod. Phys. D* **24**, 1530028 (2015).
80. LIGO Scientific Collaboration and Virgo Collaboration, B. P. Abbott *et al.*, Observation of gravitational waves from a binary black hole merger, *Phys. Rev. Lett.* **116**, 061102 (Feb.12, 2016); http://www.nytimes.com/2016/02/12/science/ligo-gravitational-waves-black-holes-einstein.html.
81. In the opinion of many, including the author, it has been established that mathematics is the precise language of Nature. The author hopes that in some small measure this paper has helped to make the point that everyone, scientists and non-scientists, can participate in the excitement and the enrichment found in

the frontiers of scientific discovery, just as everyone can enjoy music and painting without being able to play a musical instrument and to paint. This point of view has guided the author in communicating physics (particle, condensed matter, and cosmology) to non-scientists since the late 1980s. For a sampling of her approach with good results, see the web page of her 2005 Physics 10 for non-scientist undergraduates, http://www.physics.ucdavis.edu/Classes/Physics10-B-W05/.

Discovery of the First Yang–Mills Gauge Particle — The Gluon*

Sau Lan Wu

University of Wisconsin-Madison, Madison, WI 53706, USA

This article "Discovery of the First Yang–Mills Gauge Particle — The Gluon" is dedicated to Professor Chen Ning Yang. The Gluon is the first Yang–Mills non-Abelian gauge particle discovered experimentally. The Yang–Mills non-Abelian gauge field theory was proposed by Yang and Mills in 1954, sixty years ago.

The experimental discovery of the first Yang–Mills non-Abelian gauge particle — the gluon — in the spring of 1979 is summarized, together with some of the subsequent developments, including the role of the gluon in the recent discovery of the Higgs particle.

In 1977, I became a faculty member as an assistant professor of physics at the University of Wisconsin at Madison. With support from the US Department of Energy, I formed a group of three post-docs and one graduate student. The first major decisions that I had to make were: what important problem of particle physics should I work on, and which experiment should I participate in? At that time, I was interested in positron–electron colliding experiments for various reasons, one of the most important ones being that the events are quite clean. The best choices were therefore the PEP at SLAC, Stanford University in California and the PETRA in Hamburg, Germany.

After discussions with several members of the experiments at these two accelerators, I went to talk to Björn Wiik of the TASSO Collaboration at PETRA. Getting to know Wiik constitutes a major lucky break at the beginning of my career in physics as a faculty member. Wiik introduced me to his colleagues Paul Söding and Günter Wolf, and brought me into the TASSO Collaboration.

After becoming a member of the TASSO Collaboration, my group and I concentrated on the construction of the drift chamber and the Čerenkov counters, two of the most interesting components, together with the F35 group of the TASSO Collaboration.

---

*This article is based on a talk at the Conference on 60 Years of Yang–Mills Gauge Field Theories, Nanyang Technological University, Singapore, May 25–28, 2015.

1. Particle Physics in 1977

Even before becoming a member of the TASSO Collaboration, I spent a great deal of time thinking about what physics to work on. For my thinking, it is useful to review some of the historical aspects of physics.

One of the most far-reaching conceptional advances in physics during the nineteenth century is that there is no action-at-a-distance. The best example is that the interaction between two charged particles, such as two electrons, is not correctly described in general by the Coulomb potential; instead, these two electrons interact through the intermediary of an electromagnetic field. This is the essence of the Maxwell equations. Thus the Coulomb potential by itself gives a correct description only for the static limit, i.e., when the two electrons are not moving with respect to each other.

Quantum mechanics, which was developed at the beginning of the twentieth century, taught us that particles are waves and waves are particles. In one of the epoch-making 1905 papers of Einstein,[1] he predicted that electromagnetic waves are also particles, called the photons. This startling prediction was verified experimentally by Compton[2] through photon–electron elastic scattering, a process that has since borne his name. Therefore charged particles such as electrons interact with each other through the exchange of photons.

The photon is the first known particle that is the carrier of force, in this case the electromagnetic force. Such particles are called gauge particles.

The next conceptual advance was due to the genius of Yang and Mills,[3,4] where for the first time the idea of the electromagnetic field was generalized. These papers were published almost exactly sixty one years ago. While the underlying group for the electromagnetic theory is U(1), Yang and Mills considered the more complicated case of isotopic spin described by the group SU(2). Once this case of SU(2) is understood, further generalizations to other groups are immediate. While U(1) is an Abelian group, SU(2) is a non-Abelian group. From the point of physics, the difference is:

(1) For the case of the Abelian group, the electromagnetic field itself is neutral, and therefore is not the source of further electromagnetic fields; but

(2) For the case of any non-Abelian group, the corresponding "electromagnetic field", called the Yang–Mills field, is itself a source of further Yang–Mills field.

Another way of saying the same is that, while the Maxwell equations are linear in the electromagnetic field, the corresponding Yang–Mills equations are non-linear.

At that time, I was fascinated by this Yang–Mills generalization of the electromagnetic field. This was to be contrasted with the experimental situation at that time: while the photons were everywhere in the detector, no Yang–Mills particle had ever been observed in any experiment.

I therefore formulated the following problem for myself: how could I discover experimentally the first Yang–Mills gauge particle using the TASSO detector?

2. PETRA and TASSO

PETRA (Positron–Electron Tandem Ring Accelerator) was a 2.3 kilometer electron and positron storage ring located in the German National Laboratory called Deutsches Elektronen-Synchrotron (DESY) in a west suburb of Hamburg, Germany. The name, Deutsches Elektronen-Synchrotron, is derived from the first accelerator in the Laboratory, a 7-GeV alternating gradient synchrotron. This laboratory was established in 1959 under the direction of Willibald Jentschke, and has played a crucial role in the re-emergence of Germany as one of the leading countries in physics.

The proposal[5] for the project for the construction of PETRA was submitted to the German government in November 1974, and was approved one year later. The hero of the construction of PETRA was Gustav Voss; under his leadership, the electron beam was first stored on July 15, 1978, more than nine months earlier than originally scheduled — a feat that was unheard of. There are many impressive stories how Voss accomplished this feat.

PETRA started its operation as a collider in November 1978; the center-of-mass energy was initially 13 GeV, but increased to 27 GeV in the spring of 1979. PETRA, together with its five detectors in four experimental halls are shown in Fig. 1. Note that, in this figure, "DESY" refers to the original 7-GeV synchrotron, not the German laboratory.

As seen from this Fig. 1, TASSO (Two-Arm Spectrometer SOlenoid) was one of the detectors; together with two other detectors, it was moved into place in October 1978. The first hadronic event at PETRA was observed in November; since then until its eventual shutdown, PETRA had been operating regularly, reliably and well.

A feature of the TASSO detector is the two-arm spectrometer, which leads to the name TASSO. The end view of this detector, i.e., the view along the beam pipe of the completed detector, is shown in Fig. 2. When TASSO was first moved into the PETRA beams in 1978, not all of the detector components shown in Fig. 2 were in working order. In particular, the central detector, especially the drift chamber, was functioning properly.

So far as softwares are concerned, it should be emphasized that computers at that time were very slow by modern standard. As to be discussed later, this slowness of the available computers must be taken into account and played a deciding role in figuring out the practical methods of data analysis, including especially the ideas of how to discover the first Yang–Mills gauge particle.

Figure 3 shows Wiik and the author relaxing after installing some of the drift-chamber cables. This large drift chamber, which was designed and constructed under the direction of Wiik and Ulrich Kötz, had a sensitive length of 2.23 m with inner and outer diameters of 0.73 and 2.44 m. There are in total 15 layers, nine with the sense wires parallel to the axis of the chamber and six with the sense wires oriented at an angle of approximately $\pm 4°$. These six layers make it possible to measure not only the transverse momenta of the produced charged particles but also their longitudinal momenta.

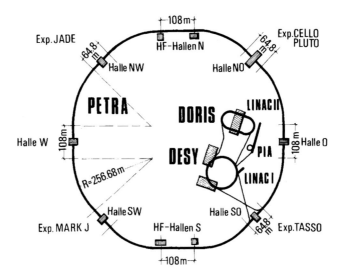

Fig. 1. PETRA (Positron-Electron Tandem Ring Accelerator).

3. Gluon Bremsstrahlung

The gluon[a] is the Yang–Mills non-Abelian gauge particle for strong interactions, i.e., the strong interactions between quarks[7,8] are mediated by the gluon, in much the same way as the electromagnetic interactions between electrons are mediated by the photon. For this reason, I concentrated on the gluon as the first Yang–Mills gauge particle, and the second gauge particle after the photon, to be discovered using the TASSO detector.

Indirect indication of gluons had been first given by deep inelastic electron scattering and neutrino scattering. The results of the SLAC-MIT deep inelastic scattering experiment[9] on the Callan–Gross sum rule were inconsistent with parton models that involved only quarks. The neutrino data from Gargamelle[10] showed that 50% of the nucleon momentum is carried by isoscalar partons or gluons. Further indirect evidence for gluons was provided by observation of scale breaking in deep inelastic scattering.[11] The very extensive neutrino scattering data from the BEBC and CDHS Collaborations[12] at CERN made it feasible to determine the distribution functions of the quark and gluon by comparison what was expected from QCD, and it was found that the gluon distribution function is sizeable. This information about the gluon is interesting but indirect, similar to that for the Z through the $\mu^+\mu^-$ asymmetry in electron–positron annihilation.[13]

At the time of PETRA turn-on in 1978–79, several groups were interested in looking for jet broadening following the suggestion of Ellis, Gaillard, and Ross.[14]

---

[a]The word "gluon" was originally introduced by Gell-Mann[6] to designate a hypothetical neutral vector field coupled strongly to the baryon current, without reference to color. Nowadays, this word is used exclusively to mean the Yang–Mills non-Abelian gauge particle for strong interactions.

Discovery of the First Yang–Mills Gauge Particle — The Gluon 115

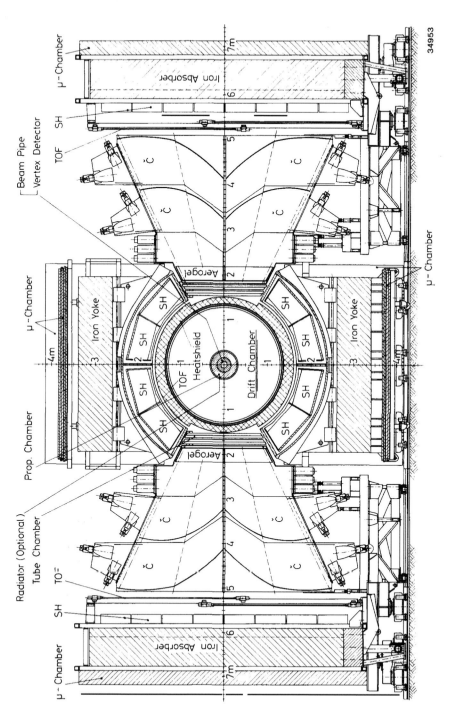

Fig. 2. End view of the TASSO detector.

Fig. 3. Björn Wiik and the author relaxing after tedious work on drift-chamber cabling. The cables in the photograph carried the timing information from the preamplifier boxes of the large central drift chamber of the TASSO detector at PETRA. They were to be connected to the read-out electronics. Dr. Ulrich Kötz of DESY/TASSO took this photograph in 1978.

The idea was to look for two-jet events where one of the jets becomes broadened due to the effect of the gluon. However, I considered this to provide just one more indirect indication of the gluon.

The discovery of the gluon requires direct observation.

Since the gluon is the gauge particle for strong interactions, the most direct way to produce a gluon is by the gluon bremsstrahlung process from a quark-antiquark pair:

$$e^+e^- \to q\bar{q}g. \qquad (1)$$

At high energy, the limited transverse-momentum distribution of the hadrons with respect to the quark direction, characteristic of strong interactions, results in back-to-back jets of hadrons[15] from the process $e^+e^- \to q\bar{q}$. In 1975, MARK I at SPEAR (SLAC) was the first to observe a two-jet structure[16] in e^+e^- annihilation as well as the spin $\frac{1}{2}$ nature of the produced quarks.[17]

How do the quark and the antiquark in the final state of process $e^+e^- \to q\bar{q}$ become jets? The heuristic picture is as follows. As the quark and the antiquark fly apart from each other, additional quark-antiquark pairs are excited; the quarks and antiquarks from these pairs then combine with the original antiquark and the original quark to form mesons (and some baryons). These mesons and baryons form the two jets. This process of jet formation is shown schematically in Fig. 4.

With this heuristic understanding of the experimental results from MARK I, I had to make my best guess as to how the gluon bremsstrahlung process (1) would look like in the PETRA detectors, including TASSO. Since the gluon is the Yang–Mills non-Abelain gauge particle for strong interactions, it is itself a source for gluon fields; see (2) of Sec. 1. It therefore seems reasonable to believe that the gluon in the final state of the process (1) would be seen in the detector as a jet, just like the quark and the antiquark.

Therefore the gluon bremsstrahlung process leads to three-jet events.

4. Three-Jet Analysis

Using the SPEAR information on the quark jets from the process, $e^+e^- \to q\bar{q}$, I convinced myself that three-jet events, if they were produced, could be detected once the PETRA energy went above three times the SPEAR energy i.e., $3 \times 7.4 \sim 22$ GeV. The arguments were as follows:

Figure 5 shows a comparison of the two-jet configuration at SPEAR with the most favorable kinematic situation of the three-jet configuration at PETRA. If the two invariant masses are taken to be the same, i.e., $\sqrt{3}E_2 \approx 7.4$ GeV, then the total energy of the three jets is $3E_2 \approx 13$ GeV, which must be further increased because each jet has to be narrower than the SPEAR jets. This additional factor is estimated to be $180°/120° = 1.5$, leading to about 20 GeV. Phase space considerations further increase this energy to about 22 GeV.

This estimate of 22 GeV was very encouraging because PETRA was expected to exceed it soon; indeed, it provided the main impetus for me

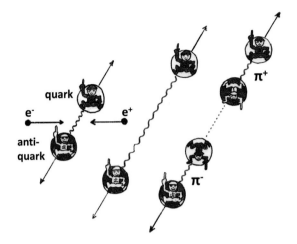

Fig. 4. Schematic picture of jet formation in $e^+e^- \to q\bar{q}$.

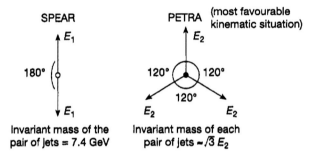

Fig. 5. Two-jet and three-jet configurations at SPEAR and PETRA respectively.

to continue the project to discover the first Yang–Mills non-Abelian gauge particle — the gluon.

At the same time, I had to address the problem: how could I find three-jet events at PETRA? I made a number of false starts until I realized the power of the following simple observation. By energy-momentum conservation, the two jets in $e^+e^- \to q\bar{q}$ must be back-to-back. Similarly, the three jets in $e^+e^- \to q\bar{q}g$ must be coplanar. Therefore, the search for the three jets can be carried out in the two-dimensional event plane, the plane formed by the momenta of $q, \bar{q}$ and g. Figure 6 shows a few pages of my notes written in June 1978.

The concept of the event plane for three-jet event deserves some discussion. As an analogous but simpler situation, consider an $\omega(782)$ at rest. Its dominant decay mode is

$$\omega \to \pi^+\pi^-\pi^0$$

followed by

$$\pi^0 \to \gamma\gamma.$$

S.L. Wu
June 1978
P8
4-8

Analysis of Jets

1. Two opposite jets
2. Two non-opposite jets
3. Planar case
4. Three jets

P10
4-10

Instead of $\vec{P}_i$, we now have a list
$$\{P_{iX}, P_{iY}\} \quad i = 1, 2, \ldots N$$

Rearrange the i's such that θ_i is in ascending order: (w.r.t. X-axis)
$$0 \leq \theta_1 \leq \theta_2 \leq \theta_3 \leq \ldots \leq \theta_N \leq 2\pi$$

where $P_{iX} + i P_{iY} = \sqrt{P_{iX}^2 + P_{iY}^2} \, e^{i\theta_i}$

B. A possible procedure for numerically treat these projected momenta is as follows

Choose three integers N_1, N_2, N_3 such that

a) $1 \leq N_1 < N_2 < N_3 \leq N$
   $N =$ no. of measured tracks

b) $\theta_{N_2-1} - \theta_{N_1} \leq \pi$
   ↑ last track of group 1   ↑ first track of group 1

   These conditions are imposed so that each jet goes more or less in a definite direction, not in both directions

c) $\theta_{N_3-1} - \theta_{N_2} \leq \pi$
   ↑ last track of group 2   ↑ first track of group 2

d) $2\pi + \theta_{N_1-1} - \theta_{N_3} \leq \pi$
   ↑ last track of group 3   ↑ first track of group 3

P9
4-9

The analysis of 3 jet events

A. With suitable χ^2 cuts, pick out the events from the region marked "Planar Distribution". For these events, T_3 is relatively small. Let $\hat{n}_3$ be eigenvector corresponding to T_3, and we project out this $\hat{n}_3$ component.

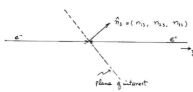

We choose n_{33} to be positive. Let the X-axis be determined by projecting the z-axis to the $\hat{n}_1 - \hat{n}_2$ plane

$$\hat{X} = \frac{\hat{z} - \hat{n}_3(\hat{z} \cdot \hat{n}_3)}{|\hat{z} - \hat{n}_3(\hat{z} \cdot \hat{n}_3)|}$$

and the Y-axis by orthogonality
$$\hat{Y} = \hat{n}_3 \times \hat{X} = \frac{\hat{n}_3 \times \hat{z}}{|\hat{z} - \hat{n}_3(\hat{z} \cdot \hat{n}_3)|}$$

P11
4-11

In this way, the N momenta have been split into three sets
$$\mathcal{A}_1 = \{N_1, N_1+1, \ldots, N_2-1\}$$
$$\mathcal{A}_2 = \{N_2, N_2+1, \ldots, N_3-1\}$$
$$\mathcal{A}_3 = \{N_3, N_3+1, \ldots, N_1-1\}$$

For each set we essentially go through the procedure of Hanson. More precisely, define
$$M_{jk}^{(\ell)} = \sum_{i \in \mathcal{A}_\ell} P_{ij} P_{ik} \quad j, k = X, Y \quad \ell = 1, 2, 3 \text{ groups}$$

Find the larger eigenvalue of the 2×2 matrix
$$M^{(\ell)} = \begin{bmatrix} M_{XX}^{(\ell)} & M_{YX}^{(\ell)} \\ M_{XY}^{(\ell)} & M_{YY}^{(\ell)} \end{bmatrix}$$

Call it $\lambda^{(\ell)}$. More precisely
$$\lambda^{(\ell)2} - (M_{XX}^{(\ell)} + M_{YY}^{(\ell)}) \lambda^{(\ell)} - (M_{XX}^{(\ell)} M_{YY}^{(\ell)} - M_{XY}^{(\ell)2}) = 0$$

choose the larger value
$$\lambda^{(\ell)} = \frac{1}{2} \left\{ M_{XX}^{(\ell)} + M_{YY}^{(\ell)} \oplus \sqrt{(M_{XX}^{(\ell)} + M_{YY}^{(\ell)})^2 - 4(M_{XX}^{(\ell)} M_{YY}^{(\ell)} - M_{XY}^{(\ell)2})} \right\}$$

$$= \frac{1}{2} \left\{ M_{XX}^{(\ell)} + M_{YY}^{(\ell)} \pm \sqrt{(M_{XX}^{(\ell)} - M_{YY}^{(\ell)})^2 + 4 M_{XY}^{(\ell)2}} \right\}$$

Fig. 6. Four of the pages from my notes of June 1978 on jet analysis.

The momenta of the π^+ and the π^- in this decay are well defined, and the momentum of π^0 is the sum of those of the two gammas. Since these momenta of π^+, π^-, π^0 must add together to zero, they in general define a plane — the event plane.

Let this consideration be applied to the gluon bremsstrahlung process $e^+e^- \to q\bar{q}g$, where the quark q, the antiquark $\bar{q}$, and the gluon g each metamorphoses into a jet. Thus the momenta of q, $\bar{q}$, and g are each given by the sum of the momenta of the particles in the respective jet. However, there is no well-defined way of assigning which of the mesons (and baryons) to which jet. Therefore, in the case of three-jet events (1), the event plane is not completely well defined. In spite of this slight ambiguity, it serves well for the purpose of three-jet analysis.

How did the event plane help me to discover the gluon? Consider first a two-jet event from $e^+e^- \to q\bar{q}$. In this case, the two jets are back-to-back in the detector. While the event, that may consist of charged tracks and signals from the ECAL and the HCAL due to neutral particles, is three-dimensional, only two-dimensional views were available in the late seventies. Because of the slowness of the computers at that time, it was not yet possible at that time to rotate the events for viewing on the computer screen.

For a two-jet event, viewing in any direction shows the back-to-back jets, the exception being views from near the direction of the two jets. Therefore, if two orthogonal views are given, at least one of them shows the two-jets structure clearly.

This is not the case for three-jet events from the gluon bremsstrahlung process $e^+e^- \to q\bar{q}g$. Contrary to the two-jet case, a three-jet event can be seen clearly as such only when viewed from certain limited directions. The best view is obtained from a direction perpendicular to the event plane. This is the underlying reason for the importance of the event plane.

Equipped with my estimate of 22 GeV center-of-mass energy as described above and the idea of viewing the events in their event planes, I was in a good position to look for three-jet events. The development of the method of data analysis for such events then proceeded rapidly. The main remaining job was to find a way of identifying the three jets; this is essential in itself, including in particular in order to compare the properties of the jets in these three-jet events with those in the two-jet events.

Projected into this two-dimensional event plane, the momenta of the particles in the detector can be naturally ordered cyclically. Thus, if the polar coordinates of N vectors $\vec{q}_j$ are (q_j, θ_j) then these $\vec{q}_j$ can be relabelled such that

$$0 \leq \theta_1 \leq \theta_2 \leq \theta_3 \leq \cdots \leq \theta_N < 2\pi. \qquad (2)$$

With this cyclic ordering, the N $\vec{q}_j$'s can be split up into three sets of contiguous vectors, and these three sets are to be identified as the three jets. There are of course a number of ways of carrying out this splitting, and, with suitable restrictions, the one with smallest average transverse momentum is chosen as the best approximation to the correct way of identifying the three jets.

This cyclic ordering (2) reduces greatly the number of combinations that need to be studied to identify the three jets. Without any ordering, the number of ways to partition the N observed tracks into three sets, each consisting of at least one track, is

$$(3^{N-1} - 2^N + 1)/2.$$

For example, this is 2.4×10^6 for $N = 15$, 5.8×10^8 for $N = 20$, 1.4×10^{11} for $N = 25$, and 3.4×10^{13} for $N = 30$. With this ordering, it is sufficient to consider three sets of contiguous vectors, and the number of such partitions is

$$\binom{N}{3} = N(N-1)(N-2)/6.$$

This is 455 for $N = 15$, 1140 for $N = 20$, 2300 for $N = 25$ and 4060 for $N = 30$; or reductions by factors of 5×10^3, 5×10^5, 6×10^7 and 8×10^9 respectively. Such is the power of using the event plane.

It should be emphasized that this large reduction should not be described merely as an improvement in efficiency, rather it shortened enormously long computer runs so that they became manageable.

This analysis of the PETRA events for three jets then proceeds as follows.[18] First, using the momentum tensor (This tensor is essentially the same as the one used previously for two-jet analysis.)

$$M_{\alpha\beta} = \sum_j p_{j\alpha} p_{j\beta}, \qquad (3)$$

the event plane is determined as the plane with the smallest transverse momentum, i.e., the plane perpendicular to the eigenvector $\hat{n}_3$ that corresponds to the smallest eigenvalue λ_3 of M. Then all the measured momenta of the produced particles are projected into this event plane. Using (2) above, rearrange these projected momenta $\vec{q}_j$ into a cyclic order. For $N \geqslant 3$, split into contiguous sets by choosing three numbers N_1, N_2 and N_3 that satisfy

$$1 \leqq N_1 < N_2 < N_3 \leqq N. \qquad (4)$$

The three sets mentioned above then consist respectively of

$$\{N_1, N_1+1, \ldots, N_2-1\}$$
$$\{N_2, N_2+1, \ldots, N_3-1\}$$
$$\{N_3, N_3+1, \ldots, N, 1, 2, \ldots, N_1-1\}. \qquad (5)$$

Each of these three sets is required to span an angle of less than π, i.e.,

$$\theta_{N_2-1} - \theta_{N_1} < \pi,$$
$$\theta_{N_3-1} - \theta_{N_2} < \pi,$$

and

$$\theta_{N_1-1} + 2\pi - \theta_{N_3} < \pi. \qquad (6)$$

For each of these sets $S^{(\tau)}$, $\tau = 1, 2, 3$, define a two-dimensional analog of the momentum tensor (3); let $\Lambda^{(\tau)}$ be the larger eigenvalue and $\hat{m}^{(\tau)}$ the corresponding normalized eigenvector.

Since each jet can contain only particles in one direction, not simultaneously in both directions, the requirement is imposed that the signs of $\hat{m}^{(\tau)}$ can be chosen so that

$$\vec{q}_j \cdot \hat{m}^{(\tau)} > 0 \tag{7}$$

for each j in the corresponding set listed in (5). These conditions (7) actually consist of the following six:

$$\vec{q}_{N_1} \cdot \hat{m}^{(1)} > 0,$$
$$\vec{q}_{N_2-1} \cdot \hat{m}^{(1)} > 0,$$
$$\vec{q}_{N_2} \cdot \hat{m}^{(2)} > 0,$$
$$\vec{q}_{N_3-1} \cdot \hat{m}^{(2)} > 0,$$
$$\vec{q}_{N_3} \cdot \hat{m}^{(3)} > 0,$$

and

$$\vec{q}_{N_1-1} \cdot \hat{m}^{(3)} > 0. \tag{8}$$

These conditions (8) are stronger than the (6) above.

For each admissible way of splitting into three contiguous sets, calculate the sum of these three largest eigenvalues:

$$\Lambda(N_1, N_2, N_3) = \Lambda^{(1)} + \Lambda^{(2)} + \Lambda^{(3)}. \tag{9}$$

This $\Lambda(N_1, N_2, N_3)$ is maximized over N_1, N_2, N_3 that satisfy (4) and (8). This maximizing partition gives the three jets and the corresponding $\hat{m}^{(1)}$, $\hat{m}^{(2)}$, $\hat{m}^{(3)}$ yield the directions of the jet axes.

In short, the event plane is used to put the projections of the measured momenta of the produced particles into cyclic order. For each way of splitting into three contiguous sets, the sum of the larger eigenvalues corresponding to the two-dimensional momentum tensors for the three sets is evaluated. The particular splitting with the largest value of this sum corresponds to the smallest average momentum transverse to the three axes, and is therefore chosen as the way to identify the three jets. It is then straightforward to study the various properties of the jets, such as the average transverse momentum of each jet in three-jet events, and compare them with the corresponding properties in two-jet events.

This procedure has a number of desirable features. First, all three jet axes are determined, and they are in the same plane. This is the feature that plays a central role in the determination of the spin of the gluon, see Sec. 6.

Secondly, particle identification is not needed, since there is no Lorentz transformation.

Thirdly, the computer time is moderate even when all the measured momenta are used.

Finally, it is not necessary to have the momenta of all the produced particles; it is only necessary to have at least one momentum from each of the three jets. Thus, for example, my procedure works well even when no neutral particles are included.

This last advantage is important, and it is the reason why this procedure is a good match to the TASSO detector at the time of the PETRA turn-on. As explained in Sec. 2, when the TASSO detector was first moved into the PETRA beams in 1978, some of the detector components were not yet in working order.

This was true not only for TASSO but also for all the other detectors at PETRA at that time. Fortunately for my three-jet analysis described here, what was needed was the TASSO central detector, which was already functioning properly.

I had at that time Georg Zobernig as my post-doc; he was excellent in working with computers. My procedure of identifying the three-jet events in order to discover the gluon, programmed by Zobernig on an IBM 370/168 computer, was ready before the turn-on of PETRA in September of 1978.

Shortly thereafter, I showed the procedure and the program to Wiik; he was very excited about them and happy that I had found and solved such an important problem. I presented my analysis method in a TASSO meeting and later had it published with Zobernig.[18]

5. Discovery of the Gluon

When we had obtained data for center-of-mass energies of 13 GeV and 17 GeV, Zobernig and I looked for three-jet events. It was not until just before the Neutrino 79 (International Conference on Neutrino, Weak Interactions and Cosmology at Bergen, Norway) in the late spring of 1979 that we started to obtain data at the higher center-of-mass energy of 27.4 GeV. We found one clear three-jet event from a total of 40 hadronic events at this center-of-mass energy. This first three-jet event of PETRA, as seen in the event plane, is shown in Fig. 7(a). When this event was found, Wiik had already left Hamburg to go to the Bergen Conference. Therefore, during the weekend before the conference, I took the display produced by my procedure for this event to Norway to meet Wiik at his house near Bergen. It turned out that John Ellis was visiting Wiik also; after seeing my event, Ellis described this event as "gold-plated". During this weekend, I also telephoned Günter Wolf, the TASSO spokesman, at his home in Hamburg and told him of the finding. Wiik showed the event in his plenary talk "First Results from PETRA", acknowledging that it was my work with Zobernig by putting our names on his transparency of the three-jet event, and referred to me for questions. Donald Perkins took this offer and challenged me by wanting to see all forty TASSO events. I showed him all forty events, and, after we had spent some time together studying the events, he was convinced.

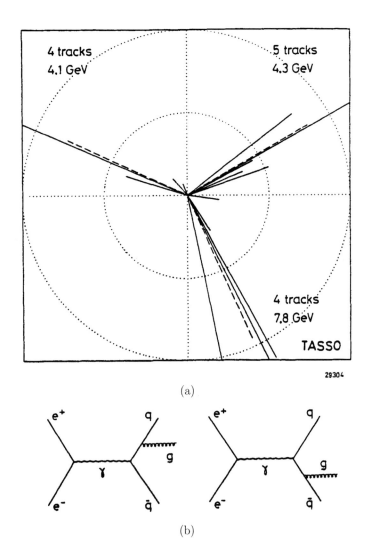

Fig. 7. (a) The first three-jet event from electron–positron annihilation, as viewed in the event plane. It has three well separated jets.[19,20] (b) Feynman diagrams for the gluon bremsstrahlung process $e^+e^- \to q\bar{q}g$.

The following is quoted from the write-up of Wiik's talk[19]:

"If hard gluon bremsstrahlung is causing the large $p_\perp$ values in the plane then a small fraction of the events should display a three jet structure. The events were analyzed for a three jet structure using a method proposed by Wu and Zobernig[27]) ... A candidate for a 3 jet event, observed by the TASSO group at 27.4 GeV, is shown in Fig. 21 viewed along the $\hat{n}_3$ direction. Note that the event has a three clear well separated jet and is just not a widening of a jet."

As soon as I returned from Bergen, I wrote a TASSO note with Zobernig on the observation of this three-jet event.[20] The first page of this TASSO note is shown in Fig. 8. Both in Wiik's talk[19] and in the TASSO note,[20] this three-jet event was already considered to be due to the hard gluon bremsstrahlung process (1), described by the Feynman diagrams of Fig. 7(b). As seen from Fig. 7(a), this first three-jet event had three clear, well separated jets, and was considered to be more convincing than a good deal of statistical analysis. Indeed, before the question of statistical fluctuation could be seriously raised, events from $E_{\text{cm}} = 27.4$ GeV rolled in and we found a number of other three-jet events.

Less than two weeks after the Bergen Conference, four of the TASSO three-jet events, as given in Fig. 9(a), were shown by Paul Söding of DESY/TASSO at the European Physical Society (EPS) Conference in Geneva.[21] Figure 9(b) gives several plots of the various transverse momentum distributions. The first one is the distribution of

$$\langle p_\perp^2 \rangle_{\text{out}} = \frac{1}{N} \sum_j (\bar{p}_j \cdot \hat{n}_3)^2 \tag{10}$$

(= square of momentum component normal to the event plane averaged over the charged particles in one event), while the second one is that of

$$\langle p_\perp^2 \rangle_{\text{in}} = \frac{1}{N} \sum_j (\bar{p}_j \cdot \hat{n}_2)^2 \tag{11}$$

(= square of momentum component in the event plane and perpendicular to the sphericity axis averaged the same way). A comparison of these two plots shows that the major difference is the absence of a tail for $\langle p_\perp^2 \rangle_{\text{out}}$ and the presence of one for $\langle p_\perp^2 \rangle_{\text{in}}$. Since three-jet events tend to have a small $\langle p_\perp^2 \rangle_{\text{out}}$ but a much larger $\langle p_\perp^2 \rangle_{\text{in}}$, this distribution $\langle p_\perp^2 \rangle_{\text{in}}$ shows a continuous transition from two-jet events to three-jet events.

Also shown in Fig. 9(b) is $\langle p_\perp^2 \rangle_{\text{in, 3 jet axes}}$, which is defined the same way as (11) but, for each jet, the jet axis found by my method is used.

TASSO Note No. 84
26.6.1979

From Sau Lan Wu and Haimo Zobernig

On: A three-jet candidate (run 447 event 13177)

We have made a three jet analysis to all the hadronic candidates (43 events for $\Sigma_i |P_i| \geq 9$ GeV) of the May 1979 data at $E_{cm} = 27.4$ GeV using our method described in DESY 79/23 (A method of three jet analysis in e^+e^- annihilation).

Fig. 1 gives the triangular plot of the normalized eigenvalues Q_1, Q_2 and Q_3 ($Q_1 \leq Q_2 \leq Q_3$) of the momentum matrix

$$M_{\alpha\beta} = \sum_j (P_{j\alpha} P_{j\beta})$$

(See equation (1) and Fig. 1 of DESY 79/23). We find two three jet candidates

    run 447   event 13177

    run 439   event 12845

We then display each event on the 3 planes

    plane 1: normal to $\hat{n}_1$, the normalized eigenvector corresponding to Q_1. $\sum_i |P_{i\perp}|^2$

with respect to this plane is minimized.

    plane 2: normal to $\hat{n}_2$, the normalized eigenvector ~~to $\hat{n}_2$~~ corresponding to Q_2

    plane 3: normal to $\hat{n}_3$, the normalized eigenvector corresponding to Q_3.

Fig. 2 displays the three jet candidate (run 447 event 13177) on planes 1, 2, 3.

Fig. 3 displays plane 1 of this event in a blow up scale.

    The axis for each of the three jets are found. Given the axes and $\Sigma_i |P_i|$ of each jet, the total energy of each jet is determined assuming the mass of each quark (or gluon) is zero.

Fig. 4 displays the event run 439 event 12845. This event looks like two charged jets and one neutral jet.

Fig. 8. The first page of TASSO Note No. 84, June 26, 1979, by Wu and Zobernig.[20]

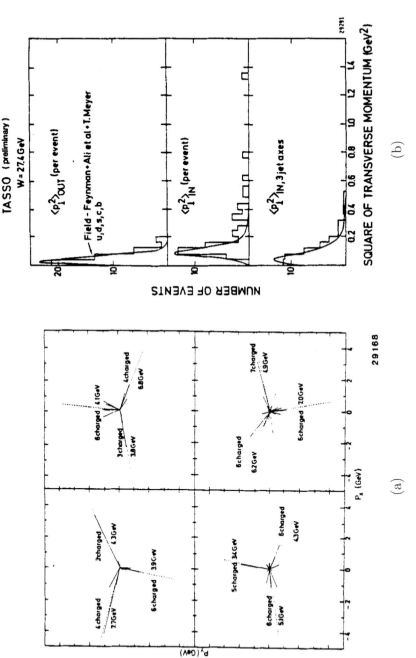

Fig. 9. (a) Four TASSO three-jet events as seen in the event plane. (b) Distribution of the average squared transverse momentum component out of the event plane (top), and in the event plane (center), for the early TASSO events at the center-of-mass energy of 27.4 GeV (averaging over charged hadrons only). The curves are for $q\bar{q}$ jets without gluon bremsstrahlung. Comparison of these distributions gives evidence that broadening (compared to $q\bar{q}$ jets) occurs in one plane. The bottom figure shows $\langle p_\perp^2 \rangle$ per jet when 3 jet axes are fitted, again compared with the 2-jet model.[21]

The absence of a tail and the similarity to the first distribution means that the jets in three-jet events are similar to those in two-jet events, justifying the use of the same word "jet" in both cases. At this time of the EPS Conference in Geneva, no other experiment at PETRA mentioned anything about three-jet events.

On July 31, 1979, at the presentations by each of the PETRA experiments at the open session of the DESY Physics Research Committee, again only TASSO (represented by Peter Schmüser of the University of Hamburg) gave evidence of three-jet events.

With these three-jet events, the question is: what are the three jets? Since quarks are fermions, and two fermions (electron and positron) cannot become three fermions, it immediately follows that these three jets cannot all be quarks and antiquarks. In other words, *a new particle has been discovered.*

Secondly, since this new particle, similar to the quarks, also hadronizes into a jet, it cannot be a color singlet. Color singlets, such as the pion, the kaon, and the proton, either leave a track (if charged), or give an energy deposition in a calorimeter, or decay into well-defined final states, but do not metamorphose into jets. Therefore, the abundance of three-jet events means that the carrier of strong forces, unlike the photon, is not an Abelian gauge particle (which must be colorless).

For these reasons, it was readily accepted by most of the high-energy physicists that the three-jet events are due to $e^+e^- \rightarrow q\bar{q}g$.

6. Confirmations and the Spin of the Gluon

As more data were obtained, by the end of August other experiments at PETRA began to have their own three-jet analysis ready. It is quite natural that a discovery is corroborated by higher statistics and by other experiments at a later date. At the Lepton–Photon Conference at FNAL in late August of 1979, all four experiments at PETRA gave more extensive data, confirming the earlier observation of TASSO. Since these experiments were run simultaneously, the amounts of data were similar and there is not much difference in their statistical significance. Since this was the highlight of the conference, Leon Lederman, Director of FNAL, called for a press conference on the discovery of the gluon. In a period of three months, between August 29 and December 7, these more extensive data were submitted for publication by TASSO,[22] MARK J,[23] PLUTO[24] and JADE,[25] in this order. The gluon remains one of the most interesting discoveries from PETRA.[26]

The early papers related to the PETRA three-jet events are the following.

(1) Sau Lan Wu and Georg Zobernig, *Z. Phys. C — Particles and Fields* **2**, 107 (1979).

(2) B. H. Wiik, *Proceedings of International Conference on Neutrinos, Weak Interactions and Cosmology*, Bergen, Norway, June 18–22, 1979, p. 113.

(3) Sau Lan Wu and Haimo Zobernig, *TASSO Note 84*, June 26, 1979.

(4) P. Söding, *Proceedings of European Physical Society International Conference on High Energy Physics*, Geneva, Switzerland, 27 June–4 July, 1979, p. 271.

(5) TASSO Collaboration (R. Brandelik *et al.*), *Phys. Lett.* B **86**, 243 (1979) [received on August 29, 1979].

(6) MARK J Collaboration (D. P. Barber *et al.*), *Phys. Rev. Lett.* **43**, 830 (1979) [received on August 31, 1979].

(7) PLUTO Collaboration (Ch. Bergen *et al.*), *Phys. Lett.* B **86**, 418 (1979) [received on September 13, 1979].

(8) JADE Collaboration (W. Bartel *et al.*), *Phys. Lett.* B **91**, 142 (1980) [received on December 7, 1979].

(9) JADE Collaboration, MARK J Collaboration, PLUTO Collaboration and TASSO Collaboration, *Proceedings of 1979 International Symposium on Lepton and Photon Interactions at High Energies*, Fermi National Accelerator Laboratory (FNAL), Batavia, Illinois, August 23–29, 1979.

In the above list, the first one provides the method of analysis used in the TASSO papers, and the others are all experimental.

Since the gluon is a Yang–Mills non-Abelian gauge particle, it must have spin 1. It is nevertheless nice to have a measurement of its spin, especially since it is the first such particle ever seen experimentally. As described in Sec. 3, my procedure gives all three jet axes, and they are in the same plane. This ability to isolate individual jets in high-energy three-jet events makes it possible to measure the correlation between the directions of the three jets. Since the quark spin[16,17] is known to be $\frac{1}{2}$ and this correlation depends on the gluon spin, it can be used to give the desired determination of the spin of the gluon.

The amount of data collected at PETRA up to the time of the FNAL Conference was not quite enough for this spin determination. As soon as there were enough data, in September 1980, the TASSO Collaboration determined the spin of the gluon.[27] In this determination, the Ellis–Karliner angle[28] was used with my three-jet analysis. Not surprisingly, the spin of the gluon was found to be indeed 1. One month later, the PLUTO Collaboration reached the same conclusion.[29] This result was confirmed subsequently by the other PETRA Collaborations.

Because of this discovery of the gluon by the TASSO Collaboration, Söding, Wiik, Wolf, and I were awarded the 1995 European Physical Society High Energy and Particle Physics Prize (Figs. 10 and 11). Because of my leading role in this discovery, I was chosen to give the acceptance speech at the EPS award ceremony.[30]

7. Recent Developments

The gluon was discovered thirty six years ago. During these thirty six years, it has become more and more important, one prominent example being its role in the

Fig. 10. The four Prize Recipients at the ceremony of the 1995 European Physical Society High Energy and Particle Physics Prize in Brussels, Belgium. Front row: Günter Wolf and Sau Lan Wu; second row: Björn Wiik and Paul Söding.

recent discovery of the Higgs particle[31] by the ATLAS Collaboration[32] and the CMS Collaboration[33] using the Large Hadron Collider (LHC) at CERN. This observation of the Higgs particle in 2012 has led to the award of the 2013 Nobel Prize in physics to Englert and Higgs. Brout had unfortunately died a year and half earlier.

Since the gluon is the Yang–Mills gauge particle for strong interactions, to a good approximation a proton consists of a number of gluons in addition to two u quarks and a d quark. Since the coupling of the Higgs particle to any elementary particle is proportional to its mass, there is little coupling between the Higgs particle and these constituents of the proton. Instead, some heavy particle needs to be produced in a proton-proton collision at LHC, and is then used to couple to the Higgs particle. Among all the known elementary particles, the top quark t, with a mass of 173 GeV/c^2, is the heaviest.[34]

The top quark is produced predominantly together with an anti-top quark or an anti-bottom quark. Since the top quark has a charge of $+2/3$ and is a color triplet, such pairs can be produced by

(a) a photon: $\gamma \to t\bar{t}$;
(b) a Z: $Z \to t\bar{t}$;
(c) a W: $W^+ \to t\bar{b}$; or
(d) a g: $g \to t\bar{t}$.

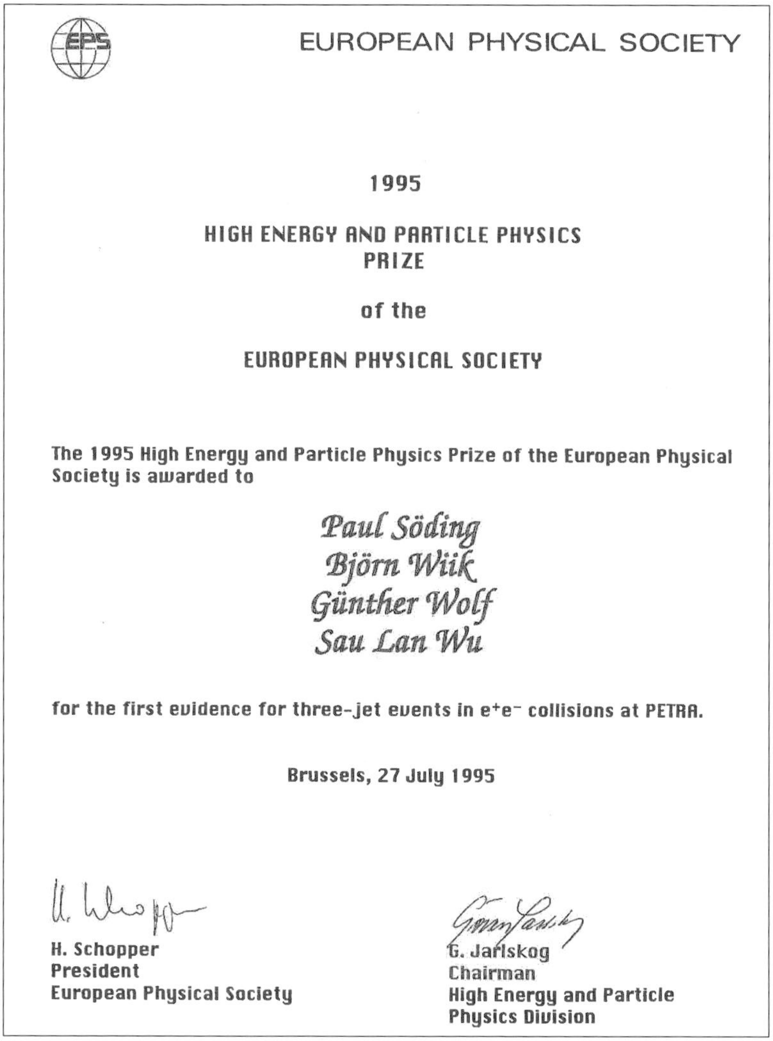

EUROPEAN PHYSICAL SOCIETY

1995

HIGH ENERGY AND PARTICLE PHYSICS
PRIZE

of the

EUROPEAN PHYSICAL SOCIETY

The 1995 High Energy and Particle Physics Prize of the European Physical Society is awarded to

Paul Söding
Björn Wiik
Günther Wolf
Sau Lan Wu

for the first evidence for three-jet events in e⁺e⁻ collisions at PETRA.

Brussels, 27 July 1995

H. Schopper
President
European Physical Society

G. Jarlskog
Chairman
High Energy and Particle
Physics Division

Fig. 11. European Physical Society High Energy and Particle Physics Prize, 1995.

As discussed in the preceding paragraph, there is no photon, or Z, or W as a constituent of the proton. Since, on the other hand, there are gluons in the proton, (d) is by far the most important production process for the top quark.

Because of color conservation — the gluon has color but not the Higgs particle — the top and anti-top pair produced by a gluon cannot annihilate into a Higgs particle. In order for this annihilation into a Higgs particle to occur, it is necessary for the top or the anti-top quark to interact with a second gluon to change its color content. It is therefore necessary to involve two gluons, one each from the protons of the two opposing beams of LHC, and we are led to the diagram of Fig. 12 for Higgs production.

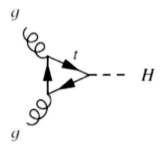

Fig. 12. Feynman diagram for the Higgs (H) production by gluon–gluon fusion (also called gluon fusion).

This production process is called "gluon–gluon fusion" (also called "gluon fusion"). As expected from the large mass of the top quark, this gluon–gluon fusion is by far the most important Higgs production process, and shows the central role played by the gluon in the discovery of the Higgs particle in 2012.

It is desirable to make this statement more quantitative. It is shown in Fig. 13 the various Higgs production cross sections from calculation as functions of the Higgs mass.[35] The top curve is for gluon–gluon fusion, and next one is for vector boson fusion (VBF); note that the vertical scale is logarithmic.

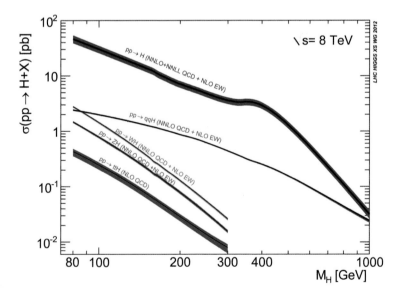

Fig. 13. Higgs production cross sections from gluon–gluon fusion (top curve), vector boson fusion, and three associated production processes at the LHC center-of-mass energy of 8 TeV.[35] It is seen that gluon–gluon fusion is the most important production process.

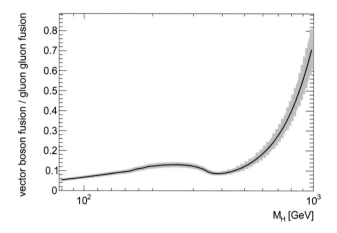

Fig. 14. Ratio of VBF cross section divided by the gluon–gluon fusion cross section.

In order to show even more clearly the importance of the gluon in the production of the Higgs particle, Fig. 14 shows the ratio of the second most important production process VBF to gluon–gluon fusion. It is seen that, for the relatively low masses of the Higgs particle, VBF cross section is less than 10% of that of gluon–gluon fusion. It is known that the Higgs particle[32,33] indeed has a mass in this range. Thus, through gluon–gluon fusion, the gluon contributes about 90% of Higgs production at the Large Hadron Collider. A more dramatic way of saying the same thing is that, if there were no gluon, the Higgs particle could not have been discovered for years!

8. Summary

In summary, since two fermions cannot turn into three fermions, the experimental observation of three-jet events in e^+e^- annihilation, first accomplished by the TASSO Collaboration in the spring of 1979 and confirmed by the other Collaborations at PETRA two months later in the summer, implies the discovery of a new particle. Similar to the quarks, this new particle hadronizes into a jet, and therefore cannot be a color singlet. These three-jet events are most naturally explained by hard non-collinear bremsstrahlung $e^+e^- \to q\bar{q}g$. One year later, the spin of the gluon was determined experimentally to be indeed 1, as it should be for a gauge particle.

Thus the 1979 discovery of the second gauge particle, the gluon, occurred more than half a century after that by Compton[2] of the first, the photon. This second gauge particle is also the first Yang–Mills non-Abelian gauge particle,[3,4] i.e., a gauge particle with self-interactions. Four years later, in 1983, the second and the third non-Abelian gauge particles, the W and the Z, were discovered at the CERN proton–antiproton collider by the UA1 Collaboration of Carlo Rubbia, Simon van der Meer et al.[36] and by the UA2 Collaboration of Pierre Darriulat et al.[37]

Since its first observation thirty six years ago, the importance of the gluon in particle physics has grown significantly. An especially noticeable example is its essential role in the discovery of the Higgs particle in 2012 by the ATLAS Collaboration[32] and the CMS Collaboration[33] at the Large Hadron Collider at CERN.

Acknowledgments

I am most grateful to the support, throughout many years, of the United States Department of Energy (Grant No. DE-FG02-95ER40896) and the University of Wisconsin through the Wisconsin Alumni Research Foundation and the Vilas Foundation.

I would very much like to thank Professor K. K. Phua for the invitation to contribute to this conference on 60 Years of Yang–Mills Gauge Field Theories.

References

1. Albert Einstein, "Über einen die Erzeugung und Verwandlung des Lichtes betreffenden heuristischen Gesichtspunkt" (On a Heuristic Point of View Concerning the Production and Transformation of Light) *Annalen der Physik* **17**, 132 (1905).
2. Arthur Holly Compton, "The Total Reflexion of X-Rays", *Philosophical Magazine* **45**, 1121 (1923).
3. Chen Ning Yang and Robert L. Mills, "Isotopic Spin Conservation and a Generalized Gauge Invariance", *Phys. Rev.* **95**, 631 (1954).
4. Chen Ning Yang and Robert L. Mills, "Conservation of Isotopic Spin and Isotopic Gauge Invariance", *Phys. Rev.* **96**, 191 (1954).
5. Deutsches Elektronen-Synchrotron, PETRA — a proposal for extending the storage-ring facilities at DESY to higher energies, DESY, Hamburg, November 1974; Deutsches Elektronen-Synchrotron, PETRA — updated version of the PETRA proposal, DESY, Hamburg, February 1976.
6. Murray Gell-Mann, "Symmetries of Baryons and Mesons", *Phys. Rev.* **125**, 1067 (1962).
7. Murray Gell-Mann, "A Schematic Model of Baryons and Mesons", *Phys. Lett.* **8**, 214 (1964).
8. George Zweig, "An SU3 Model for Strong Interaction Symmetry and its Breaking", CERN Preprints TH401 (January 17, 1964) and TH412 (February 21, 1964).
9. SLAC-MIT Collaboration (Elliott D. Bloom *et al.*), *Phys. Rev. Lett.* **23**, 930 (1969); SLAC-MIT Collaboration (Martin Breidenbach, *et al.*), *Phys. Rev. Lett.* **23**, 935 (1969); SLAC-MIT Collaboration, data presented by H. Kendall, *Symp. Electron and Photon Interactions*, 1971, Cornell Univ., Ithaca, p. 248; J. Friedman and H. Kendall, *Ann. Rev. Nucl. Sci.* **22**, 203 (1972); SLAC-MIT Collaboration (G. Miller *et al.*), *Phys. Rev. D* **5**, 528 (1972).

10. Gargamelle Collaboration (T. Eichten et al.), Phys. Lett. B **46**, 274 (1973).
11. Y. Watanabe et al., Phys. Rev. Lett. **35**, 898 (1975); C. Chang et al., Phys. Rev. Lett. **35**, 901 (1975); W. B. Atwood et al., Phys. Lett. B **64**, 479 (1976).
12. P. C. Bosetti, et al., Nucl. Phys. B **142**, 1 (1978); J. G. H. de Groot, et al., Phys. Lett. B **82**, 456 (1979); J. G. H. de Groot, et al., Z. Phys. C: Particles and Fields **1**, 143 (1979).
13. JADE Collaboration (W. Bartel, et al.), Phys. Lett. B **108**, 140 (1982); TASSO Collaboration (R. Brandelik, et al.), Phys. Lett. B **110**, 173 (1982); MARK J Collaboration (B. Adeva, et al.), Phys. Rev. Lett. **48**, 1701 (1982); CELLO Collaboration (H. J. Behrend, et al.), Z. Phys. C: Particles and Fields **14**, 283 (1982).
14. John Ellis, Mary K. Gaillard, and Graham Ross, Nucl. Phys. B **111**, 253 (1976); [Erratum ibid. **130**, 516 (1977)]. Many experimentalists, in preparing for the data analysis at the beginning of the PETRA run, were influenced by the following sentences in this paper: "The first observable effect should be a tendency for the two-jet cigars to be unexpectedly oblate, with a high large p_T cross section. Eventually, events with large p_T would have a three-jet structure, without local compensation of p_T."
15. S. D. Drell, D. J. Levy, and T. M. Yan, Phys. Rev. **187**, 2159 (1969), and Phys. Rev. D **1**, 1617 (1970); N. Cabibbo, G. Parisi, and M. Testa, Nuovo Cimento **4**, 35 (1970); J. D. Bjorken and S. J. Brodsky, Phys. Rev. D **1**, 1416 (1970); R. P. Feynman, Photon-Hadron Interactions (Benjamin, Reading, Mass., 1972), p. 166.
16. G. Hanson, et al., Phys. Rev. Lett. **35**, 1609 (1975).
17. R. F. Schwitters, et al., Phys. Rev. Lett. **35**, 1320 (1975).
18. Sau Lan Wu and Georg Zobernig, "A Method of Three-jet Analysis in e^+e^- Annihilation", Z. Phys. C: Part. Fields **2**, 107 (1979).
19. Björn H. Wiik, "First Results from PETRA", in Proceedings of Neutrino 79, International Conference on Neutrinos, Weak Interactions and Cosmology, Volume **1**, Bergen June 18–22 1979, 113–154. [see pp. 127–128 for the quotation].
20. Sau Lan Wu and Haimo Zobernig, "A three-jet candidate (run 447, event 13177)", TASSO Note No. 84 (June 26, 1979).
21. Paul Söding, "Jet Analysis", Proceedings of European Physical Society International Conference on High Energy Physics, Geneva, Switzerland, 27 June–4 July, 1979, pp. 271–281.
22. TASSO Collaboration (R. Brandelik et al.), "Evidence for Planar Events in e^+e^- Annihilation at High Energies", Phys. Lett. B **86**, 243 (1979), received on August 29, 1979.
23. MARK J Collaboration (D.P. Barber et al.), "Discovery of Three-Jet Events and a Test of Quantum Chromodynamics at PETRA", Phys. Rev. Lett. **43**, 830 (1979), received on August 31, 1979.

24. PLUTO Collaboration (Ch. Berger et al.), "Evidence for Gluon Bremsstrahlung in e^+e^- Annihilation at High Energies", *Phys. Lett. B* **86**, 418 (1979), received on September 13, 1979.
25. JADE Collaboration (W. Bartel et al.), "Observation of Planar Three-Jet Events in e^+e^- Annihilation and Evidence for Gluon Bremsstrahlung", *Phys. Lett. B* **91**, 142 (1980), received on December 7, 1979.
26. Sau Lan Wu, "e^+e^- Physics at PETRA — the First Five Years", *Phys. Rep.* **107**, 59–324 (1984).
27. TASSO Collaboration (R. Brandelik, et al.), *Phys. Lett. B* **97**, 453 (1980).
28. J. Ellis and I. Karliner, *Nucl. Phys. B* **148**, 141 (1979).
29. PLUTO Collaboration (Ch. Berger, et al.), *Phys. Lett. B* **97**, 459 (1980).
30. P. Söding, B. Wiik, G. Wolf and S. L. Wu (presented by S. L. Wu) for the 1995 EPS High Energy and Particle Physics Prize, in *Proceedings of the International Europhysics Conference on High Energy Physics*, Brussels, Belgium 27 Jul.– 2 Aug. 1995, pp. 3–10.
31. F. Englert and R. Brout, *Phys. Rev. Lett.* **13**, 321 (1964); P. W. Higgs, *Phys. Lett.* **12**, 132 (1964); P. W. Higgs, *Phys. Rev. Lett.* **13**, 508 (1964); G. S. Guralnik, C. R. Hagen and T. W. B. Kibble, *Phys. Rev. Lett.* **13**, 585 (1964); P. W. Higgs, *Phys. Rev.* **145**, 1156 (1966); T. W. B. Kibble, *Phys. Rev. Lett.* **155**, 1554 (1967).
32. ATLAS Collaboration (G. Aad et al.), "Observation of a new particle in the search for the Standard Model Higgs boson with the ATLAS detector at the LHC", *Phys. Lett. B* **716**, 1 (2012).
33. CMS Collaboration (S. Chatrchyan et al.), "Observation of a new boson at a mass of 125 GeV with the CMS experiment at the LHC", *Phys. Lett. B* **716**, 30 (2012).
34. CDF Collaboration (F. Abe et al.), "Observation of Top Quark Production in p anti-p Collisions with the Collider Detector at Fermilab", *Phys. Rev. Lett.* **74**, 2626 (1995); D0 Collaboration (S. Abachi et al.), "Observation of the Top Quark", *Phys. Rev. Lett.* **74**, 2632 (1995).
35. S. Heinemeyer, C. Mariotti, G. Passarino and R. Tanaka (eds.), CERN Yellow Report — Handbook of LHC Higgs cross sections: Books 1, 2 and 3. Report of the LHC Higgs Cross Section Working Group, arXiv:1101.0593v3 [hep-ph] 20 May 2011, arXiv:1201.3084v1 [hep-ph] 15 Jan. 2012, arXiv:1307.1347v1 [hep-ph] 4 Jul. 2013. See also The Higgs Cross Section Working Group web page: https://twiki.cern.ch/twiki/bin/view/LHCPhysics/CrossSections, 2013.
36. UA1 Collaboration (G. Arnison et al.), "Experimental Observation of Isolated Large Transverse Energy Electrons with Associated Missing Energy at $s =$ 540 GeV", *Phys. Lett. B* **122**, 103 (1983); "Experimental Observation of Lepton Pairs of Invariant Mass around 95 GeV/c^2 at the CERN SPS Collider", *Phys. Lett. B* **126**, 398 (1983).

37. UA2 Collaboration (M. Banner *et al.*), "Observation of Single Isolated Electrons of High Transverse Momentum in Events with Missing Transverse Energy at the CERN SPS Collider", *Phys. Lett. B* **122**, 476 (1983); UA2 Collaboration (P. Bagnaia *et al.*), "Evidence for $Z^0 \to e^+e^-$ at the CERN SPS Collider", *Phys. Lett. B* **129**, 130 (1983).

Yang–Mills Gauge Theory and Higgs Particle*

Tai Tsun Wu

Gordon McKay Laboratory, Harvard University,
Cambridge, MA 02138, USA

Sau Lan Wu[†]

Physics Department, University of Wisconsin-Madison,
Madison, WI 53706, USA

Motivated by the experimental data on the Higgs particle from the ATLAS Collaboration and the CMS Collaboration at CERN, the standard model, which is a Yang–Mills non-Abelian gauge theory with the group $U(1) \times SU(2) \times SU(3)$, is augmented by scalar quarks and scalar leptons without changing the gauge group and without any additional Higgs particle. Thus there is fermion–boson symmetry between these new particles and the known quarks and leptons. In a simplest scenario, the cancellation of the quadratic divergences in this augmented standard model leads to a determination of the masses of all these scalar quarks and scalar leptons. All these masses are found to be less than $100\,\text{GeV}/c^2$, and the right-handed scalar neutrinos are especially light. Alterative procedures are given with less reliance on the experimental data, leading to the same conclusions.

1. Introduction

When one of us (TTW) was a first-year graduate student, Professor Yang came to Harvard to give a colloquium on his Yang–Mills non-Abelian gauge theory. From what we have learned from Professor Yang during this conference, that was the first time that he had given a seminar on this topic. It was also the first time when TTW met Professor Yang, a meeting that has since influenced greatly the path of TTW's career.

The 1954 paper of Professor Yang and Robert L. Mills[1] is the most important theoretical paper for the second half of the twentieth century. Their non-Abelian gauge theory has permeated many branches of physics, and can be expected to be even more important for this twenty-first century.

Let us begin by quoting from the talk that Professor Yang gave at this conference, entitled "The Future of Physics — Revisited".

What have we learned in the fifty odd years since that 1961 panel? *A lot.*

---

*This paper is based on a talk presented by TTW at the Conference on 60 Years of Yang–Mills Gauge Field Theories, Nanyang Technological University, Singapore, May 25–28, 2015.
[†]Work supported in part by the U.S. Department of Energy under Grant No. DE-FG02-95ER40896.

Through very intensive collaborative efforts between theorists and experimentalists important conceptual developments were proposed and verified:
- One specific model of Symmetry Breaking
- Electroweak Theory
- Renormalizability of non-Abelian Gauge Theory
- Asymptotic Freedom and QCD

Capping these developments was the dramatic experimental discovery in 2012 of the Higgs particle.

We have now a workable

standard model, a $SU(3) \times SU(2) \times U(1)$ gauge theory.

Thus in these fifty odd years since 1961, we have reached *one more layer* of physical concepts, which is based on *all* the previous layers *plus* a large number of very large experiments.

It is the purpose of this present talk to attempt to give a first discussion of "the next layer of physical concepts".

2. Starting Point

The experimental discovery of the Higgs particle[2] was accomplished in 2012 by the ATLAS Collaboration[3] and the CMS Collaboration[4] using the Large Hadron Collider (LHC) at CERN.

This LHC is a proton–proton Collider housed in the tunnel previously used for the Large Electron Positron collider (LEP), a circular tunnel which is 17 miles in circumference. Its design center-of-mass energy is 14 TeV (=14,000 GeV); when it first began to produce experimental data in 2010, it was operating at half of the design energy, i.e., a center-of-mass energy of 7 TeV. This means that its initial operating energy was more than three and half times that of the previous highest-energy accelerator, the Tevatron Collider at Fermilab. This factor of 3.5 represents an enormous jump in energy, and, together with the expectation of finding the Higgs particle,[2] it was generally believed by the high-energy community that exciting physics was forthcoming.

A year later, in 2011, the center-of-mass energy of LHC was increased to 8 TeV. Very recently in 2015, this center-of-mass energy was further increased by another 60% to 13 TeV. LHC is expected to operate at this 13 TeV for the next years.

On July 4, 2012, the experimental discovery of the Higgs particle was officially announced at CERN. This discovery was widely celebrated and made newspaper headlines all over the world. Figure 1 shows some such examples.

The Higgs particle was the only missing particle in the aforementioned standard model, and hence its discovery completed the list of the elementary particles in the

Fig. 1. Example of newspaper headlines for the July 4, 2012 discovery.

standard model.[5] But the observation of the Higgs particle is much more than that: it is the discovery of not only a new particle, but a new type of particle, one with spin 0. No such spin 0 elementary particle has ever been seen before. Furthermore, comparing the experimental data from the ATLAS Collaboration and the CMS Collaboration with the calculation given in Fig. 26 of Ref. 6 gives a first indication of the physics beyond the standard model.

The starting point of the present first discussion of the next layer of physical concepts is therefore the experimental data from the ATLAS Collaboration and the CMS Collaboration.

3. Inputs from Experimental Data

Physics is basically an experimental science. Therefore, in order to gain some understanding of the next layer of physical concepts, it is advantageous to begin with a study of the experimental data.

The discovery of the Higgs particle, the first spin 0 elementary particle ever observed, tell us which experimental data to concentrate on — those from the ATLAS Collaboration and the CMS Collaboration.

Here is a brief summary of the experimental data from these two collaborations to be used as inputs, together with a discussion of their interpretations and possible implications. In the next section, Sec. 4, the most important theoretical input is to be introduced. In Sec. 5, other theoretical inputs will be given. These theoretical inputs are based on the present knowledge of particle physics taken together with the experimental inputs of this section.

(a) Rate of the decay $H \to \gamma\gamma$

The discovery of the Higgs particle on 2012 was accomplished[3,4] mainly through two decay channels; in order of their importance to this discovery, they are

$$H \to \gamma\gamma \qquad (1)$$

and

$$H \to ZZ \qquad (2)$$

followed by both Z decaying into a lepton pair, i.e., $Z \to e^+e^-$ or $Z \to \mu^+\mu^-$. In (2), one or both of the Z's are off mass shell; this decay is referred to as the four-lepton decay of the Higgs particle.

These two decay modes (1) and (2) of the Higgs particle are fundamentally different in the following way. The decay (2) is well described by tree diagrams, i.e., diagrams that do not include a loop. In contrast, since, the photon bring massless, there is no direct coupling of the Higgs particle to the photon, there is no tree diagram for the decay (1) and therefore this decay is described approximately by one-loop diagrams.

The decay process (1) was mentioned already in the original proposals to build the ATLAS and CMS detectors, but its importance for the discovery of the Higgs particle was not realized until much later, for example, in Ref. 7. It is the most important decay process for discovering the Higgs particle. For the purpose of trying to reach the next layer of physical concepts, the experimental data for this decay process (1) play a most important role, as to be discussed in the next section.

(b) Coupling of the Higgs boson to various particles

The data from the CDF Collaboration and the D0 Collaboration using the Tevatron Collider at Fermilab give the coupling strength of the Higgs particle to the bottom quark. The more extensive data from the ATLAS Collaboration and the CMS Collaboration give the couplings to the gauge bosons Z and W. These data also tell us the coupling to the top quark through the associated production of a Higgs particle together with a top–antitop pair. Taken together, these experimental data from CERN and Fermilab show that the couplings of the Higgs particle to

$$t, \ Z, \ W, \text{ and } b \qquad (3)$$

are in approximate agreement with the predictions of the standard model.[8] To this list of particles (3), the τ lepton may be added.

The simplest interpretation of this agreement with the standard model is that there is no second Higgs particle, i.e., the Higgs particle that was discovered in 2012 is responsible for giving masses to all the elementary particles. For the purpose of the present investigation on the next layer of physical concepts, this interpretation is accepted.

(c) Search for a second Higgs particle

Since the interpretation of the last paragraph is of central importance even beyond the present investigation, a direct search for the second Higgs particle has been carried out at the Large Hadron Collider. Such searches have been partially motivated by supersymmetry.[9] The theory of supersymmetry is based on the mathematics of graded Lie algebra, a generalization of Lie algebra so successfully used by Yang and Mills in particle physics and beyond. It is an immediate consequence of using the graded Lie algebra that there have to be at least two Higgs doublets, not one as in the case of the standard model.

The search for the second Higgs particle has been carried out mostly through the decay

$$H' \to W^+W^-. \qquad (4)$$

This decay offers major advantages for high-mass Higgs particles. No such second Higgs particle has been found up[10] to the mass of $1500\,\text{GeV}/c^2$, which is more than ten times the mass of the Higgs particle discovered in 2012. The failure of this search reinforces the conclusion of (b) above, namely that the Higgs particle discovered in 2012 is responsible for giving masses to all the elementary particles.

(d) Absence of new stable or long-lived charged particles

Even after extensive and careful search, no new stable or long-lived charged particle has been found at the Large Hadron Collider.[11] This absence of such particles is an important experimental input — see Sec. 5.

4. Input from Theory

As already mentioned in Sec. 3, the decay process $H \to \gamma\gamma$ is described approximately by one-loop diagrams. Since the Higgs boson couples to elementary particles with a strength proportional to its mass, the most important one-loop diagrams are those that involve the heaviest charged particles in the standard model, namely,

(1) the top quark, and
(2) the gauge particle W.

The contribution from the top-quark loop was calculated thirty five years ago by Rizzo,[12] while that from the W loop one year[13,14] before the experimental discovery of the Higgs particle. In both cases, as a consequence of the fact that the photon is massless and hence does not couple directly to the Higgs particle, the calculation can be carried out in a completely straightforward way without ghosts or regularization. With this understanding, the one-loop diagrams for the cases (1) and (2) above are shown in Figs. 2 and 3 respectively.

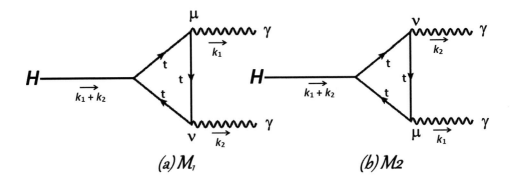

Fig. 2. The two diagrams for the decay $H \to \gamma\gamma$ via a top loop.

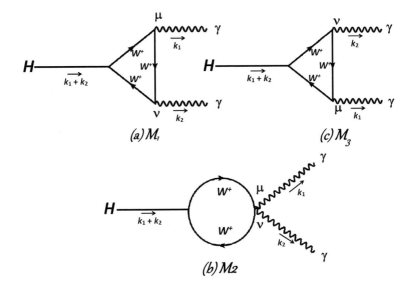

Fig. 3. The three diagrams for the decay $H \to \gamma\gamma$ via a W loop.

This calculation for the decay rate of $H \to \gamma\gamma$ in the standard model has a few noteworthy features, some of which are the following:

(i) There is significant destructive interference between the contributions from the top loop and the W loop;
(ii) The calculation for the W loop is much more complicated than that for the top loop; and
(iii) The result in the case of the W loop has been verified by Christova and Todorov[15] using a completely different method.

The importance of these theoretical considerations became evident only one year after its completions, when the decay rate for $H \to \gamma\gamma$ was experimentally measures — see the (a) of Sec. 3. It turned out that[12–15]

$$\frac{[\text{decay rate of H} \to \gamma\gamma]_{\text{ATLAS, CMS}}}{[\text{decay rate of H} \to \gamma\gamma]_{\text{standard model}}} \sim 2.5 \quad (5)$$

In other words, there is a significant disagreement of the experimental data from the ATLAS Collaboration and the CMS Collaboration from the prediction of the standard model.

In physics and especially particle physics, discrepancies between new experimental measurements and previous predictions from the existing theories are golden: it is such discrepancy that leads to fundamental developments. It is the purpose of the present paper to use this disagreement (5) between experiment and theory as the entry to discuss the next layer of physical concepts. See Sec. 12 for an alternative point of view.

5. Augmented Standard Model — Particle Content

The first task is to develop a theoretical framework: this framework must contain the standard model as a special case, and must be more general so that the discrepancy (5) is not present. In other words, this result (5) implies that the standard model must be modified, and the purpose of the theoretical framework is to narrow down which aspect of the standard model is most likely to need modification.

Here are the choices that we have followed to formulate this framework.

(a) From the standard model, the following features are retained:

- the Yang–Mills gauge group
  $U(1) \times SU(2) \times SU(3)$;
- three generations of elementary particles; and
- one Higgs doublet.

The reason for allowing only one Higgs doublet has been discussed under (b) and (c) of Sec. 3.

(b) With the discovery of the first spin 0 elementary particle — the Higgs boson — additional spin 0 elementary particles are allowed in this theoretical framework.

This point requires further discussion. The denominator on the left-hand side of (5) comes from the standard-model calculation due to the top-quark loop and the W loop, which are shown respectively in Figs. 2 and 3. If there are additional heavy charged particles, which can also contribute to this loop, then the left-hand side (5) is altered.

What can the spin for these additional heavy charged particle be? They cannot be of spin 1, because of (a) above, the Yang–Mills gauge group being the same as that of the standard model. If they are of spin $\frac{1}{2}$, then they are likely to be quarks or leptons of the fourth generation, but, even after diligent searching, no particles of the fourth generation has been seen. Therefore the spin of these additional heavy charged elementary particles is taken to the zero, implying that the Higgs particle is not the only spin 0 elementary particle. See the discussions of Sec 7.

Similar to the top-quark loop diagrams of Fig. 2 and the W loop diagrams of Fig. 3, the three diagrams for the loop due to the charged spin 0 particle, called X here, are shown in Fig. 4.

The conclusion is therefore reached that spin 0 particles, in addition to the Higgs particle already discovered, are to be added to the standard model.

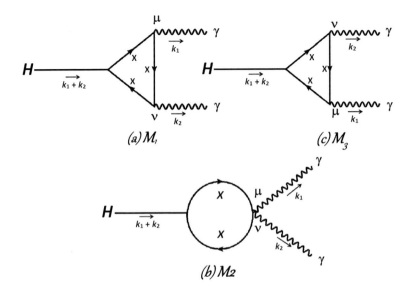

Fig. 4. The three diagrams for the decay $H \to \gamma\gamma$ via a scalar loop X.

Using this framework, we proceed to modify the standard model. The immediate question is: what can these additional spin 0 particles be?

We take the most conservative choice: besides the Higgs particle, the other spin 0 elementary particles are scalar quarks and scalar leptons. This choice is conservative because such scalar quarks and scalar leptons have been studied theoretically for many years.[16] Scalar quarks (scalar leptons) have the same quantum number as quarks (leptons) except the spin. When this is combined with the (a) above, the result is: three generations of quarks, leptons, scalar quarks, and scalar leptons. Note that scalar quarks have fractional baryon numbers and fractional charges, while the scalar leptons have lepton numbers.

How many scalar quarks and how many scalar leptons are there in each of the three generations? In order to answer this question, we rely on the idea of fermion–boson symmetry.

This fermion–boson symmetry is to play a central role in the present study of the next layer of physical concepts, and it will be investigated in greater depth in Sec. 7. For the time being, it is used merely to ascertain the numbers of scalar quarks and scalar leptons.

For each generation, there are two quarks and two leptons. For definiteness, consider the two quarks in the first generation. Since the quarks are spin $\frac{1}{2}$ fermions, there are four degrees of freedom, namely,

$$u_L, \ u_R, \ d_L, \ \text{and} \ d_R. \tag{6}$$

With the notation of a tilde to denote a scalar quark (and also a scalar lepton), the fermion–boson symmetry says that there are four scalar quarks in the first generation, to be denoted correspondingly as

$$\tilde{u}_L, \ \tilde{u}_R, \ \tilde{d}_L \ \text{and} \ \tilde{d}_R. \tag{7}$$

Since u_L and d_L from a doublet under the $SU(2)$ of the gauge group, so do $\tilde{u}_L$ and $\tilde{d}_L$. Similarly, u_R, d_R, $\tilde{u}_R$, and $\tilde{d}_R$ are all singlets. This correspondence, not only for quarks but also for leptons, is given in Fig. 5. For all three generations, there are twelve scalar quarks

$$\tilde{u}_L, \ \tilde{u}_R, \ \tilde{d}_L, \ \tilde{d}_R; \ \tilde{c}_L, \ \tilde{c}_R, \ \tilde{s}_L, \ \tilde{s}_R; \ \tilde{t}_L, \ \tilde{t}_R, \ \tilde{b}_L, \ \tilde{b}_R. \tag{8}$$

and twelve scalar leptons

$$\tilde{\nu}_{eL}, \ \tilde{\nu}_{eR}, \ \tilde{e}_L, \ \tilde{e}_R; \ \tilde{\nu}_{\mu L}, \ \tilde{\nu}_{\mu R}, \ \tilde{\mu}_L, \ \tilde{\mu}_R; \ \tilde{\nu}_{\tau L}, \ \tilde{\nu}_{\tau R}, \ \tilde{\tau}_L, \ \tilde{\tau}_R. \tag{9}$$

It should be emphasized that, while u_L and u_R are two spin states for the same u quark and therefore must have the same mass, the scalar quarks $\tilde{u}_L$ and $\tilde{u}_R$ are independent spin 0 particles and therefore there is no reason why they should have the same mass. In other words, there are two mass values for the particles listed in (6), but there are four for those of (7).

Since the standard model is augmented by the twenty four spin 0 particles listed in (8) and (9), this modified standard model is to be referred to as the augmented standard model.

The next question to be discussed is: how do these scalar quarks and scalar leptons decay? Again for definiteness, consider the case of scalar quarks of (8). Similar to the quarks, the scalar quarks have the decay

$$\text{heavier scalar quark} \to \text{lighter scalar quark} + W \tag{10}$$

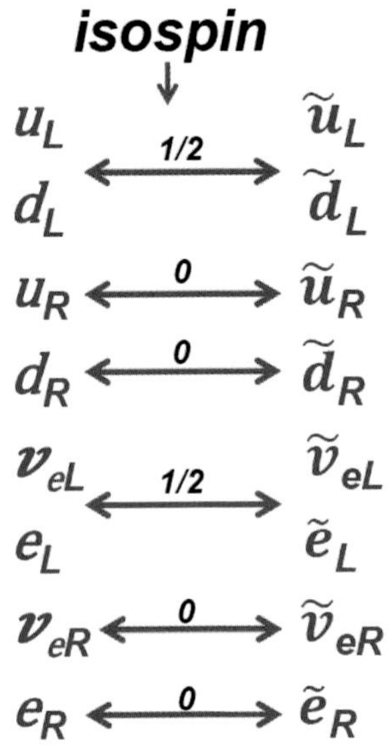

$$\begin{array}{ccc}
& \text{\textbf{\textit{isospin}}} & \\
& \downarrow & \\
\begin{matrix} u_L \\ d_L \end{matrix} & \xleftrightarrow{1/2} & \begin{matrix} \tilde{u}_L \\ \tilde{d}_L \end{matrix} \\
u_R & \xleftrightarrow{0} & \tilde{u}_R \\
d_R & \xleftrightarrow{0} & \tilde{d}_R \\
\begin{matrix} \nu_{eL} \\ e_L \end{matrix} & \xleftrightarrow{1/2} & \begin{matrix} \tilde{\nu}_{eL} \\ \tilde{e}_L \end{matrix} \\
\nu_{eR} & \xleftrightarrow{0} & \tilde{\nu}_{eR} \\
e_R & \xleftrightarrow{0} & \tilde{e}_R
\end{array}$$

Fig. 5. For the first generation, the correspondence between the quarks and leptons on the one hand and the scalar quarks and scalar leptons on the other hand.

where the W may be real or virtual; if the W is virtual, then this (10) is followed by, for example,

$$W^+ \to \bar{e} + \nu_e. \tag{11}$$

It remains to be determined whether the lightest scalar quark is stable or not.

This problem has been studied in detail.[17] Here the important experimental input is the (d) of Sec. 3: absence of new stable or long-lived charged particle.[11] By a somewhat complicated argument, the conclusion is reached that this (d) implies that the lightest scalar quark cannot be stable. Therefore, there must be another decay for the scalar quarks besides that of (10). The simplest choice for this additional decay mode is

$$\text{scalar quark} \to \text{quark} + \text{new fermion}. \tag{12}$$

The scalar leptons must behave similarity, i.e., corresponding to these (10) and (12) for the scalar quarks, the scalar leptons have the decays

$$\text{heavier scalar lepton} \to \text{lighter scalar lepton} + W. \tag{13}$$

and

$$\text{scalar lepton} \to \text{lepton} + \text{new fermion}. \tag{14}$$

In both (12) and (14), the new fermion may be real (on mass shell) or virtual (off mass shell). These new fermions complete the list of elementary particles for the augmented standard model; this list is shown in Fig. 6.

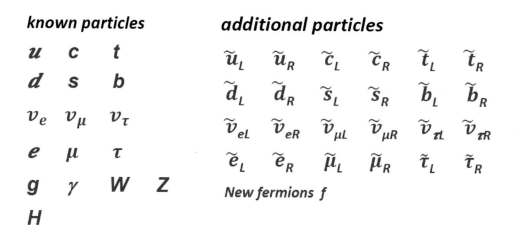

Fig. 6. The particle content of the augmented standard model.

6. Augmented Standard Model — Interactions

After deciding on the particle content as given in Fig. 6, the next step involves the determination of the interaction Lagrangian. This is a much more difficult task, and this section is devoted to some of the initial considerations.

First, only terms allowed by renormalization are to be accepted. This is obvious.

In order to go further, consider for a moment the case of the standard model. The interactions in the standard model consist of three types of terms:

(a) Yang–Mills gauge interactions;
(b) interactions of the fermions with the Higgs doublet; and
(c) self interaction of the Higgs doublet.

The Yang–Mills gauge couplings are of course dimensionless; through the Englert–Brout–Higgs mechanism, all the mass terms, except that of the Higgs particle itself, are replaced by dimensionless couplings to the Higgs particle. If we use the convention of considering a mass term to be a coupling with the mass square as the coupling constant, then we can say that, with the exception of the one mass term for the Higgs particle itself, all the coupling constants in the standard model are dimensionless. Technically, all these coupling terms are said to be of dimension 4.

As a side remark, this is also true for quantum electrodynamics — there is only one massive particle, the electron, and the interaction strength between the electron and the photon is given by the fine-structure constant, which is dimensionless.

On the basis of these two examples, it is proposed that in the case of the present augmented standard model, the interaction Lagrangian should also consist of many terms with dimensional coupling constants together with one mass term, that for the Higgs particle.

Since there are twelve scalar quarks and twelve scalar leptons in the augmented standard model, this restriction on the interaction Lagrangian is highly non-trivial and disallows numerous interaction terms, including especially all the terms that are cubic in the scalar fields.

It should also be emphasized that only one mass term is permitted in this proposal. One such mass term is necessary in order to set the scale; if there are two or more, then there is a complicated issue of how these scales can be related to each other. Furthermore, that there is one mass form implies that there can only be one Higgs doublet. Therefore, it is gratifying that the theoretical preference here and the experimental inputs (b) and (c) of Sec. 3 point in the same direction.

It may be worth stating that, since there is only one Higgs particle, the one that was discovered in 2012, in the augmented standard model all the elementary particles, including the quarks, the leptons, the scalar quarks, and the scalar leptons, get their masses from this one Higgs particle.

These are not the only restrictions on the Lagrangian for the augmented standard model, and further conditions are to be introduced in later sections. So far, only one-loop diagrams have been studied; the investigation of two-loop diagram is very difficult and is believed to be beyond reach at present. Therefore, only interaction terms that can contribute to one-loop diagrams have been considered.

7. Fermion–Boson Symmetry

Fermion–boson symmetry is just about the only fundamental symmetry that has not been studied systematically. The central question is this: since symmetry dictates interactions,[18] what does the fermion–boson symmetry imply about the nature of the interactions of elementary particles?

Fermion–boson symmetry tends to make a quantum field theory less divergent. This is the property to be exploited for the next layer of physical concepts.

The standard model, as well as the augmented standard model, is a Yang–Mills gauge theory. The gauge invariance already reduces the divergence; this is why all the spin 1 elementary particles are gauge particles. What is more natural than to combine fermion–boson symmetry with gauge invariance to reduce the divergence even further? Perhaps in this way, it is possible to get a quantum field theory that has no divergence, i.e., a finite quantum field theory in four dimensions that describes nature.

In their ways to reduce divergences, gauge invariance and fermion–boson symmetry are vastly different. In perturbation calculations, the potentially most divergent part is not allowed by gauge invariance and hence must be absent. In comparison, the effort of fermion–boson symmetry is less direct.

As noted at the end of last paragraph, the present considerations are limited to one-loop diagrams; this can be seen already from Figs. 2–4. Let us therefore limit ourselves also to one-loop diagrams in discussing fermion–boson symmetry. Because of their anti-commutation properties, there is an extra minus sign for each fermion loop. Thus the contributions from a fermion loop and a boson loop have opposite signs and tend to cancel each other. However, for the contributions to actually cancel, their absolute values have to be equal, which is not the case in general. In other words, unlike gauge invariance, the suppression of divergences from fermion–boson symmetry does not happen automatically; rather, this suppression happens only when a suitable condition is satisfied, the condition to make the aforementioned absolute values equal for the contribution from the fermion loop and that from the boson loop.

The next step is to apply these general considerations to the augmented standard model.

8. Two Fundamental Problems of Particle Physics

In particle physics, there are problems that are not often discussed. In the present study of the next layer of physical concepts, here are two such problems that we would like to address.

(1) The theory of renormalization for quantum electrodynamics is one of the important developments in the twentieth century; it was pioneered by Schwinger,[19] Tomonaga,[20] Feynman,[21] Dyson,[22] Ward,[23] Mills and Yang,[24] etc. It leads to the very successful and detailed agreement between the theoretical predictions from QED with the experimental results. Nevertheless, on a purely theoretical level, renormalization leaves something unpleasant: why would a quantum field theory such as quantum electrodynamics have ugly divergences which have to be swept under the rug?

Wouldn't it be much better not to have such divergences?

(2) A fundamental problem that is rarely discussed is: what determines the masses of elementary particles? After all, the mass, together with the baryon number, the spin and the charge, are the basic attributes of an elementary particle. For example, there is no theoretical understanding whatsoever why the muon is about two hundred times heavier than the electron.

In the present discussion of the next layer of theoretical concepts, it is proposed that these two fundamental problems of particle physics are related to each other. More precisely, the proposal is that, under suitable conditions, the divergences in the augmented standard model (or possibly a slightly modified version) may be cancelled and hence absent, and that these conditions take the form of relations between the masses of the elementary particles.

Such relations between the masses of elementary particle have been given in the literature. To the best of our knowledge, the earliest papers on this topic are those of Ferrara, Girardello and Palumbo[25] and of Veltman.[26] It is not clear whether the results are applicable to the standard model; if they are, then the result are

$$m_{\nu_e}^2 + m_{\nu_\mu}^2 + m_{\nu_\tau}^2 + m_e^2 + m_\mu^2 + m_\tau^2 + 3(m_u^2 + m_d^2 + m_c^2 + m_s^2 + m_t^2 + m_b^2)$$
$$= \frac{3}{2}m_W^2 + \frac{3}{4}m_Z^2 + \frac{1}{4}m_H^2 \quad (15)$$

and

$$m_{\nu_e}^2 + m_{\nu_\mu}^2 + m_{\nu_\tau}^2 + m_e^2 + m_\mu^2 + m_\tau^2 + 3(m_u^2 + m_d^2 + m_c^2 + m_s^2 + m_t^2 + m_b^2)$$
$$= \frac{3}{2}m_W^2 + \frac{3}{4}m_Z^2 + \frac{3}{4}m_H^2 \quad (16)$$

where the neutrino masses have been added to the left-hand side. These two formulas have the same left-hand side, and the only difference is the coefficient of the last term.

Later, the question was raised whether (15) or (16) can be obtained within the standard model, and what input is needed to arrive at these formulas. The answer turns out to be yes for the first part of this question, and the necessary input is the cancellation of quadratic divergences in the standard model.[5]

Actually, there are two quadratic divergences in the standard model, one from the Higgs tadpole and the other from Higgs self energy. The Higgs tadpole diagrams are those with only one external line, a Higgs line. It turns out that, to the order of one loop, the cancellations of the quadratic divergences in these two cases lead to the same condition, the Veltman condition (16).[27]

That the same condition is attained twice from the standard model is of supreme importance for the present study of the next layer of physical concepts.

Since the discovery of the Higgs particle in 2012, all the masses involved in (15) and (16) have been measured. [The neutrino masses are still not known, but they are so small that they do not matter.] Neither (15) nor (16) is anywhere near being satisfied. This is most easily seen by noticing that, on the left-hand sides of (15) and (16), which are the same, the largest term, $3m_t^2$, is much larger than the terms on the right-hand sides. Indeed,

$$\text{LHS} \sim 3m_t^2 \quad (17)$$

is an excellent approximation for the left-hand sides.

The immediate question is: why are the mass relations (15) and (16) so far away from being satisfied? Is it because the idea of the cancellation of divergences in quantum field theory not valid?

We believe that the idea of the cancellation of divergences is a good one, and the reason for the failure of (15) and (16) is to be sought elsewhere. In the last section, it has been stated that

"What is more natural than to combine fermion–boson symmetry with gauge invariance to reduce the divergences even further?"

The standard model has gauge invariance but not fermion–boson symmetry. It is the combination of gauge invariance with fermion–boson symmetry that can lead to a quantum field theory with less divergences.

It is therefore proposed to apply the consideration of Ref. 27 to a quantum field theory that has both gauge invariance and fermion–boson symmetry. Indeed, it is with this proposal in mind that the augmented standard model has been formulated.

9. Simplest Scenario

In Sec. 5, the particle content of the augmented standard model is determined, including especially the twelve scalar quarks and the twelve scalar leptons. In Sec. 6, constraints are formulated for the allowed interactions of this augmented standard model. Because of these twenty four scalar particles, numerous combinations are possible for ϕ^4 interactions, which are all allowed in a renormalizable theory. In other words, the fermion–boson symmetry leads to many, many possible additional terms in the Lagrangian density.

Because of these terms, it is difficult to proceed to study this augmented standard model in full generality. Therefore, in this and the next sections, no attempt is made to maintain this generality. Instead, this discussion will be limited to an especially simple special case.

The mass relation (16) is obtained from the cancelation of quadratic divergences for the Higgs tadpole and the Higgs self-energy in the standard model. These two divergences and their cancellations are only slightly different for the augmented standard model, and the calculation is almost the same. The corresponding mass relation is:

$$m_{\nu_e}^2 + m_{\nu_\mu}^2 + m_{\nu_\tau}^2 + m_e^2 + m_\mu^2 + m_\tau^2$$
$$+ 3(m_u^2 + m_d^2 + m_c^2 + m_s^2 + m_t^2 + m_b^2) + \sum m_f^2$$
$$= \frac{3}{2}m_W^2 + \frac{3}{4}m_Z^2 + \frac{3}{4}m_H^2$$
$$+ \frac{1}{2}(m_{\tilde\nu_e L}^2 + m_{\tilde\nu_e R}^2 + m_{\tilde\nu_\mu L}^2 + m_{\tilde\nu_\mu R}^2 + m_{\tilde\nu_\tau L}^2 + m_{\tilde\nu_\tau R}^2) \quad (18)$$
$$+ m_{\tilde e_L}^2 + m_{\tilde e_R}^2 + m_{\tilde\mu_L}^2 + m_{\tilde\mu_R}^2 + m_{\tilde\tau_L}^2 + m_{\tilde\tau_R}^2)$$
$$+ \frac{3}{2}(m_{\tilde u_L}^2 + m_{\tilde u_R}^2 + m_{\tilde d_L}^2 + m_{\tilde d_R}^2 + m_{\tilde c_L}^2 + m_{\tilde c_R}^2 + m_{\tilde s_L}^2$$
$$+ m_{\tilde s_R}^2 + m_{\tilde t_L}^2 + m_{\tilde t_R}^2 + m_{\tilde b_L}^2 + m_{\tilde b_R}^2),$$

where the term $\sum m_f^2$ comes from the new fermions.

It is immediately noted that this (18) has the following important advantage over the previous (16): there are many more terms on the right-hand side due to the scalar particles and these terms help the fulfillment of this mass relation by taking care of the very large term $3m_t^2$ on the left-hand side.

There are, however, many more quadratic divergences in the augmented standard model; more precisely, there are twenty four additional quadratic divergences due to the self-energies of the twenty four scalar quarks and scalar leptons. In principle, they can lead to twenty four mass relations similar to (18).

The problem is therefore to find a way to determine these mass relations without knowing the interactions of the scalar quarks and scalar leptons.

For the simplest scenario to be discussed here, the following assumption is to be introduced:

$$\text{The approximation (17) also holds for (18).} \tag{19}$$

What this assumption means is that the contribution $\sum m_f^2$ is much less than $3m_t^2$.

At the end of Sec. 6, it is pointed out that only one-loop diagrams are treated in the present study. Within this one-loop approximation and using the approximation (19), all the fermions are massless except the top quark. Therefore the left-hand sides of all the twenty four additional mass relations coming from the cancellation of quadratic divergences from the one-loop self-energy diagrams of the scalar quarks and scalar leptons must be proportional to m_t^2. This is the underlying reason why the assumption (19) is so powerful.

Once these left-hand sides are known within this one-loop approximation, further considerations lead to the strong conclusion that all twenty six $(2 + 24 = 26)$ quadratic divergences in the augmented standard model are cancelled provided that just a single mass relation is satisfied:

$$\begin{aligned}
3m_t^2 &= \frac{3}{2}m_W^2 + \frac{1}{2}m_Z^2 + \frac{3}{4}m_H^2 \\
&+ \frac{1}{2}(m_{\tilde{\nu}_e L}^2 + m_{\tilde{\nu}_e R}^2 + m_{\tilde{\nu}_\mu L}^2 + m_{\tilde{\nu}_\mu R}^2 + m_{\tilde{\nu}_\tau L}^2 + m_{\tilde{\nu}_\tau R}^2 \\
&+ m_{\tilde{e}_L}^2 + m_{\tilde{e}_R}^2 + m_{\tilde{\mu}_L}^2 + m_{\tilde{\mu}_R}^2 + m_{\tilde{\tau}_L}^2 + m_{\tilde{\tau}_R}^2) \\
&+ \frac{3}{2}(m_{\tilde{u}_L}^2 + m_{\tilde{u}_R}^2 + m_{\tilde{d}_L}^2 + m_{\tilde{d}_R}^2 + m_{\tilde{c}_L}^2 + m_{\tilde{c}_R}^2 \\
&+ m_{\tilde{s}_L}^2 + m_{\tilde{s}_R}^2 + m_{\tilde{t}_L}^2 + m_{\tilde{t}_R}^2 + m_{\tilde{b}_L}^2 + m_{\tilde{b}_R}^2).
\end{aligned} \tag{20}$$

Since this is the first venture into the next layer of physical concepts, there are a number of open questions, some of which will be mentioned in Sec. 11. In the context of the relation between masses, there is the following mystery. For the standard model, as already mentioned, there are two quadratic divergences, and the one mass relation, (16), is sufficient for the cancellation of both divergences; we do not know why. For the augmented standard model, there are 26 quadratic divergences. When the assumption (19) is used, one mass relation, (20), is sufficient for the cancellation of all 26 divergences. Again we do not know why.

10. Masses of Scalar Quarks and Scalar Leptons

That all twenty six quadratic divergences are cancelled when (20) is satisfied has numerous consequences. Let us limit ourselves to those on the masses of the scalar quarks and scalar leptons in this simplest scenario.

A first consequence is that, within the one-loop approximation, the masses of the three generations are equal, i.e.,

$$\begin{aligned} m_{\tilde{u}_L} &= m_{\tilde{c}_L} = m_{\tilde{t}_L}, & m_{\tilde{u}_R} &= m_{\tilde{c}_R} = m_{\tilde{t}_R} \\ m_{\tilde{d}_L} &= m_{\tilde{s}_L} = m_{\tilde{b}_L}, & m_{\tilde{d}_R} &= m_{\tilde{s}_R} = m_{\tilde{b}_R} \\ m_{\tilde{\nu}_{eL}} &= m_{\tilde{\nu}_{\mu L}} = m_{\tilde{\nu}_{\tau L}}, & m_{\tilde{\nu}_{eR}} &= m_{\tilde{\nu}_{\mu R}} = m_{\tilde{\nu}_{\tau R}}, \\ m_{\tilde{e}_L} &= m_{\tilde{\mu}_L} = m_{\tilde{\tau}_L}, & m_{\tilde{e}_R} &= m_{\tilde{\mu}_R} = m_{\tilde{\tau}_R} \end{aligned} \qquad (21)$$

for general Kobayashi–Maskawa matrices[28] in the case of scalar quarks and scalar leptons. This is a surprise to us.

What (21) means is that the mass hierarchies are the opposite for fermions and bosons; while the masses of the quarks and leptons are very different for the three generations, those for the scalar quarks and scalar leptons are nearly the same for the three generations.

Again within the one-loop approximation, a second consequence is the determination of these masses. The results are

$$\begin{aligned} m_{\tilde{u}_L} &= \sqrt{81.6^2 + m_0^2} \text{ GeV}/c^2 & m_{\tilde{\nu}_L} &= 85.1 \text{ GeV}/c^2 \\ m_{\tilde{d}_L} &= \sqrt{81.6^2 + m_0^2} \text{ GeV}/c^2 & m_{\tilde{l}_R} &= 85.1 \text{ GeV}/c^2 \\ m_{\tilde{u}_R} &= \sqrt{34.2^2 + m_0^2} \text{ GeV}/c^2 & m_{\tilde{\nu}_R} &= 0 \\ m_{\tilde{d}_R} &= \sqrt{17.1^2 + m_0^2} \text{ GeV}/c^2 & m_{\tilde{l}_R} &= 51.3 \text{ GeV}/c^2 \end{aligned} \qquad (22)$$

where m_0 gives a parameterization of the effect from strong interactions. Of course, these values are approximate.

These mass values have many intriguing features; here are a few.

(a) For small values of m_0 consistent with the existing experimental data, all these masses are less than 100 GeV/c^2.

(b) For these scalar particles, the masses of each $SU(2)$ doublet are the same. This is very different from the case of quarks and leptons, and is probably related to the presence of quadratic divergences in the case of spin 0 particles but not in the case of spin $\frac{1}{2}$ particles.

(c) The right-handed scalar neutrinos are found to be massless in the one-loop approximation. Thus in some sense it plays the role of the ordinary neutrinos. Just as the ordinary neutrinos are often produced in the decays of ordinary quarks and leptons, these right-handed scalar neutrinos may be expected to be produced often in the decay of the scalar quarks and scalar leptons. However, these decays of the scalar particles seem to be more complicated, depending partly on the masses of the new fermions.

(d) Many questions remain concerning the experimental observation of the scalar quarks and the scalar leptons at these masses.

11. Discussions

There are numerous open questions and puzzles; let us limit ourselves to the following.

In the calculation of (22) for the masses of the scalar quarks and the scalar leptons, the known masses of the Z and W gauge bosons and the Higgs particle are used as inputs but not the mass 173 GeV/c^2 of the top quark. Therefore, it is a highly non-trivial check to substitute the values of the masses in (22) into the right-hand side of (20) to find out whether the result is $3m_t^2$. This turns out to be roughly the case provided that the parameter m_0 is relatively small.

A most important next step in the present program is to determine the properties of the new fermions in the augmented standard model. Some progress has been made in this direction, but much more is still needed before a systematic study becomes possible for the decay modes of the scalar quarks and the scalar leptons. For example, at the moment it is not clear how many of the new fermions should there be, and roughly what their masses are.

12. Conclusion

The central theme of the present work is to have a first peek at the next layer of physical concepts. This next layer has a number of salient features that are not emphasized in the existing understanding of elementary particle physics, including especially the following two.

(a) This next layer of physical concepts should provide a basis understanding the masses of elementary particles; and

(b) The theory that describes the interaction of elementary particles should be free of divergences.

At present, we are having only a first peek, and are far from being able to achieve either of these two highly desirable features.

The theoretical determination of the masses of elementary particles is clearly an important one, but is a topic that is very rarely discussed. This is to be contrasted with the possibility of a quantum field theory without divergences, a topic very much in the mind of many physicists. However, such theories without divergences are often studied by introducing large Yang–Mills gauge groups. From the present point of view, the introduction of such large gauge groups has consequences that may be considered to be undesirable, such as gauge particles that have never been seen, and very heavy new particles.

It is our purpose here to keep the gauge group as $U(1) \times SU(2) \times SU(3)$ and to refrain from introducing very heavy particles. It is a fact that experimentally the top quark has remained the heaviest known elementary particle for twenty years, a very long time.

There are many ways to look at the present attempt to investigate the next layer of physical concepts. We have here given the presentation following the path of our own work:

(a) The starting point is the disagreement between theory and experiment;
(b) New massive charged particles are added to get out of this disagreement;
(c) Fermion–boson symmetry is used to classify these new particles; and
(d) The resulting augmented standard model is used to study the two fundamental problems described at the beginning of this section.

In this way, the experimental results from the ATLAS Collaboration and the CMS Collaboration at the Large Hadron Collider play a central role.

This is of course not the only way to present our work. One of the possible alternative sequences is as follows:

(a′) The starting point is the two fundamental problems described at the beginning of this section;
(b′) Fermion–boson symmetry, together with gauge invariance, is used to reduce divergences in quantum field theory;
(c′) For this purpose of having fermion–boson symmetry, scalar quarks and scalar leptons are added to the standard model; and
(d′) This results in the formulation of the augmented standard model.

In this way, the experimental results from the ATLAS Collaboration and the CMS Collaboration at the Large Hadron Collider play only a secondary role.

The final results are of course the same.

Acknowledgments

A deep sense of gratitude is owed to Professor Chen Ning Yang, who introduced one of us (TTW) to particle physics together with numerous subsequent discussions. We thank Professor Kok Khoo Phua for inviting us to this exciting conference. The hospitality of CERN, where part of this work was carried out, is much appreciated.

References

1. C. N. Yang and R. L. Mills, *Phys. Rev.* **96**, 191 (1954).

2. F. Englert and R. Brout, *Phys. Rev. Lett.* **13**, 321 (1964); P. W. Higgs, *Phys. Lett.* **12**, 132 (1964); G. S. Guralnik, C. R. Hagen, and T. W. B. Kibble, *Phys. Rev. Lett.* **13**, 585 (1964).
3. ATLAS Collaboration, *Phys. Lett. B* **716**, 1 (2012).
4. CMS Collaboration, *Phys. Lett. B* **716**, 30 (2012).
5. S. Glashow, *Nucl. Phys.* **22**, 579 (1961); S. Weinberg, *Phys. Rev. Lett.* **19**, 1264 (1967); A. Salam, *Proc. 8th. Nobel Symp.* (Stockholm, 1968), ed. N. Svartholm (Almgvist, 1968), p. 367.
6. S. L. Wu, *Modern Phys. Lett. A* **29**, 1330027 (2014).
7. K. Cranmer, B. Mellado, W. Quelle, and S. L. Wu, ATLAS Note, ATL-PHYS-2003-036, arXiv:hep-ph/0401088 (2003).
8. ATLAS Collaboration, *Phys. Lett. B* **726**, 88 (2013); CDF and D0 Collaborations, *Phys. Rev. D* **88**, 052014 (2013); CMS Collaboration, *Eur. Phys. J. C* **75**, 212 (2015).
9. Y. Goifand and E. Likhtman, *JETP Lett.* **13**, 323 (1971); D. Volkov and V. Akulov, *Phys. Lett. B* **46**, 109 (1973); J. Wess and B. Zumino, *Nucl. Phys. B* **70**, 39 (1974); A. Salam and J. Strathdee, *ibid.*, **76**, 477 (1974).
10. ATLAS Collaboration (to be published).
11. CMS Collaboration, *J. High Energy Phys. C* **75**, 362 (2015).
12. T. G. Rizzo, *Phys. Rev. D* **22**, 178 (1980).
13. R. Gastmans, S. L. Wu, and T. T. Wu, CERN preprint CERN-PH-TH/2011-200, arXiv:1108.5322 (2011).
14. R. Gastmans, S. L. Wu, and T. T. Wu, CERN preprint CERN-PH-TH/2011-201, arXiv:1108.5872 (2011).
15. E. Christova and I. Todorov, arXiv: 1410.7061v1 [hep-ph] 26 Oct., 2014.
16. Some of the papers on fermionic mesons are: D. Stanley and D. Robson, *Phys. Rev. D* **21**, 3180 (1980); J. Carlson, J. B. Kogut, and V. Pandharipande, *ibid. D* **27**, 233 (1983); L. Basdevant and S. Boukraa, *Z. Phys. C* **28**, 413 (1985); S. Godfrey and N. Isgur, *Phys. Rev. D* **34**, 899 (1986); E. Hiyama, Y. Kino, and M. Kamimura, *Prog. Part. Nucl. Phys.* **51**, 223 (2003).
17. J. M. Richard and T. T. Wu, CERN preprint CERN-PH-TH/2013-200 (2013).
18. C. N. Yang, *Phys. Today* **33**, 42 (1980); C. N. Yang, *Oskar Klein Memorial Lectures*, Vol. 1, ed. G. Ekspong (World Scientific, 1991).
19. J. Schwinger, *Phys. Rev.* **73**, 416 (1948).
20. S. Tomonaga, *Phys. Rev.* **74**, 224 (1948).
21. R. P. Feynman, *Rev. Mod. Phys.* **20**, 367 (1948).
22. F. J. Dyson, *Phys. Rev.* **75**, 486 and 1736 (1949).
23. J. C. Ward, *Phys. Rev.* **78**, 182 (1950); *Proc. Phys. Soc. A* **64**, 54 (1951).
24. R. L. Mills and C. N. Yang, *Suppl. Progr. Theor. Phys.* **37** and **38**, 507 (1966).

25. S. Ferrara, L. Girardello, and F. Palumbo, *Phys. Rev. D* **20**, 403 (1979).
26. M. Veltman, *Acta Phys. Pol. B* **12**, 437 (1981).
27. P. Osland and T. T. Wu, *Z. Phys. C — Particles and Fields* **55**, 569 (1992) and **55**, 585 (1992).
28. M. Kobayashi and T. Maskawa, *Prog. Theor. Phys.* **49**, 652 (1973).

Scenario for the Renormalization in the 4D Yang–Mills Theory

L. D. Faddeev

St. Petersburg Department of V.A. Steklov Institute of Mathematics, Russia
St. Petersburg State University, Russia

The renormalizability of the Yang–Mills quantum field theory in four-dimensional space–time is discussed in the background field formalism.

Yang–Mills quantum field theory has a unique character, allowing a self-consistent formulation in the four-dimensional space–time. Two important properties — asymptotic freedom and dimensional transmutation — which are the characteristic features that distinguish it from other theories. I think that the typical textbook exposition of this theory, which based on general paradigm of QFT, still does not underline these specifics. In my talk, I propose a scheme for the description of the Yang–Mills theory which exactly does this. I do not claim finding anything new; my proposal has simply a methodological value.

As a main tool for my presentation, I have chosen the object called the effective action which is defined via the background field. Following Feynman's ideas, I consider the functional of a background field as the generating functional for the S-matrix, whereas the Schwinger functional of external current, generating the Green functions, needs LSZ reduction formulas to define the S-matrix. Moreover, the latter functional is not manifestly gauge invariant.

Ironically, the standard description of the background field method[1–4] uses the external current and Legendre transformation. Alternative formulation began in Ref. 5 and entering Ref. 6 was improved in Ref. 7. In what follows I use the latter approach.

An effective action $W(B)$ is a functional of a classical Yang–Mills field $B_\mu(x)$ given by a series in a dimensionless coupling α,

$$W(B) = \frac{1}{\alpha} W_{-1}(B) + W_0(B) + \sum_{n \geq 1} \alpha^n W_n(B),$$

where $W_{-1}(B)$ is a classical action, $W_0(B)$ is a one-loop correction, defined via the determinants of vector and scalar operators M_1 and M_0,

$$M_0 = \nabla_\mu^2, \quad M_1 = \nabla_\sigma^2 \delta_{\mu\nu} + 2[F_{\mu\nu}, \cdot\,],$$

where

$$\nabla_\mu = \partial_\mu + B_\mu,$$
$$F_{\mu\nu} = \partial_\mu B_\nu - \partial_\nu B_\mu + [B_\mu, B_\nu],$$
$$W_0 = -\frac{1}{2} \ln \det M_1 + \ln \det M_0$$

and W_k, $k = 1, 2, \ldots$ are defined as the contribution of strongly connected vacuum diagrams with $k+1$ loops, constructed via Green functions M_0^{-1} and M_1^{-1} and vertices defined by the forms of vector and scalar fields $a_\mu(x)$, $\bar{c}(x)$, $c(x)$

$$\Gamma_3(B) = g \int \operatorname{tr} \nabla_\mu a_\nu [a_\mu, a_\nu] d^4x,$$
$$\Gamma_4(B) = g^2 \int \operatorname{tr}[a_\mu, a_\nu]^2 d^4x,$$
$$\Omega(B) = g \int \operatorname{tr} \nabla_\mu \bar{c}[a_\mu, c] d^4x,$$

where $g = \frac{1}{2}\sqrt{\alpha}$, taking into account anticommuting properties of ghosts $\bar{c}(x)$, $c(x)$.

The divergences of the diagrams should be regularized. I believe that there exists a regularization defined by the cutoff momentum Λ such that all infinities are powers in

$$L = \ln \frac{\Lambda}{\mu},$$

where μ is some normalization mass. Unfortunately, at present I do not know a satisfactory procedure for such a regularization. That is why I call my exposition a "scenario."

The renormalizability of the Yang–Mills theory means there exists a dependence of the coupling constant α on cutoff Λ such that the full action $W(B)$ is finite.

In the case of one loop, everything is clear. The functional $W_0(B)$ can be defined via the proper time method of Fock[8] giving the formula

$$W_0(B) = \int_0^\infty \frac{ds}{s} T(B, s),$$

where the functional $T(B, s)$ has the following behavior for small s,

$$T(B, s) = T_0(B) + sT_1(B, s)$$

and

$$T_0(B) = \frac{1}{2}\beta_1 W_{-1},$$

where β_1 is a famous negative constant. So the only divergence is proportional to the classical action and can be compensated by the renormalization of the coupling constant α. In more detail, we regularize $W_0(B)$ as

$$W_0^{\text{reg}}(B) = \int_0^{1/\Lambda^2} ds\, T_1(B,s) + \int_{1/\Lambda^2}^{\infty} \frac{ds}{s} T(B,s)$$

and choose the dependence of the coupling constant α on Λ as

$$\frac{1}{\alpha(\Lambda)} = -\beta_1 \ln \frac{\Lambda}{m},$$

where m is a new parameter with the dimension of mass. It is clear that the regularized one-loop $W(B)$ does not depend on Λ and so is finite. More explicitly, we rewrite $W_0^{\text{reg}}(B)$ as

$$W_{00}(B) = W_{00} + W_{01} L,$$

where

$$W_0^{\text{reg}}(B) = \int_0^{1/\mu^2} ds\, T_1(B,s) + \int_{1/\mu^2}^{\infty} \frac{ds}{s} T(B,s),$$

and

$$W_{01} = \beta_1 W_{-1}$$

and define the renormalized running coupling constant

$$\frac{1}{\alpha(\mu)} = -\beta_1 \ln \frac{\mu}{m},$$

so that

$$W_{\text{1-loop}}^{\text{reg}}(B) = \frac{1}{\alpha(\mu)} W_{-1} + W_{00}.$$

Thus we have traded the dimensionless α for a dimensional parameter m. However, m enters trivially, defining the scale. Observe that $\alpha(\Lambda)$ and $\alpha(\mu)$ are the values of the same function for two values of the argument. This function satisfies the first approximation to Gell-Mann–Low equation,

$$x \frac{d}{dx} \alpha(x) = \beta(\alpha) = \beta_1 \alpha^2,$$

where the R.H.S. does not depend on x and m plays the role of the conserved integral. The shift of x can be interpreted as an Abelian group action, this is the famous renormalization group. The same is true for the whole one-loop functional

$$W_{\text{1-loop}}^{\text{reg}}(B,\Lambda) = W_{\text{1-loop}}(B,\mu)$$

and the renormalized action $W(B,\mu)$ does not depend on the running momentum μ.

My scenario is based on the assumption that the whole $W(B)$ has the same properties. I believe that all infinities in $W(B)$ can be combined into the form

$$W(B,\Lambda) = \frac{1}{\alpha}W_{-1} + W_{00} + W_{01}L$$
$$+ \alpha(W_{10} + W_{11}L) + \cdots$$
$$+ \alpha^n(W_{n0} + W_{n1}L + \cdots W_{nn}L^n) + \cdots,$$

where the functionals $W_{k0}(B)$, $k = 0, \ldots$ are finite and depend on μ. The coupling constant $\alpha(\Lambda)$ should satisfy the full Gell-Mann–Low equation

$$\Lambda\frac{d\alpha}{d\Lambda} = \beta(\alpha) = \beta_1\alpha^2 + \beta_2\alpha^3 + \cdots + \beta_n\alpha^n + \cdots$$

and the full $W(B)$ should be independent of L and so it is finite.

The equation

$$\frac{dW}{dL} = 0$$

immediately gives

$$W_{11} = \beta_2 W_{-1}$$

and leads to an equation expressed via double series in powers of α and L. The condition that the corresponding coefficients vanish leads to the recurrent relations expressing W_{nm} via $W_{n-1,m}$, $n = 2, 3, \ldots$, $n \geq m$.

Here are examples of such equations at the lowest orders:

$$\beta_1 W_{10} + W_{21} = \beta_3 W_{-1},$$
$$\beta_2 W_{10} + 2\beta_1 W_{20} + W_{31} = \beta_4 W_{-1},$$
$$\beta_1 W_{11} + 2W_{22} = 0,$$
$$\beta_2 W_{11} + 2\beta_1 W_{21} + 2W_{32} = 0.$$

One can solve some of these equations exactly. For instance, for the highest coefficients W_{nn} we get relation

$$(n-1)\beta_1 W_{n-1,n-1} + nW_{nn} = 0$$

and as W_{11} is proportional to W_{-1} the same is true for all W_{nn}. Thus their contribution $\sum \alpha^n L^n W_{nn}$ is summed up to

$$\frac{\beta_2}{\beta_1}\ln(1 + \beta_1 \alpha L)W_{-1}$$

and this gives the next correction to the coefficient in front of W_{-1} in $W(B)$, namely

$$\frac{1}{\alpha} + \beta_1 L + \frac{\beta_2}{\beta_1}\left[\ln\alpha + \ln\left(\frac{1}{\alpha} + \beta_1 L\right)\right],$$

leading to the next approximation for the renormalized coupling constant

$$\frac{1}{\alpha_r} = \frac{1}{\alpha} + \beta_1 L + \frac{\beta_2}{\beta_1}\ln L,$$

consistent with the Gell-Mann–Low equation.

We see that all coefficients W_{nm}, $n \geq m$, $n \geq 1$ are expressed via the finite ones W_{n0}. More detailed investigation of these equations[9] shows that the full expression for $W(B)$ can be rewritten in terms of W_{n0} and powers of renormalized coupling constant

$$W(B,\mu) = \frac{1}{\alpha_r}W_{-1} + W_{00} + \sum \alpha_r^n W_{n0}.$$

This is consistent with the equation

$$W_{\text{reg}}(B, \Lambda) = W(B, \mu)$$

and

$$L|_{\Lambda=\mu} = 0.$$

Let us comment that the recurrence relations make sense when the only nonzero coefficients in β-function are β_1 and β_2. This enables a speculation that there exist the regularizations in which $\beta_3 = \beta_4 = \cdots = 0$.

Acknowledgments

This work is partially supported by RFBR grant 12-01-00207 and the programme "Mathematical problems of nonlinear dynamics" of RAS. I thank S. Derkachov and A. Ivanov for collaboration in development of my programme.

References

1. B. S. DeWitt, Quantum theory of gravity. 2. The manifestly covariant theory, *Phys. Rev.* **162**, 1195 (1967).
2. B. S. DeWitt, Quantum theory of gravity. 3. Applications of the covariant theory, *Phys. Rev.* **162**, 1239 (1967).
3. G. 't Hooft, The background field method in gauge field theories, in *Karpacz 1975, Proceedings, Acta Universitatis Wratislaviensis No. 368*, Vol. 1, Wroclaw, 1976, pp. 345–369.
4. L. F. Abbott, Introduction to the background field method, *Acta Phys. Pol. B* **13**, 33 (1982).
5. I. Y. Arefeva, L. D. Faddeev and A. A. Slavnov, Generating functional for the s matrix in gauge theories, *Theor. Math. Phys.* **21**, 1165 (1975) [*Teor. Mat. Fiz.* **21**, 311 (1974)].

6. L. D. Faddeev and A. A. Slavnov, Gauge fields. Introduction to quantum theory, *Front. Phys.* **50**, 1 (1980) [*Front. Phys.* **83**, 1 (1990)].
7. L. D. Faddeev, Mass in quantum Yang–Mills theory: Comment on a clay millenium problem, arXiv:0911.1013 [math-ph].
8. V. Fock, Proper time in classical and quantum mechanics, *Phys. Z. Sowjetunion* **12**, 404 (1937).
9. S. Derkachov, L. Faddeev and A. Ivanov, to be published.

Statistical Physics in the Oeuvre of Chen Ning Yang

Michael E. Fisher

*Institute for Physical Science and Technology and Department of Physics,
University of Maryland, College Park, MD 20742, USA*

Starting from his early years selected contributions of C. N. Yang to statistical mechanics are high-lighted and the physics to which they led is briefly reviewed.

1. First Publications

From his earliest days as a student Chen Ning Yang — later known to friends in the West as Frank Yang — had been aware of mathematics; his father, Wu-Chih Yang, was, indeed, a professor of the subject. Not surprisingly, perhaps, the first paper published by Cheng-Ning Yang (see Fig. 1) appeared in the *Bulletin of the American Mathematical Society* in 1944. Nevertheless, as a graduate student at the Southwest Associated University in Kunming in 1942–44, he became intrigued by statistical mechanics. His advisor, Professor J. S. Wang (who had studied in Britain with R. H. Fowler in the 1930's) supervised his Master's thesis under the title: *"Contributions to the Statistical Theory of Order-Disorder Transitions."*

Through his studies of order-disorder phenomena — one of the most notable examples being the transition in beta-brass which is an alloy of copper and zinc

---

ON THE UNIQUENESS OF YOUNG'S DIFFERENTIALS

CHENG-NING YANG

Introduction. The differential of a function of several variables may be defined in a variety of ways, of which the one given by Young[1] renders the best parallelism with the case of a single variable. Stated in the way given below, his definition is applicable to a function defined in a set S of points containing limiting points at which the function is to have differentials. The question of the uniqueness of the differentials, however, arises. In this paper we shall first define, and prove two theorems concerning, the "limiting directions" which describe the directional distribution of the points of S near a limiting point. Then we proceed to show that properties of these limiting directions determine whether the differential is unique or not.

Fig. 1. C. N. Yang's first published paper based on Young's paper in *Proc. London Math. Soc.* **7**, 157 (1908). The article appeared in *Bull. Amer. Math. Soc.* **50**, 373–375 (1944).

168 M. E. Fisher

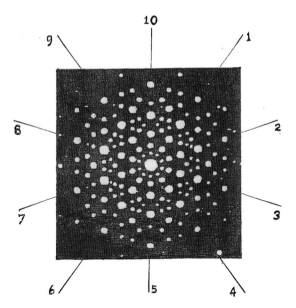

Fig. 2. Scattering observations illustrating the evidence for order in a crytal that can undergo an order-disorder transition. [After D. Shectman, I. Blech, D. Gratias, and J. W. Cahn, *Phys. Rev. Lett.* **53**, 1951 (1984).]

close to a 50:50 composition — the young Mr. Yang surely became familiar with X-ray scattering observations such as that illustrated in Fig. 2. But here there is a bit of a cheat: the crystal studied in Fig. 2 by Dan Shechtman and coworkers is actually Al_6Mn which, as might be concluded from the 10-fold symmetry, is actually a *quasicrystal*! Be that as it may, Yang's first paper in statistical physics was based on his Master's thesis and published in *The Journal of Chemical Physics* in 1945: see Fig. 3. This was submitted while he was a Research Fellow of the China

THE JOURNAL OF CHEMICAL PHYSICS   VOLUME 13, NUMBER 2   FEBRUARY, 1945

A Generalization of the Quasi-Chemical Method in the Statistical Theory of Superlattices

C. N. Yang*
National Tsing Hua University, Kunming, China
(Received November 17, 1944)

The quasi-chemical method in the investigation of the equilibrium distribution of atoms in the pairs of neighboring sites in a superlattice is generalized by considering groups containing large numbers of sites. The generalized method may be used to obtain successive approximations of the free energy of the crystal. The labor of integration is avoided by the introduction of a Legendre transformation. In order to analyze the fundamental assumption underlying the method more closely, the number of arrangements of the atoms for given long-distance order is calculated and the hypothesis of the non-interference of local configurations discussed. The method is applied to the calculation of the free energy in the different approximations discussed in this paper, including Bethe's second approximation and a simple approximation for the face-centered cubic crystal Cu_3Au.

Fig. 3. The second paper published. Note that the author is now identified as "C. N. Yang."

Foundation at the National Tsing Hua University (which had been evacuated to Kunming owing to the Japan–China war, 1937–1945). Notable, and characteristic, is that the paper develops a new formulation of the quasi-chemical method: "which is capable of yielding successively higher approximations."

2. The Ising Model and Onsager

Doubtless already as a graduate student working on his Master's thesis, Yang had come across the Ising model of a ferromagnet. In its simplest form at each site, i, of a square lattice, in $d = 2$ dimensions (or a simple cubic lattice, if $d = 3$), there is a "spin variable," s_i, which takes only the values $+1$ or -1. Nearest neighbor spins, at sites i and j are coupled yielding a favorable energy $\Delta \mathcal{H} = -Js_i s_j$ while the magnetic field, H, contributes $-mHs_i$ for each spin. To model a binary AB alloy, the values $s_i = +1$ or -1 are merely associated with an atom A or B.

Ernst Ising solved the model explicitly for $d = 1$ in 1924. Too soon, however, he became a refugee from Hitler's Germany. Much later he worked for almost three decades, 1948–1976, as a Professor of Physics in Bradley University in Peoria, Illinois. But to solve the Ising model for $d > 1$ proved intractable for many years. Furthermore, as we now know well, the available approximate solutions were quite misleading!

But Yang has related how one day in 1944–45, while still a Master's student, his normally quiet and reserved advisor, Professor J. S. Wang, became quite excited, almost agitated. It transpired that he had just learned that Lars Onsager (shown below with C. N. Yang in Fig. 4) had exactly solved the square-lattice Ising model in

Fig. 4. Chen Ning Yang and Lars Onsager together much later in March 1965.

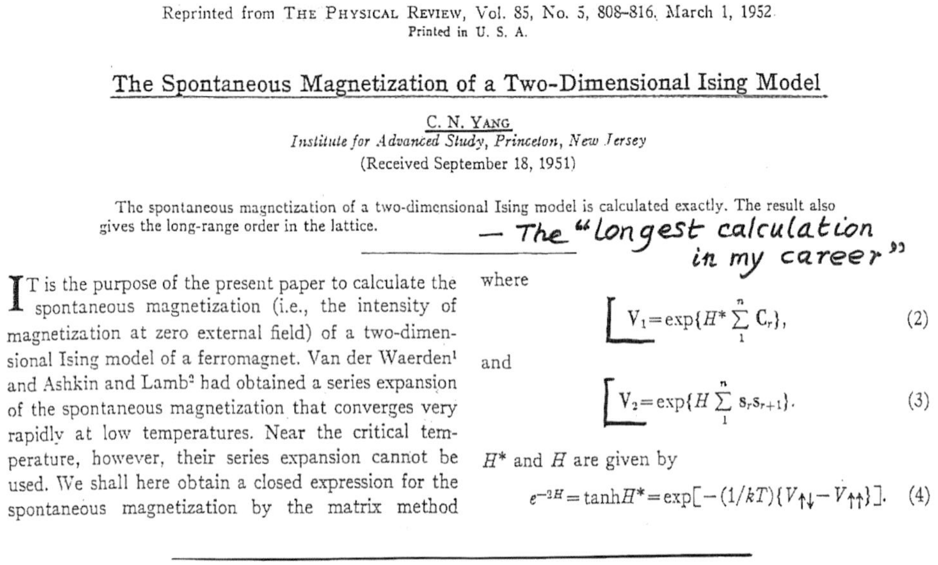

Reprinted from THE PHYSICAL REVIEW, Vol. 85, No. 5, 808–816, March 1, 1952
Printed in U. S. A.

The Spontaneous Magnetization of a Two-Dimensional Ising Model

C. N. YANG
Institute for Advanced Study, Princeton, New Jersey
(Received September 18, 1951)

The spontaneous magnetization of a two-dimensional Ising model is calculated exactly. The result also gives the long-range order in the lattice.

— The "longest calculation in my career"

IT is the purpose of the present paper to calculate the spontaneous magnetization (i.e., the intensity of magnetization at zero external field) of a two-dimensional Ising model of a ferromagnet. Van der Waerden[1] and Ashkin and Lamb[2] had obtained a series expansion of the spontaneous magnetization that converges very rapidly at low temperatures. Near the critical temperature, however, their series expansion cannot be used. We shall here obtain a closed expression for the spontaneous magnetization by the matrix method

where

$$V_1 = \exp\{H^* \sum_1^n C_r\}, \quad (2)$$

and

$$V_2 = \exp\{H \sum_1^n s_r s_{r+1}\}. \quad (3)$$

H^* and H are given by

$$e^{-2H} = \tanh H^* = \exp[-(1/kT)\{V_{\uparrow\downarrow} - V_{\uparrow\uparrow}\}]. \quad (4)$$

Fig. 5. The 1951 article providing the exact solution of the spontaneous magnetization of the square-lattice Ising model and establishing the critical exponent $\beta = \frac{1}{8}$. Note the author's later remark! The basic transfer matrix is $\mathbf{V}_1 \mathbf{V}_2$ where (in the notation of Bruria Kaufman[1] who used $H = J/k_B T$) $\mathbf{V}_1$ and $\mathbf{V}_2$ appear here in Eqs. (2) and (3).

zero magnetic field finding a logarithmic divergence of the specific heat at criticality! He told Yang about the paper: *Phys. Rev.* **65**, 117 (1944); but Yang, like so many others, could not see how to go further: no "strategic plan" seemed evident. A little while later, when already in Chicago, significant progress still eluded him.

However, in November 1949, with a Chicago Ph.D. safely in hand, Yang had a chance conversation with J. M. Luttinger at the Institute for Advanced Study in Princeton. Through that he learned of Bruria Kaufman's simplification of Onsager's approach in terms of a system of anticommuting hermitian matrices. Kaufman's work[1] opened the door to understanding Onsager's solution and to an appreciation of how much had been learned about the Ising model on a square lattice, albeit in zero magnetic field.

In January 1951 Yang realized that the spontaneous magnetization, $M_0(T)$ as a function of the temperature T, could be derived from an off-diagonal matrix element between the two eigenvectors of the underlying transfer matrix that had the largest eigenvalues. Following that idea — which entailed months of hard labour — led to the submission in mid-September of the remarkable paper displayed in Fig. 5. Following a series of "miraculous cancellations" (including elaborate factors denoted, with bold originality, **I**, **II**, **III**, and **IV**), the calculation led to the surprisingly

---

[1] See: B. Kaufman, *Phys. Rev.* **76**, 1232 (1949).

simple expression

$$M_0(T) = \left[\frac{1+x^2}{(1-x^2)^2}(1-6x^2+x^4)^{\frac{1}{2}}\right]^{2\beta}, \quad (1)$$

where $x = \exp(-2J/k_B T)$, while k_B is Boltzmann's constant. The crucial and *universal critical exponent* proved to be $\beta = \frac{1}{8}$, in strong contrast with nearly all previous approximate theoretical treatments[2] which yielded $\beta = \frac{1}{2}$ — often called the "mean-field value."

The problem of a rectangular Ising model with distinct values of J_1 and J_2 — already broached by Onsager — was suggested to C. H. Chang (of the University of Washington, Seattle). His results (guided by Yang!) could be expressed in terms of the modulus, k, of the crucial elliptic integral as

$$M_0(T) = (1-k^2)^\beta, \quad (2)$$

where, with $x_i = \exp(-2J_i/k_B T)$ for $i = 1, 2$,

$$k = \frac{2x_1}{1-x_1^2}\frac{2x_2}{1-x_2^2}. \quad (3)$$

As expected, Chang found $\beta = \frac{1}{8}$ for all values of $n = J_2/J_1$; the striking behavior of $M_0(T; n)$ is shown in Fig. 6.

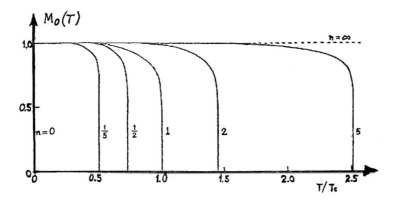

Fig. 6. The spontaneous magnetization of a two-dimensional rectangular Ising model with vertical and horizontal couplings, J_1 and J_2, related by $n = J_2/J_1$. Note that T_c is the critical point of the square lattice with $J = J_1 = J_2$ and $x_c = \sqrt{2}-1$. [After C. H. Chang, *Phys. Rev.* **88**, 1422 (1952).]

---

[2] It is worth recalling, however, that Cyril Domb, using an extrapolation of the exact low-temperature series expansions for $M_0(T)$, had, in 1949, obtained the estimate $\beta \lesssim 0.16$ only 20% higher than Yang's exact result: see *Proc. Roy. Soc. A* **199**, 199–221 (1949).

3. Lars Onsager and C. N. Yang

A year later Yang encountered Onsager at a 1953 *Tokyo–Kyoto International Conference*. But while Onsager spoke in his characteristic, somewhat vague style, punctuated by broad smiles, he did not address the Ising model. Thus it was only some twelve years later in March 1965 that Yang was able to learn, directly from Onsager himself, the reasons for all the unobvious commutator calculations in the 1944 paper. The background occasion, recorded in Fig. 7, was the *University of Kentucky Centennial Conference on Phase Transitions* commemorating the founding of

Fig. 7. Newly commisioned Kentucky Colonels with their host at the Conference on Phase Transitions.

the university in 1865. While waiting after the meeting at the airport,[3] Onsager explained how, in succession, he had undertaken to diagonalize the transfer matrix by hand: first the $2 \times \infty$ matrix, then the $3 \times \infty$, and so on! Eventually, by the $6 \times \infty$ case, he confirmed that the $64 = 2^6$ eigenvalues were (essentially) all of the form $\exp(\pm\gamma_1 \pm \cdots \pm \gamma_6)$. That suggested an underlying product algebra which, in turn, had led him to the elaborate structure of his original derivation.

One may wonder what brought Yang and Onsager together. It transpired that Onsager was an old friend of the host at the Conference, W. C. de Marcus, who approached him for advice. Thus, indeed, it was Onsager's approval that brought Yang to the meeting! He also suggested Professor Mark Kac (by then some years at the Rockefeller University in New York) and, in addition, as a junior speaker the Author of this article. All four, as shown in Fig. 7, were awarded a special honor of the *Commonwealth of Kentucky*, that is to say, we were commisioned as Kentucky Colonels. A document recording the author's commision is reproduced in Fig. 8. In accepting the commisions on behalf of all four of us, Mark Kac expressed his belief that this "Southern Honor" was particularly appropriate in that Lars Onsager came from Southern Norway, Frank Yang hailed from Southern China, Michael Fisher was brought up in Southern England, while he himself originated in Southern Poland!

4. Two Dimensions in the Real World

The critical exponent, $\beta = \frac{1}{8}$, found for the Ising model by Yang (and confirmed more broadly by Chang) was for a "manifestly artificial model" — as the Ising model was generally regarded in the middle of the last century. Beyond other aspects, the "two-dimensionality" added to the sense of artificiality. But that was no serious obstacle to another Chinese immigrant to the United States, namely, the talented experimentalist M. H. W. Chan. Moses, as he is known in the West, was born in November 1946 in Xi'an but moved with his family to Hong Kong in 1949 and came to America as a student, gaining his Ph.D. at Cornell University.

It was known that if one deposited a submonolayer of methane, CH_4, on graphite at a temperature below about 70 K it phase separated into a low density (or gaseous) "vapor" and a higher density "liquid." At first sight this appeared to be an ideal lattice gas and so, reasonably described by an Ising model. But previous attempts to measure the coexistence curve as a function of temperature and hence determine the exponent β experimentally had, at best, proved inconclusive. It was realized by Moses Chan, however, that the difficulty was closely associated with the steepness of the anticipated "two-dimensional" coexistence curve, clearly evident in Fig. 6. He realized that an optimal route is to determine the phase diagram by carefully

---

[3] As related in Chen Ning Yang, *Selected Papers 1945–1980 With Commentary* (W.H. Freeman & Co., San Francisco, 1983).

Fig. 8. Document announcing the commissioning of Michael E. Fisher as a Kentucky Colonel on 17 March 1965 in the 173rd year of the Commonwealth of Kentucky.

measuring the specific heat at a range of constant coverages, n: see the specific-heat traces in Fig. 9. On crossing the coexistance curve at constant n from the two-phase (vapor + liquid) region below T_c to the supercritical fluid phase a clear break would be evident in temperature plots of the specific heat.

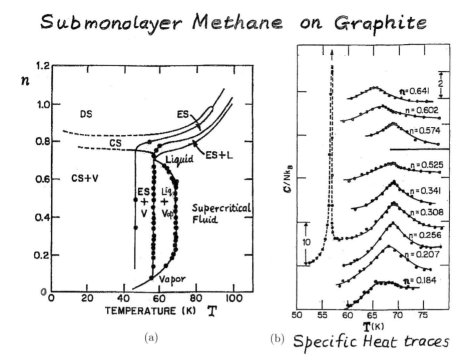

Fig. 9. (a) The coverage, n, vs. temperature, T, phase diagram of methane deposited on graphite at low coverages as determined by (b) specific heat plots, C/Nk_B, versus T. (After H. K. Kim and M. H. W. Chan, *Phys. Rev. Lett.* **53**, 170–173 (1984).

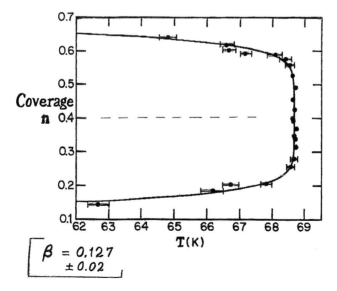

Fig. 10. Coexistence curve for temperatures from 62 K to 69 K of submonolayer methane on graphite to determine the critical exponent β. (After H. K. Kim and M. H. W. Chan, *Phys. Rev. Lett.* **53**, 170–173 (1984).

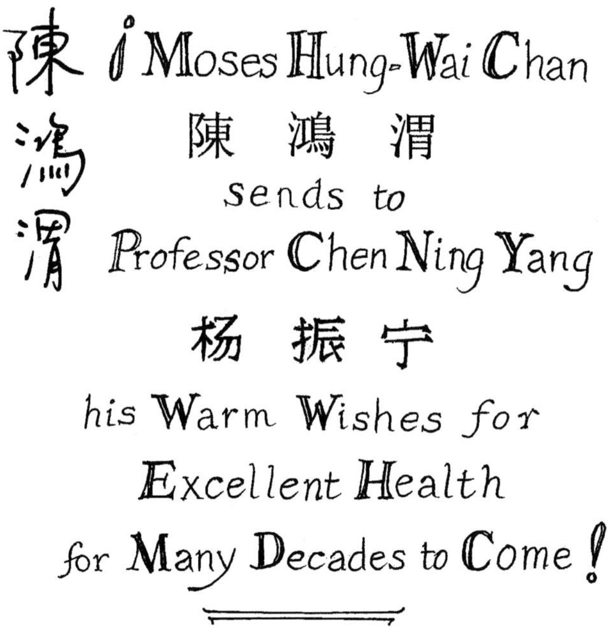

陳 / Moses Hung-Wai Chan
鴻　陳　鴻　渭
渭　sends to
　　Professor Chen Ning Yang
　　　杨　振　宁
　his Warm Wishes for
　　Excellent Health
　for Many Decades to Come !

Fig. 11. A good-will message from Moses Hung-Wai Chan to Professor Chen Ning Yang (as displayed on 28 May 2015 in the author's talk to the Conference at the Institute of Advanced Studies in Nanyang Technological University).

Using this approach, Chan (with his student H. K. Kim) was able, in 1984, to determine the coexistence curve displayed in Fig. 10. The fitted value, $\beta = 0.127 \pm 0.02$, agrees remarkably well with Yang's result of $\beta = \frac{1}{8} = 0.125$ — now clearly seen as a "prediction" thirty two years before the observation! In celebration of these theoretical and experimental achievements the greeting shown in Fig. 11 was displayed.

5. Condensation and Circles in the Complex Plane

In March 1982 a Sanibel Symposium was organized by Per-Olav Löwdin in Florida, in honor of Joseph Mayer (who died only a year later). In an interesting historical essay[4] C. N. Yang has related how Mayer, in 1937, approached the problem of *condensation* of a fluid from a vapor to a liquid. Mayer developed systematic cluster expansions that, as Yang observes, "started an analysis of the mathematics[5] and

---

[4] C. N. Yang, *Int. J. Quant. Chem.: Quantum Chemistry Symposium*, **16**, 21–24 (1982). See also pages 43–46 in Chen Ning Yang, *Selected Papers II With Commentaries* (World Scientific Publishing Co., Singapore, 2013).

[5] Among authors inspired to address the issues mathematically, Yang cites L. van Hove, *Physica* **15**, 951 (1949). Unhappily, however, as orginally pointed out by N. G. van Kampen and as I reported in *Arch. Ratl. Mech. Anal.* **17**, 377–410 (1964), the second (but unnumbered!) equation in the appendix of van Hove's article is deeply flawed. This fact actually invalidates van Hove's

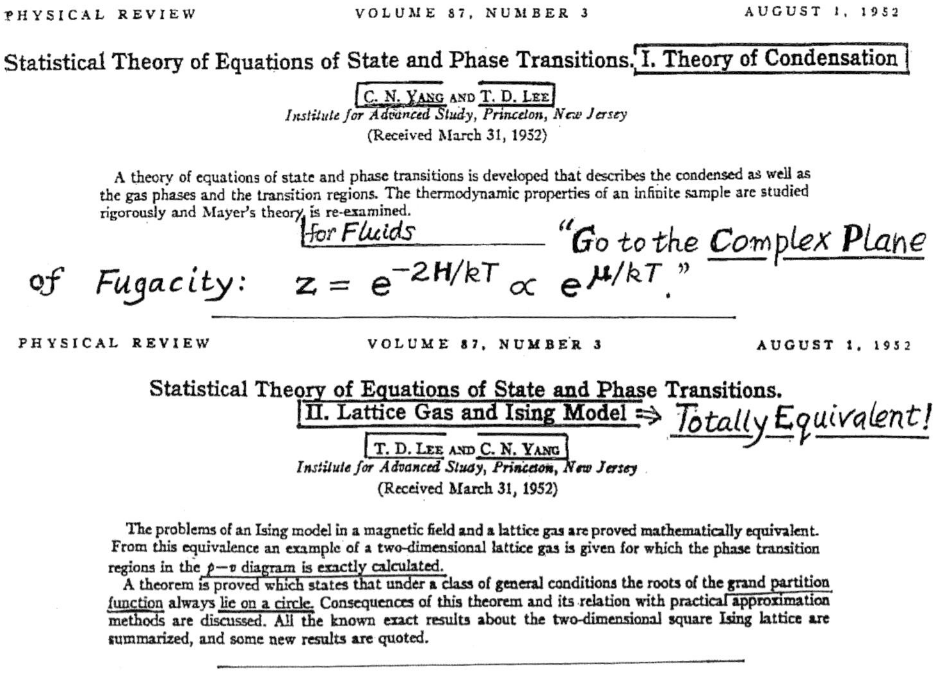

Statistical Theory of Equations of State and Phase Transitions. I. Theory of Condensation

C. N. YANG AND T. D. LEE
Institute for Advanced Study, Princeton, New Jersey
(Received March 31, 1952)

A theory of equations of state and phase transitions is developed that describes the condensed as well as the gas phases and the transition regions. The thermodynamic properties of an infinite sample are studied rigorously and Mayer's theory is re-examined.

[handwritten annotations:] for Fluids — "Go to the Complex Plane of Fugacity: $z = e^{-2H/kT} \propto e^{\mu/kT}$."

Statistical Theory of Equations of State and Phase Transitions. II. Lattice Gas and Ising Model

[handwritten: ⇒ Totally Equivalent!]

T. D. LEE AND C. N. YANG
Institute for Advanced Study, Princeton, New Jersey
(Received March 31, 1952)

The problems of an Ising model in a magnetic field and a lattice gas are proved mathematically equivalent. From this equivalence an example of a two-dimensional lattice gas is given for which the phase transition regions in the $p-v$ diagram is exactly calculated.

A theorem is proved which states that under a class of general conditions the roots of the grand partition function always lie on a circle. Consequences of this theorem and its relation with practical approximation methods are discussed. All the known exact results about the two-dimensional square Ising lattice are summarized, and some new results are quoted.

Fig. 12. The two-part 1952 article by C. N. Yang and T. D. Lee addressing the statistical mechanical theory of phase transitions. This work established the significance of the complex plane for understanding how a sharp first-order phase transition could appear in a large system.

physics of such phenomena." In fact, Mayer expected some sort of mathematical singularity at the condensation point (with, as a consequence, the failure of any analytical continuation of an isotherm beyond condensation). But conceptual puzzles remained: "How can the gas models 'know' when they have to coagulate to form a liquid or solid?" asked Born and Fuchs in 1938."[4] This was the topic that attracted Yang's attention in 1952. It led to a collaboration with T. D. Lee and to the two-part articles shown in Fig. 12. Their recipe, as the figure states, was: "Go into the Complex Plane!"

As indicated in Fig. 12, Yang and Lee chose to examine the complex plane of the fugacity z. For a ferromagnet subject to a magnetic field, H, this is simply the corresponding Boltzmann factor; but for a fluid the *chemical potential*, μ, is involved. This variable is much beloved by chemists but anathema to many physicists! However, in terms of z the grand canonical partition function, $\Xi(z;\Omega)$, for a finite domain Ω that contains N particles is merely a polynomial of degree N. Since $\Xi(z)$ must be real and positive for positive real z and T (or μ/T or H/T) all zeroes of the polynomial $\Xi(z)$ must lie in the complex plane (or on the negative real axis).

---

claim of a proof of the existence of a well defined *thermodynamic limit* of the statistical mechanical expression for the free energy of a system.

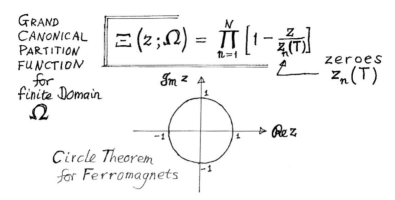

Fig. 13. Illustration of the Lee–Yang circle theorem for lattice gases or for ferromagnetic Ising models where $z = \exp(2H/k_{\rm B}T)$: all zeroes must lie on the unit circle. In the thermodynamic limit this implies that the only possible singularities in the free energy — indicative of a phase transition — must be located in zero magnetic field ($H = 0$) or, for a lattice gas, on a unique locus $\mu_\sigma(T)$.

As a result, knowing the distribution of the zeroes amounts to a full knowledge of the thermodynamics. Consequently if the zeroes approach the real axis as $N \to \infty$, a singularity must appear in the thermodynamic limit at the corresponding value of μ and T.

In Part II of their article Lee and Yang explicitly considered the Ising model and the lattice gas and found them to be completely equivalent mathematically. But the great surprise was the *circle theorem*! Specifically, as illustrated in Fig. 13, the zeroes of the grand canonical partition function for any pairwise ferromagnetic interactions whatsoever (of arbitrary range or structure, etc.) were proved to always lie on the unit circle in the complex z-plane. Equivalently, the zeroes are restricted to lie *on* the pure imaginary axis in the complex plane of μ, the chemical potential.

It is worth stressing that, building on the work of Yang and Lee, we now know that Mayer's vision of a singularity at condensation is, in fact, correct. At a first-order transition one must expect[6,7] an *essential singularity*. Thus, all the nth derivatives remain bounded but their behavior as $n \to \infty$ is insufficient to make the associated Taylor series converge.

For many years it was felt that the Lee–Yang zeroes must basically remain unobservable — even though the *density of zeroes*, $g(\theta)$ near the real axis must

---

[6] See M. E. Fisher, IUPAP Conf. Stat. Mech. (Brown University, 1962) as recorded by S. Katsura, *Adv. Phys.* **12**, 416 (1963); and *Physics* **3**, 255–283 (1967). Note that the 1967 article is the text of the talk presented at the Centennial Conference on Phase Transitions held at the University of Kentucky, 18–20 March 1965 (as illustrated in Figs. 7 and 8, above).

[7] The concept of an essential singularity at condensation was independently advanced by A. F. Andreev, *Sov. Phys, JETP* **18**, 1415 (1964). For a lattice gas the result was proved with full mathematical rigor by S. N. Isakov, *Commun. Math. Phys.* **95**, 427 (1984): see also the discussion in M. E. Fisher, *Proc. Gibbs Symp.* (Yale University, 15–17 May 1989; 1990 American Mathematical Society), pp. 47–50.

influence the thermodynamics at condensation. Recently, however, Ren-Bao Liu,[8] of the Chinese University of Hong Kong, and his coworkers have demonstrated how Lee–Yang zeroes may be investigated by "measuring quantum coherence of a probe spin coupled to a Ising-type spinbath." Indeed, Dr. Ren-Bao Liu presented his results at the Yang–Mills Conference.

Of course, as the saying has it: "What's good for the goose is good for the gander!" Thus, as pointed out some time ago,[9] the complex plane is also valuable for other variables, perhaps most notably the temperature. This leads to what have been called[8] "Fisher zeroes."

For an Ising model on a rectangular lattice one knows from Kaufman[1] that the canonical partition function for an $m \times n$ torus can be expressed as[9]

$$Z_N^{(1)} = \prod_{r=1}^{m} \prod_{s=1}^{n} \left\{ \frac{1+v^2}{1-v^2} \frac{1+v'^2}{1-v'^2} - \frac{2v}{1-v^2} \cos \frac{2\pi r}{m} - \frac{2v'}{1-v'^2} \cos \frac{2\pi s}{n} \right\}, \quad (1)$$

together with three other quite similar products. Here the natural thermal variables, vanishing like $1/T$ as $T \to \infty$, are

$$v = \frac{1-x_1}{1+x_1} = \tanh\left(\frac{J_1}{k_B T}\right), \quad v' = \frac{1-x_2}{1+x_2} = \tanh\left(\frac{J_2}{k_B T}\right). \quad (2)$$

For the symmetric case $v = v'$ (or $J_1 = J_2$) it is not hard to see, as illustrated in Fig. 14, that all the zeros of (5.1) lie on two circles in the complex v plane.

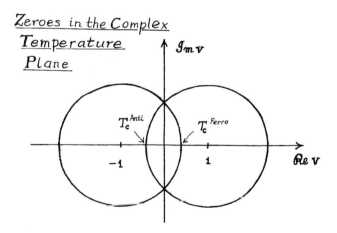

Fig. 14. Loci of zeroes of the canonical partition function of the square lattice Ising model in the complex $v = v' = \tanh(J/k_B T)$ plane. The critical points of the ferromagnetic and antiferromagnetic models are indicated.

---

[8]X. Ping, H. Zhou, B.-B. Wei, J. Cui, J. Du, and R.-B. Liu, *Phys. Rev. Lett.* **114**, 010601 (2015); see also B.-B. Wei and R.-B. Liu, *Phys. Rev. Lett.* **109**, 185701 (2012).
[9]M. E. Fisher, "*The Nature of Critical Points*," in *Lectures in Theoretical Physics, Vol. VIIC* (Univ. of Colorado Press, Boulder, Colorado, 1965), Secs. 13 and 19.

Where one of the circles crosses the real positive axis at $\text{Re}(v) < 1$ corresponds to the ferromagnetic critical point. Conversely, where the other circle crosses the negative v axis [at $\text{Re}(v) > -1$] locates the critical point of the antiferromagnet (for which $J < 0$). If θ is an angle describing the circle centered at $\text{Re}(v) = -1$, the density of zeroes varies as

$$g(\theta) = |\sin\theta| F(\theta), \tag{3}$$

where $F(\theta)$ is an analytic function periodic in θ. It then readily follows[9] that the specific heat varies simply as $A \ln|T - T_c|$ — just as orginally found by Onsager!

6. The Universal Yang–Lee Edge Singularity

Let us identify $\mu_\sigma(T)$ as the locus of condensation in the plane of real chemical potential vs. temperature. We may, for convenience, write

$$\Delta\mu(T) \equiv \mu - \mu_\sigma(T) = h'(T) + ih''(T). \tag{1}$$

We then know, thanks to the circle theorem, that the Lee–Yang zeroes are confined to the imaginary plane $\text{Re}\{\Delta\mu\} \equiv h'(T) = 0$; hence their density, $g(T; h'')$, is a function only of $h'' \equiv \text{Im}\{\Delta\mu\}$.

Because of the essential singularity at $h'(T) = h''(T) = 0$, always to be expected below the critical temperature, T_c, the density $g(T; h'')$ must be nonvanishing for small h'' when $T < T_c$. On the other hand, when T increases above T_c there must open up a gap, say $h_{\text{YL}}(T) > 0$, that is free of zeroes (in the thermodynamic limit). The gap function, $h_{\text{YL}}(T)$, serves to define the *Yang–Lee edge* as illustrated in Fig. 15. On grounds of symmetry we expect two equivalent edges located at $h'' = \pm h_{\text{YL}}(T)$.

Now, as noted by P. J. Kortman and R. B. Griffiths, *Phys. Rev. Lett.* **27**, 1439, (1971) these edges must be branch points of the free energy $F(T, \mu)$. We will call such branch points *Yang–Lee Edge Singularities*[10] and, following Kortman and Griffiths, ask about their nature. It will be argued that they should be treated as a potentially new class of critical points and analyzed accordingly!

To this end, let us posit that the density of zeroes near the edge varies as

$$g(T, h'') \sim [h_{\text{YL}}(T) - h'']^\sigma, \tag{2}$$

when $h'' \to h_{\text{YL}}(T)$ from below and enquire as to the value of the critical exponent σ.

---

[10] M. E. Fisher, *Phys. Rev. Lett.* **40**, 1610–1613 (1978). See also D. A. Kurtze and M. E. Fisher, *J. Stat. Phys.* **19**, 205–218 (1978) and *Phys. Rev. B* **20**, 2785–2796 (1979).

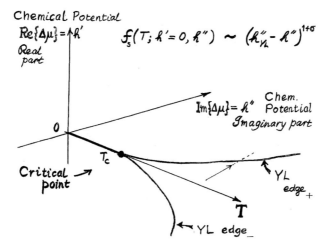

Fig. 15. Depiction of the Yang–Lee edges above the critical temperature, T_c, when there is a real condensation locus, $\mu_\sigma(T)$. The real and imaginary parts, (h', h''), of $\Delta\mu \equiv \mu - \mu_\sigma(T)$ are defined in the text; the Lee–Yang zeroes occur only on the plane $h' = 0$.

It follows that there is a singular contribution to the total free energy $F(T,h)$, near the Yang–Lee edge which behaves as

$$f_s(T; h' = 0, h'') \approx A_\pm |h_{\text{YL}}(T) - h''|^{1+\sigma}, \tag{3}$$

where the amplitudes A_+ and A_- may, in general, depend on the sign of $h'' - h_{\text{YL}}$. This implies that the susceptibility, $\chi \propto (\partial^2 F/\partial h^2)$, diverges with an exponent $\gamma_{\text{YL}} = 1 - \sigma$, where we have anticipated that σ is always less than unity.

Indeed, it is natural to suppose that $\sigma(d)$ depends only on the dimensionality, d, of the system and is, thus, *universal*. This is, in fact, upheld by further investigation.[10] For $d = 1$ the results can be found exactly yielding $\sigma(1) = -\frac{1}{2}$. Furthermore, one can check other aspects of criticality such as *scaling*: specifically, that suggests the correlation function near criticality for an Ising (or more general) system — say with spins s_0 at the origin and $s_\mathbf{R}$ at separation $\mathbf{R}$ — should behave as

$$G(R; T, h) \equiv \langle s_0 s_\mathbf{R}\rangle - \langle s_0\rangle\langle s_\mathbf{R}\rangle$$
$$\approx D(Rh^{\nu_c})/R^{d-2+\eta}, \tag{4}$$

when $h, 1/R \to 0$. This can be checked precisely[10] for $d = 1$ yielding the Yang–Lee values $\eta_{\text{YL}}(1) = -1$ and $\nu_c(1) = \frac{1}{2}$. Similarly, hyperscaling relations such as

$$\sigma = \frac{d - 2 + \eta_{\text{YL}}}{d + 2 - \eta_{\text{YL}}} \quad \text{and} \quad \gamma_c = \frac{2}{d + 2 - \eta_{\text{YL}}}, \tag{5}$$

are verified up to a borderline dimension that turns out to be[10] $d_{\text{YL}} = 6$.

Beyond that, consideration of mean-field or Landau-type theories demonstrates that Yang–Lee criticality should, indeed, be universal. However, in place of the standard field theoretic φ^4 analysis, it transpires that an $i\varphi^3$ theory is needed! That being understood, a renormalization-group treatment is possible and leads to

$$\sigma = \tfrac{1}{2} - \tfrac{1}{12}\epsilon \quad \text{and} \quad \eta_{\text{YL}} = -\tfrac{1}{9}\epsilon, \tag{6}$$

to first order in $\epsilon = 6 - d(> 0)$. (In second order the $\tfrac{1}{9}$ for η is replaced by[10] $\tfrac{43}{81}$. A third order calculation has been undertaken by Bonfim, Kirkham, and McKane, J. Phys. A **14**, 2391 (1981); see also Ref. 11 and Fig. 17 below.)

It turns out, however, that this is not the end of the story! In 1984 a physical chemist at Johns Hopkins University, D. Poland, proposed [in J. Stat. Phys. **35**, 341 (1984)] that both lattice and continuum *hard-core fluids* are characterized, in the Mayer fugacity expansion of the pressure, by a *universal* dominant singularity on the *negative* fugacity axis, say at $z = z_0 < 0$. On introducing an exponent $\phi(d)$ for this singularity, the pressure displays a contribution of the form

$$p_s(T, z) \sim P(z - z_0)^{\phi(d)}. \tag{7}$$

A subsequent study[11] led to the exponent values $\phi(1) = \tfrac{1}{2}$, $\phi(2) = \tfrac{5}{6} = 0.833\ldots$, (derived from Baxter's well known 1980 exact solution of the hard-hexagon model) $\phi(3) \simeq 1.06$, and $\phi(4) \simeq 1.2$. When d became large, it appeared that $\phi(d)$ approached $\tfrac{3}{2}$.

A natural first question is: "How crucial is the 'hard core' — implying an interaction potential, $u(\mathbf{r})$, which is actually *infinite* over a finite range of particle separation $\mathbf{r}$?" An answer can be found by studying models with a 'soft repulsive core' for which $u(r)$ is always finite. A suitable model, for which low-density series of significant length could be generated for arbitrary d turned out to be the *Gaussian Molecule Mixture* (GMM); on this my student Sheng-Nan Lai labored.[11] This consists of similar A and B particles with interaction potentials

$$\begin{aligned} u_{\text{AA}}(\mathbf{r}) = u_{\text{BB}}(\mathbf{r}) = 0 \quad &\text{but} \\ \exp[-u_{\text{AB}}(\mathbf{r})/k_{\text{B}}T] &= 1 - e^{-r^2/r_0^2}, \end{aligned} \tag{8}$$

where r_0 merely sets the scale of the repulsive potential. (Evidently the Mayer f-function is Gaussian.) The singularities arising in the (z_A, z_B) plane for this GMM model are shown in Fig. 16 (with $d > 1$).

The numerical values found for $\phi(d)$ as d varied strongly supported the natural hypothesis, namely, that the intrinisic 'hardness' (or otherwise!) did not matter at all! In other words, universality extended to whatever the nature of the particle repulsive cores that generated the singularity. (It is worth remarking that some form of repulsive cores are essential to maintain thermodynamic stability.[12])

---

[11] S.-N. Lai and M. E. Fisher, J. Chem. Phys. **103**, 8144–8155 (1995) and references therein.
[12] See, e.g., M. E. Fisher, Arch. Ratl. Mech. Anal. **17**, 377–410 (1964).

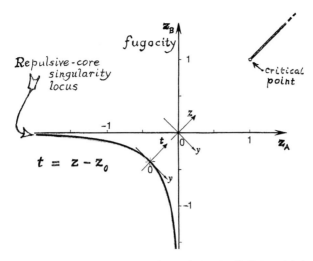

Fig. 16. Schematic view of the fugacity plane, (z_A, z_B), for the GMM model showing, in the first quadrant, a first-order phase boundary on the axis of symmetry, $z_A = z_B > 0$, that terminates at a critical point. In the third quadrant is a locus of *repulsive-core singularities* (which, in general, may extend into the second and fourth quadrants).

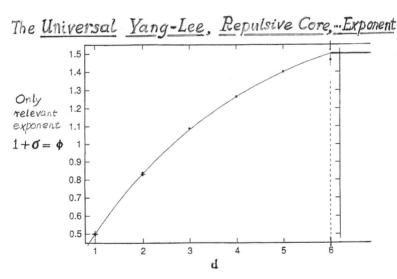

Fig. 17. Variation with dimensionality, d, of the universal repulsive-core singularity exponent, $\phi(d)$, now identified with the Yang–Lee edge exponent, $\sigma(d) = \phi(d) - 1$. The borderline dimensionality is $d = 6$; beyond that $\phi = \frac{3}{2}$ and $\sigma = \frac{1}{2}$. The continuous plot embodies an $O(\epsilon^3)$ expansion where $\epsilon = 6 - d > 0$.

In the same way, the similarity of the repulsive core exponents as a function of d to the Yang–Lee exponents led[11] to the explicit proposal

$$\sigma(d) = \phi(d) - 1. \tag{9}$$

A theoretical justification for this identification was originally lacking. However, with the aid of a field-theoretic approach which employed separate representations for the repulsive and attractive parts of the pairwise interactions, such a demonstration was eventually constructed.[13] (Note that particle–particle *attractions* are needed to yield condensation, criticality, and a Yang–Lee gap, $h''_{\rm YL}(T)$, above T_c.) Accepting (6.9), enables us to present, in Fig. 17 a combined plot of the dimensional dependence of both the Yang–Lee edge exponent, σ, and the repulsive core exponent ϕ.

Is that then the end of the story? Not really! Indeed, separate investigations by many authors[11] have shown not only that $\phi(d)$ describes assorted square and triangular lattice models of hard squares, hard hexagons (as mentioned), and dimers (with holes) but have also established the relations

$$\phi_{\rm D}(d) = \phi(d-1) \quad \text{and} \quad \phi_{\rm I}(d) = 1 + \phi(d-2) \tag{10}$$

for *Directed Site* and *Bond Animals*[14,15] and *Isotropic Site and Bond Animals* or *Branched Polymers*.[16] Note that in the case of animals, etc., of size n one expects the numbers of animals to vary asymptotically as

$$a_n \approx A\lambda^n n^{-\phi_D}, \tag{11}$$

(and similarly for isotropic animals, etc.) where A and λ are lattice-dependent, nonuniversal constants. The corresponding singularities now lie on the *positive* real fugacity axis.[11]

7. The Quantal Many-Body Problem at low T

In Fall 1955 Kerson Huang became a member of the Institute for Advanced Study at Princeton and introduced C. N. Yang to Fermi's *pseudopotential* approach. This soon led to joint work on the quantal many-body problem for Bose and Fermi systems. Ultimately Yang published some 15 or so papers on this topic, many involving a close collaboration with T. D. Lee. The first of these, together with four other early articles are shown in Fig. 18. However, unanticipated issues arose — as related in Yang's much later commentary[3] — and, in retrospect, these prove instructive illustrating the vicissitudes to which, even for the most talented scientist, research can lead in the real world!

In the original paper ([A] in Fig. 18) the results for the ground-state energy, E_0, of a Fermi system proved satisfactory; but for a Bose system of dilute hard spheres

---

[13] Y. Park and M. E. Fisher, *Phys. Rev. E* **60**, 6223–6228 (1999).
[14] J. L. Cardy, *J. Phys. A* **15**, L593 (1982).
[15] See D. Dhar, *Phys. Rev. Lett.* **51**, 853 (1983), *ibid.* **49**, 959 (1982), and H. E. Stanley, S. Redner, and Z.-R. Yang, *J. Phys. A* **15**, L569 (1982) (and references therein).
[16] See D. S. Gaunt, *J. Phys. A* **13**, L97 (1980) and G. Parisi and N. Sourlas, *Phys. Rev. Lett.* **46**, 871 (1981).

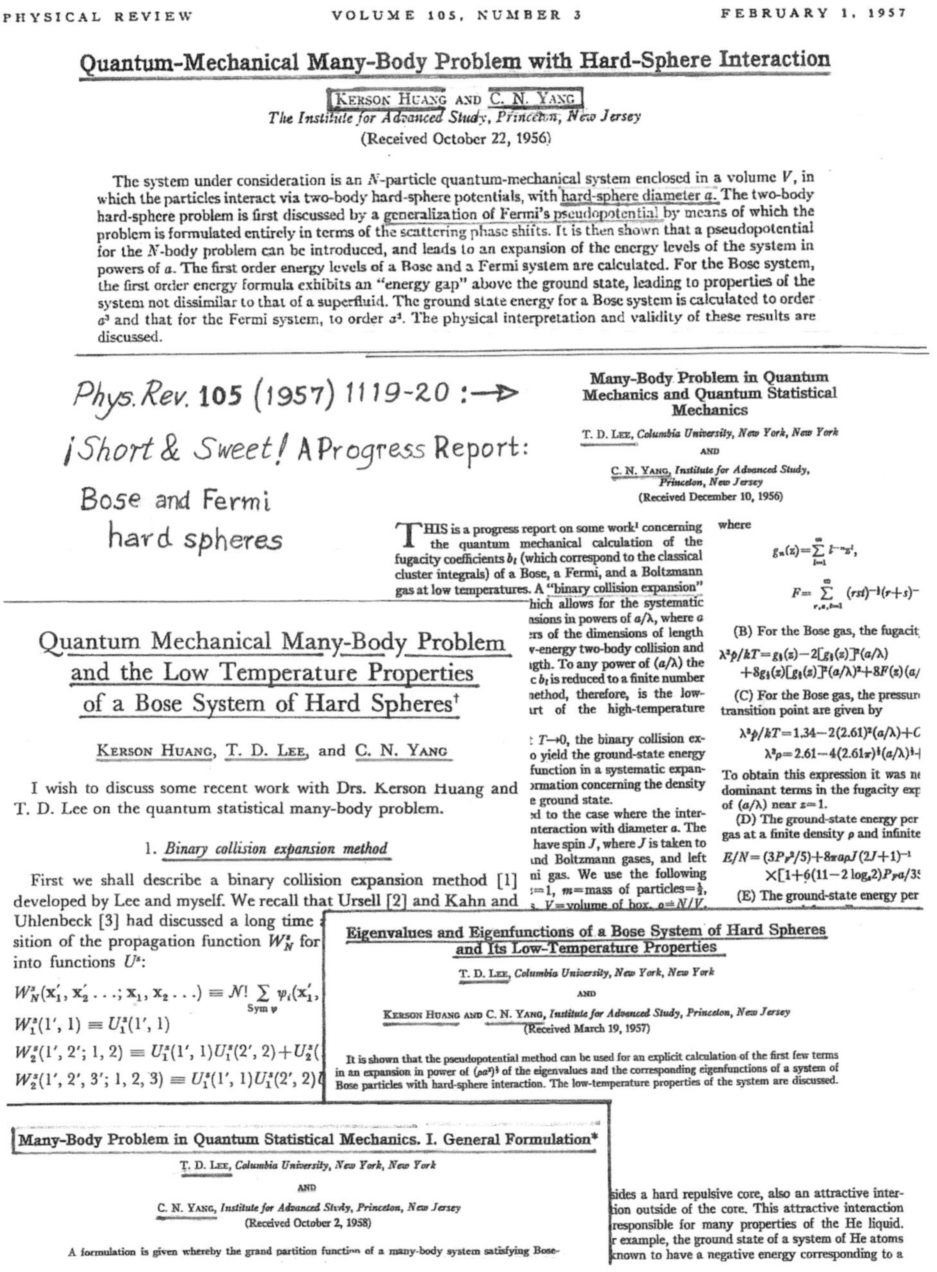

PHYSICAL REVIEW VOLUME 105, NUMBER 3 FEBRUARY 1, 1957

Quantum-Mechanical Many-Body Problem with Hard-Sphere Interaction

KERSON HUANG AND C. N. YANG
The Institute for Advanced Study, Princeton, New Jersey
(Received October 22, 1956)

The system under consideration is an N-particle quantum-mechanical system enclosed in a volume V, in which the particles interact via two-body hard-sphere potentials, with hard-sphere diameter a. The two-body hard-sphere problem is first discussed by a generalization of Fermi's pseudopotential by means of which the problem is formulated entirely in terms of the scattering phase shifts. It is then shown that a pseudopotential for the N-body problem can be introduced, and leads to an expansion of the energy levels of the system in powers of a. The first order energy levels of a Bose and a Fermi system are calculated. For the Bose system, the first order energy formula exhibits an "energy gap" above the ground state, leading to properties of the system not dissimilar to that of a superfluid. The ground state energy for a Bose system is calculated to order a^2 and that for the Fermi system, to order a^2. The physical interpretation and validity of these results are discussed.

Phys. Rev. 105 (1957) 1119-20 :→
¡Short & Sweet! A Progress Report:
Bose and Fermi
hard spheres

Many-Body Problem in Quantum Mechanics and Quantum Statistical Mechanics

T. D. LEE, *Columbia University, New York, New York*
AND
C. N. YANG, *Institute for Advanced Study, Princeton, New Jersey*
(Received December 10, 1956)

THIS is a progress report on some work[1] concerning the quantum mechanical calculation of the fugacity coefficients b_l (which correspond to the classical cluster integrals) of a Bose, a Fermi, and a Boltzmann gas at low temperatures. A "binary collision expansion" ...hich allows for the systematic ...nsions in powers of a/λ, where a ...rs of the dimensions of length ...v-energy two-body collision and ...gth. To any power of (a/λ) the ...c b_l is reduced to a finite number ...nethod, therefore, is the low- ...urt of the high-temperature

... $T\to 0$, the binary collision ex- ...o yield the ground-state energy function in a systematic expan- ...ormation concerning the density ...e ground state. ...d to the case where the inter- ...nteraction with diameter a. The ...have spin J, where J is taken to ...nd Boltzmann gases, and left ...ni gas. We use the following ...=1, m=mass of particles=½, ... V=volume of box, $\rho = N/V$.

where

$$g_n(z)=\sum_{l=1}^{\infty} l^{-n}z^l,$$

$$F=\sum_{r,s,t=1}^{\infty} (rst)^{-\frac{1}{2}}(r+s)^-$$

(B) For the Bose gas, the fugacit...

$$\lambda^3 p/kT = g_{\frac{5}{2}}(z) - 2[g_{\frac{3}{2}}(z)]^2(a/\lambda) + 8g_{\frac{1}{2}}(z)[g_{\frac{3}{2}}(z)]^2(a/\lambda)^2 + 8F(z)(a/$$

(C) For the Bose gas, the pressur... transition point are given by

$$\lambda^3 p/kT = 1.34 - 2(2.61)^2(a/\lambda) + C$$
$$\lambda^3 \rho = 2.61 - 4(2.61\pi)^{\frac{1}{2}}(a/\lambda)^{\frac{3}{2}} + $$

To obtain this expression it was ne... dominant terms in the fugacity exp... of (a/λ) near $z=1$.

(D) The ground-state energy per ... gas at a finite density ρ and infinite

$$E/N = (3P_F^2/5) + 8\pi a\rho J(2J+1)^{-1} \times[1+6(11-2\log_e 2)P_F a/35$$

(E) The ground-state energy per

Quantum Mechanical Many-Body Problem and the Low Temperature Properties of a Bose System of Hard Spheres†

KERSON HUANG, T. D. LEE, and C. N. YANG

I wish to discuss some recent work with Drs. Kerson Huang and T. D. Lee on the quantum statistical many-body problem.

1. *Binary collision expansion method*

First we shall describe a binary collision expansion method [1] developed by Lee and myself. We recall that Ursell [2] and Kahn and Uhlenbeck [3] had discussed a long time ... sition of the propagation function W_N^s for ... into functions U^s:

$$W_N^s(x_1', x_2' \ldots; x_1, x_2 \ldots) \equiv N! \sum_{\text{Sym }\psi} \psi_i(x_1',$$
$$W_1^s(1', 1) \equiv U_1^s(1', 1)$$
$$W_2^s(1', 2'; 1, 2) \equiv U_1^s(1', 1)U_1^s(2', 2) + U_2^s($$
$$W_3^s(1', 2', 3'; 1, 2, 3) \equiv U_1^s(1', 1)U_1^s(2', 2)$$

Eigenvalues and Eigenfunctions of a Bose System of Hard Spheres and Its Low-Temperature Properties

T. D. LEE, *Columbia University, New York, New York*
AND
KERSON HUANG AND C. N. YANG, *Institute for Advanced Study, Princeton, New Jersey*
(Received March 19, 1957)

It is shown that the pseudopotential method can be used for an explicit calculation of the first few terms in an expansion in power of $(\rho a^3)^{\frac{1}{2}}$ of the eigenvalues and the corresponding eigenfunctions of a system of Bose particles with hard-sphere interaction. The low-temperature properties of the system are discussed.

Many-Body Problem in Quantum Statistical Mechanics. I. General Formulation*

T. D. LEE, *Columbia University, New York, New York*
AND
C. N. YANG, *Institute for Advanced Study, Princeton, New Jersey*
(Received October 2, 1958)

A formulation is given whereby the grand partition function of a many-body system satisfying Bose-

...sides a hard repulsive core, also an attractive inter- ...tion outside of the core. This attractive interaction ...responsible for many properties of the He liquid. ...r example, the ground state of a system of He atoms ...known to have a negative energy corresponding to a

Fig. 18. Papers published by Yang on the quantal many-body problem, initially with Kerson Huang and later with T. D. Lee. From the first, at the top, and downwards, the specific references are: [A] *Phys. Rev.* **105**, 767 (1957); [B] *Phys. Rev.* **105**, 1119 (1957); [C] pp. 165–175 in *The Many-Body Problem*, ed. J. K. Percus (Wiley-Interscience, New York, 1963); [D] *Phys. Rev.* **106**, 1135 (1957); and [E] *Phys. Rev.* **113**, 1165 (1959).

the pseudopotential approach resulted in

$$E_0 = N \frac{4\pi a(N-1)}{L^3} \left\{ 1 + 2.37 \frac{a}{L} + \frac{a^2}{L^2} \left[(2.37)^2 + \frac{\xi}{\pi^2}(2N-5) \right] \right\} + \cdots, \quad (1)$$

where a is the hard-sphere diameter, N is the number of spheres, $L \times L \times L$ is the periodic box containing the system, while ξ is a convergent sum over three integers, (l,m,n). For fixed $\rho = N/L^3$, as $L \to \infty$ in the thermodynamic limit, however, the last term in (7.1) does not make sense: it behaves as $\xi N a^2/L^2 = \xi L a^2 \rho$ and so diverges! This observation disturbed the authors but they eventually decided to publish anyway.

Of course, the pressure was then on to find an alternative approach: this led to the *binary collision expansion* method which was presented (under the names of all three authors) as a lecture by Yang at the *Steven Conference on the Many-Body Problem* in January 1957. This is displayed as [C] in Fig. 18; but publication was unhappily delayed for six years. The technique employs a summation of the most divergent terms: applying this Lee and Yang found that the troublesome term identified in (7.1) dropped out after summation leading to

$$\frac{E_0}{N} = 4\pi \rho a \left[1 + \frac{128}{15} \sqrt{\frac{\rho a^3}{\pi}} + \cdots \right]. \quad (2)$$

The correction term of order $(\rho a^3)^{1/2}$ can be found merely by summing the most divergent terms (as shown in [C] of Fig. 18). The next term, of order $\rho a^3 \ln(\rho a^3)$, was published first by T. T. Wu, *Phys. Rev.* **115**, 1390 (1959) but soon confirmed by others. But T. T. Wu also found that in some cases summation of the most divergent terms — always a risky enterprise! — actually fails: see *Phys. Rev.* **149**, 380 (1960).

Meanwhile a defect had been found in the original application of the pseudopotential method; that invalidated (7.1). Once it was understood that a careful technique of subtractions was needed, it turned out that the pseudopotential method could, in fact, yield the now well-established result (7.2). The details were published in [D], the penultimate article exhibited in Fig. 18. While the various calculations by different methods required thorough-exposition, once the results were clear, it was wisely decided that a very short — less than two pages — "*Progress Report*" was called for! That is the second article, [B], in Fig. 18. As more-or-less visible in the figure, this article essentially consists of a series, (A) to (E), of explicit formulae for Fermi, Bose, and for the ground state, Boltzmann systems. It was noted that by a classic argument — which I first learned from the writings of Freeman Dyson — all the low-temperature expansions are likely to be of asymptotic character [allowing contributions like $\exp(1/\rho a^3)$].

Finally, in Fig. 18 the last article [E], by Lee and Yang, is the first in a series of five papers, published in 1959–60. The last two articles represented the hope[3] of revealing the true nature of the lambda transition to a superfluid in helium-4. But that was unsuccessful; progress in that direction had to await more general findings regarding critical phenomena, their universality and scaling, the renormalization group and the $\epsilon = 4 - d$ expansion, and so on.

8. Off-Diagonal Long-Range Order

Ever since Einstein accepted the ideas sent to him in 1924 by S. N. Bose and applied them to an ideal gas, it has been hard to avoid the picture presented in Fig. 19. But how might it relate to the actual phenomenon of superfluidity in liquid helium? And, perhaps, also have some connection to superconductivity in metallic systems? Quantum mechanics clearly matters; but in helium as in real metals, the interactions between atoms are crucial and an ideal gas cannot be an acceptable model.

Personally, my first encounter with this central issue for many-body quantum mechanics was in 1954. Having rejoined King's College, London now as a doctoral student after a couple of years National Service in Her Majesty's Royal Air Force), I was sent to Cambridge by Cyril Domb, newly arrived as Professor of Theoretical Physics. The reason for the visit was, I believe, related to my duties as Student Librarian in King's for the Physics Department. At the end of the afternoon in Cambridge I looked up Oliver Penrose, a close friend of my brother-in-law-to-be, David Castillejo (the younger brother of Leonardo Castillejo of some fame in high-energy physics arising from the "CDD" paper by Castillejo, Dalitz and Dyson). To my fascination Oliver Penrose told me of his thoughts, (already published three years previously) regarding the intrinsically "wave-mechanical" nature of the 'order' underlying superfluidity in real helium! Not long afterwards Oliver went, as a postdoctoral scholar, to work with Lars Onsager in Yale. Together in 1956 they published a basic paper in which, to quote C. N. Yang,[3] they "were the first to give a precise definition of Bose condensation in the case of interacting Bosons."

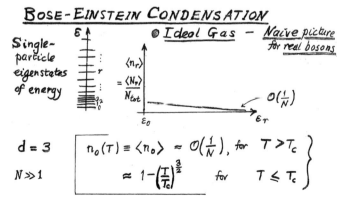

Fig. 19. The theoretical picture of condensation in an ideal gas of N bosons. Above T_c the occupancy of the lowest eigenstate of energy remains negligible; but below T_c a finite fraction of the bosons are "condensed" into the ground state, the fraction approaching 100% as $T \to 0$.

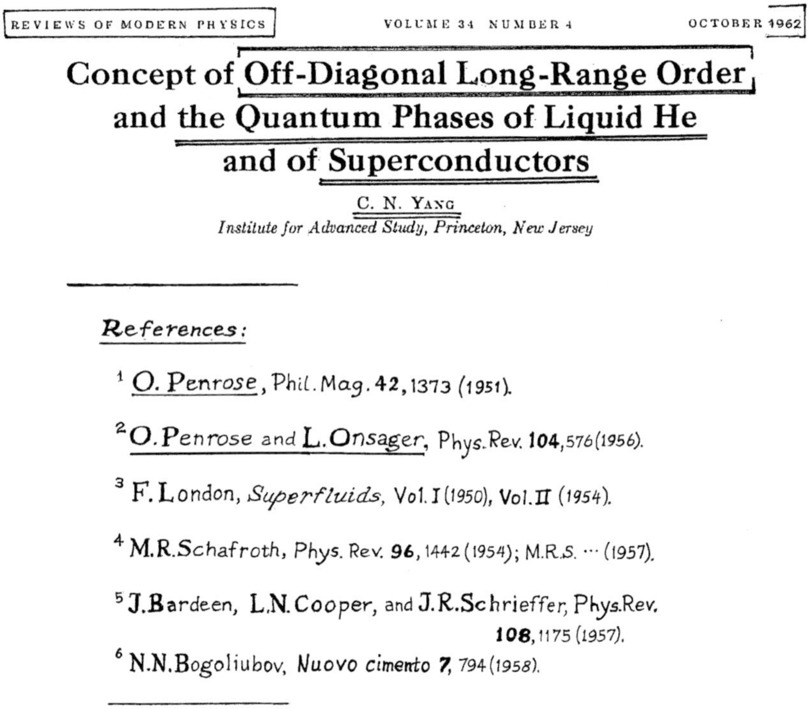

REVIEWS OF MODERN PHYSICS VOLUME 34 NUMBER 4 OCTOBER 1962

Concept of Off-Diagonal Long-Range Order and the Quantum Phases of Liquid He and of Superconductors

C. N. YANG

Institute for Advanced Study, Princeton, New Jersey

References:

[1] O. Penrose, Phil. Mag. **42**, 1373 (1951).

[2] O. Penrose and L. Onsager, Phys. Rev. **104**, 576 (1956).

[3] F. London, *Superfluids*, Vol. I (1950), Vol. II (1954).

[4] M. R. Schafroth, Phys. Rev. **96**, 1442 (1954); M. R. S. ··· (1957).

[5] J. Bardeen, L. N. Cooper, and J. R. Schrieffer, Phys. Rev. **108**, 1175 (1957).

[6] N. N. Bogoliubov, Nuovo cimento **7**, 794 (1958).

Fig. 20. Yang's fundamental paper on ODLRO, [*Rev. Mod. Phys.* **34**, 694–704 (1962)], showing the first six references visible as footnotes on the title page.

Yang then introduced the name "*Off-Diagonal Long-Range Order*" (or ODLRO) for the essential concept and, in the article shown in Fig. 20, he presented resulting general features for both Bose–Einstein and Fermi–Dirac particles. True to his customary meticulous regard for proper credit, the first six references, reproduced in Fig. 20, refer to the previous studies of Penrose and Onsager as well as the less precise ideas of London for superfluids and, for superconductors, of Schafroth and others, including "BCS" and Bogoliubov.

The essential step in defining ODLRO is to recognize that a quantum mechanical system of N particles must be described by a wave function,

$$\Psi_N = \Psi_N(\mathbf{r}_1, \ldots, \mathbf{r}_N; \mathbf{q}) \cong \hat{\psi}(\mathbf{r}), \tag{1}$$

of appropriate symmetry in the positional co-ordinates $\mathbf{r}_1, \ldots, \mathbf{r}_N$, while $\mathbf{q}$ denotes the remaining co-ordinates of the closed system. Following Lev Landau the overall density matrix ρ is then

$$\rho(\mathbf{r}_i; \mathbf{r}'_j) = \int d\mathbf{q} \Psi^*_N(\mathbf{r}_i; \mathbf{q}) \Psi_N(\mathbf{r}'_j; \mathbf{q}), \tag{2}$$

and reduced single-particle, two-particle, ... density matrices can be defined by integrating over the co-ordinates of $(N-1), (N-2), \ldots$ particles.

In terms of the density operator, $\hat{\psi}(\mathbf{r})$ in (8.1), the off-diagonal matrix element of the single-particle density matrix behaves as

$$\rho_1 \equiv \langle \hat{\psi}^\dagger(\mathbf{r})\hat{\psi}(\mathbf{r}')\rangle \to 0 \qquad \text{for} \quad T > T_c,$$
$$\approx \Psi_0^*(T)\Psi_0(T) \neq 0 \quad \text{for} \quad T < T_c, \tag{3}$$

when $|\mathbf{r}' - \mathbf{r}| \to \infty$. Clearly T_c locates the transition at which ODLRO, embodied in $n_0(T) = |\Psi_0(T)|^2$, and Bose–Einstein condensation sets in. For liquid helium it is natural to identify $T_c \equiv T_\lambda$ as the lambda point at which superfluidity appears and the specific heat exhibits its extremely sharp spike.

More generally, to include superconductivity one should, following Yang, consider at least ρ_2 the two-particle reduced density matrix. Then large eigenvalues, $\geq O(N)$, and corresponding ODLRO, originate from pairs of fermions, e.g., Cooper pairs of electrons, engaging in Bose–Einstein degeneracy. Thus, the smallest value of n for which ODLRO appears in a reduced density matrix $\rho_m (m \geq n)$ identifies the "collection of n particles that, in a sense, forms the *basic group* exhibiting the long-range correlation." In fact, the article notes that although it is customary to regard liquid helium as a collection of helium atoms obeying Bose statistics, a "much better description is a collection of electrons and He nuclei."

In the 1980's Yang expressed[3] his fondness for this paper but remarked that "it is clearly unfinished." Indeed, there is little doubt that in Yang's mind were also the questions: "How best to define the observable *superfluid density*, $\rho_s(T)$, in microscopic terms?" and, ultimately related: "What is the value of the corresponding critical exponent?" As regards ODLRO it became clear that, in general, this is analogous to $[M_0(T)]^2$ for a ferromagnet; thus the appropriate exponent is 2β. It remained, however, for Josephson[17,18] to convincingly argue that for $\rho_s(T)$ the exponent is slightly different, namely,

$$2\beta - \eta\nu = 2 - \alpha - 2\nu = (d-2)\nu, \tag{4}$$

where the last equality is restricted to dimensionalities $d \leq 4$. In Eq. (8.4) the exponents α and ν describe, respectively, the specific heat singularity and the divergence of the correlation length, while $\eta < 0.05$ is defined via the critical point decay of the pair correlation function. see Eq. (6.4) above. And, finally, Yang himself drew attention, 35 years ago, to his conjectured formula for the *penetration depth* for superconductors in magnetic fields.

---

[17] B. D. Josephson, *Phys. Lett.* **21**, 608 (1966).
[18] See also M. E. Fisher, M. N. Barber, and D. Jasnow, *Phys. Rev. A* **8**, 1111–1124 (1973), where $\rho_s(T)$ is identified as proportional to the so-called *helicity modulus*, $\Upsilon(T)$, whose definition requires the difference of free energies in *finite-size* systems with *anti*periodic and periodic boundary conditions on the *phase* of the wavefunction (or its operator).

9. Yang–Yang Thermodynamics and the Scaling Axes

When C. N. Yang visited Stanford in 1963 William Little told him of the striking experiments by A. V. Voronel and workers in Russia. This led to the article with his brother, Chen Ping Yang, shown in Fig. 21. Voronel's observations on argon and oxygen made it clear that the specific heats of ordinary fluids exhibit very sharp peaks at criticality well described by the form

$$C_V(T;\rho_c) \approx A^{\pm}/|t|^{\alpha} + B, \quad t = (T-T_c)/T_c. \tag{1}$$

This was in direct contradiction to van der Waals (or mean field) predictions. Rather, as Yang and Yang noted, the data were strongly reminiscent of the close-to logarithimically divergent specific heat peak for the superfluid transition in helium found by Fairbank, Buckingham, and Kellers in 1957. We now know, however, that in argon and similar fluids, the exponent α introduced in (9.1) is close to 0.109 whereas in superfluid helium it is actually slightly negative with $\alpha \simeq -0.013$. (Note that a logarithmic divergence corresponds, via a simple limit, to $\alpha = 0$.)

Yang and Yang then made a series of remarks of which the first — just visible in Eq. (3) in Fig. 21 — was a misleadingly simple thermodynamic formula, which we rewrite as

$$C_V^{\text{total}} = VT \left(\frac{\partial^2 p}{\partial T^2}\right)_V - NT \left(\frac{\partial^2 \mu}{\partial T^2}\right)_V. \tag{2}$$

As they emphasized, this relation, based simply on $SdT = Vdp - Nd\mu$, must hold in the two-phase region below T_c. Hence it must apply to the observations of Voronel and coworkers in (9.1). Consequently, p and μ in (9.2) can be replaced by the condensation loci, $p_\sigma(T)$ and $\mu_\sigma(T)$; thus one reaches the question posed pictorially in Fig. 22 (and stated verbally[19] in the caption).

Now, on the one hand, it is known that $\mu_\sigma''(T)$ always remains finite in the usual lattice gases, in simple droplet models,[20] and, in other exactly soluble systems.[21] On the other hand in all these cases $p_\sigma''(T)$ becomes singular at T_c! Experimentally, this as also true for H_2O, CO_2 and propane, C_3H_8.[19] For the hard-core square-well fluid this is likewise confirmed by simulation. But as Yang and Yang remark: "For real gases, it is more reasonable to expect that both $(d^2\mu_\sigma/dT^2)$ and (d^2p_σ/dT^2) become ∞" — supposing $\alpha \geq 0$.

---

[19] M. E. Fisher and G. Orkoulas, *Phys. Rev. Lett.* **85**, 696–699 (2000); G. Orkoulas, M. E. Fisher, and C. Üstün, *J. Chem. Phys.* **113**, 7530–7545 (2000).

[20] M. E. Fisher, *Physics* **3**, 255–283 (1967). More elaborate cluster-interaction models are in M. E. Fisher and B. U. Felderhof, *Ann. Phys. (N.Y.)* **58**, 176–216, 217–267 (1970); B. U. Felderhof and M. E. Fisher, *ibid.* pp. 268–280. None-the-less, some of the models described in Secs. 7 and 8 of these articles do yield a divergence of $\mu_\sigma''(T_c)$.

[21] It is worth noting here, that for ferromagnets (and other systems with a corresponding symmetry) the vanishing of $(d^2\mu_\sigma/dT^2)$ cannot be questioned. Thus the chemical potential is replaced by the magnetic field, H, subject to the symmetry, $H \to -H$, and the phase boundary, $\mu_\sigma(T)$, becomes simply $H = 0$.

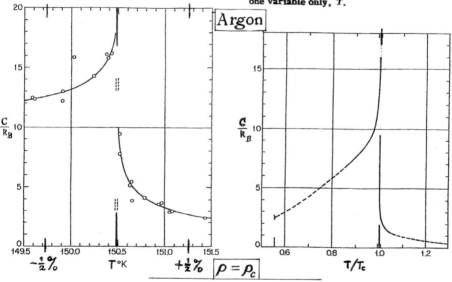

Fig. 21. Title and introductory paragraphs of a paper by C. N. Yang and C. P. Yang stimulated by specific heat measurements by Voronel and coworkers [M. I. Bagatskii, A. V. Voronel, and V. G. Gusak, *Zh. Eksperimi. i Teor. Fiz.* **43**, 728 (1962)] on argon, shown in the graphs, and on oxygen a year later.

But why might this "detail" be significant? The answer is: "Because the orientations of the basic *scaling axes* depend crucially on the behavior of $(d^2\mu_\sigma/dT^2)$." To understand this we define the deviations from critically in the "*space of fields*" (p, μ, T), as

$$\check{p} = \frac{p - p_c}{\rho_c k_B T_c}, \quad \check{\mu} = \frac{\mu - \mu_c}{k_B T_c}, \quad t = \frac{T - T_c}{T_c}. \tag{3}$$

Next notice that the full thermodynamics[22] follows from the functional relation between p, μ and T. Consequently, the presence of asymptotic scaling when $\tilde{p}, \tilde{\mu}$, and $t \to 0$, can be expressed as

$$\Phi\left(\frac{\tilde{p}}{\tilde{t}^{2-\alpha}}, \frac{\tilde{\mu}}{\tilde{t}^{\beta+\gamma}}\right) = 0, \tag{4}$$

where, up to second-order corrections, the three *scaling fields* have the form

$$\tilde{p} = \check{p} - k_0 t - l_0 \check{\mu} + O_2(\cdots), \tag{5}$$

$$\tilde{\mu} = \check{\mu} - k_1 t, \tag{6}$$

$$\tilde{t} = t. \tag{7}$$

Note that the dimensionless constants, k_0, k_1, and l_0 (and, to come below, l_1, j_0 and j_1) serve to specify the *nonuniversal orientation* of the scaling axes for $(\tilde{p}, \tilde{\mu}, \tilde{t})$ in the space of fields, (9.2). Recall, however, that the exponents $\alpha, \beta,$ and γ — for the specific heat [as in (9.1) and (8.4)], for the magnetization or coexistence curve [as in (2.1), (2.2), and (8.4)], and for the susceptibility or compressibility — should be universal.

The coefficient k_1 in (9.6) clearly serves to specify the slope $(d\mu_\sigma/dT_c)$: see dashed line Fig. 22. Studies of models, e.g., by Mermin,[23] established the need for a term $-l_1\check{\mu}$ in (9.7); but, until the question raised by Yang and Yang was considered, that seemed adequate. However, a careful analysis[19] of extensive data

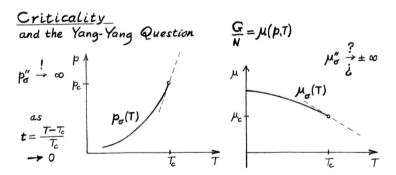

Fig. 22. The Yang–Yang Question posed graphically! In words: "Given that the well-defined pressure locus of condensation $p_\sigma(T)$, exhibits a divergent curvature (i.e., second derivative) at the critical point, is the same true (or not?) for the corresponding chemical potential locus, $\mu_\sigma(T)$? And, if true, what is the appropriate[19] dimensionless Yang–Yang ratio, R_μ, measuring the relative strength of the divergence of $\mu_\sigma''(T)$? Might R_μ even be negative? [The dashed lines at criticality indicate a well-defined slope, i.e., first derivative.]

---

[22] See, e.g., E. A. Guggenheim, *Thermodynamics* (2nd Edn., North Holland Publishing, Co., Amsterdam, 1950), Chap. II and, especially, p. 58; R. A. Sack, *Mol. Phys.* **2**, 8 (1959). The corresponding partition function might be described as "great grand canonical."
[23] N. D. Mermin, *Phys. Rev. Lett.* **26**, 169 (1971).

for propane provided by Abdulagatov et al.[24] — but not available until the late '90s — revealed the existence of a clear "Yang–Yang anomaly," i.e., a divergence of $\mu''_\sigma(T)$ at T_c! Indeed, the Yang–Yang ratio was estimated[19] as $R_\mu = 0.56 \pm 0.04$ (propane). Examination of restricted data[24] suggests a *negative* ratio for CO_2 (with $R_\mu \simeq -0.4$).

Either way it is clear that "pressure mixing" is needed.[19] Thus the previously accepted equations (9.6) and (9.7) must be supplemented to read

$$\tilde{\mu} = \check{\mu} - k_1 t - j_2 \check{p}, \qquad (8)$$

$$\tilde{t} = t - l_1 \check{\mu} - j_1 \check{p}, \qquad (9)$$

with two further amplitudes j_1 and j_2 [as well as higher order terms as in (9.5)].

It is then straightforward to see that the condensation boundaries have the form

$$\mu_\sigma(T) = \mu_c + \mu_1 t + A_\mu |t|^{2-\alpha} + \mu_2 t^2 + \cdots, \qquad (10)$$

$$p_\sigma(T) = p_c + p_1 t + A_p |t|^{2-\alpha} + p_2 t^2 + \cdots. \qquad (11)$$

On returning to the Yang–Yang expression (9.2), one sees that both amplitudes A_μ and A_p specify the divergence (for $\alpha > 0$) of the specific heat at constant volume. The Yang–Yang ratio then follows from $R_\mu = -j_2/(1 - j_2)$.

But a further surprise remains! Sufficiently accurate measurements of the liquid and vapor phase boundaries, $\rho_{\text{liq}}(T)$ and $\rho_{\text{vap}}(T)$, can determine the coexistence curve diameter. Scaling now predicts

$$\rho_{\text{diam}}(T) = \rho_c [1 + c_{1-\alpha} l_1 |t|^{1-\alpha} + c_{2\beta} j_2 |t|^{2\beta} + c_1 t + \cdots], \qquad (12)$$

with unsuspected power-law correction terms, with amplitudes $c_1, c_{2\beta}$, and $c_{1-\alpha}$. The correction proportional to l_1 is of order $|t|^{0.89}$; the surprise is that pressure mixing induces a new term, of order $j_2 |t|^{0.65}$, which dominates that previously known! Clearly, both power-law corrections dominate the anticipated term $c_1 t$, embodied in the *Law of the Rectilinear Diameter*.

10. Yang–Yang Anomaly and "Compressible" Fluid Models

It is natural to ask if there are models that might reveal something about the origins of pressure mixing in the scaling fields as is demanded by Yang–Yang anomalies both observed[19] and subsequently found in precise simulations.[25] To this end consider, first, the standard nearest-neighbor lattice gas (equivalent to a basic Ising model).

---

[24] I. M. Abdulagatov et al., *Fluid Phase Equilib.* **127**, 205 (1997); data for CO_2 by I. M. Abdulagatov et al., *J. Chem. Thermodyn.* **26**, 1031 (1994).

[25] Y. C. Kim, M. E. Fisher, and E. Luijten, *Phys. Rev. Lett.* **91**, 065701 (2003). This work led to $R_\mu \simeq -0.044$ for the hard-core square-well fluid (with $b/a = 1.50$ in Fig. 23) and to $R_\mu \simeq 0.26$ for the restricted primitive model of an electrolyte. See also: G. Orkoulas, M. E. Fisher, and A. Z. Panagiotopoulos, *Phys. Rev. E* **63**, 051507 (2001); Y. C. Kim, M. E. Fisher, and G. Orkoulas, *Phys. Rev. E* **67**, 061506 (2003).

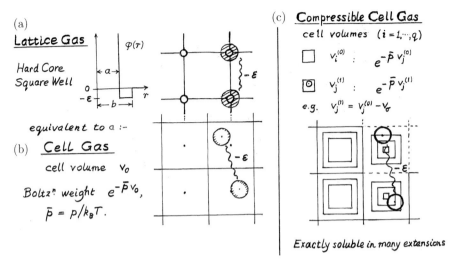

Fig. 23. Ilustration of (a) the standard *lattice gas* and (b) its reinterpretation as a *cell gas* which, in turn, suggests (c) the *compressible cell gas* model which, in general, yields a Yang–Yang anomaly.

As illustrated in Fig. 23(a), this can be regarded as representing a fluid of hard spheres of diameter a interacting via an attractive square well of range b. Provided the lattice spacing is less than b, which, in turn, must be less than the next-nearest-neighbor lattice distance, the only approximation is the restriction of the particles to lattice sites.

But the lattice gas can be viewed instead[26] as a "cell gas" [illustrated in Fig. 23(b)] in which one identifies adjacent lattice *cells* of volume v_0. A cell may be empty or (due to the hard cores) occupied by no more than a single particle that can move freely in its cell. This motion contributes a Boltzmann factor $\exp(-pv_0/k_B T)$ for each particle. The main approximation is that the attractive well comes into play only between particles in nearest-neighbor cells and does not depend on their actual positions within the cells.

How might this model be improved? Can one relax the "rigidity" imposed by the lattice structure? The alternative (c) in Fig. 23 presents one such extension: Even in the absence of particles the lattice volume may fluctuate and respond to pressure: thus individual cells are allowed to assume distinct volumes, $v_i^{(0)}$ ($i = 1, 2, \ldots, q$), that lead to corresponding (but independent) Boltzmann factors $\exp(-pv_i^{(0)}/k_B T)$. The same idea may be used for singly-occupied cells: thus the model supposes that particle motion is restricted to a (possibly) different set of cell volumes, $v_i^{(1)}$. It is natural, but not necessary, to suppose that each $v_i^{(1)}$ is smaller than the associated $v_i^{(0)}$; simplest is to take $v_i^{(1)} = v_i^{(0)} - v_\sigma$ for fixed $v_\sigma > 0$.

---

[26] This way of regarding a lattice gas was learnt from Benjamin Widom (private communication).

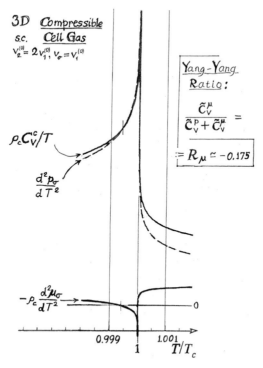

Fig. 24. The components of the Yang–Yang relation (9.2) for a simple compressible cell gas on the simple cubic lattice yielding a negative ratio R_μ.

Now when the original lattice gas can be analyzed (say, numerically) this *Compressible Cell Gas* (or CCG) proves equally soluble! And the same applies to many of its extensions and different versions.[27] As an illustration Fig. 24 presents the simple case where $q = 2$ and $v_2^{(0)} = 2v_1^{(0)}$ while $v_\sigma = v_1^{(0)}$ (the underlying lattice being simple cubic). Evidently the Yang–Yang ratio, R_μ, is now unmistakably *negative*[27] (as, one may recall,[19] seems so for carbon dioxide). However, positive ratios, R_μ, are found when the cell volumes, $v_i^{(1)}$, are coupled with distinct interaction energies, in particular when smaller cells are associated with larger energies.[27]

11. The Richness of Statistical Mechanics

Following his first paper in 1964, C. N. Yang wrote half a dozen further articles with his brother Chen Ping Yang, in the next five years. Many addressed anisotropic Heisenberg spin systems, particularly in one-dimensional chains. Three papers, in

---

[27] R. T. Willis and M. E. Fisher [poster], *Thermo 2005* Meeting, University of Maryland, 29–30 April 2005; C. A. Cerdeiriña, G. Orkoulas, and M. E. Fisher [abstract and poster], *XV Congress of Stat. Phys.*, Royal Spanish Phys. Soc., Salamanca, 27–29 March 2008; [abstract] *Proc. 7th Liquid Matter Conf.*, Lund, June 2008; to be published.

particular; *Phys. Rev.* **150**, 321, 327; **151**, 258 (1966), discussed, extended, and exploited the mathematics originally introduced for Heisenberg spins by Hans Bethe over three decades earlier in 1931.[28] A notable by-product, published with B. Sutherland as first author, reported an: "Exact Solution of a Model of a Two-Dimensional Ferroelectric in an Arbitrary External Electric Field," *Phys. Rev. Lett.* **19**, 588 (1967).

From a more mathematical point of view the analysis also led specifically to the so-called *Yang–Baxter equations*.[29] The first publication where these equations appeared was by C. N. Yang in: "Some Exact Results for the Many Body Problem in One Dimension with Repulsive Delta-Function Interaction," *Phys. Rev. Lett.* **19**, 1312 (1967). But space did not allow proofs of some of the crucial equations; complete details, however, were supplied by[30] M. K. Fung in *J. Math. Phys.* **22**, 2017 (1981). A fuller presentation of the 1D model had to wait over twenty years until the appearance in *Commun. Math. Phys.* **122**, 105 (1989) with first author Gu Chao-Hao, who visited the State University of New York at Stony Brook from the Chinese University of Science and Technology in Hefei. Gu and Yang presented a 1D Fermionic model with an explicitly factorized S matrix. The Yang–Baxter equations for the related matrices X_{ij}^{ab}, namely,

$$X_{ij}^{ab} X_{kj}^{cb} X_{ki}^{ca} = X_{ki}^{ca} X_{kj}^{cb} X_{ij}^{ab}, \tag{1}$$

appeared as consistency conditions for the validity of Bethe's hypothesis.

An exposition of the essential steps in using the hypothesis had been presented in 1970 at a Winter School of Physics in Kapacz in Poland.[3] Much later, on reviewing his "Journey through Statistical Mechanics" in 1987 at the Nankai Institute of Mathematics,[30] Yang pointed out the similarity of the Yang–Baxter equations to the braid group relation ABA = BAB used in the classification of knots. The reader was invited to conclude that perhaps "these cubic equations embody some fundamental structure still to be explored."

After 33 years at SUNY in Stony Brook, Chen Ning Yang retired in 1999. Four years later, in December 2003, he moved to Tsinghua University in Beijing. There he had the opportunity of developing his passion for the history of physics, especially as regards its philosophical and conceptual roots. Recall that, years earlier in 1982, while still in the United States, he[4] had drawn attention to Joseph Mayer's 1937 attack on the statistical mechanics of condensation in fluids. But still the stimulus of new experiments and the challenge of understanding them led once

---

[28] H. A. Bethe, *Zeit. Physik.* **71**, 205–226 (1931).
[29] R. J. Baxter, *Phys. Rev. Lett.* **26**, 832–833 (1971), solved exactly the so-called eight-vertex model: see also *Exactly Solved Models in Statistical Mechanics* (Academic Press, London, New York, 1982), Chaps. 9 and 10; and "Some Academic and Personal Reminiscences of Rodney James Baxter", *J. Phys. A Math. Theor.* **48**, 254001 (2015).
[30] See C. N. Yang, *Selected Papers II, With Commentaries* (World Scientific Publishing Co., Singapore, 2013).

again to work in statistical physics or, essentially indistinguishable at the theoretical level, "condensed matter physics." Note in particular, a paper alone in 2009 and further articles in 2009–2010 concerning fermions in one-dimensional traps with Ma Zhong-Qi and, concerning multicomponent fermions and bosons, with You Yi-Zhuang.[30]

The main focus of the career of Chen Ning Yang — as also true of this meeting — has clearly been on quantum field theory for the understanding of fundamental particles or "high-energy physics." It may, hence, be appropriate to draw attention to the proceedings[31] of a meeting held in Boston University on 1–3 March 1996 concerning the foundations of quantum field theory. The two-day symposium, followed by a workshop, was sponsored by the Center for Philosophy and History of Science at Boston University. Among the speakers were David J. Gross, who featured here in Singapore, and I myself. There were lively discussions and, as the *Preface*[31] put it: "interesting material about the tension between two groups of scholars."

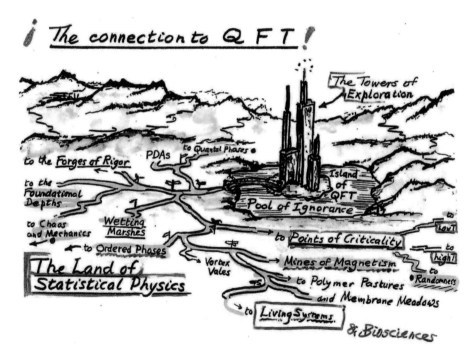

Fig. 25. The wide "Land of Statistical Physics" connected by a now sturdy bridge to the "Island of Quantum Field Theory".

---

[31] T. Y. Cao (editor), *Conceptual Foundations of Quantum Field Theory* (Cambridge University Press, Cambridge, 1999), Presentations 1 to 26 in parts I to X.

C. N. Yang is almost unique in having contributed fundamentally to both the broader aspects of statistical mechanics and to central issues in quantum field theory. In concluding my present contribution, I would like to draw attention to the article by David Nelson entitled: *"What is Fundamental Physics? A Renormalization Group Perspective"* and the subsequent discussion (Sec. 18, pages 264–267, in the proceedings[31]). Finally, I offer Fig. 25 as a pictorial representation of the richness of statistical physics and the connections to quantum field theory.

Quantum Vorticity in Nature

Kerson Huang

*Physics Department, Massachusetts Institute of Technology,
Cambridge, MA 02139, USA*

Quantum vorticity occurs in superfluidity, which arises from a spatial variation of the quantum phase. As such, it can occur in diverse systems over a wide range of scales, from the electroweak sector and QCD of the standard model of particle theory, through the everyday world, to the cosmos. I review the observable manifestations, and their unified description in terms of an order parameter that is a complex scalar field.

1. Overview

Quantum vorticity is a manifestation of superfluidity, which arises from a spatial variation of the quantum phase. This can occur in different types of physical systems, but a unified description can be given in terms of an order parameter that is a complex scalar field:

$$\phi(x) = F(x)e^{i\sigma(x)}. \tag{1}$$

Its existence signals the spontaneous breaking of a $U(1)$ gauge symmetry.

In atomic systems with Bose–Einstein condensation,[1] the broken gauge symmetry is global. In this case, ϕ represents the condensate wave function, which can be described by the Gross–Pitaevskii equation, a nonlinear Schrödinger equation (NLSE):

$$-\frac{\hbar^2}{2m}\nabla^2\phi + (g\phi^*\phi - \mu_0)\phi = i\hbar\frac{\partial\phi}{\partial t}, \tag{2}$$

where m is the mass of the atoms, g an interaction parameter, and μ_0 is the chemical potential.

In superconductivity,[2] the broken gauge symmetry is local, and ϕ obeys a nonlinear Schrödinger equation with gauge coupling:

$$-\frac{\hbar^2}{2m}\left(\nabla - \frac{i\hbar q}{c}\mathbf{A}\right)^2\phi + (g\phi^*\phi - \mu_0)\phi = i\hbar\frac{\partial\phi}{\partial t}, \tag{3}$$

where $q = 2e$ is the charge of Cooper pairs, and $\mathbf{A}$ is a magnetic field. This is called the Ginsburg–Landau (GL) equation.

In relativistic systems,[3] these equations generalize to a nonlinear Klein–Gordon (NLKG) equation:

$$\Box \phi + V'\phi = J, \qquad (4)$$

where $\Box$ is the d'Alembertian in curved spacetime, which may contain gauge couplings, and J denotes a possible external current. The self-interaction potential may be chosen to be a phenomenological ϕ^4 potential:

$$V = \frac{\lambda}{2}(\phi^*\phi - F_0)^2 + V_0. \qquad (5)$$

Thus

$$V' = \lambda(\phi^*\phi - F_0), \qquad (6)$$

where a prime denotes differentiation with respect to $\phi^*\phi$. Physically, ϕ may correspond to a component of the Higgs field in the standard model, the order parameter associated with chiral symmetry breaking in QCD, or Higgs-like fields in grand-unified or supersymmetric theories.

All these systems exhibit superfluidity, with superfluid velocity given by[4]

$$\mathbf{v}_s = \kappa \nabla \sigma, \qquad (7)$$

where $\kappa = -c(\partial \sigma/\partial t)^{-1}$. In the relativistic case, the requirement $|\mathbf{v}_s| < c$ is guaranteed, when $\partial^\mu \sigma$ is timelike, i.e. $\partial^\mu \sigma \partial_\mu \sigma < 0$. In the nonrelativistic limit, $\sigma \to -(mc^2/\hbar)t + \nu$, and $\kappa \to \hbar/m$.

The equations of motions can be rewritten in terms of F and σ, and the equation for σ yields a equation for v_s similar to the classical Euler equation of hydrodynamics, with quantum corrections. However, for many purposes it is simpler to stay with the original forms, particularly numerical analysis.

Quantum vorticity arises from a quantization of the circulation, by virtue of the continuity of ϕ:

$$\oint_C d\mathbf{x} \cdot \nabla \sigma = 2\pi n \quad (n = 0, \pm 1, \pm 2, \ldots), \qquad (8)$$

where C is a closed loop. If $n \neq 0$, the loop encircles a vortex line on which $\phi = 0$. As illustrated in Fig. 1, the field rises from zero at the line to an asymptotic value, with a characteristic healing length ξ. The vortex line can thus be thought of as a tube of radius ξ in which symmetry is restored, with expulsion of the order parameter. The superfluid velocity v_s drops off with distance r from the line center like r^{-1}. Since the energy density of the superfluid flow is proportional to v_s^2, the energy per unit length of a single straight vortex line diverges with the radius of the container. The existence of vorticity depends on a vacuum field, hence on the nonlinearity; it is absent in the usual Schrödinger equation, or Klein–Gordon equation, where the field goes to zero at infinity.

A vortex line cannot terminate, except on boundaries, or upon itself, forming a closed curve. The simplest example of the latter is vortex ring of radius R, as

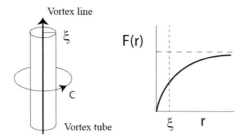

Fig. 1. The vortex tube. The superfluid velocity decreases inversely with distance from the central vortex line.

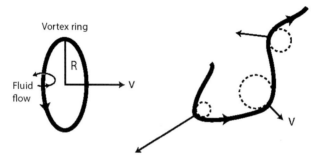

Fig. 2. Left panel: The vortex ring has a translational velocity v approximately inversely proportional to its radius R. Right panel: A general vortex line has local velocity normal to the tangential vortex ring, with magnitude inversely proportional to the radius of curvature.

illustrated in Fig. 2. The ring moves with a translational velocity normal to the plane of the ring $v \sim R^{-1} \ln R$, with energy $E \sim R \ln R$. Thus, roughly speaking, $E \sim v^{-1}$. A vortex line of arbitrary shape will move in such a manner that, at any point of the line, the local velocity is normal to the tangent circle, with velocity inversely proportional to the local radius of curvature. Thus, the line will generally execute a self-driven writhing motion, as illustrated in Fig. 2.

When two vortex lines cross, they will reconnect, as illustrated in Fig. 3. In the final configuration, there appear two cusps, which spring away from each other at great speed, because of the smallness of the radii of curvature. A reconnection thus creates two jets, which represents an efficient way to convert potential energy into kinetic energy in a short time. The reconnection of magnetic flux lines in the sun's corona is believed to create jets of solar flares, as shown in Fig. 3.

Vortex rings can be created in a superfluid with a heat source. They would expand, intersect and reconnect, and reach a steady-state vortex tangle called quantum turbulence. Figure 4[6] shows a computer simulation of the process. The vortex tangle is a new geometrical object, with a fractal of dimension 1.6.[7] When the heat source is removed, the vortex tangle will decay. The dynamics of growth and decay

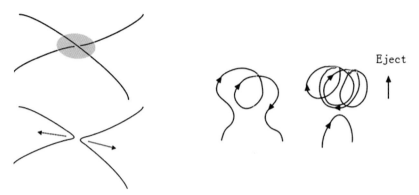

Fig. 3. Left panel: Reconnection of vortex lines creates two cusps that spring away from each other at high velocity, creating high-velocity jets of flow. Right panel: Jets in magnetic reconnections in the sun's corona creates solar flares.

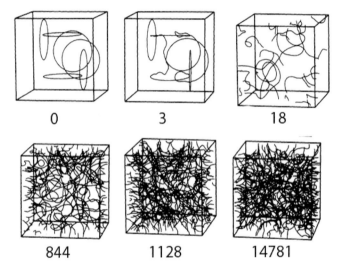

Fig. 4. Emergence of quantum turbulence in computer simulation. Numbers under each picture indicate the number of reconnections (from Ref. 6).

can be described phenomenologically by Vinen's equation[5]

$$\frac{\partial \ell}{\partial t} = A\ell^{3/2} - B\ell^2, \tag{9}$$

where ℓ is the average vortex-line density ℓ (length per unit volume), and A and B are phenomenological parameters, with A governing the growth, and B the decay of the vortex tangle.

In the following, we briefly survey quantum vorticity in diverse physical systems.

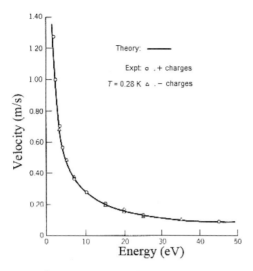

Fig. 5. Velocity-energy curve of vortex ring created by ions shot into liquid helium. The ion (alpha particle or electron) gets trapped in the ring it created. The theory curve is that for a vortex with one unit of quantized circulation.

2. Liquid Helium and Cold Trapped Atomic Gases

Quantized vortices in superfluid helium and atomic gases can be described by the NLSE (2), with superfluid density given by $n = \mu_0/g$, and healing length given by

$$\xi = \frac{\sqrt{g}}{\mu_0}. \tag{10}$$

By shooting alpha particles or electrons into superfluid helium, vortex rings can be created, which trap the projectiles.[8] The velocity-energy curve of the composite object can be obtained by dragging it cross the liquid against an electric field. The results fit that of a quantized vortex ring, as shown in Fig. 5.

In superfluid helium and cold condensed atomic gases, vortices can be created by rotating the container. When the angular velocity of the container exceeds a critical value, the superfluid responds by developing vortex lines parallel to the axis of rotation.[9,10] Experimental results in a trapped atomic gas are shown in Fig. 6, showing the development of a vortex lattice as the angular frequency increases.[11]

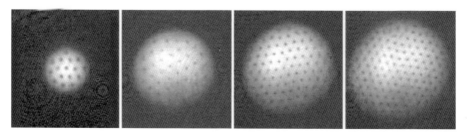

Fig. 6. Lattice of vortices in rotating cold trapped atomic gas, at increasing angular velocity of rotation (from Ref. 11).

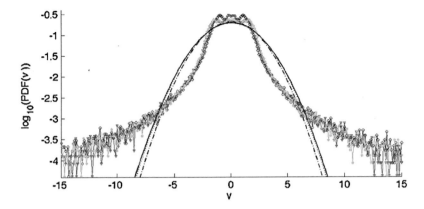

Fig. 7. Velocity distribution in quantum turbulence in superfluid helium (data points), as compared with the Gaussian distribution of classical turbulence (solid curves). The former has a v^{-3} tail due to the jets of large velocities accompanying vortex reconnections, which are essential for the maintenance of quantum turbulence (from Ref. 13).

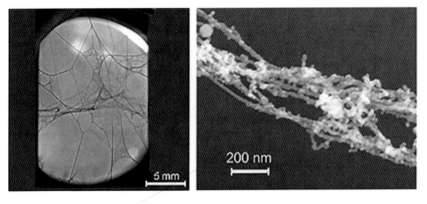

Fig. 8. Vortices in superfluid helium made visible by coating the surface of the vortex with metallic dust of nano size. The adhesion is due to Bernoulli pressure (from Ref. 15).

Quantum turbulence,[13] as well as vortex reconnections,[12] have been studied in superfluid helium. The velocity distribution in quantum turbulence is found to have a power-law tail v^{-3}, as shown in Fig. 7. This is quite different from the Gaussian distribution of classical turbulence, and is due to the occurrence of large velocities in vortex reconnections. Computer simulations using the NLSE have reproduced this distribution.[14]

Vortex lines in liquid helium have been made visible by coating them with nano-sized metallic dust, as shown in Fig. 8.[15] The metallic dust adheres to the surface of the vortex tube because of the Bernoulli pressure arising from a decrease of superfluid velocity away from the core.

The computer simulations in Fig. 9[16] demonstrates that a Bose condensate far from equilibrium creates quantum turbulence, and relaxes through its decay.

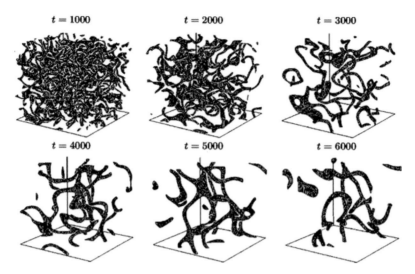

Fig. 9. Computer simulation of the evolution of a Bose condensate far from equilibrium, showing that it equilibrates through the creation and decay of quantum turbulence. These pictures show the decay sequence (from Ref. 16).

3. Superconductivity

In superconductivity, as described by the GL equation (3), there is another length besides the healing length ξ, the penetration depth d. This arises from the conserved current

$$\mathbf{J} = \frac{\hbar q}{2mi}(\phi\nabla\phi^* - \phi^*\nabla\phi) - \frac{q^2}{mc}\phi^*\phi\mathbf{A}. \tag{11}$$

When this is substituted into Maxwell's equation $\nabla \times \nabla \times \mathbf{A} = \frac{4\pi}{c}\mathbf{J}$, we obtain, in Coulomb gauge $\nabla \cdot \mathbf{A} = 0$,

$$\left(\nabla^2 + \frac{4\pi n q^2}{mc^2}\right)\mathbf{A} = 0 \tag{12}$$

for uniform $\phi^*\phi = n$. The second term in the brackets is the photon's squared mass inside a superconductor. An external magnetic field can penetrate a superconductor only to a depth d corresponding to the inverse mass:

$$d = \sqrt{\frac{mc^2}{4\pi n q^2}}. \tag{13}$$

This is the Meissner effect, a simple example of the Higgs mechanism, i.e. spontaneous breaking of local gauge invariance gives mass to the gauge particle.

The presence of two characteristic lengths leads two types of superconductors: If $\xi > d$, we have type I behavior: the magnetic field is completely repelled from the interior. If $\xi < d$ we have type II behavior: the magnetic field penetrates the body in vortex tubes containing quantized magnetic flux hc/e, arranged in a regular lattice know as the Abrikosov lattice. We can understand these behavior by referring to

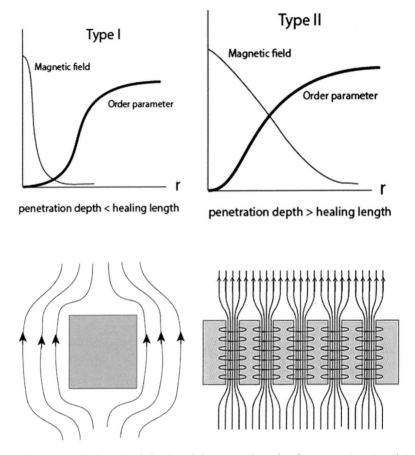

Fig. 10. Upper panels show the behavior of the magnetic and order parameter at an interface, for type I and type II superconductors. Lower panels illustrate the expulsion of magnet flux in type I superconductors, and the formation of a quantized-flux lattice in type II case.

Fig. 10, which depicts the interface between magnetic field and order parameter. In the type I case, there is little overlap between magnetic field and order parameter. If a flux tube were formed, the system can lower the energy by expelling it outside, and filling in the hole in the order parameter. For type II, an overlap between magnetic field and order parameter lowers the energy, creating a negative surface tension. Thus, flux tubes are formed, penetrating the superconducting body, and maintained by solenoidal supercurrents. If the medium is infinite, the flux cannot be expelled, and finite-energy configurations are flux tube of finite length, terminated by magnetic monopoles. In this regard, see Fig. 11 for a comparison between the magnetic field due to two monopoles in vacuum and in a superconductor. If the tube is cut, one does not get free monopoles, but more flux tubes terminating in N and S. In this sense, a superconductor is a medium of magnetic-monopole confinement.

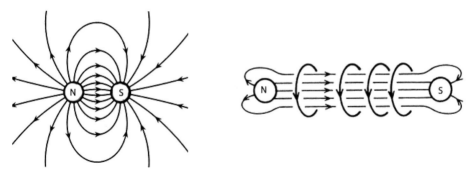

Fig. 11. The magnetic field of a magnetic dipole in empty space (left) and in a superconductor (right). In the latter the magnetic flux is squeezed into a flux tube by circulation supercurrents. The situation suggests that a superconductor would exhibit magnetic monopole confinement.

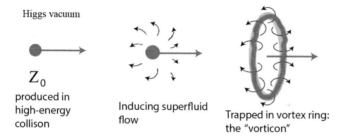

Fig. 12. A Z_0 vector boson created in a high-energy collision induces superfluid flow in the background Higgs field, creating, and is trapped by, of a vortex ring, resulting in a unstable particle dubbed the "vorticon," with a mass of approximately 3 TeV, and a lifetime of the order of 3×10^{-25} s.

4. "Vorticons" in the Higgs Field

In the electroweak sector of the standard model, the local gauge symmetry is spontaneously broken by of the multi-component Higgs field, which gives mass to the gauge bosons $W_\pm$ and Z_0, while leaving the photon massless. We can envisage a "vorticon,"[17] a microscopic vortex ring in the Higgs field, in the shape of a donut, containing a gauge boson that is massless, as illustrated in Fig. 12. It is a microscopic donut-shaped waveguide, which could be created when an energetic Z_0, say, tears through the Higgs field, much as a projectile can create and be trapped by a vortex ring in liquid helium.

The mass of a vorticon can be estimated by constructing normal modes of the Z_0 field inside a torus cut out from the background Higgs field, and minimizing the energy of the lowest mode.[17] The standard-model Hamiltonian in the Z_0 sector is, with $\hbar = c = 1$,

$$H = \int d^3x \left[\frac{1}{2}(\mathbf{B}^2 + \mathbf{E}^2) + |(\nabla - iq\mathbf{Z})\phi|^2 + V(\phi)\right]. \qquad (14)$$

Here, ϕ is the Higgs field, $\mathbf{Z}$ the vector potential in Coulomb gauge ($\nabla \cdot \mathbf{Z} = 0$), and $\mathbf{B} = \nabla \times \mathbf{Z}$, $\mathbf{E} = -\partial \mathbf{Z}/\partial t$. The Higgs potential is given by

$$V(\phi) = \frac{\lambda}{2}(\phi^*\phi - F_0^2)^2 \qquad (15)$$

with

$$\begin{aligned} F_0 &= 174 \text{ GeV}, \\ \sqrt{2\lambda}F_0 &= m_H = 125 \text{ GeV}, \end{aligned} \qquad (16)$$

where m_H is the Higgs mass. Thus $\lambda = 0.256$. The gauge coupling constant q and Z_0 mass are given by

$$q = -\frac{e}{\sin 2\theta_W}, \qquad (17)$$

$$m_Z = \frac{gF_0}{\sqrt{2}\cos\theta_W}, \qquad (18)$$

where e is given through the fine-structure constant $e^2/4\pi \approx 1/137$, and θ_W is the Weinberg angle given through $\sin^2\theta_W \approx 1/4$.

There are two types of vorticons: magnetic and electric, with the magnetic (electric) field pointing along the toroidal direction. As in electromagnetic wave guides, there is no completely transverse mode. The masses are found to be

$$\begin{aligned} \frac{M_{\text{mag}}}{m_Z} &= 35.3 + 6.42\left(\frac{m_H}{m_Z}\right)^{2/3} - 1.03\left(\frac{m_H}{m_Z}\right)^{1/2}, \\ \frac{M_{\text{elec}}}{m_Z} &= 27.7 + 5.66\left(\frac{m_H}{m_Z}\right)^{2/3} + 0.504\left(\frac{m_H}{m_Z}\right)^{1/2}. \end{aligned} \qquad (19)$$

With the experimental values $m_Z = 91$ GeV and $m_H = 125$ GeV, we have

$$\begin{aligned} M_{\text{mag}} &= 3.47 \text{ TeV}, \\ M_{\text{elec}} &= 2.88 \text{ TeV}. \end{aligned} \qquad (20)$$

The size of these vorticons are of order $m_Z^{-1} \approx 10^{-12}$ cm. They are unstable, with lifetimes of the order the Z_0 lifetime 3×10^{-25} s.

5. QCD Strings

A phenomenological description of quark confinement in QCD can be modeled after the magnetic monopole confinement in a superconductor. The difference is that quarks generate Yang–Mills color-electric fields instead of magnetic fields, so the QCD vacuum confines electric flux instead of magnetic flux. (In the nonlinear Yang–Mills gauge theory, there is no electric–magnetic duality as in Maxwell theory.) A meson consists of a quark and antiquark pair, belonging respectively to the representations **3** and **3̄** of color $SU(3)$, connected by a flux tube. This can be represented by the picture in Fig. 11, if we substitute **3** for N, **3̄** for S. The order

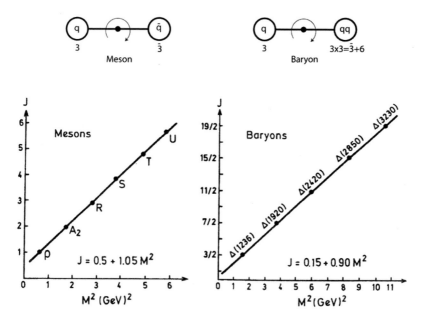

Fig. 13. Hadrons modeled as strings terminated in quarks. The strings are idealization of vortex tubes containing color-electric flux. Their rotational angular momentum J and squared mass M^2 bear a linear relation that agrees with the observe Regge trajectory of particles, and show the string has a tension of about 16 tons.

parameter corresponds to a condensate of quark–antiquark pairs, and the electric flux tube is maintained by solenoidal color-magnetic currents. A baryon made of three quarks can be represented by a flux tube with one quark at one end, at two at the other. The two-quark system has color representations $\mathbf{3} \times \mathbf{3} = \mathbf{\bar{3}} + \mathbf{6}$, and we can use the $\mathbf{\bar{3}}$ irreducible combination. Alternatively, the baryon could be represented by Y-shaped flux tubes each terminated by $\mathbf{3}$.

One can approximate the flux tube by a massless string terminated at both ends by quarks, with the string rotating about a perpendicular axis, under tension T_0,[18] as illustrated in the upper panel of Fig. 13. In the limit of small quark masses, one obtains a relation between J and M:

$$J = \alpha_0 + \alpha' M^2, \qquad (21)$$

where α_0 is a model parameter, and

$$\alpha' = \frac{1}{2\pi T_0}. \qquad (22)$$

As shown in Fig. 13, J–M^2 plots of meson and baryons do yield straight lines known as Regge trajectories, with a universal slope $\alpha' \approx 0.9 \, \text{GeV}^{-2}$. This corresponds to a string tension of approximately 16 tons.

It would be desirable to set up an analog of the GL equation for QCD, but the situation here is more complicated than superconductivity, because of the intrinsic nonlinearity of Yang–Mills gauge fields. Besides quark confinement, there is chiral symmetry breaking. The condensate could involve topological object such as monopoles. We refer the reader to Refs. 19–21, for works related to that goal.

6. Cosmology

The Higgs field of particle theory has found experiment support with the discovery of the Higgs boson.[22,23] Grand-unified or supersymmetric theories call for more Higgs-like fields. Any long-range order, such as the phase coherence of these fields, will persist to infinite distances in spatial dimension greater than two. In two or fewer dimensions, the long-range order will be destroyed by the long-wavelength fluctuations of the Goldstone modes.[24] The Higgs field, and other possible Higgs-like fields, thus make the whole universe a superposition of different types of superfluid, (like a mixture of ^{4}He and ^{3}He below 10^{-3} K.) We discuss the manifestations of superfluidity in terms of a generic complex scalar field, in three different epochs of the universe: the big-bang era, the CMB (cosmic microwave background) era, and the present.

The scalar field presumably emerges during the big bang, but how it does this depends on model. A quantum field needs a high-momentum cutoff Λ, which should be infinite at the big bang. If the potential emerges from zero at that moment, it must be asymptotically free. This rules out all polynomial forms, and admits only the Halpern–Huang potential,[25] with exponential behavior for large fields. A big-bang model has been constructed based on this potential,[26,27] with uniform Robertson–Walker metric whose scale is tied to the cutoff: $a = \Lambda^{-1}$. Solving Einstein equation numerically, as an initial-value problem, shows that the length scale expands like

$$a(t) \sim \exp(c_0 t^{1-p}), \tag{23}$$

where c_0 and $p < 1$ depend on model parameters and initial data. The result is equivalent to having a cosmological constant that decays in time according to a power law. If this behavior persists to later times, it could explain dark energy without "fine-tuning." Phenomenological studies so based are referred to as "intermediate inflation."[28]

During the big bang, the scalar field would emerge far from equilibrium, and, like the Bose gas illustrated in Fig. 9, equilibrate by going through a period of quantum turbulence. The vortex reconnections that maintain the vortex tangle produce jets of energy that can create matter, like the solar flares created by magnetic reconnections illustrated in Fig. 3. This provides a new scenario for the inflation era, in which all matter in the present universe were created in quantum turbulence.

A detailed analysis of the turbulent era based on the coupled Einstein-scalar equations, unfortunately, faces the formidable problem of a nonuniform metric. We

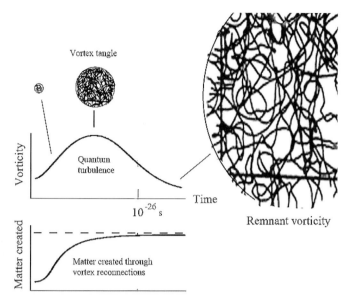

Fig. 14. The inflation era: creation of matter all matter the vortex reconnections in quantum turbulence (Ref. 27).

can bypass this difficulty with a phenomenological approach using Vinen's equation, which deals with a uniform distribution of vortex lines. It is then possible to construct a scenario for the growth and decay of a vortex tangle, as illustrated in Fig. 14. Parameters can be so chosen that the lifetime of the vortex tangle is of the order of 10^{-26} s, during which time the universe expanded by a factor of 10^{27}, and a total amount of matter was created equal to what is presently observed, roughly 10^{22} suns. This presents a new picture of inflation.[27]

The big-bang era, including inflation and the creation of all matter through quantum turbulence, lasted about 10^{-26} s. After this, nonuniformities in the universe become important, and the Robertson–Walker metric ceases to be valid. The Halpern–Huang potential, which is a high-cutoff approximation, may need corrections. Also, while there was only one scale in the big-bang era, the Planck scale of 10^{22} GeV, a new scale appears with the emergence of matter, the QCD scale of 1 GeV. Vortices created during the big-bang era, including those in the vortex tangle, have core sizes of the order of the Planck length of 10^{-33} cm, and they will co-expand with the universe. Those created after the emergence of the matter, however, will have much smaller fixed core sizes that do not expand with the universe.

After the big-bang era, our model will be replaced by a cosmic perturbation theory, which must include the scalar field, and this will govern the creation of the CMB. The widely used ΛCDM model would be modified with inclusion of a vacuum complex scalar field, which supplies a cosmological constant (Λ) via its energy–momentum tensor, and generates cold dark matter (CDM) through density

212   K. Huang

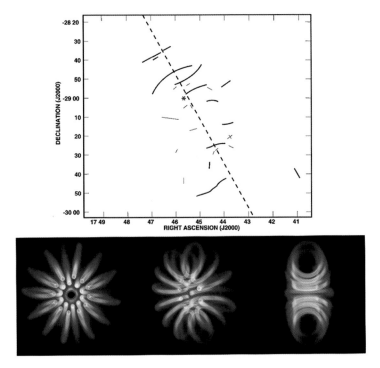

Fig. 15.   Upper panel: The "nonthermal filaments" observed near the center of the Milky Way, Ref. 34, could be vortex lines surrounding black holes. Lower panel: 3D simulations of a vortex-ring assembly arising from a rotating body at the center, based on the NLKG, in various perspectives, Ref. 3.

variations of the scalar field. A complete reformulation is yet to be done, but we expect that the results of the usual cosmic perturbation theories will be largely preserved. The only new element would be quantum vorticity, which will make contributions to the tensor mode, in competition with gravitational waves.

In the present universe, galaxies will attract superfluid to form halos of denser-than-vacuum superfluid around it, and these will be perceived as dark matter, through gravitational lensing.[4] When galaxy clusters collide, their haloes undergo distortions according to superfluid hydrodynamics, as observed in the "bullet cluster."[29] Computer simulations based on the NLKG may be found in Ref. 4. The literature contains studies of cosmic vorticity under the name "cosmic strings."[32,33]

A fast-rotating body such as a black hole will drag the surrounding superfluid into rotation through the creation of vortices. The so-called "nonthermal" filaments[34] observed near the center of the Milky Way could be vortex lines, as shown in the upper panel of Fig. 15. The lower panels shows 3D simulations based on the NLKG.[3]

The core sizes of remnant vortices from quantum turbulence in the big bang era, originally of order of the Planck size of 10^{-33} cm, will co-expand with the universe, and in the 13.7 billion years since, they could reach sizes of the order of

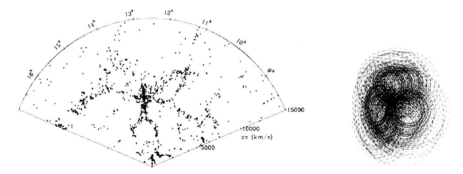

Fig. 16. Left: Voids in galaxy distribution in the "stick man" configuration, Ref. 35. Right: Simulation by the superposition of three primordial vortex tubes, which have co-expanded with the universe to gigantic sizes. The inside of the tubes are devoid of galaxies, which cling to the outside dues to hydrodynamic pressure, like the metallic dust clinging to vortex tubes in superfluid helium shown in Fig. 8.

Fig. 17. Lathrop's suggestion for galaxy formation: dust particles accrete onto a vortex ring, gravitate and clump, squeezing the ring into a spiral shape with a central mass, which would become a black hole.

10^7 light years. Since matter was created in the scalar field, these cores are devoid of matter, and show up as voids in the galactic distribution. The so-called "stick man" configuration[35] is shown in Fig. 16, with a simulation by superposition of vortex tubes.

7. Galaxy Formation

The CMB was formed about 10^5 yrs after the big bang. Between that and the present, there was a long period of galactic formation, in which quantum vorticity may play an important role. A speculation of Lathrop[36] is that a galaxy can form from a large vortex ring with accretion of dust, which gravitates to form a central mass, squeezing the vortex ring into a shape with spiral arms, as illustrated in Fig. 17. Can one find any hint of this mechanism in currently-observed galactic properties?

The central mass would correspond to the black hole observed at the center of all galaxies, whose mass M bears power-law relations to other galactic properties[37]:

$$M \sim X^\beta \qquad (24)$$

as indicated in the following:

	X	β
Stellar mass	m	1.05 ± 0.11
Luminosity	L	1.11 ± 0.13
Stellar velocity	v	5.57 ± 0.33

(25)

Assume that the dust particles initially accrete onto the vortex ring uniformly, and then clump up under gravitation. Assume further that a fixed fraction forms the central mass, which becomes a black hole and goes dark, while the rest remains luminous. This would mean $m \propto M$, and $L \propto m$, which are consistent with the relevant exponents being unity.

The initial vortex ring may be generated by self-avoiding random walk (SAW). (The fact that it is a close ring matters little for the arguments here.) Thus, the dust particles, which initially adhere uniformly to the vortex ring, form a SAW sequence, with the relation $N \sim R^{5/3}$, where R is the spatial extension, and N the number of steps. In our case, R corresponds to the size of the galaxy, and N is proportional to $M + m$, hence to M. This means that M scales like $R^{5/3}$. Now assume that the total angular momentum J, which is a constant of the motion, scales as the galactic volume:

$$M \sim R^{5/3},$$
$$J \sim R^3.$$

(26)

Defining the stellar velocity v through $J = Rmv \propto RMv$, we obtain

$$M \sim v^5$$

(27)

which is not inconsistent with observations. As an interesting note, the SAW exponent 5/3 is the same as the Kolmogorov exponent in classical turbulence, and the Flory exponent for polymers.[38]

References

1. F. Dafolvo, S. Giorgini, L. D. Pitaevskii and S. Stringari, *Rev. Mod. Phys.* **71**, 463 (1999).
2. P. G. de Gennes, *Superconductivity of Metals and Alloys* (Westview Press, Boulder, 1999).
3. C. Xiong, M. Good, X. Liu and K. Huang, *Phys. Rev. D* **90**, 125019 (2014), arXiv:1408.0779.
4. K. Huang, C. Xiong and X. Zhao, *Int. J. Mod. Phys. A* **29**, 1450074 (2014), arXiv:1304.1595.
5. S. K. Nemirovskii and W. Fizdon, *Rev. Mod. Phys.* **67**, 37 (1995) 013603 (2002).
6. K. W. Schwarz, *Phys. Rev. B* **38**, 2398 (1988).
7. D. Kivotides, C. F. Barenghi and D. C. Samuels, *Phys. Rev. Lett.* **87**, 155301 (2001).

8. G. W. Rayfield and F. Reif, *Phys. Rev.* **136**, A1194 (1964).
9. R. J. Donelly, *Quantized Vortices in Helium II* (Cambridge University Press, 1991).
10. A. L. Fetter, *Rev. Mod. Phys.* **81**, 647 (2009).
11. Abo-Shaeer, I. Raman and W. Ketterle, *Phys. Rev. Lett.* **88**, 070409 (2002).
12. M. S. Paoletti, M. E. Fisher and D. P. Lathrop, *Physica D* **239**, 1367 (2010).
13. M. S. Paoletti, M. E. Fisher, K. R. Sreenivasan and D. P. Lathrop, *Phys. Rev. Lett.* **101**, 154501 (2008).
14. A. C. White, C. F. Barenghi, N. P. Proukakis, A. L. Youd and D. H. Hawks, *Phys. Rev. Lett.* **104**, 075301 (2010).
15. V. Lebedev, P. Moroshkin, B. Grobey and A. Weis, *J. Low Temp. Phys.* **165**, 166 (2011).
16. N. G. Berlof and B. V. Svistunov, *Phys. Rev. B* **66**, 013603 (2002).
17. K. Huang and R. Tipton, *Phys. Rev.* **23**, 3050 (1981).
18. K. Johnson and C. Nohl, *Phys. Rev. D* **19**, 291 (1979).
19. K. Huang, *Quarks, Leptons and Gauge Fields*, 2nd edn. (World Scientific, Singapore, 1992), Chap. 14.
20. V. Gusynin and V. A. Miranskii, *Zh. Eksp. Teor. Fiz.* **101**, 414 (1992) [*Sov. Phys. JETP* **74**, 216 (1992)].
21. C. Xiong, *Phys. Rev. D* **88**, 025042 (2013).
22. ATLAS Collab., *Phys. Lett. B* **716**, 1 (2012).
23. CMS Collab., *Phys. Lett. B* **716**, 30 (2012).
24. K. Huang, *Statistical Mechanics*, 2nd edn. (Wiley, New York, 1987), Sec. 16.7.
25. K. Halpern and K. Huang, *Phys. Rev.* **53**, 3252 (1996).
26. K. Huang, H. B. Low and R. S. Tung, *Class. Quantum Grav.* **29**, 155014 (2012), arXiv:1011.4012.
27. K. Huang, H. B. Low and R. S. Tung, *Int. J. Mod. Phys. A* **27**, 1250154 (2012), arXiv:1106.5283.
28. J. Barrow, *Phys. Rev. D* **51**, 2729 (1995).
29. D. Crowe *et al.*, *Astrophys. J.* **648**, 806 (2006).
30. B. Kain and H. Y. Ling, *Phys. Rev. D* **82**, 064042 (2010).
31. V. C. Rubin, Thornnard and W. K. J. Ford, *Astrophys. J.* **238**, 471 (1980).
32. B. Gradwohl, G. Kalbermann, T. Piran and E. Bertschinger, *Nucl. Phys. B* **338**, 371 (1990).
33. A. Villenkin and E. P. S. Shellard, *Cosmological Strings and Other Topological Defects* (Cambridge University Press, 1994).
34. T. N. LaRosa *et al.*, *Astrophys. J.* **607**, 302 (2004).
35. V. de Lapparent, M. J. Geller and J. P. Huchra, *Astrophys. J.* **302**, L1 (2004).
36. D. P. Lathrop, to be published.
37. J. M. Nicolas and C. P. Ma, arXiv:1211.2816.
38. K. Huang, *Lectures on Statistical Physics and Protein Folding* (World Scientific, Singapore, 2005), Chaps. 14 and 15.

Yang–Mills Theory and Fermionic Path Integrals

Kazuo Fujikawa

RIKEN Nishina Center, Wako 351-0198, Japan
k.fujikawa@riken.jp

The Yang–Mills gauge field theory, which was proposed 60 years ago, is extremely successful in describing the basic interactions of fundamental particles. The Yang–Mills theory in the course of its developments also stimulated many important field theoretical machinery. In my talk I discuss the path integral techniques, in particular, the fermionic path integrals which were developed together with the successful applications of quantized Yang–Mills field theory. I start with the Faddeev–Popov path integral formula with emphasis on the treatment of fermionic ghosts as an application of Grassmann numbers. I then discuss the ordinary fermionic path integrals and the general treatment of quantum anomalies. The contents of this talk are mostly pedagogical except for a recent analysis of path integral bosonization.

1. Yang–Mills Theory

We start with the quantization of Yang–Mills theory.[1] The quantization of Yang–Mills theory stimulated the developments of basic machinery in quantum field theory such as the path integral,

$$\int d\mu \exp\left[\frac{i}{\hbar}\int d^4x \mathcal{L}_{\text{eff}}\right] \quad (1)$$

with the effective Lagrangian $\mathcal{L}_{\text{eff}}$

$$\mathcal{L}_{\text{eff}} = -\frac{1}{4}F^a_{\mu\nu}F^{a\mu\nu} + B^a(x)\partial^\mu A^a_\mu - i\bar{c}^a(x)\partial^\mu$$
$$\times [\partial_\mu c^a(x) + gf^{abc}A^b_\mu(x)c^c(x)] + \frac{\xi}{2}(B^a(x))^2 \quad (2)$$

and the integration measure

$$d\mu = \mathcal{D}A^a_\mu \mathcal{D}B^a \mathcal{D}\bar{c}^a \mathcal{D}c^a. \quad (3)$$

We adopted the ξ-*gauge* which is a generalization of Lorentz gauge

$$\partial^\mu A^a_\mu(x) = 0. \quad (4)$$

Fermionic scalars $c^a(x)$ and $\bar{c}^a(x)$ are the Faddeev–Popov **ghost** and **antighost**,[2] respectively, expressed by **Grassmann** numbers.[3]

The BRST symmetry of $\mathcal{L}_{\text{eff}}$ is defined using a Grassmann parameter λ

$$A_\mu^a(x) \to A_\mu^a(x) + i\lambda(\partial_\mu c^a(x) + gf^{abc}A_\mu^b(x)c^c(x)),$$

$$c^a(x) \to c^a(x) - i\lambda \frac{g}{2} f^{abc} c^b(x) c^c(x),$$

$$\bar{c}^a(x) \to \bar{c}^a(x) + \lambda B^a(x),$$

$$B^a(x) \to B^a(x). \tag{5}$$

Note that $c^a(x)^\dagger = c^a(x)$, $\bar{c}^a(x)^\dagger = \bar{c}^a(x)$ and $\lambda^\dagger = \lambda$. See Refs. 4 and 5.

Quick derivation of path integral

To understand the above path integral formula, I start with a definition of the path integral following the idea of *Schwinger's action principle*:

$$\frac{\hbar}{i} \frac{\delta}{\delta J(x)} \langle 0, +\infty | 0, -\infty \rangle_J = \langle 0, +\infty | \hat{\varphi}(x) | 0, -\infty \rangle_J \tag{6}$$

where $\langle 0, +\infty | 0, -\infty \rangle_J$ is the vacuum-to-vacuum transition amplitude in the presence of the source $J(x)$, which is an (infinitesimal) c-number quantity. The solution of this functional equation is Feynman's path integral,

$$\langle 0, +\infty | 0, -\infty \rangle_J = \int \mathcal{D}\varphi \exp\left\{ \frac{i}{\hbar} \int d^4 x \mathcal{L}_J \right\} \tag{7}$$

with

$$\mathcal{L}_J = \mathcal{L} + \varphi(x) J(x). \tag{8}$$

The basic requirement of the path integral measure is

$$\mathcal{D}(\varphi + \epsilon) = \mathcal{D}\varphi \tag{9}$$

for an arbitrary infinitesimal function $\epsilon(x)$.

To understand this formal solution, one may start with an identity

$$\int \mathcal{D}\varphi \exp\left\{ \frac{i}{\hbar} \int d^4 y \mathcal{L}_J(\varphi) \right\}$$

$$= \int \mathcal{D}(\varphi + \epsilon) \exp\left\{ \frac{i}{\hbar} \int d^4 y \mathcal{L}_J(\varphi + \epsilon) \right\}$$

$$= \int \mathcal{D}\varphi \exp\left\{ \frac{i}{\hbar} \int d^4 y \mathcal{L}_J(\varphi) \right\}$$

$$+ \int \mathcal{D}\varphi \frac{i}{\hbar} \int d^4 x \epsilon(x) \left(\frac{\delta}{\delta \varphi(x)} \int d^4 y \mathcal{L}_J \right) \exp\left\{ \frac{i}{\hbar} \int d^4 y \mathcal{L}_J(\varphi) \right\} \tag{10}$$

where we used (9) in the third line. Since $\epsilon(x)$ is arbitrary

$$\left\langle 0, +\infty \left| \frac{\delta}{\delta\hat{\varphi}(x)} \int d^4y \hat{\mathcal{L}}_J \right| 0, -\infty \right\rangle J = \frac{\hbar}{i} \int \mathcal{D}\varphi \frac{\delta}{\delta\varphi(x)} \exp\left\{ \frac{i}{\hbar} \int d^4y \mathcal{L}_J \right\} = 0, \tag{11}$$

namely, an integration of a total derivative vanishes. Thus the *quantized equation of motion*

$$\frac{\delta}{\delta\hat{\varphi}(x)} \int d^4y \hat{\mathcal{L}}_J = 0 \tag{12}$$

is satisfied.

Coming back to the basic requirement of the translation invariant measure in functional space (9), we can satisfy the condition using the ordinary complex field and ordinary integral for a Bose field at all the space–time points

$$\mathcal{D}\varphi = \prod_x d\varphi(x). \tag{13}$$

For a Fermi field, one can use the *left-derivative* with respect to Grassmann-valued function

$$\mathcal{D}\psi = \prod_x \frac{\delta}{\delta\psi(x)}. \tag{14}$$

Here *Grassmann numbers* are defined as anti-commuting "numbers" at all space–time points

$$\psi(x)\psi(y) + \psi(y)\psi(x) = 0, \tag{15}$$

in particular,

$$\psi(x)\psi(x) = 0 \tag{16}$$

and thus no magnitude. Grassmann number valued fields become Fermi particles after quantization, but they are not defined as *classical waves* before quantization.

Spin-statistics theorem

To understand the naming "ghost", we need to know the spin-statistics theorem which states that a particle with a spin half-odd integer follows the Fermi statistics and a particle with a spin integer follows the Bose statistics. This theorem is usually discussed in the operator formalism, but it is also analyzed in the framework of path integral. I briefly explain the basic idea following Ref. 6.

In a Lorentz invariant local field theory, the theorem holds if following conditions are satisfied (W. Pauli):

1. Positive energy condition,
2. Causality,
3. No negative probability.

(1) In the path integral, the positive energy condition is satisfied by *Feynman's* $m - i\epsilon$ *prescription*, which dictates that positive energy solutions propagate forward in time and negative energy solutions propagate backward in time, and thus the net energy flow in any fixed time slice is always positive. This prescription thus ensures the positive energy condition regardless of the assignment of statistics.

(2) Causality, namely, fields either commute or anti-commute at space-like separation, is ensured in Lorentz invariant theory by any Green's functions defined in terms of time ordered products of complex number valued fields or Grassmann number valued fields. This is explicitly checked using the Bjorken–Johnson–Low prescription.

(3) Thus the spin-statistics relation is determined by the positive norm condition, which is examined by looking at the residue at the pole of the propagator. This analysis then excludes the *abnormal* spin-statistics relation, namely, Fermi statistics for both spin-0 scalar (and similarly spin-1 vector) particles and Bose statistics spin-1/2 Dirac particles.

In short, the Dirac equation leads to Fermi statistics, and the Klein–Gordon-type equation leads to Bose statistics independent of space-time dimensionality. The Faddeev–Popov particle, which is described by the Klein–Gordon-type equation and yet by anti-commuting Grassmann numbers, is thus classified as "ghost".

2. Quantum Anomalies

As is well known, the notion of anomalies was established by Bell and Jackiw[7] and Adler[8] by clarifying the physical meanings of quantum anomalies. As for reviews of quantum anomalies, see Refs. 9–13.

We here briefly review the history of the subtle phenomenon of quantum (triangle or chiral) anomaly. The anomalous behavior in the neutral pion decay was first recognized by Fukuda and Miyamoto[14] soon after the establishment of modern renormalizable quantum field theory. This anomalous behavior was further analyzed by Steinberger[15] and Tomonaga[16] and his associates. In the footnote 11) in the paper of Steinberger, he mentions that he learned the paper of Fukuda and Miyamoto through H. Yukawa, who was staying at Princeton at that time. Tomonaga apparently learned the anomalous behavior more directly from Fukuda and Miyamaoto, who belonged to the research group of Tomonaga at that time. This issue of gauge invariance was also examined by Schwinger,[17] but its real physical significance was not clear for about 20 years.

At about the same time of Bell and Jackiw and Adler, the mathematicians, Atiyah and Singer,[18] formulated the famous Atiyah–Singer index theorem. In the history of quantum anomalies, physicists were thus ahead of mathematicians for about 20 years. In 1982, I encountered a topological anomaly related to the ghost number symmetry in the first quantization of the bosonic string theory.[19] This anomaly was later recognized to be related to the Riemann–Roch theorem in the theory of Riemann surfaces.[20] If one includes the ghost number anomaly into anomalies, it may be said that mathematicians in the 19th century knew the quantum anomaly.

2.1. *Path integrals and anomalies*

Following the suggestion of organizers, I would like to describe the essence of the path integral formulation of quantum anomalies.[21] We consider the QCD-type theory (in Euclidean space)

$$\mathcal{L} = \bar{\psi}[i\slashed{D} - m]\psi + \mathcal{L}_{\text{YM}} \qquad (17)$$

but we concentrate on fermionic path integral;

$$Z = \int \mathcal{D}\bar{\psi}\mathcal{D}\psi \exp\left\{\int \bar{\psi}[i\slashed{D} - m]\psi\right\}. \qquad (18)$$

For chiral transformation,

$$\psi'(x) = e^{i\alpha(x)\gamma_5}\psi(x), \quad \bar{\psi}'(x) = \bar{\psi}(x)e^{i\alpha(x)\gamma_5} \qquad (19)$$

the general Ward–Takahashi identity is written as

$$\int \mathcal{D}\bar{\psi}\mathcal{D}\psi \exp\left\{\int \bar{\psi}[i\slashed{D} - m]\psi\right\} = \int \mathcal{D}\bar{\psi}'\mathcal{D}\psi' \exp\left\{\int \bar{\psi}'[i\slashed{D} - m]\psi'\right\}. \qquad (20)$$

Naive Noether's theorem states that the variation of action vanishes

$$\langle \delta S \rangle = \int d^4x \alpha(x) \langle \partial_\mu(\bar{\psi}\gamma^\mu\gamma_5\psi) - 2im\bar{\psi}\gamma_5\psi \rangle = 0. \qquad (21)$$

The *anomalous* identity means to take into account the Jacobian factor

$$\mathcal{D}\bar{\psi}'\mathcal{D}\psi' = J(\alpha)\mathcal{D}\bar{\psi}\mathcal{D}\psi. \qquad (22)$$

The evaluation of Jacobian proceeds as follows; we formally define an exact fermionic path integral by expanding the field variables

$$\psi(x) = \sum_n b_n \varphi_n(x),$$

$$\bar{\psi}(x) = \sum_n \bar{b}_n \varphi_n(x)^\dagger, \qquad (23)$$

where b_n and $\bar{b}_n$ are Grassmann numbers, with

$$\slashed{D}\varphi_n = \lambda_n \varphi_n, \quad \int \varphi_n(x)^\dagger \varphi_n(x) d^4x = \delta_{nm}, \qquad (24)$$

and thus

$$Z = \lim_{N\to\infty} \int \prod_{n=1}^N d\bar{b}_n db_n \exp\left[\sum_{n=1}^N (i\lambda_n - m)\bar{b}_n b_n\right]. \qquad (25)$$

The chiral transformation is defined by the change of field variables,
$$\psi(x) \to \psi(x)' = e^{i\alpha(x)\gamma_5}\psi(x),$$
$$\bar{\psi}(x) \to \bar{\psi}(x)' = \bar{\psi}(x)e^{i\alpha(x)\gamma_5}. \quad (26)$$

The Jacobian for this transformation is calculated by (in the space defined by $\{\varphi_n(x)\}$),

$$\ln J(\alpha) = -2i \lim_{N\to\infty} \sum_{n=1}^{N} \int \alpha(x)\varphi_n(x)^\dagger \gamma_5 \varphi_n(x) d^4x$$

$$= -2i \lim_{M\to\infty} \sum_{n=1}^{\infty} \int \alpha(x)\varphi_n(x)^\dagger \gamma_5 f\left(\left(\frac{\lambda_n}{M}\right)^2\right) \varphi_n(x) d^4x, \quad (27)$$

where we replaced the mode cut-off by a smooth eigenvalue cut-off, namely, we adopt the *gauge invariant mode cut-off* with any smooth function

$$f(0) = 1, \quad f(\infty) = 0, \quad xf'(x)|_{x=0} = xf'(x)|_{x=\infty} = 0. \quad (28)$$

We then evaluate the quantity for large M,

$$\sum_n \varphi_n(x)^\dagger \gamma_5 f(\lambda_n^2/M^2)\varphi_n(x)$$

$$= \sum_n (\gamma_5 f(\slashed{D}(A_\mu)^2/M^2)\varphi_n(x))\varphi_n(x)^\dagger$$

$$= \sum_n \text{tr}\langle x|\gamma_5 f(\slashed{D}(A_\mu)^2/M^2)|n\rangle\langle n|x\rangle$$

$$\equiv \text{tr}\langle x|\gamma_5 f(\slashed{D}(A_\mu)^2/M^2)|x\rangle$$

$$= \text{tr}\int \frac{d^4k}{(2\pi)^4}\{\gamma_5 f\left(\slashed{D}(A)^2/M^2\right)\langle x|k\rangle\}\langle k|x\rangle$$

$$= \text{tr}\int \frac{d^4k}{(2\pi)^4}e^{-ikx}\gamma_5 f\left(\slashed{D}(A)^2/M^2\right)e^{ikx}$$

$$= \text{tr}M^4 \int \frac{d^4k}{(2\pi)^4}\gamma_5 f\left((ik_\mu + D_\mu/M)^2 - i\frac{1}{M^2}\frac{1}{4}F_{\mu\nu}[\gamma^\mu,\gamma^\nu]\right)$$

$$= \text{tr}\int \frac{d^4k}{(2\pi)^4}\gamma_5 f''\left((ik_\mu)^2\right)\frac{1}{2!}\left(-i\frac{1}{4}F_{\mu\nu}[\gamma^\mu,\gamma_\nu]\right)^2$$

$$= \frac{1}{16\pi^2}\text{tr}\gamma_5 \frac{1}{2!}\left(-i\frac{1}{4}F_{\mu\nu}[\gamma^\mu,\gamma_\nu]\right)^2$$

$$= \frac{1}{32\pi^2}\text{tr}\epsilon^{\mu\nu\alpha\beta}F_{\mu\nu}F_{\alpha\beta}(A), \quad (29)$$

for $M \to \infty$ and the final result is independent of the choice of specific $f(x)$. In the fourth line of (29), all the differential operators in $\mathrm{tr}\langle x|\gamma_5 f(\slashed{D}(A_\mu)^2/M^2)|x\rangle$ act on $\langle x|k\rangle = e^{ikx}$ following Dirac's prescription

$$\langle x|\hat{p}_\mu = i\partial_\mu^x \langle x|.$$

Inside an abstract operator notation such as $\mathrm{tr}\langle x|\gamma_5 f(\slashed{D}(A_\mu)^2/M^2)|x\rangle$, the covariant derivative is understood as $i\slashed{D}(A) = \gamma^\mu(\hat{p}_\mu + A_\mu(\hat{x}))$. We also used $\slashed{D}(A)^2 = D^\mu D_\mu(A) - \frac{i}{4}F_{\mu\nu}[\gamma^\mu, \gamma^\nu]$, and $f''(x)$ indicates the second derivative. In the seventh line of (29), we made a scale transformation $k_\mu/M \to k_\mu$. The remaining trace in (29) is over the internal gauge freedom, and we defined

$$\epsilon^{1234} = 1. \tag{30}$$

For global γ_5-transformation with $\alpha = \mathrm{const}$, the above evaluation is consistent with the index theorem in the context of field theory,

$$\ln J = -2i\alpha \sum_{n=1}^{\infty} \int \varphi_n(x)^\dagger \gamma_5 f(\lambda_n^2/M^2)\varphi_n(x) d^4 x$$

$$= -2i\alpha \sum_{n \text{ with } \lambda_n = 0} \int \varphi_n(x)^\dagger \gamma_5 \varphi_n(x) d^4 x$$

$$= -2i\alpha(n_+ - n_-), \tag{31}$$

namely,

$$\sum_{n=1}^{\infty} \int \varphi_n(x)^\dagger \gamma_5 f(\lambda_n^2/M^2)\varphi_n(x) d^4 x = (n_+ - n_-),$$

where $n_\pm =$ stands for the number of zero-modes with $\gamma_5 \varphi_n = \pm \varphi_n$, since

$$\int \varphi_n(x)^\dagger \gamma_5 \varphi_n(x) d^4 x = 0 \quad \text{for} \quad \lambda_n \neq 0 \tag{32}$$

by noting $\{\slashed{D}, \gamma_5\} = 0$. The Atiya–Singer index theorem states that the index $n_+ - n_-$ is related to the Chern–Pontryagin number

$$n_+ - n_- = \int d^4 x \frac{1}{32\pi^2} F\tilde{F}. \tag{33}$$

The chiral identity in an operator statement thus becomes

$$\langle \partial_\mu(\bar{\psi}\gamma^\mu \gamma_5 \psi) \rangle - 2im\bar{\psi}\gamma_5\psi \rangle = \left\langle \frac{i}{16\pi^2} F\tilde{F} \right\rangle. \tag{34}$$

As an application of this formulation, we mention the relation

$$\int \mathcal{D}\bar{\psi}\mathcal{D}\psi \exp\left\{\int \bar{\psi}[i\slashed{D}-m]\psi\right\}$$
$$= \int \mathcal{D}\bar{\psi}\mathcal{D}\psi \exp\left\{\int \left[\bar{\psi}(i\slashed{D}-me^{2i\alpha\gamma_5})\psi - \frac{i\alpha}{16\pi^2}F\tilde{F}\right]\right\} \quad (35)$$

for a global chiral transformation $\psi(x) \to e^{i\alpha\gamma_5}\psi(x)$, which is useful in the analysis of strong CP breaking in QCD.

2.2. Weyl anomaly

The Weyl transformation is defined by

$$g^{\mu\nu} \to e^{2\alpha(x)}g^{\mu\nu}, \quad (36)$$

which changes the length but preserves the local angle. The dilatation is a part of general coordinate transformation which has no anomaly in QCD type theory in $d=4$. It is thus more precise to say that the conformal anomaly arises from the Weyl symmetry.

In a curved space–time without torsion, the action

$$\int d^4x \sqrt{g}\bar{\psi}[ie_k^\mu \gamma^k D_\mu - m]\psi + S_{YM} \quad (37)$$

is invariant under Weyl transformation

$$g^{\mu\nu}(x) \to e^{2\alpha(x)}g^{\mu\nu}(x), \quad g_{\mu\nu}(x) \to e^{-2\alpha(x)}g_{\mu\nu}(x),$$
$$\sqrt{g} \to e^{-4\alpha(x)}\sqrt{g}, \quad e_k^\mu(x) \to e^{\alpha(x)}e_k^\mu(x),$$
$$\psi(x) \to e^{\frac{3}{2}\alpha(x)}\psi(x), \quad A_\mu^a(x) \to A_\mu^a(x), \quad (38)$$

except for the mass term. The spin connection, which is included in D_μ, is expressed in terms of $e_k^\mu(x)$.

In a curved space, the general coordinate invariant path integral measure is defined by[22]

$$\mathcal{D}\tilde{\bar{\psi}}\mathcal{D}\tilde{\psi} \quad (39)$$

with

$$\tilde{\psi} = g^{1/4}\psi, \quad \tilde{\bar{\psi}} = g^{1/4}\bar{\psi} \quad (40)$$

which is confirmed by evaluating the Jacobian for general coordinate transformation and asking its vanishing, or more intuitively by noting

$$\int \mathcal{D}\tilde{\bar{\psi}}\mathcal{D}\tilde{\psi} \exp\left[\int d^4x \sqrt{g} m\bar{\psi}\psi\right] = \text{constant}, \quad (41)$$

and the general coordinate invariance of $\int d^4x \sqrt{g} m \bar{\psi}\psi$. The basic Weyl transformation of integration variables is thus

$$\tilde{\psi}(x) \to e^{-\frac{1}{2}\alpha(x)}\tilde{\psi}(x), \quad \tilde{\bar{\psi}}(x) \to e^{-\frac{1}{2}\alpha(x)}\tilde{\bar{\psi}}(x) \tag{42}$$

which defines the transformation law even in the *flat-space limit*. Note that a naive Weyl transformation of fermionic variables is $\psi(x) \to e^{\frac{3}{2}\alpha(x)}\psi(x)$.

The Jacobian for the Weyl transformation of $\mathcal{D}\tilde{\bar{\psi}}\mathcal{D}\tilde{\psi}$ is thus given by

$$\begin{aligned}
J_W(\alpha) &= \int \alpha(x) \lim_{M \to \infty} \sum_n \varphi_n(x)^\dagger f(\lambda_n^2/M^2)\varphi_n(x) \\
&= \int \alpha(x) \lim_{M \to \infty} \sum_n \varphi_n(x)^\dagger f(\slashed{D}^2/M^2)\varphi_n(x) \\
&= \int \alpha(x) \lim_{M \to \infty} \mathrm{tr} \int \frac{d^4k}{(2\pi)^4} e^{-ikx} f(\slashed{D}^2/M^2) e^{ikx} \\
&= \int \alpha(x) \frac{1}{48\pi^2} F^a_{\mu\nu} F^{\mu\nu a}
\end{aligned} \tag{43}$$

independent of the choice of $f(x)$. The actual evaluation procedure is the same as for the chiral anomaly in Eq. (29).[22]

For QCD-type theory, the coefficient of Weyl anomaly gives the fermion contribution to *one-loop* β-function,

$$\frac{\beta(g)}{2g^3} = \frac{f}{48\pi^2} \tag{44}$$

where f stands for the number of fermionic variables.[23]

When one analyzes the Weyl anomaly for Yang–Mills theory, the basic transformation law of path integral variables is

$$g^{1/4} e_a^\mu A_\mu(x) \to e^{-\alpha(x)} g^{1/4} e_a^\mu A_\mu(x) \tag{45}$$

and the Weyl anomaly defined as the Jacobian for the path integral measure $\mathcal{D}(g^{1/4} e_a^\mu A_\mu)$, which is explicitly evaluated in the one-loop level, leads to the well-known asymptotic freedom; in this definition of integration variables, one first forms a world-scalar quantity $e_a^\mu A_\mu(x)$, which is analogous to $\psi(x)$, and then multiplies the weight factor $g^{1/4}$ as for $\psi(x)$. Note that $A_\mu(x)$ itself is Weyl scalar. In the actual evaluation of Weyl anomaly for Yang–Mills theory, one needs to include the contributions of Faddeev–Popov ghosts and thus the evaluation is slightly more involved.[13]

Lattice regularization and chiral anomaly

The fermionic path integral is Gaussian in renormalizable theory such as QCD, and thus one can define the path integral exactly. Consequently, one can hope for the exact evaluation of the Jacobian and thus chiral anomaly. (In the case of conformal anomaly, one cannot hope for the exact evaluation of the Jacobian for non-linear Yang–Mills theory, at least in the present scheme of evaluation.)

The recent development in the definition of fermions on the lattice is based on the *Ginsparg–Wilson fermion*.[24–26] In this formalism, one starts with the path integral

$$Z = \int \mathcal{D}\bar{\psi}\mathcal{D}\psi e^{\int \bar{\psi} D \psi} \tag{46}$$

where the lattice Dirac operator is defined to satisfy the Ginsparg–Wilson relation,

$$\gamma_5 D + D\gamma_5 = a D \gamma_5 D,$$

$$(\gamma_5 D)^\dagger = \gamma_5 D, \tag{47}$$

or equivalently

$$\Gamma_5 \gamma_5 D + \gamma_5 D \Gamma_5 = 0,$$

$$\Gamma_5 \equiv \gamma_5 \left(1 - \frac{a}{2} D\right) = \Gamma_5^\dagger. \tag{48}$$

Essentially, this construction satisfies the following correspondence with the continuum definition of Dirac operators;

$$\text{continuum} \longleftrightarrow \text{lattice}$$

$$\gamma_5 i \slashed{D} \longleftrightarrow \gamma_5 D$$

$$\gamma_5 \longleftrightarrow \Gamma_5$$

$$Tr\gamma_5 \longleftrightarrow Tr\Gamma_5. \tag{49}$$

One may note that the eigenfunctions of the lattice Dirac operator

$$\gamma_5 D \phi_n = \lambda_n \phi_n \tag{50}$$

satisfy

$$\phi_n^\dagger(\Gamma_5 \phi_n) \equiv \sum_x \phi_n^\dagger(x)(\Gamma_5 \phi_n)(x) = 0, \quad \text{for } \lambda_n \neq 0, \tag{51}$$

since one has $\gamma_5 D(\Gamma_5 \phi_n) = -\lambda_n (\Gamma_5 \phi_n)$ because of $\Gamma_5 \gamma_5 D + \gamma_5 D \Gamma_5 = 0$.

Starting with the path integral

$$Z = \int \mathcal{D}\bar{\psi}\mathcal{D}\psi e^{\int \bar{\psi} D \psi} \tag{52}$$

one may consider the chiral transformation[27]

$$\psi \to \psi' = (1 + i\epsilon \Gamma_5)\psi,$$

$$\bar{\psi} \to \bar{\psi}' = \bar{\psi}(1 + i\epsilon \left(1 - \frac{a}{2} D\right)\gamma_5) \tag{53}$$

which keeps the lattice fermionic action invariant for global ϵ. The Jacobian for this transformation is given by

$$\ln J = -2i\text{Tr}\epsilon\Gamma_5. \tag{54}$$

For a global chiral transformation, ϵ = constant, one has

$$\text{Tr}\Gamma_5 = \sum_n \phi_n^\dagger \Gamma_5 \phi_n$$

$$= \sum_n \phi_n^\dagger \Gamma_5 f((\gamma_5 D)^2/M^2) \phi_n \tag{55}$$

independent of $f(x)$ for $f(0) = 1$, since the non-vanishing eigenvalues of $\gamma_5 D$ does not contributes to the index because of (51).

For a suitable $f(x)$ such as $f(x) = e^{-x}$, one can evaluate the index as [28]

$$\sum_n \phi_n^\dagger \Gamma_5 f((\gamma_5 D)^2/M^2)\phi_n$$

$$= \sum_n \phi_n^\dagger \gamma_5 \left(1 - \frac{a}{2}D\right) f((\gamma_5 D)^2/M^2)\phi_n$$

$$\to_{a\to 0} \sum_n \phi_n^\dagger \gamma_5 f((\gamma_5 D)^2/M^2)\phi_n$$

$$= \int d^4x \,\text{tr}\int \frac{d^4k}{(2\pi)^4} e^{-ikx} \gamma_5 f((\gamma_5 D)^2/M^2) e^{ikx}$$

$$\to_{a\to 0} \int d^4x \,\text{tr}\int \frac{d^4k}{(2\pi)^4} e^{-ikx} \gamma_5 f(\slashed{D}^2/M^2) e^{ikx} \tag{56}$$

which agrees with the continuum result for sufficiently smooth gauge fields.

To summarize, the above lattice derivation is based on the basic properties of D,

1. Ginsparg–Wilson relation,
2. $\gamma_5 D \to -\gamma_5 i \slashed{D}$ for $a \to 0$,
3. Free of species doublers.

An explicit form of lattice D, which satisfies the above properties, is given by H. Neuberger.[25] This analysis suggests that chiral anomaly in path integral is non-perturbatively defined, to be consistent with the so-called Adler–Bardeen theorem.[29]

One may summarize the path integral analysis of anomalies by saying that *the path integral contains all the ingredients to understand quantum anomalies*. As an important implications of quantum anomalies, I mention the fermion number anomaly first noted by 't Hooft;[30] the lepton and quark number currents in the Standard Model contain anomaly, but $B - L$ is free of anomaly. This is known to be essential to understand the "baryon number asymmetry" in the universe. This is one of the profound implications implied by a better understanding of quantum theory, which can even modify our idea about the origin of the matter in the universe. One my hope that a deeper understanding of quantum theory in the future may also lead to an even more precise understanding of the beginning of the universe.

3. Bosonization in the Path Integral Formulation

In this section, an application of chiral anomaly to the bosonization of two-dimensional fermions is discussed.[31,32] See Refs. 33, 34 for reviews with extensive references. We first briefly describe the past formulation and then describe a recent work on the "direct bosonization" in path integral formulation.[35]

We start with the path integral

$$e^{iW(v_\mu)} = \int \mathcal{D}\bar\psi \mathcal{D}\psi \exp\left[i\int d^2x\left(\bar\psi i\gamma^\mu \partial_\mu \psi + v_\mu \bar\psi \gamma^\mu \psi\right)\right]. \quad (57)$$

We then observe that the vector field in two-dimensional space–time is decomposed into two arbitrary real functions α and β as

$$v_\mu(x) = \partial_\mu \alpha(x) + \epsilon_{\mu\nu}\partial^\nu \beta(x). \quad (58)$$

We can then write

$$e^{iW(v_\mu)} = \int \mathcal{D}\bar\psi \mathcal{D}\psi \exp\left\{i\int d^2x\left[\bar\psi i\gamma^\mu(\partial_\mu - i\partial_\mu \alpha - i\partial_\mu \beta \gamma_5)\psi\right]\right\} \quad (59)$$

by noting $\epsilon_{\mu\nu}\gamma^\mu = \gamma_\nu \gamma_5$.

The vector part $\partial_\mu \alpha(x)$ is transformed away by a suitable gauge transformation without generating any Jacobian factor. The axial-vector part $\partial_\mu \beta(x)$ is also transformed away by a repeated application of an infinitesimal chiral transformation $\psi(x) \to e^{i\delta\beta(x)\gamma_5}\psi(x)$ and $\bar\psi(x) \to \bar\psi(x)e^{i\delta\beta(x)\gamma_5}$ but with a non-trivial Jacobian factor. The α and β dependences in the action are thus extracted as

$$e^{iW(v_\mu)} = \exp[i\Gamma(v_\mu)] \int \mathcal{D}\bar\psi \mathcal{D}\psi \exp\left[i\int d^2x\left(\bar\psi i\gamma^\mu \partial_\mu \psi\right)\right] \quad (60)$$

where $\Gamma(v_\mu)$ stands for the integrated Jacobian (or anomaly) written in Minkowski metric

$$i\Gamma(v_\mu) = \frac{i}{\pi}\int d^2x\left(-\frac{1}{2}\partial^\mu \beta \partial_\mu \beta\right).$$

We can also write

$$e^{iW(v_\mu)} = \int \mathcal{D}\xi \exp\left[i\int d^2x\left(\frac{1}{2}\partial^\mu \xi \partial_\mu \xi - \frac{v_\mu}{\sqrt{\pi}}\epsilon^{\mu\nu}\partial_\nu \xi\right)\right], \quad (61)$$

where we used $\partial^\mu \partial_\mu \beta = \epsilon^{\mu\nu}\partial_\mu v_\nu$.

The theory of a free Dirac fermion

$$e^{iW(v_\mu)} = \int \mathcal{D}\bar\psi\mathcal{D}\psi \exp\left[i\int d^2x \left(\bar\psi i\gamma^\mu\partial_\mu\psi + v_\mu\bar\psi\gamma^\mu\psi\right)\right] \tag{62}$$

and the theory of a free real Bose field

$$e^{iW(v_\mu)} = \int \mathcal{D}\xi \exp\left[i\int d^2x \left(\frac{1}{2}\partial^\mu\xi\partial_\mu\xi - \frac{v_\mu}{\sqrt\pi}\epsilon^{\mu\nu}\partial_\nu\xi\right)\right], \tag{63}$$

define the identical generating functional $W(v_\mu)$ of Green's functions, which is called *bosonization*. This observation implies, for example,

$$\langle T^*\bar\psi(x)\gamma^\mu\psi(x)\bar\psi(y)\gamma^\nu\psi(y)\rangle = \left(\frac{1}{\pi}\right)\langle T^*\epsilon^{\mu\alpha}\partial_\alpha\xi(x)\epsilon^{\nu\beta}\partial_\beta\xi(y)\rangle. \tag{64}$$

Direct bosonization

In the operator formulation of bosonization, the direct $d=2$ on-shell bosonization

$$\psi_L(x_+) = e^{i\xi(x_+)},$$
$$\psi_R^\dagger(x_-) = e^{i\xi(x_-)}, \tag{65}$$

with $x_\pm = x_0 \pm x_1$ is commonly used.[33,34] This direct bosonization is not possible in the path integral approach since we cannot define the path integral in terms of the on-shell variables such as $\xi(x_+)$ and $\xi(x_-)$, which satisfy $\Box\xi(x_\pm) = 0$. To understand the direct bosonization in path integral, we first derive the off-shell relations

$$\psi_L(x)\psi_R^\dagger(x) = \exp[i\xi(x)],$$
$$\psi_R(x)\psi_L^\dagger(x) = \exp[-i\xi(x)], \tag{66}$$

which are defined in the path integral, and then consider the on-shell limit.

Similarly, the on-shell bosonizations of the bosonic commuting spinor,

$$\phi_L(x_+) = ie^{-i\xi(x_+)}\partial^+ e^{-i\chi(x_+)},$$
$$\phi_R^\dagger(x_-) = e^{-i\xi(x_-)-i\chi(x_-)}, \tag{67}$$

and

$$\phi_R(x_-) = ie^{i\xi(x_-)}\partial^- e^{+i\chi(x_-)},$$
$$\phi_L^\dagger(x_+) = e^{i\xi(x_+)+i\chi(x_+)}, \tag{68}$$

are established in path integral formulation by first deriving the off-shell relations

$$\phi_L(x)\phi_R^\dagger(x) = ie^{-i\xi(x)}\partial^+ e^{-i\chi(x)},$$
$$\phi_R(x)\phi_L^\dagger(x) = ie^{i\xi(x)}\partial^- e^{i\chi(x)}, \tag{69}$$

which are well-defined in path integral.

A brief sketch of the derivation

We explain the derivation of (65). We start with

$$e^{iW(v_\mu, j_R, j_L)} = \int \mathcal{D}\bar{\psi}\mathcal{D}\psi \exp\left\{i\left(\frac{1}{2\pi}\right)\int d^2x \mathcal{L}\right\} \tag{70}$$

where, following the procedure in (60),

$$\mathcal{L} = \bar{\psi}i\gamma^\mu\partial_\mu\psi + v_\mu\bar{\psi}\gamma^\mu\psi - j_L(x)\psi_L\psi_R^\dagger - j_R(x)\psi_R\psi_L^\dagger$$
$$\Rightarrow \bar{\psi}i\gamma^\mu\partial_\mu\psi - j_L(x)e^{-2i\beta}\psi_L\psi_R^\dagger - j_R(x)e^{2i\beta}\psi_R\psi_L^\dagger - \partial_\mu\beta\partial^\mu\beta \tag{71}$$

and then using **Cauchy's lemma**, which converts the ferminonic path integral to the bosonic path integral, we have

$$e^{iW(v_\mu, j_R, j_L)} = \int \mathcal{D}\xi \exp\left\{i\left(\frac{1}{2\pi}\right)\int d^2x \mathcal{L}\right\} \tag{72}$$

with

$$\mathcal{L} = \frac{1}{2}\partial^\mu\xi(x)\partial_\mu\xi(x) - 2j_L(x)\Lambda e^{-2i\beta+i\xi} - 2j_R(x)\Lambda e^{2i\beta-i\xi} - 2\partial_\mu\beta\partial^\mu\beta$$
$$\Rightarrow \frac{1}{2}\partial^\mu\xi(x)\partial_\mu\xi(x) - 2v_\mu\epsilon^{\mu\nu}\partial_\nu\xi(x) - 2j_L\Lambda e^{i\xi} - 2j_R\Lambda e^{-i\xi}, \tag{73}$$

where a shift of integration variable $\xi \to \xi + 2\beta$ is made. Namely, we have

$$e^{iW(v_\mu, j_R, j_L)} = \int \mathcal{D}\xi \exp\left\{i\left(\frac{1}{4\pi}\right)\int d^2x \mathcal{L}\right\} \tag{74}$$

with

$$\mathcal{L} = \frac{1}{2}\partial^\mu\xi(x)\partial_\mu\xi(x) - 2v_\mu\epsilon^{\mu\nu}\partial_\nu\xi(x) - 2j_L\Lambda e^{i\xi} - 2j_R\Lambda e^{-i\xi}. \tag{75}$$

Comparing (75) with (71), we have established path integral bosonization rules,

$$\bar{\psi}\gamma^\mu\gamma_5\psi = \partial^\mu\xi,$$
$$\psi_L(x)\psi_R^\dagger(x) = \Lambda\exp[i\xi(x)],$$
$$\psi_R(x)\psi_L^\dagger(x) = \Lambda\exp[-i\xi(x)]. \tag{76}$$

In the limit of *on-shell fields* with $\partial^+\partial^-\xi(x) = 0$ and $\xi(x) = \xi(x_+) + \xi(x_-)$, the last two relations in (76) imply the on-shell direct bosonization rules in (65),

$$\psi_L(x_+) = \Lambda^{1/2}e^{i\xi(x_+)} =: e^{i\xi(x_+)}:,$$

$$\psi_R^\dagger(x_-) = \Lambda^{1/2}e^{i\xi(x_-)} =: e^{i\xi(x_-)}: . \qquad (77)$$

Similarly, one can establish the direct bosonization rules in (66) and (67).

4. Conclusion

Yang–Mills gauge field theory, together with Maxwell field and Einstein's general relativity, is remarkably successful in describing the fundamental interactions. It also stimulated many interesting field theoretical machinery including the development of path integral formulation briefly described here.

Acknowledgments

I thank Professor C. N. Yang for letting me to visit Institute for Theoretical Physics (now C. N. Yang Institute) at Stony Brook where some of the ideas discussed in my talk were worked out. I also thank P. van Nieuwenhuizen and W. Weissberger, among others, for encouragement.

References

1. C. N. Yang and R. Mills, *Phys. Rev.* **96**, 191 (1954).
2. L. D. Faddeev and V. N. Popov, *Phys. Lett. B* **25**, 29 (1967);
   L. D. Faddeev, *Theor. Math. Phys.* **1**, 1 (1970).
3. F. A. Berezin, *The Method of Second Quantization* (Academic Press, 1966).
4. C. Becchi, A. Rouet and R. Stora, *Comm. Math. Phys.* **42**, 127 (1975); *Ann. of Phys*, **98**, 287 (1976).
5. G. 't Hooft, *Nucl. Phys. B* **33**, 173 (1971); *Nucl. Phys. B* **35**, 167 (1971).
6. K. Fujikawa, *Int. J. Mod. Phys. A* **16**, 425 (2001).
7. J. S. Bell and R. Jackiw, *Nuovo Cim. A* **60**, 47 (1969).
8. S. L. Adler, *Phys. Rev.* **177**, 2426 (1969).
9. S. L. Adler, Perturbation theory anomalies, in: *Lectures on Elementary Particles and Quantum Field Theory*, Vol. 1., S. Deser et al. (eds.), (MIT Press, 1970).
10. S. B. Treiman, R. Jackiw, B. Zumino and E. Witten, *Current Algebra and Anomalies* (World Scientific, 1985).
11. P. van Nieuwenhuizen, *Anomalies in Quantum Field Theory: Cancellation of Anomalies in D = 10 Supergravity* (Leuven University Press, 1988).
12. R. Bertlemann, *Anomalies in Quantum Field Theory* (Oxford University Press, 1996).

13. K. Fujikawa and H. Suzuki, *Path Integrals and Quantum Anomalies* (Oxford University Press, 2004).
14. H. Fukuda and Y. Miyamoto, *Prog. Theor. Phys.* **4**, 347 (1949).
15. J. Steinberger, *Phys. Rev.* **76**, 1180 (1949).
16. S. Tomonaga et al., *Prog. Theor. Phys.* **4**, 477 (1949).
17. J. Schwinger, *Phys. Rev.* **82**, 664 (1951).
18. M. Atiyah, R. Bott and V. Patodi, *Invent. Math.* **19**, 279 (1973).
19. K. Fujikawa, *Phys. Rev. D* **25**, 2584 (1982)
20. O. Alvarez, *Nucl. Phys. B* **216**, 125 (1983);
    D. Friedan, S. Shenker and P. Matinec, *Nucl. Phys. B* **271**, 93 (1986).
21. K. Fujikawa, *Phys. Rev. Lett.* **42**, 1195 (1979);
    K. Fujikawa, *Phys. Rev. D* **21**, 2848 (1980) [*Phys. Rev. D* **22**, 1499 (1980)].
22. K. Fujikawa, *Phys. Rev. Lett.* **44**, 1733 (1980).
23. S. L. Adler, J. C. Collins and A. Dunkan, *Phys. Rev. D* **15**, 1712 (1977).
24. P. H. Ginsparg and K. G. Wilson, *Phys. Rev. D* **25**, 2649 (1982).
25. H. Neuberger, *Phys. Lett. B* **417**, 141 (1998);
    H. Neuberger, *Phys. Lett. B* **427**, 353 (1998).
26. P. Hasenfratz, *Nucl. Phys. B* **525**, 401 (1998);
    P. Hasenfratz, V. Laliena and F. Niedermayer, *Phys. Lett. B* **427**, 125 (1998).
27. M. Lüscher, *Phys. Lett. B* **428**, 342 (1998).
28. K. Fujikawa, *Nucl. Phys. B* **546**, 480 (1999).
29. S. L. Adler and W. A. Bardeen, *Phys. Rev.* **182**, 1517 (1969).
30. G. 't Hooft, *Phys. Rev. Lett.* **37**, 8 (1976).
31. S. R. Coleman, Phys. Rev. D **11**, 2088 (1975).
32. R. Roskies and F. Schaposnik, *Phys. Rev. D* **23**, 558 (1981).
33. E. Abdalla, M. C. B. Abdalla and K. D. Rothe, *Nonperturbative Methods in Two-Dimensional Quantum Field Theory* (World Scientific, 1991).
34. M. Stone, *Bosonization* (World Scientific, 1994).
35. K. Fujikawa and H. Suzuki, *Phys. Rev. D* **91**, 065010 (2015), and references therein.

Yang–Mills Gauge Theory and the Higgs Boson Family

Ngee-Pong Chang

Physics Department, City College of CUNY, New York, NY 10031, USA[*]
The Graduate Center, CUNY, New York, NY 10016, USA
and
Institute of Advanced Studies, Nanyang Technological University, Singapore[†]
npccc@sci.ccny.cuny.edu

The gauge symmetry principles of the Yang–Mills field of 1954 provide the solid rock foundation for the Standard Model of particle physics. To give masses to the quarks and leptons, however, SM calls on the solitary Higgs field using a set of mysterious complex Yukawa coupling matrices.

We enrich the SM by reducing the Yukawa coupling matrices to a single Yukawa coupling constant, and endowing it with a family of Higgs fields that are degenerate in mass.

The recent experimental discovery of the Higgs resonance at 125.09 ± 0.21 GeV does not preclude this possibility. Instead, it presents an opportunity to explore the interference effects in background events at the LHC.

We present a study based on the maximally symmetric Higgs potential in a leading hierarchy scenario.

Keywords: Enriched Standard Model; *r*-symmetry; Yukawa coupling; family of Higgs fields.

1. Introduction

On the occasion of this memorable conference to honor and celebrate the 60th anniversary of the seminal paper by Yang and Mills[1] on non-Abelian local gauge fields, it is a good time to step back and marvel on the remarkable implications and achievements of local non-Abelian gauge symmetries in physics. For it has blossomed to become the theory of everything, describing the strong, electromagnetic and weak interactions of all matter in the universe. Yet from the day of conception, the Yang–Mills (YM) gauge field was faced with the difficult challenge posed by Pauli on the question of mass. When the YM field was applied to the broken isospin symmetry in nuclear interactions, the gauge fields involved clearly had to be massive. But how?

This question was fully resolved first in the context of electroweak unification of Glashow–Weinberg–Salam,[2–4] with the local gauge group $SU(2)_W \times U(1)_Y$. Out of the four gauge bosons, W_μ^a, $a = 1, 2, 3$ and B_μ, one linear combination

---
[*]Permanent address
[†]Visiting Professor

is identified with the ubiquitous massless photon, while the other three become the massive vector bosons mediating weak interactions. The weak vector bosons acquire mass as a result of the spontaneous symmetry breaking[5] (SSB) of the electroweak gauge group $SU(2)_W \times U(1)_Y$ down to the usual $U(1)_Q$ electric charge gauge symmetry.

This SSB comes about through the introduction of the $SU(2)_W$ doublet Higgs[6-8] scalar field, ϕ^α, $\alpha = +, 0$. Unlike a normal scalar field, the Higgs potential is minimal around a nonzero vacuum expectation value, $\langle \phi^o \rangle = v$, so that the vacuum state breaks the local gauge symmetry. The gauge covariant kinetic energy term in the Higgs field Lagrangian then gives rise to masses, m_W and m_Z, for the charged $W^\pm$ and neutral Z bosons.

It was not, however, until the ground-breaking proof by 't Hooft[9] and Veltman[10] of the renormalizability of massive YM gauge theories that the Yang–Mills theory reached adolescence. In this, the pioneering work by Faddeev and Popov[11] laid the groundwork for a Feynman diagram analysis of the renormalization program.

This breakthrough in the regularization and renormalization of YM gauge theories quickly led to the birth of Quantum Chromodynamics (QCD) through Fritzsch and Gell-Mann[12] and the discovery of asymptotic freedom by Gross–Wilczek[13] and Politzer.[14]

In QCD, the strong interactions of colored quarks is described in terms of the YM gauge fields of $SU(3)_c$. The gluon fields lend themselves to an intuitive picture of the strong binding between quarks through the trilinear and quadrilinear self-couplings of the gluon.[15] Computer simulations in lattice-QCD have shown that the flux tube linking quarks leads to confinement of quarks.[16]

With QCD, we now have a theory of everything that includes strong, electromagnetic and weak interactions. This is the Standard Model (SM), with the complete gauge group, $SU(3)_c \times SU(2)_W \times U(1)_Y$, that acts on the 3-family of quarks and leptons, and the solitary Higgs doublet, ϕ^α.

2. Yukawa Coupling

In the SM, the fermions acquire mass through the spontaneous symmetry breaking in the vacuum expectation value of the neutral Higgs scalar field, $\langle \phi^o \rangle = v$. This is accomplished through the Yukawa coupling

$$\mathcal{L}_Y = -Y^u_{ij} \bar{u}^i_R q^{j\alpha}_L \phi^\beta \epsilon_{\alpha\beta} + \text{h.c.} - Y^d_{ij} \bar{d}^i_R q^{j\alpha}_L \phi^\dagger_\alpha + \text{h.c.}, \qquad (1)$$

where I have suppressed the color indices but display explicitly the family indices $i, j = 1, 2, 3$, while $\alpha = 1, 2$ are the $SU(2)_W$ doublet indices.

As a solitary Higgs that is to produce masses for all the quarks, the Yukawa couplings are necessarily complex 3×3 matrices. In the world of particle physics we have become used and inured to work with such complex Yukawa coupling matrices.

In this talk, however, I would like to entertain the possibility that the Higgs not be a lone ranger in this universe, but that the Higgs is part of a larger family. In simple terms, I propose to transfer the complexity of the coupling to a family tree of Higgs bosons:

$$\mathcal{L}_Y = -h_u \bar{u}_R^i q_L^{j\alpha} \phi_{ij}^\beta \epsilon_{\alpha\beta} + \text{h.c.} - h_d \bar{d}_R^i q_L^{j\alpha} \hat{\phi}_{\alpha ij} + \text{h.c.} \qquad (2)$$

Here I have introduced the ϕ_{ij}^α to be associated with the u-quark mass, and the $\hat{\phi}^{ij,\alpha}$ for the d-quark mass. These Higgs fields are, as usual, color-neutral. Here, I am using the notation

$$\hat{\phi}_{\alpha ij} \equiv (\hat{\phi}^{\alpha ij})^\dagger \qquad (3)$$

with the vacuum expectation values for the neutral scalar fields

$$\langle \phi_{ij}^o \rangle = v_{ij}, \qquad (4)$$

$$\langle \hat{\phi}^{o\,ij} \rangle = \hat{v}^{ij}. \qquad (5)$$

3. r-Symmetry

To make the model more predictive, we make the simple requirement

$$h_u = h_d \qquad (6)$$

so that

$$\mathcal{L}_Y = -h_q(\bar{u}_R^i q_L^{j\alpha} \phi_{ij}^\beta \epsilon_{\alpha\beta} + \bar{d}_R^i q_L^{j\alpha} \hat{\phi}_{\alpha ij}) + \text{h.c.} \qquad (7)$$

This requirement may look odd, as everyone knows that the down quark in each family is much lighter than the up quark. But, for this Enriched Standard Model, with the two families of Higgs fields, ϕ_{ij}^α, and $\hat{\phi}_{ij,\alpha}$, the difference in the physical up and down quark masses may be attributed to the difference in vacuum expectation values of the corresponding Higgs fields.

To implement this requirement we extend the enriched Standard Model to the gauge group $SU(2)_L \times U(1)_Y \times U(1)_R$ and impose a new r-symmetry on the full Lagrangian.

	Y_R	Y'	$(I_3)_L$		Y_R	Y'	$(I_3)_L$
ϕ_{ij}^+	+1/2		+1/2	u_L^j		+1/6	+1/2
ϕ_{ij}^o	+1/2		−1/2	d_L^j		+1/6	−1/2
$\hat{\phi}^{+ij}$	+1/2		+1/2	u_{Ri}	+1/2	+1/6	
$\hat{\phi}^{o\,ij}$	+1/2		−1/2	d_{Ri}	−1/2	+1/6	

The covariant derivatives for the quark and Higgs fields now read as[a]

$$D_\mu q_L^{j\alpha} = \partial_\mu q_L^{j\alpha} + i\frac{g}{2}(\boldsymbol{\tau}\cdot\mathbf{W})^\alpha_\beta q_L^{j\beta} + i\frac{g'}{6}B'_\mu q_L^{j\alpha}, \tag{8}$$

$$D_\mu q_{Ri}^{a} = \partial_\mu q_{Ri}^{a} + i\frac{g_R}{2}(\tau_3 B_{R\mu})^a_b q_{Ri}^{b} + i\frac{g'}{6}B'_\mu q_{Ri}^{a}, \tag{9}$$

$$D_\mu \phi_{ij}^{\alpha} = \partial_\mu \phi_{ij}^{\alpha} + i\frac{g}{2}(\boldsymbol{\tau}\cdot\mathbf{W})^\alpha_\beta \phi_{ij}^{\beta} + i\frac{g_R}{2}B_{R\mu}\phi_{ij}^{\alpha}, \tag{10}$$

$$D_\mu \hat{\phi}_{\alpha ij} = \partial_\mu \hat{\phi}_{\alpha ij} - i\frac{g}{2}(\boldsymbol{\tau}\cdot\mathbf{W})^\beta_\alpha \hat{\phi}_{\beta ij} - i\frac{g_R}{2}B_{R\mu}\hat{\phi}_{\alpha ij}. \tag{11}$$

The full Lagrangian is invariant under the r-symmetry

$$\left.\begin{array}{c} u_{Ri} \to d_{Ri}, \qquad d_{Ri} \to -u_{Ri} \\ \phi_{ij}^{\alpha} \to -\epsilon^{\alpha\beta}\hat{\phi}_{\beta ij}, \quad \hat{\phi}_{\alpha ij} \to -\epsilon_{\alpha\beta}\phi_{ij}^{\beta} \end{array}\right\} \tag{12}$$

and for the gauge fields

$$\left.\begin{array}{c} B_{R\mu} \to -B_{R\mu} \\ \mathbf{W}_\mu \to +\mathbf{W}_\mu \\ B'_\mu \to +B'_\mu \end{array}\right\}. \tag{13}$$

This extension of the Standard Model parallels the $SU(2)_L \times SU(2)_R \times U(1)_Y$ of Mohapatra and Senjanovic.[17] Instead of the full set of $SU(2)_R$ gauge bosons, however, we have only the neutral B_R gauge boson. To have the correct neutrino phenomenology, we introduce a set of very heavy Higgs fields, $\Delta_{ab}^{R,ij}$. As shown in a previous paper,[18] they serve to give rise to the heavy mass for Z_R, whose energy scale is far above the Higgs mass scale. For this paper, we take advantage of the Georgi–Weinberg[19] decoupling theorem and focus our attention on the Higgs mass scale, and thus work only with the Higgs potential, V_ϕ, involving the ϕ_{ij}^{α} and $\hat{\phi}^{\alpha ij}$ fields.

4. Mass Hierarchy

Transforming to the basis where the physical quark mass terms are diagonal through the unitary transforms

$$\bar{u}_R^i = \bar{\mathbf{u}}_R^I (V_{uR}^*)^i_I, \tag{14}$$

$$u_L^j = (V_{uL})^j_J \mathbf{u}_L^J, \tag{15}$$

$$\bar{d}_R^i = \bar{\mathbf{d}}_R^I (V_{dR}^*)^i_I, \tag{16}$$

$$d_L^j = (V_{dL})^j_J \mathbf{d}_L^J, \tag{17}$$

---

[a]Here $a, b = 1, 2$ refer to the flavor of the right-handed quark families q_{Ri}^a, while $\alpha = 1, 2$ refer as usual to the flavor of the left-handed q_L^j quark families, and $i, j = 1, 2, 3$ refer to the families.

the Higgs fields take the form

$$\phi^{(o)}_{ij} = (\tilde{V}_u)^I_i (V_u^\dagger)^J_j \Phi^{(o)}_{IJ}, \tag{18}$$

$$\phi^{(+)}_{ij} = (\tilde{V}_u)^I_i (V_d^\dagger)^J_j \Phi^{(+)}_{IJ}, \tag{19}$$

$$\hat{\phi}_{(o)ij} = (\tilde{V}_d)^I_i (V_d^\dagger)^J_j \hat{\Phi}_{(o)IJ}, \tag{20}$$

$$\hat{\phi}_{(-)ij} = (\tilde{V}_d)^I_i (V_u^\dagger)^J_j \hat{\Phi}_{(-)IJ} \tag{21}$$

AND the vacuum expectation values are diagonal

$$\langle \Phi^{(o)}_{IJ} \rangle = \begin{pmatrix} v_1 & 0 & 0 \\ 0 & v_2 & 0 \\ 0 & 0 & v_3 \end{pmatrix}_{IJ} \tag{22}$$

and the mass hierarchy is expressed in terms of

$$v_3 \gg v_2 \gg v_1. \tag{23}$$

5. Maximally Symmetric Higgs Potential

In principle, there are as many parameters available in this Enriched Standard Model as in the SM with the solitary Higgs. We will, however, take advantage of the r-symmetry and work with the scenario of a maximally symmetric Higgs potential. In this scenario we work with the simplest possible Higgs potential with a fully degenerate ϕ^α_{ij} and $\hat{\phi}^{ij,\alpha}$ Higgs family:

$$V_\phi = +\frac{\lambda_1}{2}(\epsilon_{\alpha\beta}\phi^\alpha_{a,ij}\phi^\beta_{b,k\ell})(\epsilon^{\alpha'\beta'}\phi^{a,ij}_{\alpha'}\phi^{b,k\ell}_{\beta'})$$
$$+\frac{\lambda_2}{2}(\phi^\alpha_{a,ij}\phi^{b,k\ell}_\alpha)(\phi^\beta_{b,k\ell}\phi^{a,ij}_\beta) - \frac{\lambda_3}{4}(\phi^\alpha_{a,ij}\phi^{a,ij}_\alpha)^2. \tag{24}$$

Here we introduce the compact notation for the Higgs fields

$$\phi^\beta_{a,ij} = \begin{pmatrix} \hat{\phi}_{(o)ij} & \phi^{(+)}_{ij} \\ -\hat{\phi}_{(-)ij} & \phi^{(o)}_{ij} \end{pmatrix}^\beta_a. \tag{25}$$

The Higgs potential about the new vacuum takes the form

$$V_\phi = +\frac{\lambda_1}{2}(\epsilon_{\alpha\beta}[\phi^\alpha_{a,ij}\phi^\beta_{b,k\ell} - v^\alpha_{a,ij}v^\beta_{b,k\ell}])(\epsilon^{\alpha'\beta'}[\phi^{a,ij}_{\alpha'}\phi^{b,k\ell}_{\beta'} - v^{a,ij}_\alpha v^{b,k\ell}_\beta])$$
$$+\frac{\lambda_2}{2}([\phi^\alpha_{a,ij}\phi^{b,k\ell}_\alpha - v^\alpha_{a,ij}v^{b,k\ell}_\alpha])([\phi^\beta_{b,k\ell}\phi^{a,ij}_\beta - v^\beta_{b,k\ell}v^{a,ij}_\beta])$$
$$-\frac{\lambda_3}{4}(\phi^\alpha_{a,ij}\phi^{a,ij}_\alpha - v^2 - \hat{v}^2)^2, \tag{26}$$

where

$$\langle \phi^\beta_{bij} \rangle = \begin{pmatrix} \hat{v}_{ij} & 0 \\ 0 & v_{ij} \end{pmatrix}^\beta_b \qquad (27)$$

and

$$v^2 = v^\alpha_{ij} v^{ij}_\alpha, \qquad (28)$$

$$\hat{v}^2 = \hat{v}^{\alpha ij} \hat{v}_{\alpha ij} \qquad (29)$$

with the hierarchy

$$v \gg \hat{v}.$$

6. Meet the Higgs Family

In the leading hierarchy where $v_3 \gg v_2 \gg v_1$, we consider only the $j = 3$ family to be massive. The physical Higgs family that results ($v_t \equiv \sqrt{v^2 + \hat{v}^2}$)

Table 1. Higgs boson family.

M^2	Higgs
$(\lambda_1 + \lambda_2 - \lambda_3)v_t^2$	$h_A = (v h_3 + \hat{v}\hat{h}_3)/v_t$
$\lambda_2 v_t^2$	$\begin{cases} h_B = (\hat{v} h_3 - v\hat{h}_3)/v_t \\ z_B = (\hat{v} z_3 - v\hat{z}_3)/v_t \\ \phi_B^+ = (\hat{v}\phi_{33}^+ - v\hat{\phi}^{+33})/v_t \end{cases}$
$\lambda_1 \hat{v}^2 + \lambda_2 v^2$	$h_{ij}, z_{ij}, \hat{\phi}^{+ij}$ for $ij \neq 33$
$\lambda_1 v^2 + \lambda_2 \hat{v}^2$	$\hat{h}_{ij}, \hat{z}_{ij}, \phi^{+ij}$ for $ij \neq 33$

with the Goldstone bosons

Table 2. Goldstone bosons.

M^2	Higgs
0	$z_A = (v z_3 + \hat{v}\hat{z}_3)/v_t$
0	$\phi_A^+ = (v\phi_{33}^+ + \hat{v}\hat{\phi}^{+33})/v_t$

Here we have introduced the scalar field decompositions for the neutral Higgs fields

$$\phi^o_{33} = \frac{h_3 - iz_3}{\sqrt{2}}, \tag{30}$$

$$\hat{\phi}^{o\,33} = \frac{\hat{h}_3 - i\hat{z}_3}{\sqrt{2}}. \tag{31}$$

For the special case, with

$$\lambda_1 = \lambda_2 = \lambda_3 \equiv \lambda$$

we have the interesting scenario of a fully degenerate Higgs boson family.

Table 3. Degenerate Higgs boson family.

M^2	Higgs
λv_t^2	$h_A = (vh_3 + \hat{v}\hat{h}_3)/v_t$
λv_t^2	$\begin{cases} h_B = (\hat{v}h_3 - v\hat{h}_3)/v_t \\ z_B = (\hat{v}z_3 - v\hat{z}_3)/v_t \\ \phi_B^+ = (\hat{v}\phi_{33}^+ - v\hat{\phi}^{+33})/v_t \end{cases}$
λv_t^2	$h_{ij}, z_{ij}, \hat{\phi}^{+ij}$ for $ij \neq 33$
λv_t^2	$\hat{h}_{ij}, \hat{z}_{ij}, \phi^{+ij}$ for $ij \neq 33$

7. Phenomenological Implications

To date, only one Higgs has been observed,[20] and there is no indication of any structure in the Higgs mass plots around 125.09 GeV $\pm$ 0.21 (stat) $\pm$ 0.11 (sys).

Does this not refute the possibility of a big Higgs family?

And yet the members of this Higgs family do not all interact in the same way with the electroweak bosons. And as you will see in the ensuing discussion, much of the data for the other members of the Higgs family may still be hidden in data. So do bear with me as I point out the subtleties of this scenario.

7.1. h_A emulates the SM Higgs field

In the Standard Model, the production processes for Higgs may be classified in terms of (a) Gluon Fusion Production, (b) Associative Production with W, (c) Vector Boson Fusion and (d) Associative Production with top quark pair. Of these, the gluon fusion production cross-section is the highest, arising from the top quark loop. In all of these processes, the SM Higgs is emitted by W, Z or the top quark loop.

In our scenario, h_A field has the same couplings to the electroweak gauge bosons as the SM Higgs scalar ($c \equiv \cos\theta_W$, $m_W = gv_t/\sqrt{2}$)

$$\mathcal{L}^g_{h_A} = -gm_W h_A W^+_\mu W^-_\mu - \frac{gm_Z}{2c} h_A Z_\mu Z_\mu$$
$$-\frac{g^2}{4} h_A^2 W^+_\mu W^-_\mu - \frac{g^2}{8c^2} h_A^2 Z_\mu Z_\mu. \tag{32}$$

7.2. h_B and z_B have only higher order quartic coupling to W and Z

Unlike h_A, however, the cousins h_B and z_B decouple from the trilinear coupling with the W and Z. Their gauge interactions are of second order in the weak coupling constant, g,

$$\mathcal{L}^g_{h_B, z_B} = -\frac{g^2}{4} h_B^2 W^+_\mu W^-_\mu - \frac{g^2}{8c^2} h_B^2 Z_\mu Z_\mu$$
$$-\frac{g^2}{4} z_B^2 W^+_\mu W^-_\mu - \frac{g^2}{8c^2} z_B^2 Z_\mu Z_\mu. \tag{33}$$

Thus, to leading order, the h_B and z_B Higgs are not produced in the Associative Production with W, nor in the Vector Boson Fusion process.

7.3. h_A, h_B and z_B interference in gluon fusion

The dominant Higgs production process at LHC is through Gluon Fusion, mediated by the top quark loop. All three Higgs are emitted by the top quark loop according to the Lagrangian:

$$\mathcal{L}^t_{h_A, h_B, z_B} = -\frac{gm_t}{2m_W} h_A \bar{t}t - \frac{gm_b}{2m_W} h_A \bar{b}b - \frac{gm_b}{2m_W} h_B \bar{t}t$$
$$+ \frac{gm_t}{2m_W} h_B \bar{b}b + i\frac{gm_b}{2m_W} z_B \bar{t}\gamma_5 t + i\frac{gm_t}{2m_W} z_B \bar{b}\gamma_5 b. \tag{34}$$

In the fully degenerate scenario, where all the members of the Higgs family have the same mass, we have the interesting possibility of a quantum interference among the three Higgs through their subsequent decays, say, into the bottom quarks.

For on resonance, the amplitude for $b\bar{b}$ production is given by

$$\langle b\bar{b}|gg\rangle = \frac{\mathcal{M}^S_{gg}(v)(\hat{v})(\bar{b}b)}{v_t^2(-im_H\Gamma_A)} + \frac{\mathcal{M}^S_{gg}(-\hat{v})v(\bar{b}b)}{v_t^2(-im_H\Gamma_B)}$$
$$+ \frac{\mathcal{M}^P_{gg}(\hat{v})(v)(\bar{b}\gamma_5 b)}{v_t^2(-im_H\Gamma_{z_B})}. \tag{35}$$

In this scenario, therefore, at $s = m_H^2$, the observed Higgs peak in Gluon Fusion consists of three peaks, with different widths. However, with a tree-level estimate of the decay widths

$$\Gamma_{h_A \to b\bar{b}} = \frac{3g^2 m_b^2}{32\pi m_W^2} \left(1 - \frac{4m_b^2}{m_H^2}\right)^{3/2}, \qquad (36)$$

$$\Gamma_{h_B \to b\bar{b}} = \frac{3g^2 m_t^2}{32\pi m_W^2} \left(1 - \frac{4m_b^2}{m_H^2}\right)^{3/2}, \qquad (37)$$

$$\Gamma_{z_B \to b\bar{b}} = \frac{3g^2 m_t^2}{32\pi m_W^2} \left(1 - \frac{4m_b^2}{m_H^2}\right)^{1/2}, \qquad (38)$$

what we find is that the narrow h_A resonance dominates over the other two broader resonance by three orders of magnitude, so that

$$\langle b\bar{b}|gg\rangle = \frac{\mathcal{M}^S_{gg}(v)(\hat{v})(\bar{b}b)}{v_t^2(-im_H \Gamma_A)}[1 + O(10^{-3})]. \qquad (39)$$

Off resonance, for $s \gg m_H^2$, we have

$$\langle b\bar{b}|gg\rangle|_{h_A, h_B} = \frac{\mathcal{M}^S_{gg}(v\hat{v})(\bar{b}b)}{v_t^2(s - m_H^2 - im_H \Gamma_A)}$$

$$+ \frac{\mathcal{M}^S_{gg}(-\hat{v}v)(\bar{b}b)}{v_t^2(s - m_H^2 - im_H \Gamma_B)} \xrightarrow{s \gg m_H^2} 0$$

so that the h_A and h_B states interfere destructively off resonance and the LHC signal above SM background come from the z_B field, given by

$$\langle b\bar{b}|gg\rangle|_{z_B} = \frac{\mathcal{M}^P_{gg}(\hat{v}v)(\bar{b}\gamma_5 b)}{v_t^2(s - m_H^2 - im_H \Gamma_{z_B})}. \qquad (40)$$

7.4. The observation of Higgs decay into 2γ

The question of the parity of the SM Higgs was settled by the precision measurement of the Higgs decay into 2γ, demonstrating that it is a scalar field, rather than a pseudoscalar. In our scenario, as Eq. (34) shows, both h_A and h_B fields are scalar, while z_B is a pseudoscalar. All three are produced on resonance, but h_A is a narrow peak that dominates, and thus our scenario is not ruled out, *per se*, by the 2γ data.

7.5. h_{ij}, z_{ij}, $\hat{\phi}^{+ij}$ and $\hat{h}_{ij}$, $\hat{z}_{ij}$, ϕ^{+ij} with $ij \neq 33$

These Higgs fields have the Yukawa interactions of the form

$$\mathcal{L}_Y = -h_t \sum_{(ij)\neq(33)} \left\{ \bar{u}_R^i u_L^j \frac{(h_{ij} - iz_{ij})}{\sqrt{2}} + \bar{d}_R^i d_L^j \frac{(\tilde{h}_{ij} + i\tilde{z}_{ij})}{\sqrt{2}} \right.$$

$$\left. - \bar{u}_R^i d_L^j \phi_{ij}^+ + \bar{d}_R^i u_L^j \tilde{\phi}_{-ij} \right\} + \text{h.c.} \qquad (41)$$

They are produced with associative production through $q\bar{q}$ pairs. As these production cross-sections are much smaller than the dominant gluon fusion cross-sections, more precision data is needed to look for their signals.

8. Conclusion

What I have sketched here in this talk is but a broad outline of the subtleties contained in the Enriched Standard Model. The experts in the field can easily see that I have not gone beyond tree level and simple leading order (LO) considerations, while the SM machinery is up to NNLO and beyond. And yet, in the hunger and thirst for new discoveries, it may be an interesting exercise for us to look beyond and under the troves of LHC data to spot the quantum mechanics of degenerate states, their production and interference in a rich family of Higgs bosons. Thank you for your attention.

Note Added in Proof

New LHC data at 13 TeV show possible new Higgs at 750 GeV. If confirmed, the Higgs family tree in Table 1 would indicate scenario with $\lambda_2 = \lambda_3 \sim 36\lambda_1$, and consequently a rich structure of h_B, z_B, ϕ_B^+ and others in that higher mass region.

References

1. C. N. Yang and R. L. Mills, *Phys. Rev.* **96**, 191 (1954).
2. S. Glashow, *Nucl. Phys.* **22**, 579 (1961).
3. S. Weinberg, *Phys. Rev. Lett.* **19**, 1264 (1967).
4. A. Salam, in *Elementary Particle Theory*, ed. N. Svartholm (Almqvist and Wiksells, 1969), p. 367.
5. Y. Nambu, *Phys. Rev. Lett.* **4**, 380 (1960).
6. P. W. Higgs, *Phys. Lett.* **12**, 132 (1964); *Phys. Rev. Lett.* **13**, 508 (1964); *Phys. Rev.* **145**, 1156 (1966).
7. R. Englert and R. Brout, *Phys. Rev. Lett.* **13**, 321 (1964).
8. G. S. Guralnik, C. R. Hagen and T. W. B. Kibble, *Phys. Rev. Lett.* **13**, 585 (1964); *Adv. Phys.* **2**, 567 (1967).
9. G. 't Hooft, *Nucl. Phys.* B **35**, 167 (1971).
10. G. 't Hooft and M. Veltman, *Nucl. Phys.* B **44**, 189 (1972).
11. L. D. Faddeev and V. N. Popov, *Phys. Lett.* B **35**, 29 (1967).
12. H. Fritzsch and M. Gell-Mann, *ICHEP*, Chicago (1972), eConf C 720906 v2 p. 135 (1972).
13. D. Gross and F. Wilczek, *Phys. Rev. Lett.* **30**, 1343 (1973).
14. D. Politzer, *Phys. Rev. Lett.* **30**, 1346 (1973).
15. F. Wilczek, *Phys. Today* **53**, 22 (2000).
16. M. Lüscher, CERN-TH/2002-321.
17. R. Mohapatra and G. Senjanovic, *Phys. Rev.* D **23**, 165 (1981).

18. N. P. Chang, *Int. J. Mod. Phys. A* **29**, 1444008 (2014); preliminary version presented in N. P. Chang, What if the Higgs has Brothers?, *Proc. of the Conf. in Honor of the 90th Birthday of Freeman Dyson* (World Scientific, 2014).
19. H. Georgi and S. Weinberg, *Phys. Rev. D* **17**, 275 (1978).
20. ATLAS Collab. (G. Aad *et al.*), *Phys. Lett. B* **716**, 1 (2012); CMS Collab. (S. Chatrchyan *et al.*), *ibid.* **716**, 30 (2012); ATLAS Collab. and CMS Collab. (G. Aad *et al.*), *Phys. Rev. Lett.* **114**, 191803 (2015).

On the Physics of the Minimal Length: The Question of Gauge Invariance

Lay Nam Chang[1], Djordje Minic[1], Ahmed Roman[1,2],
Chen Sun[1] and Tatsu Takeuchi[1]

[1] *Center for Neutrino Physics, Department of Physics, Virginia Tech,
Blacksburg VA 24061, USA*
[2] *Department of Organismic and Evolutionary Biology, Harvard University,
Cambridge MA 02138, USA*
laynam@vt.edu, dminic@vt.edu, mido@vt.edu,
sunchen@vt.edu, takeuchi@vt.edu

In this note we discuss the question of gauge invariance in the presence of a minimal length. This contribution is prepared for the celebration of the 60th anniversary of the Yang–Mills theory.

Keywords: Minimal length; gauge invariance.

1. Introduction

Yang–Mills theory[1] represents one of the most remarkable achievements of theoretical physics in the second half of the 20th century.[a] Its fundamental importance is found in high energy physics as well as condensed matter physics and mathematics.[b] In this note we address the issue of gauge invariance which underlies the structure of Yang–Mills theory in the context of theories endowed with a minimal length.

One of the ubiquitous features of quantum gravity[11] is its possession of a fundamental length scale. In a theory of quantum gravity such as string theory,[12–16] the fundamental scale determines the typical spacetime extension of a fundamental string.[17] In canonical string theory this is $\ell_S = \sqrt{\alpha'}$, where $\hbar c/\alpha'$ is the string tension. Such a feature is to be expected of any candidate theory of quantum gravity, since gravity itself is characterized by the Planck length $\ell_P = \sqrt{\hbar G_N/c^3}$. Moreover, $\ell_P \sim \ell_S$ is understood to be the *minimal length* below which spacetime distances cannot be resolved:[18–22]

$$\delta s \gtrsim \ell_P \sim \ell_S. \qquad (1)$$

Local quantum field theory, on the other hand, is completely oblivious to the presence of such a scale, despite being the putative infrared limit of some more

---

[a] For a historical account of the concept of gauge invariance that lead to Yang–Mills theory, see Ref. 2. See also the recent account by Yang in Ref. 3.
[b] For a recent discussion of the mass gap problem in the pure Yang–Mills theory[4] see, for example, Refs. 5–10.

fundamental theory, such as string theory. A natural question to ask is, therefore, whether the formalism of quantum mechanics (and thus of an effective local quantum field theory) can be deformed or extended in such a way as to consistently incorporate the minimal length. If it is at all possible, the precise manner in which quantum mechanics must be modified may point to solutions of yet unresolved mysteries such as the cosmological constant problem,[23–27] which is essentially quantum gravitational in its origin. Such a structure should also illuminate the nature of quantum gravity, including string theory.[28–32]

The very idea of the minimal length in the context of quantum theory has a fascinating and long history. It was used by Heisenberg in 1930[33,34] to address the infinities of the then newly formulated theory of quantum electrodynamics.[35] Over the years, the idea has been picked up by many authors in a plethora of contexts, e.g. Refs. 36–44 to list just a few. (For a comprehensive bibliography consult Ref. 22.) Various ways to deform or extend quantum mechanics have also been suggested.[45–47] In this note, we focus our attention on how a minimal length can be introduced into quantum mechanics by modifying its algebraic structure.[48–50] In particular, we follow our previous work in Refs. 51–58.

The starting point of our discussion is the minimal length uncertainty relation (MLUR),[59,60] also known as the generalized uncertainty principle (GUP)

$$\delta x \sim \left(\frac{\hbar}{\delta p} + \alpha' \frac{\delta p}{\hbar} \right), \qquad (2)$$

which is suggested by a re-summed perturbation expansion of the string-string scattering amplitude in a flat spacetime background.[61–64] This is essentially a Heisenberg microscope argument[65] in the S-matrix language[66–69] with fundamental strings used to probe fundamental strings. The first term inside the parentheses on the right-hand side is the usual Heisenberg term coming from the shortening of the probe-wavelength as momentum is increased, while the second-term can be understood as due to the lengthening of the probe string as more energy is pumped into it:

$$\delta p = \frac{\delta E}{c} \sim \frac{\hbar}{\alpha'} \delta x. \qquad (3)$$

Equation (2) implies that the uncertainty in position, δx, is bounded from below by the string length scale,

$$\delta x \gtrsim \sqrt{\alpha'} = \ell_S, \qquad (4)$$

where the minimum occurs at

$$\delta p \sim \frac{\hbar}{\sqrt{\alpha'}} = \frac{\hbar}{\ell_S} \equiv \mu_S. \qquad (5)$$

Thus, ℓ_S is the minimal length below which spatial distances cannot be resolved, consistent with Eq. (1). In fact, the MLUR can be motivated by fairly elementary general relativistic considerations independent of string theory, which suggests that it is a universal feature of quantum gravity.[18–21]

Note that in the trans-Planckian momentum region $\delta p \gg \mu_S$, the MLUR is dominated by the behavior of Eq. (3), which implies that large δp (UV) corresponds to large δx (IR), and that there exists a correspondence between UV and IR physics. Such UV/IR relations have been observed in various string dualities,[12–16] and in the context of AdS/CFT correspondence[70,71] (albeit between the bulk and boundary theories). Thus, the MLUR captures another distinguishing feature of string theory.[c]

Note that the recent reformulation of quantum gravity in the guise of metastring theory[29–32] based on such novel concepts as relative locality[85] and dynamical energy-momentum space,[d] strongly suggests that the fundamental length scale always goes hand-in-hand with the fundamental energy scale, and that the two appear in a symmetric manner, suggested by Born reciprocity.[e] In this note, we concentrate only on the implications of the fundamental length scale. In particular, we make a few observations on the question of gauge invariance in the presence of a minimal length. This seems to be an appropriate topic for the celebration of the 60th anniversary of Yang–Mills theory.[1]

2. Local Gauge Invariance in the Presence of a Minimal Length

The question of gauge invariance in the presence of a minimal length has been addressed in the context of non-commutative field theories (NCFT).[97–103] In that case the minimal length can be related, in a particular realization, to an effective magnetic length. Here we want to investigate this question from a different point of view, following our previous work on the minimal length as motivated by the foundations of quantum gravity and string theory.[17] As we will note in what follows, the algebra of NCFT, in which the question of gauge invariance is fully understood, is a limiting case of a more general discussion involving a minimal length motivated by quantum gravity and string theory.

---

[c]In addition to the MLUR, another uncertainty relation has been pointed out by Yoneya as characteristic of string theory. This is the so-called spacetime uncertainty relation (STUR) $\delta x\, \delta t \sim \ell_S^2/c$, which can be motivated in a somewhat hand-waving manner by combining the usual energy-time uncertainty relation $\delta E\, \delta t \sim \hbar$ with Eq. (3).[72–74] However, it can also be supported via an analysis of D0-brane scattering in certain backgrounds in which δx can be made arbitrary small at the expense of making the duration of the interaction δt arbitrary large.[75–84] While the MLUR pertains to dynamics of a particle in a non-dynamic spacetime, the STUR can be interpreted to pertain to the dynamics of spacetime itself in which the size of a quantized spacetime cell is preserved.

[d]There are many examples of fundamental physics structures that stay unchanged under the interchange of conjugate variables, such as spatial and momentum coordinates: $x \to p$ and $p \to -x$. See Refs. 86 and 87. By combining the uncertainty relation between x and p and the fact that x is endowed with a metric structure in general relativity, Born reciprocity would suggest a dynamical momentum and thus a dynamical phase space. See Ref. 88.

[e]Some of the historically interesting follow-up work on Born reciprocity and dynamical energy-momentum space can be found in Refs. 89–96.

2.1. Commutation relations

To place the MLUR, Eq. (2), onto a firmer footing, we begin by rewriting it as

$$\delta x\, \delta p \geq \frac{\hbar}{2}\left(1 + \beta\, \delta p^2\right), \tag{6}$$

where we have introduced the parameter $\beta = \alpha'/\hbar^2$. This uncertainty relation can be reproduced by deforming the canonical commutation relation between $\hat{x}$ and $\hat{p}$ to:

$$\frac{1}{i\hbar}[\hat{x},\hat{p}] = 1 \quad \longrightarrow \quad \frac{1}{i\hbar}[\hat{x},\hat{p}] = A(\hat{p}^2), \tag{7}$$

with $A(p^2) = 1 + \beta p^2$. Indeed, we find

$$\delta x\, \delta p \geq \frac{1}{2}\left|\langle[\hat{x},\hat{p}]\rangle\right| = \frac{\hbar}{2}\left(1 + \beta\langle \hat{p}^2\rangle\right) \geq \frac{\hbar}{2}\left(1 + \beta\, \delta p^2\right), \tag{8}$$

since $\delta p^2 = \langle \hat{p}^2\rangle - \langle \hat{p}\rangle^2$. The function $A(p^2)$ can actually be more generic, with βp^2 being the linear term in its expansion in p^2.

When we have more than one spatial dimension, the above commutation relation can be generalized to[17]

$$\frac{1}{i\hbar}[\hat{x}_i,\hat{p}_j] = A(\hat{\mathbf{p}}^2)\,\delta_{ij} + B(\hat{\mathbf{p}}^2)\,\hat{p}_i\hat{p}_j, \tag{9}$$

where $\hat{\mathbf{p}}^2 = \sum_i \hat{p}_i^2$. The right-hand side is the most general form that depends only on the momentum and respects rotational symmetry. Assuming that the components of the momentum commute among themselves,

$$[\hat{p}_i,\hat{p}_j] = 0, \tag{10}$$

the Jacobi identity demands that

$$\frac{1}{i\hbar}[\hat{x}_i,\hat{x}_j] = -\left\{2(\hat{A} + \hat{B}\hat{\mathbf{p}}^2)\hat{A}' - \hat{A}\hat{B}\right\}\hat{\ell}_{ij}, \tag{11}$$

where we have used the shorthand $\hat{A} = A(\hat{\mathbf{p}}^2)$, $\hat{A}' = \dfrac{dA}{dp^2}(\hat{\mathbf{p}}^2)$, $\hat{B} = B(\hat{\mathbf{p}}^2)$, and $\hat{\ell}_{ij} = (\hat{x}_i\hat{p}_j - \hat{x}_j\hat{p}_i)/\hat{A}$. Here, $\hat{\ell}_{ij}$ is the angular momentum operator which generates rotations, as can be seen from its commutation relations which are

$$\frac{1}{i\hbar}[\hat{\ell}_{ij},\hat{x}_k] = \delta_{ik}\hat{x}_j - \delta_{jk}\hat{x}_i,$$

$$\frac{1}{i\hbar}[\hat{\ell}_{ij},\hat{p}_k] = \delta_{ik}\hat{p}_j - \delta_{jk}\hat{p}_i,$$

$$\frac{1}{i\hbar}[\hat{\ell}_{ij},\hat{\ell}_{mn}] = \delta_{im}\hat{\ell}_{jn} - \delta_{in}\hat{\ell}_{jm} + \delta_{jn}\hat{\ell}_{im} - \delta_{jm}\hat{\ell}_{in}. \tag{12}$$

Note that the non-commutativity of the components of position can be interpreted as a reflection of the dynamic nature of space itself, as would be expected in quantum gravity.

Various choices for the functions $A(\mathbf{p}^2)$ and $B(\mathbf{p}^2)$ have been considered in the literature.[37,48,50,104–109] Here, we choose[f]

$$A(\mathbf{p}^2) = 1 - \beta \mathbf{p}^2, \quad B(\mathbf{p}^2) = 2\beta, \tag{13}$$

which leads to the algebra

$$[\hat{p}_i, \hat{p}_j] = 0,$$

$$\frac{1}{i\hbar}[\hat{x}_i, \hat{p}_j] = (1 - \beta \hat{\mathbf{p}}^2)\delta_{ij} + 2\beta \hat{p}_i \hat{p}_j,$$

$$\frac{1}{i\hbar}[\hat{x}_i, \hat{x}_j] = 4\beta \hat{\ell}_{ij}. \tag{14}$$

In 1D this reduces to the form we assumed in Eq. (7):

$$\frac{1}{i\hbar}[\hat{x}, \hat{p}] = (1 - \beta \hat{p}^2) + 2\beta \hat{p}^2 = 1 + \beta \hat{p}^2. \tag{15}$$

For higher dimensions, the diagonal commutators read

$$\frac{1}{i\hbar}[\hat{x}_i, \hat{p}_i] = (1 + \beta \hat{p}_i^2) - \beta \sum_{j \neq i} \hat{p}_j^2, \tag{16}$$

(no summation of repeated indices) which implies

$$\delta x_i \, \delta p_i \geq \frac{1}{2}\left|\langle[\hat{x}_i, \hat{p}_i]\rangle\right| = \frac{\hbar}{2}\left|1 + \beta \langle \hat{p}_i^2 \rangle - \beta \sum_{j \neq i} \langle \hat{p}_j^2 \rangle\right|. \tag{17}$$

If $\langle \hat{p}_j^2 \rangle = 0$ for $j \neq i$, this in turn implies $\delta x_i \sim \delta p_i$ for $\delta p_i \gg 1/\sqrt{\beta}$. Despite this observation, summing over i gives us

$$\sum_{i=1}^{D} \delta x_i \, \delta p_i \geq \frac{\hbar}{2} \sum_{i=1}^{D} \left|\langle[\hat{x}_i, \hat{p}_i]\rangle\right|$$

$$\geq \frac{\hbar}{2}\left|\sum_{i=1}^{D} \langle[\hat{x}_i, \hat{p}_i]\rangle\right| = \frac{\hbar}{2}\left|D - \beta(D-2)\langle \hat{\mathbf{p}}^2\rangle\right|, \tag{18}$$

which suggests that for $D \geq 3$, unlike the $D = 1$ case, we should take $\beta < 0$ for a minimal length to exist. For the borderline $D = 2$ case, the lower bound on $\sum_{i=1}^{D} \delta x_i \, \delta p_i$ is independent of β or $\langle \hat{\mathbf{p}}^2 \rangle$, despite the deformation of the algebra, suggesting that β can be assumed to be of either sign, but the existence of the minimal length in this particular case is obscure. We have nevertheless introduced a new length scale into the algebra and continue to refer to $\hbar\sqrt{|\beta|}$ as the minimal length in this case also.

---

[f]Another popular choice in the literature has been

$$A(\mathbf{p}^2) = 1 + \beta \mathbf{p}^2, \quad B(\mathbf{p}^2) = 2\beta,$$

which leads to $[\hat{x}_i, \hat{x}_j] = O(\beta^2)$, rendering this commutator negligible if only the leading order corrections in β are considered.[105,106]

The operators which satisfy the algebra of Eq. (14) can be represented by differential operators in D-dimensional p-space as

$$\hat{p}_i = p_i,$$

$$\hat{x}_i = i\hbar \left[(1 - \beta \mathbf{p}^2) \frac{\partial}{\partial p_i} + 2\beta p_i p_j \frac{\partial}{\partial p_j} + \beta(D - \delta) p_i \right],$$

$$\hat{\ell}_{ij} = -i\hbar \left(p_i \frac{\partial}{\partial p_j} - p_j \frac{\partial}{\partial p_i} \right), \quad (19)$$

(repeated index is summed) with the inner product of the p-space wave-functions given by

$$\langle f | g \rangle_\delta = \int \frac{d^D \mathbf{p}}{(1 + \beta \mathbf{p}^2)^\delta} f^*(\mathbf{p}) g(\mathbf{p}). \quad (20)$$

This form is required to render the $\hat{x}_i$ operators symmetric, where the parameter δ appearing in the representation of $\hat{x}_i$ and the integration measure is arbitrary. By choosing $\delta = 0$ we can simplify the integration measure at the expense of adding the extra term $\beta D p_i$ to the representation of $\hat{x}_i$, which suggests that the singularity in the integration measure at $\mathbf{p}^2 = -1/\beta$ when β is negative is removable, and taking β negative would not cause any intrinsic problems in any number of dimensions as far as the differential operator representation is concerned.

Our choice of the functions $A(\mathbf{p}^2)$ and $B(\mathbf{p}^2)$ in Eq. (13) has been motivated by the fact that it leads to a particularly simple form for the commutator of the position operators, Eq. (14), allowing it to close on the angular momentum operator. This suggests that the position operators of this algebra may be realizable as angular momentum ($\beta > 0$) or boost ($\beta < 0$) operators in a higher dimension, as already pointed out in the pioneering papers of Snyder[37] and Yang.[39] Indeed, we show in the appendix that the $\hat{x}_i$ operators in D-dimensions can be realized as rotation/boost operators in $(D+1)$-dimensions, while the $\hat{p}_i$ operators can be realized as operators obtained by stereographic projection of a $(D+1)$-dimensional hypersphere ($\beta > 0$) or hyperboloid ($\beta < 0$) down to D-dimensions, $1/\sqrt{\beta}$ corresponding to the radius of the hypersphere.[39]

Taking the limit $\beta \to 0$, $\ell_{ij} \to \infty$ while keeping $4\beta \ell_{ij} \equiv \theta_{ij}$ fixed to constants leads to the algebra

$$[\hat{p}_i, \hat{p}_j] = 0,$$

$$[\hat{x}_i, \hat{p}_j] = i\hbar \delta_{ij},$$

$$[\hat{x}_i, \hat{x}_j] = i\hbar \theta_{ij}. \quad (21)$$

This limit can be understood geometrically as taking both the radius-squared of the hypersphere (hyperboloid) $1/\beta$ and the angular momenta ℓ_{ij} to infinity, while keeping their ratio fixed. This is the algebra assumed in NCFT[97–103] but with an important distinction. When θ_{ij} are c-numbers, the $\hat{p}_i$ operator can be identified with the adjoint operator

$$\hat{p}_i = [-\omega_{ij}\hat{x}_j, *], \tag{22}$$

(repeated index is summed) where

$$\theta_{ik}\omega_{kj} = \omega_{ik}\theta_{kj} = \delta_{ij}. \tag{23}$$

Indeed

$$[-\omega_{ik}\hat{x}_k, -\omega_{j\ell}\hat{x}_\ell] = -i\hbar\omega_{ij},$$
$$[\hat{x}_i, -\omega_{jk}\hat{x}_k] = i\hbar\delta_{ij}, \tag{24}$$

and we find

$$[\hat{p}_i, \hat{p}_j] * = [-\omega_{ik}\hat{x}_k, [-\omega_{j\ell}\hat{x}_\ell, *]] - [-\omega_{j\ell}\hat{x}_\ell, [-\omega_{ik}\hat{x}_k, *]]$$
$$= [[-\omega_{ik}\hat{x}_k, -\omega_{j\ell}\hat{x}_\ell], *]$$
$$= [-i\hbar\omega_{ij}, *] = 0,$$
$$[\hat{x}_i, \hat{p}_j] * = \hat{x}_i[-\omega_{jk}\hat{x}_k, *] - [-\omega_{jk}\hat{x}_k, \hat{x}_i*]$$
$$= [\hat{x}_i, -\omega_{jk}\hat{x}_k] * = i\hbar\delta_{ij}*, \tag{25}$$

for any operator $*$. Thus, the momentum operators $\hat{p}_i$ can be expressed in terms of the coordinate operators $\hat{x}_i$, and this facilitates the introduction and understanding of gauge fields in NCFT. This identification cannot be performed in our algebra; our 'momentum space' stays distinct from 'coordinate space,' and though the NCFT algebra is a particular limiting case of ours, the question of gauge invariance takes on a different nature. In particular, our formulation requires us to work in the full phase space instead of coordinate space.

2.2. *Introduction of the gauge field*

Let us now consider the introduction of the gauge field. For simplicity we consider the case of the constant magnetic field. We restrict our attention to the 2D case, which is the lowest dimension which would allow us to introduce a uniform magnetic field, or its analog. The introduction of a uniform magnetic field will also introduce a length scale, the magnetic length $\ell_M = \sqrt{\hbar/eB}$,[g] and its ratio to the minimal length $\ell_S = \hbar\sqrt{|\beta|}$ becomes an important parameter in what follows. The algebra of the position, momentum, and angular momentum in 2D reads:

$$[\hat{p}_i, \hat{p}_j] = 0,$$
$$[\hat{x}_i, \hat{p}_j] = i\hbar\{(1-\beta\hat{\mathbf{p}}^2)\delta_{ij} + 2\beta\hat{p}_i\hat{p}_j\},$$
$$[\hat{x}_i, \hat{x}_j] = 4i\hbar\beta\hat{\ell}_{ij} \equiv 4i\hbar\beta\epsilon_{ij}\hat{\ell},$$

---

[g]We set $c = 1$.

$$[\hat{\ell}, \hat{x}_i] = i\hbar \epsilon_{ij} \hat{x}_j,$$
$$[\hat{\ell}, \hat{p}_i] = i\hbar \epsilon_{ij} \hat{p}_j, \tag{26}$$

where $i, j = 1, 2$. As argued above, we allow β to take on either sign.

The covariant derivative is introduced as a shift of the momentum operator:
$$\hat{D}_i = \hat{p}_i + \hat{A}_i. \tag{27}$$

For a gauge transformation on the state vector $|\psi\rangle$,
$$|\psi\rangle \rightarrow |\psi'\rangle = \hat{U}|\psi\rangle, \tag{28}$$

the covariant derivative must also transform as
$$\hat{D}_i |\psi\rangle \rightarrow \hat{U}\hat{D}_i |\psi\rangle = \underbrace{\hat{U}\hat{D}_i \hat{U}^{-1}}_{\hat{D}'_i} \underbrace{\hat{U}|\psi\rangle}_{|\psi'\rangle}. \tag{29}$$

Therefore,
$$\begin{aligned} \hat{D} \rightarrow \hat{D}'_i &= \hat{U}\hat{D}_i \hat{U}^{-1} \\ &= \hat{U}(\hat{p}_i + \hat{A}_i)\hat{U}^{-1} \\ &= \hat{p}_i + \hat{U}[\hat{p}_i, \hat{U}^{-1}] + \hat{U}\hat{A}_i \hat{U}^{-1} \\ &= \hat{p}_i + \hat{A}'_i, \end{aligned} \tag{30}$$

that is
$$\hat{A}'_i = \hat{U}\hat{A}_i \hat{U}^{-1} + \hat{U}[\hat{p}_i, \hat{U}^{-1}]. \tag{31}$$

The replacement
$$\hat{H} = \frac{\hat{\mathbf{p}}^2}{2m} \rightarrow \hat{H}' = \frac{\hat{\mathbf{D}}^2}{2m} \tag{32}$$

then gives us a gauge covariant (not invariant) Hamiltonian: $\hat{H} \rightarrow \hat{H}' = \hat{U}\hat{H}\hat{U}^{-1}$. The eigenvalues of $\hat{H}$ are invariant and physical since it shares the same eigenvalues with $\hat{H}'$:
$$\hat{H}|\psi\rangle = E|\psi\rangle \rightarrow \underbrace{(\hat{U}\hat{H}\hat{U}^{-1})}_{\hat{H}'} \underbrace{\hat{U}|\psi\rangle}_{|\psi'\rangle} = E \underbrace{\hat{U}|\psi\rangle}_{|\psi'\rangle}. \tag{33}$$

In analogy with the $\beta = 0$ case, we introduce the vector potential, using 3D notation, as
$$\hat{\mathbf{A}} = \frac{e}{2} \mathbf{B} \times \hat{\mathbf{r}} = \frac{eB}{2} \begin{bmatrix} -\hat{x}_2 \\ \hat{x}_1 \end{bmatrix}, \tag{34}$$

where $\mathbf{B}$ is the putative uniform magnetic field in the z-direction. The magnetic field, defined in the usual fashion as a commutator of the covariant derivative, becomes a gauge covariant operator
$$\hat{f} = \frac{1}{i\hbar}[\hat{D}_1, \hat{D}_2] = eB + (eB)^2 \beta \hat{\ell}, \tag{35}$$

which transforms as $\hat{f} \to \hat{f}' = \hat{U}\hat{f}\hat{U}^{-1}$. We define the magnetic field strength as its trace:

$$f \equiv \text{Tr}\hat{f} = \text{Tr}\left[eB + (eB)^2 \beta \hat{\ell}\right] = eB, \tag{36}$$

where the trace is over the Hilbert space, and we have assumed

$$\text{Tr}\,\hat{1} = 1, \quad \text{Tr}\,\hat{\ell} = 0. \tag{37}$$

So the magnetic field strength is the same as the $\beta = 0$ case.

Note that the vector potential of Eq. (34) satisfies the Coulomb gauge condition

$$\sum_{i=1}^{2}[\hat{p}_i, \hat{A}_i] = \frac{eB}{2}\left([\hat{p}_1, -\hat{x}_2] + [\hat{p}_2, \hat{x}_1]\right)$$

$$= i\hbar eB\beta\left(\hat{p}_2\hat{p}_1 - \hat{p}_1\hat{p}_2\right) = 0, \tag{38}$$

so $\hat{\mathbf{p}} \cdot \hat{\mathbf{A}} = \hat{\mathbf{A}} \cdot \hat{\mathbf{p}}$. We find

$$\hat{H}' = \frac{1}{2m}\left(\hat{\mathbf{p}} + \hat{\mathbf{A}}\right)^2$$

$$= \frac{1}{2m}\left(\hat{\mathbf{p}}^2 + 2\hat{\mathbf{A}} \cdot \hat{\mathbf{p}} + \hat{\mathbf{A}}^2\right)$$

$$= \frac{\hat{\mathbf{p}}^2}{2m} + \frac{eB}{2m}(\hat{x}_1\hat{p}_2 - \hat{x}_2\hat{p}_1) + \frac{e^2}{8m}\hat{\mathbf{r}}^2 B^2$$

$$= \frac{\hat{\mathbf{p}}^2}{2m} + \frac{eB}{2m}(1 - \beta\hat{\mathbf{p}}^2)\hat{\ell} + \frac{e^2}{8m}\hat{\mathbf{r}}^2 B^2$$

$$= \frac{\hat{\mathbf{p}}^2}{2m} + \hat{\mu}B - \frac{1}{2}\hat{\alpha}B^2, \tag{39}$$

where we have identified the magnetic dipole moment and the magnetic polarizability with

$$\hat{\mu} = \frac{e}{2m}(1 - \beta\hat{\mathbf{p}}^2)\hat{\ell}, \quad \hat{\alpha} = -\frac{e^2}{4m}\hat{\mathbf{r}}^2. \tag{40}$$

The dependence of the magnetic dipole moment on $\beta\hat{\mathbf{p}}^2$ is intriguing. It implies that, for $\beta > 0$, the magnetic dipole moment vanishes when $\langle\hat{\mathbf{p}}^2\rangle = 1/\beta$ and can even point in the opposite direction from the angular momentum when $\langle\hat{\mathbf{p}}^2\rangle > 1/\beta$. (Note that $\hat{\mathbf{p}}^2$ and $\hat{\ell}$ commute so they can be simultaneously diagonalized.)

Now, let us perform a gauge transformation. Consider the transformation

$$\hat{U} = e^{-i\epsilon(\hat{x}_1\hat{x}_2 + \hat{x}_2\hat{x}_1)/2\hbar}. \tag{41}$$

To leading order in ϵ we find

$$\hat{U}\hat{p}_1\hat{U}^{-1} = \hat{p}_1 - \frac{i\epsilon}{2\hbar}[\hat{x}_1\hat{x}_2 + \hat{x}_2\hat{x}_1, \hat{p}_1] + \cdots$$

$$= \hat{p}_1 + \epsilon\hat{x}_2 + \frac{\epsilon\beta}{2}\left\{2(\hat{x}_1\hat{p}_1\hat{p}_2 + \hat{p}_1\hat{p}_2\hat{x}_1) + \hat{x}_2\left(\hat{p}_1^2 - \hat{p}_2^2\right)\right.$$

$$\left. + \left(\hat{p}_1^2 - \hat{p}_2^2\right)\hat{x}_2\right\} + \cdots,$$

$$\hat{U}\hat{A}_1\hat{U}^{-1} = -\frac{eB}{2}\hat{U}\hat{x}_2\hat{U}^{-1}$$
$$= -\frac{eB}{2}\left[\hat{x}_2 - 2\epsilon\beta\left(\hat{\ell}\hat{x}_2 + \hat{x}_2\hat{\ell}\right) + \cdots\right], \quad (42)$$

with similar expressions for $\hat{U}\hat{p}_2\hat{U}^{-1}$ and $\hat{U}\hat{A}_2\hat{U}^{-1}$. When $\beta = 0$, we have

$$\hat{U}(\hat{p}_1 + \hat{A}_1)\hat{U}^{-1} \xrightarrow{\beta=0} \hat{p}_1 - \left(\frac{eB}{2} - \epsilon\right)\hat{x}_2,$$

$$\hat{U}(\hat{p}_2 + \hat{A}_2)\hat{U}^{-1} \xrightarrow{\beta=0} \hat{p}_2 + \left(\frac{eB}{2} + \epsilon\right)\hat{x}_1, \quad (43)$$

with $\epsilon = \pm\frac{eB}{2}$ leading to Landau gauge. For the $\beta \neq 0$ case, the components of the gauge transformed vector potential $\hat{A}'_i$ ($i = 1, 2$) become infinite series involving both position and momentum operators. They are intrinsically non-local operators.

Though the gauge transformation operator, Eq. (41), depends only on the position operators, the transformation itself is highly non-local. This suggests a generalization of gauge transformations to those that are non-local (in x-space) even in the $\beta = 0$ limit, such as:

$$\hat{U} = e^{-i\epsilon\hat{p}_1\hat{p}_2/\hbar}, \quad \hat{U} = e^{-i\epsilon(\hat{x}_1\hat{p}_1 + \hat{x}_2\hat{p}_2)/\hbar}, \quad \text{etc.} \quad (44)$$

The rotation operator $\hat{U} = \exp[-i\theta\hat{\ell}/\hbar]$ will then be just a particular "gauge transformation" which leaves the Hamiltonian invariant. This type of generalization may be required of us in the presence of a minimal length, and it also merges nicely with the ideas of UV/IR correspondence[12-16,70,71], Born reciprocity[86,87], etc. which are expected in that context.

The traceless term in $\hat{f}$, Eq. (35), is dependent on the ratio of the minimal and magnetic lengths:

$$\ell_S = \hbar\sqrt{|\beta|}, \quad \ell_M \equiv \sqrt{\frac{\hbar}{eB}}, \quad (45)$$

and we can write

$$\hat{f} = eB\left[1 \pm \left(\frac{\ell_S}{\ell_M}\right)^2 \hat{L}\right], \quad \hat{L} = \frac{\hat{\ell}}{\hbar}, \quad (46)$$

where the $\pm$ refers to the sign of β. The Hamiltonian of a particle of mass m coupled to the gauge field is

$$\hat{H}' = \frac{\hat{D}_1^2 + \hat{D}_2^2}{2m} = \frac{\hat{D}_+\hat{D}_- + \hat{D}_-\hat{D}_+}{2m}, \quad (47)$$

where

$$\hat{D}_\pm \equiv \frac{\hat{D}_1 \mp i\hat{D}_2}{\sqrt{2}}, \quad \hat{D}_1 = \hat{p}_1 + \frac{eB}{2}\hat{x}_2, \quad \hat{D}_2 = \hat{p}_2 - \frac{eB}{2}\hat{x}_1, \quad (48)$$

and

$$\frac{1}{i\hbar}[\hat{D}_1, \hat{D}_2] = \frac{1}{\hbar}[\hat{D}_-, \hat{D}_+] = \hat{f} = eB\left[1 \pm \left(\frac{\ell_S}{\ell_M}\right)^2 \hat{L}\right]. \tag{49}$$

Consider a subspace of the Hilbert space spanned by the eigenvectors of the angular momentum operator $\hat{L}$ with eigenvalue L. In that space, the field strength operator $\hat{f}$ can be replaced by the number

$$\langle \hat{f} \rangle = f_L \equiv eB\left[1 + (eB\hbar\beta)L\right] = eB\left[1 \pm \left(\frac{\ell_S}{\ell_M}\right)^2 L\right]. \tag{50}$$

Note that f_L can be negative depending on the value and sign of βL. Let us define

$$\hat{a}_L \equiv \frac{\hat{D}_-}{\sqrt{\hbar f_L}}, \quad \hat{a}_L^\dagger \equiv \frac{\hat{D}_+}{\sqrt{\hbar f_L}}, \tag{51}$$

when $f_L > 0$, and

$$\hat{a}_L \equiv \frac{\hat{D}_+}{\sqrt{\hbar |f_L|}}, \quad \hat{a}_L^\dagger \equiv \frac{\hat{D}_-}{\sqrt{\hbar |f_L|}}, \tag{52}$$

when $f_L < 0$. In both cases we have

$$[\hat{a}_L, \hat{a}_L^\dagger] = 1, \tag{53}$$

and the Hamiltonian restricted to the L-subspace, $\hat{H}'_L$, can be written as

$$\hat{H}'_L = \hbar\omega_L \left(\frac{\hat{a}_L \hat{a}_L^\dagger + \hat{a}_L^\dagger \hat{a}_L}{2}\right) = \hbar\omega_L \left(\hat{a}_L^\dagger \hat{a}_L + \frac{1}{2}\right), \tag{54}$$

where

$$\omega_L = \frac{|f_L|}{m} = \omega_c \left|1 \pm \left(\frac{\ell_S}{\ell_M}\right)^2 L\right|, \quad \omega_c = \frac{eB}{m}. \tag{55}$$

Thus, the energy eigenvalues are

$$E_{L,n} = \hbar\omega_L \left(n + \frac{1}{2}\right), \quad n = 0, 1, 2, \ldots. \tag{56}$$

They are evenly spaced in each L-subspace with the separation depending on the ratio of the minimal to magnetic lengths ℓ_S/ℓ_M, the angular momentum L, and the sign of β. And interesting case is when $f_L = 0$. In that case, $\hat{D}_1$ and $\hat{D}_2$ commute, and the Hamiltonian $\hat{H}'_L$ for that value of L is the free particle Hamiltonian.

2.3. Introduction of the gauge field — alternative approach

The algebra involving β is covariant under changes in parametrization in p-space, so to a large extent, we may choose the functions $A(\mathbf{p}^2)$ to be anything we like. The situation is different however if we choose to change the parametrization of both p and x. Indeed, for some singular cases, we can recover the Heisenberg algebra for restricted domains over which the operators are defined.

We begin by introducing:

$$\hat{\Pi}_{\pm} \equiv \frac{1}{1+\beta\hat{\mathbf{p}}^2}\left(\frac{\hat{p}_1 \mp i\hat{p}_2}{\sqrt{2}}\right),$$

$$\hat{\Xi}_{\pm} \equiv \frac{1}{2}\left\{\left(\frac{1+\beta\hat{\mathbf{p}}^2}{1-\beta\hat{\mathbf{p}}^2}\right)\left(\frac{\hat{x}_1 \pm i\hat{x}_2}{\sqrt{2}}\right) + \left(\frac{\hat{x}_1 \pm i\hat{x}_2}{\sqrt{2}}\right)\left(\frac{1+\beta\hat{\mathbf{p}}^2}{1-\beta\hat{\mathbf{p}}^2}\right)\right\}. \quad (57)$$

It is straightforward to show that these operators satisfy the canonical commutation relations

$$[\hat{\Xi}_+, \hat{\Xi}_-] = [\hat{\Pi}_+, \hat{\Pi}_-] = 0,$$

$$[\hat{\Xi}_\pm, \hat{\Pi}_\pm] = i\hbar,$$

$$[\hat{\Xi}_\pm, \hat{\Pi}_\mp] = 0. \quad (58)$$

Note that there is no β dependence in this algebra, which is identical to the conventional Heisenberg algebra, with $\beta = 0$. From the perspective of the approach discussed in the appendix, these operators arise by projection of a paraboloid, instead of a sphere ($\beta > 0$), or hyperboloid ($\beta < 0$). In this context, the domains of the operators are restricted to $\mathbf{p}^2 < 1/|\beta|$.

The angular momentum operator can be expressed using these operators as

$$\hat{\ell} = i\left(\hat{\Xi}_+\hat{\Pi}_+ - \hat{\Xi}_-\hat{\Pi}_-\right). \quad (59)$$

Define the deformed free-particle Hamiltonian as

$$\hat{H}_\beta = \frac{1}{2m}\left(\hat{\Pi}_+\hat{\Pi}_- + \hat{\Pi}_-\hat{\Pi}_+\right) = \frac{1}{2m}\frac{\hat{p}_1^2 + \hat{p}_2^2}{(1+\beta\hat{\mathbf{p}}^2)^2}, \quad (60)$$

which commutes with the $\hat{p}_i$'s, as is required for the particle to be 'free,' and converges to the usual Hamiltonian in the limit $\beta \to 0$.

The vector potential is introduced as

$$\hat{A}_+ = -i\xi_1 eB\,\hat{\Xi}_-, \quad \hat{A}_- = +i\xi_2 eB\,\hat{\Xi}_+, \quad (61)$$

where $\xi_1 + \xi_2 = 1$, and the covariant derivative as

$$\hat{D}_+ = \hat{\Pi}_+ + \hat{A}_+ = \hat{\Pi}_+ - i\xi_1 eB\,\hat{\Xi}_-,$$

$$\hat{D}_- = \hat{\Pi}_- + \hat{A}_- = \hat{\Pi}_- + i\xi_2 eB\,\hat{\Xi}_+. \quad (62)$$

Then
$$\hat{f} = \frac{1}{\hbar}[\hat{D}_-, \hat{D}_+] = (\xi_1 + \xi_2)eB = eB. \tag{63}$$

Thus, a more generic gauge can be written down compactly in this approach, and the field strength is gauge invariant without the taking of any trace. Gauge transformation from one gauge to another is effected by the operator

$$\hat{U} = e^{\epsilon(\hat{\Xi}_+^2 - \hat{\Xi}_-^2)/2\hbar}, \tag{64}$$

which shifts the gauge parameters: $\xi_1 \to \xi_1 - \epsilon$, $\xi_2 \to \xi_2 + \epsilon$. For the symmetric $\xi_1 = \xi_2 = \frac{1}{2}$ case, the Hamiltonian in the presence of the gauge field can be written as

$$\begin{aligned}
\hat{H}'_\beta &= \frac{1}{2m}(\hat{D}_+\hat{D}_- + \hat{D}_-\hat{D}_+) \\
&= \frac{(\hat{\Pi}_+\hat{\Pi}_- + \hat{\Pi}_-\hat{\Pi}_+)}{2m} + \frac{eB}{2m}\hat{\ell} + \frac{e^2B^2}{8m}(\hat{\Xi}_+\hat{\Xi}_- + \hat{\Xi}_-\hat{\Xi}_+).
\end{aligned} \tag{65}$$

Since the $\hat{\Xi}$'s and $\hat{\Pi}$'s are canonically commuting, this Hamiltonian is exactly the same as the $\beta = 0$ case. But it is defined over a restricted domain where $|\beta|\mathbf{p}^2 < 1$, and so the corresponding eigenvalues are expected to be different. The presence of the minimal length is encoded in this restriction.

Nonetheless, this Hamiltonian will exhibit the infinite degeneracy present when $\beta = 0$. Consider the following operators:

$$\begin{aligned}
\hat{Q}_+ &\equiv \hat{\Pi}_+ + i\xi_2 eB\,\hat{\Xi}_-, \\
\hat{Q}_- &\equiv \hat{\Pi}_- - i\xi_1 eB\,\hat{\Xi}_+.
\end{aligned} \tag{66}$$

Note the different coefficients of the $\hat{\Xi}$'s from Eq. (62). These operators satisfy the commutation relations

$$\begin{aligned}
\frac{1}{\hbar}[\hat{Q}_-, \hat{Q}_+] &= (\xi_1 + \xi_2)eB = eB, \\
[\hat{Q}_\pm, \hat{D}_\pm] &= [\hat{Q}_\pm, \hat{D}_\mp] = 0.
\end{aligned} \tag{67}$$

That is, $\hat{Q}_\pm$ commute with the Hamiltonian but not with each other. The result is the infinite degeneracy already present in the normal case when $\beta = 0$.

We are yet to understand the full implication of this: the extent to which the minimal length has been 'gauged away,' so to speak, in this approach, or whether the singularities in p-space are physical and lead to observable consequences at the Planck scale. It is also unclear whether the existence of the canonically commuting operators is restricted to 2D. These issues will be explored in an upcoming paper.[110]

3. Summary and Discussion

Note that our discussion generalizes the notion of gauge invariance found in the context of non-commutative field theory (NCFT).[97–103] Given the geometric formulation presented in the appendix, it should be possible to deform the usual discussion from NCFT based on the Moyal product by realizing gauge transformations in terms of the isometries of the curved momentum space, associated with either a spherical or hyperbolic geometry. Here we give a scenario for a more general formulation that would include the geometry of the momentum space.

The first lesson of our discussion is that, in the presence of a minimal length, we should formulate quantum field theory (and thus gauge field theory) in momentum space. Then we should take into account the curved geometry of the momentum space, the curvature being directly tied to the minimal length parameter β. Thus the classical action for such a momentum space field theory (for a scalar field ϕ in $3+1$ dimensions) would read (see Ref. 111)

$$S_m = \int dt \, \frac{d^3\mathbf{p}}{(1+\beta\mathbf{p}^2)^3} \left[\pi(\mathbf{p},t) \, \dot{\phi}(\mathbf{p},t) - H\left(\phi(\mathbf{p},t), \pi(\mathbf{p},t)\right) \right] \quad (68)$$

where $\pi(\mathbf{p},t)$ is the conjugate momentum and H the Hamiltonian density. The gauge field would appear in momentum space as a generalized 1-form:

$$\pi(\mathbf{p},t) \rightarrow \pi(\mathbf{p},t) + A(\mathbf{p},t) \equiv D(\mathbf{p},t). \quad (69)$$

This suggests that the Yang–Mills theory would have to be formulated in momentum space, the curvature being given by the commutators of the above $D(\mathbf{p},t)$. Finally, the quantum theory would be defined in terms of the phase space path integral over the curved momentum space

$$\int \mathcal{D}\phi(\mathbf{p},t) \, \mathcal{D}\pi(\mathbf{p},t) \, e^{\frac{i}{\hbar} S_m} \quad (70)$$

where the measure in the path integral includes the geometry of the curved momentum space

$$\mathcal{D}\phi(\mathbf{p},t) \, \mathcal{D}\pi(\mathbf{p},t) \equiv \prod_i \prod_p \frac{d^3\mathbf{p}}{(1+\beta\mathbf{p}^2)^3} \, d\phi(\mathbf{p},t_i) \, d\pi(\mathbf{p},t_i). \quad (71)$$

In principle, such a path integral defines the required generalization of the Moyal product in our more general case.

The discussion of this note was entirely non-relativistic. Obviously the next step has to involve the issue of relativistic covariance. The usual NCFT based on the Moyal product[97,98,100–103] has some outstanding features, such as the essential non-Abelian nature of even a $U(1)$ formulation, as well as the existence of a two-scale renormalization group that was required for the proper definition of the continuum limit.[112] As we have mentioned, this canonical NCFT can be understood as a particular case of the algebraic structure that involves the minimal length β and the angular momentum L_{ij} found in our discussion. Note that in our case the non-Abelian nature is also present quite generically, but the existence of a two-scale renormalization is not clear. For that we would have to make a more precise sense of the above path integral with a Wilsonian cut-off and follow the Wilson–Polchinski-like discussion of Ref. 112. We note that the existence of a curved and dynamical momentum space as well as the two-scale renormalization is also expected on more fundamental grounds in the recent discussion of quantum gravity in the context of metastring theory,[29–32] which might be underlying the more effective analysis presented in this note. Nevertheless, the considerations of this note point to an interesting interplay of spatial non-commutativity, curved momentum space and gauge invariance in the context of the minimal length physics.

In conclusion, in this paper, we have attempted to shed new light on the issue of the minimal length. Starting from the well-understood concept of the magnetic length, which also serves as the basis for the minimal length in non-commutative field theory (NCFT), we developed a new realization of the spatial minimal length without and with the presence of an external magnetic field. The intuition here is that a uniform magnetic field through an infinite plane can be mapped to a magnetic field of a magnetic monopole through a sphere of a fixed radius. The sphere and the infinite plane are related through a stereographic projection. Thus, in order to realize the minimal spatial length, we start with the symmetries of a sphere and then induce the algebra of position and momentum operators with a minimal length, by implementing stereographic projection from that sphere onto an infinite plane. The same procedure can be realized by a stereographic projection from a hyperboloid onto an infinite plane. (This approach can in principle be covariantized, even though we do not discuss that topic in the present paper.) This procedure provides us with a new view on the origin of the minimal length and points to a formulation of physics that relies on a curved (spherical or hyperbolical) momentum-like space. By gauging this construction in the presence of a uniform magnetic field, we find an interesting interplay between the minimal spatial and magnetic lengths. The overall approach gives us a possible new handle on the experimental searches for the minimal length, which have become more topical and much more realistic given the recent remarkable breakthroughs achieved by the LIGO collaboration,[113] the gravitational interferometry being the natural experimental playground in this arena.

Acknowledgments

The work of DM is supported in part by the U.S. Department of Energy, grant DE-FG02-13ER41917, task A.

Appendix A. Projection from a Higher-Dimensional Momentum Space

In order to simply our notation, let us first introduce dimensionless operators by

$$\hat{X}_i = \frac{\hat{x}_i}{\hbar\sqrt{|\beta|}}, \quad \hat{P}_i = \sqrt{|\beta|}\,\hat{p}_i, \quad \hat{L}_{ij} = \frac{\hat{\ell}_{ij}}{\hbar}. \tag{A.1}$$

Here, we allow for the possibility that β is negative. The algebra in terms of these dimensionless operators is

$$[\hat{P}_i, \hat{P}_j] = 0,$$
$$\frac{1}{i}[\hat{X}_i, \hat{P}_j] = (1 \mp \hat{\mathbf{P}}^2)\delta_{ij} \pm 2\hat{P}_i\hat{P}_j,$$
$$\frac{1}{i}[\hat{X}_i, \hat{X}_j] = \pm 4\hat{L}_{ij}, \tag{A.2}$$

where the upper(lower) sign corresponds to positive(negative) β.

First, consider a unit D-dimensional sphere embedded in $(D+1)$-dimensional momentum space:

$$\hat{\eta}_0^2 + (\hat{\eta}_1^2 + \hat{\eta}_2^2 + \cdots + \hat{\eta}_D^2) = 1, \tag{A.3}$$

where we assume

$$[\hat{\eta}_\alpha, \hat{\eta}_\beta] = 0,$$
$$[\hat{L}_{\alpha\beta}, \hat{\eta}_\gamma] = i(\delta_{\alpha\gamma}\hat{\eta}_\beta - \delta_{\beta\gamma}\hat{\eta}_\alpha). \tag{A.4}$$

Note also

$$[\hat{L}_{0i}, \hat{L}_{0j}] = i\hat{L}_{ij},$$
$$[\hat{L}_{ij}, \hat{L}_{0k}] = i(\delta_{ik}\hat{L}_{0j} - \delta_{jk}\hat{L}_{0i}). \tag{A.5}$$

Stereographic projection from the North(South) pole of the sphere to the $\eta_0 = 0$ hyperplane yields

$$\hat{P}_i = \frac{\hat{\eta}_i}{1 \mp \hat{\eta}_0}, \quad \hat{\mathbf{P}}^2 = \sum_{i=1}^{D}\hat{P}_i^2 = \frac{1 \pm \hat{\eta}_0}{1 \mp \hat{\eta}_0}, \tag{A.6}$$

the upper(lower) sign corresponding to the North(South)-pole projection. See Fig. A1(a). It is straightforward to show that

$$[\hat{L}_{0i}, \hat{P}_j] = \pm\frac{i}{2}\left\{(1 - \hat{\mathbf{P}}^2)\delta_{ij} + 2\hat{P}_i\hat{P}_j\right\},$$
$$[\hat{L}_{ij}, \hat{P}_k] = i(\delta_{ik}\hat{P}_j - \delta_{jk}\hat{P}_i), \tag{A.7}$$

thus we can identify

$$\hat{X}_i = \pm 2\hat{L}_{0i}. \tag{A.8}$$

Next, consider a unit D-dimensional hyperboloid in $(D+1)$-dimensional momentum space:

$$\hat{\eta}_0^2 - (\hat{\eta}_1^2 + \hat{\eta}_2^2 + \cdots + \hat{\eta}_D^2) = 1. \tag{A.9}$$

The generators which keep this hyperboloid invariant are

$$\frac{1}{i}[\hat{M}_{\alpha\beta}, \hat{\eta}_\gamma] = g_{\alpha\gamma}\hat{\eta}_\beta - g_{\beta\gamma}\hat{\eta}_\alpha,$$
$$\frac{1}{i}[\hat{M}_{\alpha\beta}, \hat{M}_{\gamma\delta}] = g_{\alpha\gamma}\hat{M}_{\beta\delta} - g_{\alpha\delta}\hat{M}_{\beta\gamma} + g_{\beta\delta}\hat{M}_{\alpha\gamma} - g_{\beta\gamma}\hat{M}_{\alpha\delta}, \tag{A.10}$$

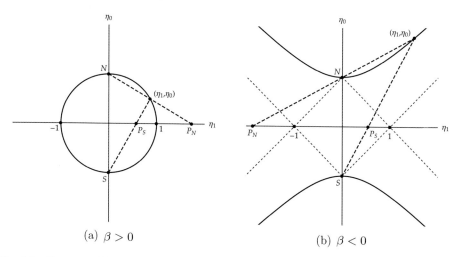

Fig. A1. Stereographic projection from a $(D+1)$-dimensional (a) sphere, and (b) hyperboloid down to D dimensions, shown for the case $D=1$. The north pole projection maps the northern hemisphere (or the northern branch of the hyperboloid) to the outside of the unit circle (sphere), and the southern hemisphere (southern branch of the hyperboloid) to the inside of the unit circle (sphere). For the south pole projection it is the other way around. A switch between the north and south pole projections corresponds to an inversion with respect to the unit circle (sphere) in D dimensions.

where

$$g_{\alpha\beta} = \mathrm{diag}(-1,\underbrace{1,1,\ldots,1}_{n}). \tag{A.11}$$

Writing $\hat{L}_{ij} = \hat{M}_{ij}$ and $\hat{K}_i = \hat{M}_{0i}$, we have

$$\begin{aligned}
{[\hat{L}_{ij}, \hat{\eta}_k]} &= i(\delta_{ik}\hat{\eta}_j - \delta_{jk}\hat{\eta}_i), & [\hat{K}_i, \hat{\eta}_j] &= -i\delta_{ij}\hat{\eta}_0, \\
{[\hat{L}_{ij}, \hat{\eta}_0]} &= 0, & [\hat{K}_i, \hat{\eta}_0] &= -i\hat{\eta}_i, \\
{[\hat{L}_{ij}, \hat{L}_{k\ell}]} &= i(\delta_{ik}\hat{L}_{j\ell} - \delta_{i\ell}\hat{L}_{jk} + \delta_{j\ell}\hat{L}_{ik} - \delta_{jk}\hat{L}_{i\ell}), & & \\
{[\hat{L}_{ij}, \hat{K}_k]} &= i(\delta_{ik}\hat{K}_j - \delta_{jk}\hat{K}_i), & & \\
{[\hat{K}_{0i}, \hat{K}_{0j}]} &= -i\hat{L}_{ij}. & &
\end{aligned} \tag{A.12}$$

Stereographic projection from the North(South) pole (points $\eta_0 = \pm 1$, $\eta_i = 0$) of the hyperboloid to the $\eta_0 = 0$ hyperplane yields

$$\hat{P}_i = \frac{\hat{\eta}_i}{1 \mp \hat{\eta}_0}, \quad \hat{\mathbf{P}}^2 = \sum_{i=1}^{D} \hat{P}_i^2 = \frac{\hat{\eta}_0 \pm 1}{\hat{\eta}_0 \mp 1}, \tag{A.13}$$

the upper(lower) sign corresponding to the North(South)-pole projection. See Fig. A1(b). Note that $\hat{\eta}_0^2 \geq 1$, which leads to the difference in sign for the expression for $\hat{\mathbf{P}}^2$. Again, it is straightforward to show that

$$[\hat{K}_i, \hat{P}_j] = \pm \frac{i}{2}\left\{(1+\hat{\mathbf{P}}^2)\delta_{ij} - 2\hat{P}_i\hat{P}_j\right\},$$

$$[\hat{L}_{ij}, \hat{P}_k] = i(\delta_{ik}\hat{P}_j - \delta_{jk}\hat{P}_i), \qquad (A.14)$$

thus we can identify

$$\hat{X}_i = \pm 2\hat{K}_i. \qquad (A.15)$$

Thus, our algebra can be understood in terms of a projection down from a higher dimension.

References

1. C.-N. Yang and R. L. Mills, Conservation of isotopic spin and isotopic gauge invariance, *Phys. Rev.* **96**, 191 (1954).
2. L. O'Raifeartaigh, *The Dawning of Gauge Theory* (Princeton University Press, 1997).
3. C.-N. Yang, The conceptual origins of Maxwell's equations and gauge theory, *Physics Today* **67**, 45 (2014).
4. A. M. Jaffe and E. Witten, Quantum Yang–Mills theory, in *Clay Mathematics Institute Millennium Problem: Yang–Mills and Mass Gap, Official Problem Description*, 2000. http://www.claymath.org/sites/default/files/yangmills.pdf.
5. D. Karabali and V. P. Nair, A Gauge invariant Hamiltonian analysis for non-Abelian gauge theories in (2+1)-dimensions, *Nucl. Phys. B* **464**, 135 (1996).
6. D. Karabali, C.-J. Kim and V. P. Nair, On the vacuum wave function and string tension of Yang–Mills theories in (2+1)-dimensions, *Phys. Lett. B* **434**, 103 (1998).
7. R. G. Leigh, D. Minic and A. Yelnikov, Solving pure QCD in 2+1 dimensions, *Phys. Rev. Lett.* **96**, p. 222001 (2006).
8. R. G. Leigh, D. Minic and A. Yelnikov, On the Glueball Spectrum of Pure Yang–Mills Theory in 2+1 Dimensions, *Phys. Rev. D* **76**, 065018 (2007).
9. L. Freidel, R. G. Leigh and D. Minic, Towards a solution of pure Yang–Mills theory in 3+1 dimensions, *Phys. Lett. B* **641**, 105 (2006).
10. L. Freidel, R. G. Leigh, D. Minic and A. Yelnikov, On the spectrum of pure Yang–Mills theory, *AMS/IP Stud. Adv. Math.* **44**, 109 (2008).
11. C. Kiefer, *Quantum gravity*, International series of monographs on physics, Vol. 155 (Oxford University Press, 2012).
12. M. B. Green, J. H. Schwarz and E. Witten, *Superstring Theory, Vol. 1: Introduction* (Cambridge University Press, 1988).
13. M. B. Green, J. H. Schwarz and E. Witten, *Superstring Theory, Vol. 2: Loop Amplitudes, Anomalies and Phenomenology* (Cambridge University Press, 1988).

14. J. Polchinski, *String Theory, Vol. 1: An Introduction to the Bosonic String* (Cambridge University Press, 2007).
15. J. Polchinski, *String Theory, Vol. 2: Superstring Theory and Beyond* (Cambridge University Press, 2007).
16. K. Becker, M. Becker and J. H. Schwarz, *String Theory and M-Theory: A Modern Introduction* (Cambridge University Press, 2006).
17. L. N. Chang, Z. Lewis, D. Minic and T. Takeuchi, On the minimal length uncertainty relation and the foundations of string theory, *Adv. High Energy Phys.* **2011**, 493514 (2011).
18. J. A. Wheeler, On the Nature of quantum geometrodynamics, *Annals Phys.* **2**, 604 (1957).
19. C. A. Mead, Possible connection between gravitation and fundamental length, *Phys. Rev.* **135**, B849 (1964).
20. M. Maggiore, A generalized uncertainty principle in quantum gravity, *Phys. Lett. B* **304**, 65 (1993).
21. L. J. Garay, Quantum gravity and minimum length, *Int. J. Mod. Phys. A* **10**, 145 (1995).
22. S. Hossenfelder, Minimal length scale scenarios for quantum gravity, *Living Rev. Rel.* **16**, 2 (2013).
23. S. Weinberg, The Cosmological Constant Problem, *Rev. Mod. Phys.* **61**, 1 (1989).
24. S. M. Carroll, The cosmological constant, *Living Rev. Rel.* **4**, 1 (2001).
25. E. Witten, The cosmological constant from the viewpoint of string theory, in *Sources and Detection of Dark Matter and Dark Energy in the Universe. Proceedings of the 4th International Symposium, DM 2000, Marina del Rey, USA, February 23–25, 2000* (2000).
26. N. Straumann, The history of the cosmological constant problem, in *18th IAP Colloquium on the Nature of Dark Energy: Observational and Theoretical Results on the Accelerating Universe Paris, France, July 1–5, 2002*, 2002.
27. S. Nobbenhuis, Categorizing different approaches to the cosmological constant problem, *Found. Phys.* **36**, 613 (2006).
28. J. Polchinski, What is string theory?, in *NATO Advanced Study Institute: Les Houches Summer School, Session 62: Fluctuating Geometries in Statistical Mechanics and Field Theory Les Houches, France, August 2–September 9, 1994* (1994).
29. L. Freidel, R. G. Leigh and D. Minic, Born reciprocity in string theory and the nature of spacetime, *Phys. Lett. B* **730**, 302 (2014).
30. L. Freidel, R. G. Leigh and D. Minic, Quantum gravity, dynamical phase space and string theory, *Int. J. Mod. Phys. D* **23**, 1442006 (2014).
31. L. Freidel, R. G. Leigh and D. Minic, Metastring theory and modular spacetime, *JHEP* **06**, 006 (2015).
32. L. Freidel, R. G. Leigh and D. Minic, Modular spacetime, *Int. J. Mod. Phys. D* **24**, 1544028 (2015).

33. H. Kragh, Arthur march, werner heisenberg, and the search for a smallest length, *Revue d'histoire des science* **48**, 401 (1995).
34. B. Carazza and H. Kragh, Heisenberg's lattice world: The 1930 theory sketch, *Am. J. Phys.* **63**, 595 (1995).
35. W. Heisenberg and W. Pauli, On quantum field theory (in German), *Z. Phys.* **56**, 1 (1929).
36. M. Born, Modified field equations with a finite radius of the electron, *Nature* **132**, 282 (1933).
37. H. S. Snyder, Quantized space-time, *Phys. Rev.* **71**, 38 (1947).
38. H. S. Snyder, The electromagnetic field in quantized space-time, *Phys. Rev.* **72**, 68 (1947).
39. C. N. Yang, On quantized space-time, *Phys. Rev.* **72**, 874 (1947).
40. C. A. Mead, Observable consequences of fundamental-length hypotheses, *Phys. Rev.* **143**, 990 (1966).
41. T. Padmanabhan, Physical significance of Planck length, *Annals Phys.* **165**, 38 (1985).
42. T. Padmanabhan, Limitations on the operational definition of space-time events and quantum gravity, *Class. Quant. Grav.* **4**, L107 (1987).
43. T. Padmanabhan, Duality and zero point length of space-time, *Phys. Rev. Lett.* **78**, 1854 (1997).
44. M. Kato, Particle theories with minimum observable length and open string theory, *Phys. Lett. B* **245**, 43 (1990).
45. S. Weinberg, Precision tests of quantum mechanics, *Phys. Rev. Lett.* **62**, 485 (1989).
46. S. Weinberg, Testing quantum mechanics, *Annals Phys.* **194**, 336 (1989).
47. C. M. Bender, D. C. Brody and H. F. Jones, Complex extension of quantum mechanics, *Phys. Rev. Lett.* **89**, 270401 (2002), [Erratum: *ibid.* **92**, 119902 (2004)].
48. M. Maggiore, Quantum groups, gravity and the generalized uncertainty principle, *Phys. Rev. D* **49**, 5182 (1994).
49. M. Maggiore, The algebraic structure of the generalized uncertainty principle, *Phys. Lett. B* **319**, 83 (1993).
50. A. Kempf, G. Mangano and R. B. Mann, Hilbert space representation of the minimal length uncertainty relation, *Phys. Rev. D* **52**, 1108 (1995).
51. L. N. Chang, D. Minic, N. Okamura and T. Takeuchi, Exact solution of the harmonic oscillator in arbitrary dimensions with minimal length uncertainty relations, *Phys. Rev. D* **65**, 125027 (2002).

52. L. N. Chang, D. Minic, N. Okamura and T. Takeuchi, The Effect of the minimal length uncertainty relation on the density of states and the cosmological constant problem, *Phys. Rev. D* **65**, 125028 (2002).
53. S. Benczik, L. N. Chang, D. Minic, N. Okamura, S. Rayyan and T. Takeuchi, Short distance versus long distance physics: The Classical limit of the minimal length uncertainty relation, *Phys. Rev. D* **66**, 026003 (2002).
54. S. Benczik, L. N. Chang, D. Minic, N. Okamura, S. Rayyan and T. Takeuchi, Classical implications of the minimal length uncertainty relation, arXiv:hep-th/0209119 (2002).
55. S. Benczik, L. N. Chang, D. Minic and T. Takeuchi, The hydrogen atom with minimal length, *Phys. Rev. A* **72**, 012104 (2005).
56. L. N. Chang, D. Minic and T. Takeuchi, Quantum gravity, dynamical energy-momentum space and vacuum energy, *Mod. Phys. Lett. A* **25**, 2947 (2010).
57. Z. Lewis and T. Takeuchi, Position and momentum uncertainties of the normal and inverted harmonic oscillators under the minimal length uncertainty relation, *Phys. Rev. D* **84**, 105029 (2011).
58. Z. Lewis, A. Roman and T. Takeuchi, Position and momentum uncertainties of a particle in a V-shaped potential under the minimal length uncertainty relation, *Int. J. Mod. Phys. A* **30**, 1550206 (2015).
59. D. Amati, M. Ciafaloni and G. Veneziano, Can space-time be probed below the string size?, *Phys. Lett. B* **216**, 41 (1989).
60. E. Witten, Reflections on the fate of space-time, *Phys. Today* **49**(4), 24 (1996).
61. D. J. Gross and P. F. Mende, The high-energy behavior of string scattering amplitudes, *Phys. Lett. B* **197**, 129 (1987).
62. D. J. Gross and P. F. Mende, String theory beyond the Planck scale, *Nucl. Phys. B* **303**, 407 (1988).
63. D. Amati, M. Ciafaloni and G. Veneziano, Superstring collisions at Planckian energies, *Phys. Lett. B* **197**, 81 (1987).
64. D. Amati, M. Ciafaloni and G. Veneziano, Classical and quantum gravity effects from Planckian energy superstring collisions, *Int. J. Mod. Phys. A* **3**, 1615 (1988).
65. W. Heisenberg, *The Physical Principles of Quantum Theory* (University of Chicago Press, 1930).
66. J. A. Wheeler, On the mathematical description of light nuclei by the method of resonating group structure, *Phys. Rev.* **52**, 1107 (1937).
67. W. Heisenberg, Die beobachtbaren größen in der theorie der elementarteilchen, *Z. Phys.* **120**, 513 (1943).

68. W. Heisenberg, Die beobachtbaren größen in der theorie der elementarteilchen. II, *Z. Phys.* **120**, 673 (1943).
69. W. Heisenberg, Die beobachtbaren größen in der theorie der elementarteilchen. III, *Z. Phys.* **123**, 93 (1944).
70. L. Susskind and E. Witten, The holographic bound in anti-de Sitter space, arXiv:hep-th/9805114 (1998).
71. A. W. Peet and J. Polchinski, UV/IR relations in AdS dynamics, *Phys. Rev. D* **59**, 065011 (1999).
72. L. I. Mandelshtam and I. G. Tamm, The uncertainty relation between energy and time in nonrelativistic quantum mechanics, *J. Phys. (USSR)* **9**, 249 (1945).
73. E. P. Wigner, On the time-energy uncertainty relation, in *Aspects of Quantum Theory* (Cambridge University Press, 1972).
74. M. Bauer and P. A. Mello, The time-energy uncertainty relation, *Annals Phys.* **111**, 38 (1978).
75. T. Yoneya, On the interpretation of minimal length in string theories, *Mod. Phys. Lett. A* **4**, 1587 (1989).
76. T. Yoneya, Schild action and space-time uncertainty principle in string theory, *Prog. Theor. Phys.* **97**, 949 (1997).
77. T. Yoneya, D particles, D instantons, and a space-time uncertainty principle in string theory, in *Recent developments in nonperturbative quantum field theory. Proceedings of the APCTP-ICTP Joint International Conference, Seoul, Korea, May 26–30, 1997* (1997).
78. T. Yoneya, String theory and space-time uncertainty principle, *Prog. Theor. Phys.* **103**, 1081 (2000).
79. T. Yoneya, Space-time uncertainty and noncommutativity in string theory, *Int. J. Mod. Phys. A* **16**, 945 (2001), [305 (2000)].
80. M. Li and T. Yoneya, D particle dynamics and the space-time uncertainty relation, *Phys. Rev. Lett.* **78**, 1219 (1997).
81. M. Li and T. Yoneya, Short distance space-time structure and black holes in string theory: A Short review of the present status, *Chaos Solitons Fractals* **10**, 423 (1999).
82. A. Jevicki and T. Yoneya, Space-time uncertainty principle and conformal symmetry in D particle dynamics, *Nucl. Phys. B* **535**, 335 (1998).
83. H. Awata, M. Li, D. Minic and T. Yoneya, On the quantization of Nambu brackets, *JHEP* **02**, 013 (2001).
84. D. Minic, On the space-time uncertainty principle and holography, *Phys. Lett. B* **442**, 102 (1998).
85. G. Amelino-Camelia, L. Freidel, J. Kowalski-Glikman and L. Smolin, The principle of relative locality, *Phys. Rev. D* **84**, 084010 (2011).
86. M. Born, Quantised field theory and the mass of the proton, *Nature* **136**, 952 (1935).

87. M. Born, Reciprocity theory of elementary particles, *Rev. Mod. Phys.* **21**, 463 (1949).
88. M. Born, A suggestion for unifying quantum theory and relativity, *Proc. Roy. Soc. Lond. A* **165**, 291 (1938).
89. H. Yukawa, Quantum theory of nonlocal fields. 1. Free fields, *Phys. Rev.* **77**, 219 (1950).
90. Y. A. Gol'fand, On the Introduction of an "Elementary Length" in the relativistic theory of elementary particles, *Sov. Phys. JETP* **10**, 356 (1960).
91. Y. A. Gol'fand, Quantum field theory in constant curvature p-space, *Sov. Phys. JETP* **16**, 184 (1963).
92. Y. A. Gol'fand, On the properties of displacements in p-space of constant curvature, *Sov. Phys. JETP* **17**, 842 (1963).
93. G. Veneziano, A stringy nature needs just two constants, *Europhys. Lett.* **2**, 199 (1986).
94. I. A. Batalin, E. S. Fradkin and T. E. Fradkina, Another version for operatorial quantization of dynamical systems with irreducible constraints, *Nucl. Phys. B* **314**, 158 (1989).
95. L. Freidel and E. R. Livine, Effective 3-D quantum gravity and noncommutative quantum field theory, *Phys. Rev. Lett.* **96**, 221301 (2006).
96. I. Bars, Gauge symmetry in phase space, consequences for physics and spacetime, *Int. J. Mod. Phys. A* **25**, 5235 (2010).
97. M. R. Douglas and N. A. Nekrasov, Noncommutative field theory, *Rev. Mod. Phys.* **73**, 977 (2001).
98. R. J. Szabo, Quantum field theory on noncommutative spaces, *Phys. Rept.* **378**, 207 (2003).
99. A. P. Polychronakos, Non-commutative fluids, *Prog. Math. Phys.* **53**, 109 (2007).
100. R. J. Szabo, Quantum gravity, field theory and signatures of noncommutative spacetime, *Gen. Rel. Grav.* **42**, 1 (2010).
101. D. N. Blaschke, E. Kronberger, R. I. P. Sedmik and M. Wohlgenannt, Gauge theories on deformed spaces, *SIGMA* **6**, 062 (2010).
102. D. D'Ascanio, P. Pisani and D. V. Vassilevich, Renormalization on noncommutative torus, arXiv:1602.01479 [hep-th] (2016).
103. A. Borowiec, T. Juric, S. Meljanac and A. Pachol, Noncommutative tetrads and quantum spacetimes, arXiv:1602.01292 [hep-th] (2016).
104. F. Scardigli, Generalized uncertainty principle in quantum gravity from micro — black hole Gedanken experiment, *Phys. Lett. B* **452**, 39 (1999).
105. F. Brau, Minimal length uncertainty relation and hydrogen atom, *J. Phys. A* **32**, 7691 (1999).
106. F. Brau and F. Buisseret, Minimal length uncertainty relation and gravitational quantum well, *Phys. Rev. D* **74**, 036002 (2006).

107. A. F. Ali, S. Das and E. C. Vagenas, Discreteness of space from the generalized uncertainty principle, *Phys. Lett. B* **678**, 497 (2009).
108. S. Das, E. C. Vagenas and A. F. Ali, Discreteness of space from GUP II: Relativistic wave equations, *Phys. Lett. B* **690**, 407 (2010) [Erratum: *ibid.* **692**, 342 (2010)].
109. A. F. Ali, S. Das and E. C. Vagenas, A proposal for testing quantum gravity in the lab, *Phys. Rev. D* **84**, 044013 (2011).
110. L. N. Chang, D. Minic, A. Roman, C. Sun and T. Takeuchi, in preparation.
111. T. Matsuo and Y. Shibusa, Quantization of fields based on generalized uncertainty principle, *Mod. Phys. Lett. A* **21**, 1285 (2006).
112. H. Grosse and R. Wulkenhaar, Renormalization of ϕ^4 theory on noncommutative R^4 in the matrix base, *Commun. Math. Phys.* **256**, 305 (2005).
113. B. P. Abbott *et al.*, Observation of gravitational waves from a binary black hole merger, *Phys. Rev. Lett.* **116**, 061102 (2016).

Generalization of the Yang–Mills Theory

G. Savvidy

Institute of Nuclear and Particle Physics,
Demokritos National Research Centre,
Athens, GR-15310, Greece
savvidy@inp.demokritos.gr
www.inp.demokritos.gr/~savvidy

We suggest an extension of the gauge principle which includes tensor gauge fields. In this extension of the Yang–Mills theory the vector gauge boson becomes a member of a bigger family of gauge bosons of arbitrary large integer spins. The proposed extension is essentially based on the extension of the Poincaré algebra and the existence of an appropriate transversal representations. The invariant Lagrangian is expressed in terms of new higher-rank field strength tensors. It does not contain higher derivatives of tensor gauge fields and all interactions take place through three- and four-particle exchanges with a dimensionless coupling constant. We calculated the scattering amplitudes of non-Abelian tensor gauge bosons at tree level, as well as their one-loop contribution into the Callan–Symanzik beta function. This contribution is negative and corresponds to the asymptotically free theory. Considering the contribution of tensorgluons of all spins into the beta function we found that it is leading to the theory which is conformally invariant at very high energies. The proposed extension may lead to a natural inclusion of the standard theory of fundamental forces into a larger theory in which vector gauge bosons, leptons and quarks represent a low-spin subgroup. We consider a possibility that inside the proton and, more generally, inside hadrons there are additional partons — tensorgluons, which can carry a part of the proton momentum. The extension of QCD influences the unification scale at which the coupling constants of the Standard Model merge, shifting its value to lower energies.

1. Introduction

It is well understood that the concept of local gauge invariance formulated by Yang and Mills[1] allows to define the non-Abelian gauge fields, to derive their dynamical field equations and to develop a universal point of view on matter interactions as resulting from the exchange of gauge quanta of different forms. The fundamental forces — electromagnetic, weak and strong interactions are successfully described by the non-Abelian Yang–Mills fields. The vector-like gauge particles — the photon, $W^{\pm}, Z$ and gluons mediate interaction between smallest constituents of matter — leptons and quarks.

The non-Abelian local gauge invariance, which was formulated by Yang and Mills,[1] requires that all interactions must be invariant under independent rotations of internal charges at all space–time points. The gauge principle allows very little arbitrariness: the interaction of matter fields, which carry non-commuting internal charges, and the nonlinear self-interaction of gauge bosons are essentially fixed by the requirement of local gauge invariance, very similarly to the self-interaction of gravitons in general relativity.[2-8]

It is therefore appealing to extend the gauge principle, which was elevated by Yang and Mills to a powerful constructive principle, so that it will define the interaction of matter fields which carry not only non-commutative internal charges, but also arbitrary large spins.[a] It seems that this will naturally lead to a theory in which fundamental forces will be mediated by integer-spin gauge quanta and that the Yang–Mills vector gauge boson will become a member of a bigger family of tensor gauge bosons.[9-11,47,48]

The proposed extension of Yang–Mills theory is essentially based on the extension of the Poincaré algebra and the existence of an appropriate transversal representations of that algebra. The tensor gauge fields take value in extended Poincaré algebra. The invariant Lagrangian is expressed in terms of new higher-rank field strength tensors. The Lagrangian does not contain higher derivatives of tensor gauge fields and all interactions take place through three- and four-particle exchanges with a dimensionless coupling constant.[9-11,47,48]

It is important to calculate the scattering amplitudes of non-Abelian tensor gauge bosons at tree level, as well as their one-loop contribution into the Callan–Symanzik beta function. This contribution is negative and corresponds to the asymptotically free theory. The proposed extension may lead to a natural inclusion of the standard theory of fundamental forces into a larger theory in which vector gauge bosons, leptons and quarks represent a low-spin subgroup.[51-53]

In the line with the above development we considered a possible extension of QCD. In so extended QCD the spectrum of the theory contains new bosons, *the tensorgluons*, in addition to the quarks and gluons. The tensorgluons have zero electric charge, like gluons, but have a larger spin. Radiation of tensorgluons by gluons leads to a possible existence of tensorgluons inside the proton and, more generally, inside the hadrons. Due to the emission of tensorgluons part of the proton momentum which is carried by the neutral constituents can be shared between gluons and tensorgluons. The density of neutral partons is therefore given by the sum: $G(x,t) + T(x,t)$, where $T(x,t)$ is the density of the tensorgluons.[51-53] To disentangle these contributions and to decide which piece of the neutral partons is the contribution of gluons $G(x,t)$ and which one is of the tensorgluons one should measure the helicities of the neutral components, which seems to be a difficult task.

---

[a]The research in high spin field theories has long and rich history. One should mention the early works of Majorana,[23] Dirac,[24] Fierz,[25] Pauli,[26] Schwinger,[30] Singh and Hagen,[31] Fronsdal,[32] Weinberg,[33] Minkowski,[27] Brink et al.[45] Berends, Burgers and Van Dam,[40] Ginzburg and Tamm,[34,36] Nambu,[37] Ramond,[35] Fradkin,[38] Vasiliev,[39] Sagnotti, Sezgin and Sundell,[41] Metsaev,[42] Manvelyan et al.[43] and many other works (see also the references in Refs. 36, 39 and 44).

The extension of QCD influences the unification scale at which the coupling constants of the Standard Model merge. We observed that the unification scale at which standard coupling constants are merging is shifted to lower energies telling us that it may be that a new physics is round the corner. Whether all these phenomena are consistent with experiment is an open question.

The paper is organized as follows. In Section 2 we shall define the composite gauge field $\mathcal{A}_\mu(x, e)$, which depends on the space–time coordinates x_μ and the new space-like vector variable e^λ. The high-rank tensor gauge fields $A^a_{\mu\lambda_1...\lambda_s}(x)$ appear in the expansion of $\mathcal{A}_\mu(x, e)$ over the vector variable. We introduce a corresponding extension of the Poincaré algebra $L_G(\mathcal{P})$ and consider the high-rank fields as tensor gauge fields taking value in algebra $L_G(\mathcal{P})$. In Section 3 we shall describe the transversal representation $L^\perp$ of the generators of the algebra $L_G(\mathcal{P})$, their helicity content and their invariant scalar products. The fact that the representation of the generators is transversal plays an important role in the definition of the gauge field $\mathcal{A}_\mu(x, e)$. In transversal representation the tensor gauge fields are projecting out into the plane transversal to the momentum and contain only positive space-like components of a definite helicity.

In Section 4 we shall define the gauge transformation of the gauge fields, the field strength tensors and the invariant Lagrangian. The kinetic term describes the propagation of positive definite helicity states. The helicity spectrum of the propagating modes is consistent with the helicity spectrum which appears in the projection of the tensor gauge fields into transversal generators $L^\perp$. The Lagrangian defines not only a free propagation of tensor gauge bosons, but also their interactions. The interaction diagrams for the lower-rank bosons are presented in Figs. 1 and 2. The high-rank bosons interact through the triple and quartic interaction vertices with a dimensionless coupling constant. In Section 5 we shall calculate and study the scattering amplitudes of the vector and tensor gauge bosons and their splitting amplitudes by using spinor representation of the momenta and polarization tensors.

In Section 6 we shall consider a possibility that inside the proton and, more generally, inside hadrons there are additional partons — tensorgluons, which can carry a part of the proton momentum. We generalize the DGLAP equation which includes the splitting probabilities of the gluons into tensorgluons and calculated the one-loop Callan-Symanzik beta function. This contribution is negative and corresponds to the asymptotically free theory. Considering the contribution of tensorgluons of all spins into the beta function we found that it is leading to the theory which is *conformally invariant* at very high energies. In Section 7 we observed that the unification scale at which standard coupling constants are merging is shifted to lower energies. In conclusion we summarize the results and discuss the challenges of the experimental verification of the suggested model.

2. Tensor Gauge Fields and Extended Poincaré Algebra

The gauge fields are defined as rank-$(s+1)$ tensors[9-11]

$$A^a_{\mu\lambda_1...\lambda_s}(x), \quad s = 0, 1, 2, ... \tag{1}$$

and are totally symmetric with respect to the indices $\lambda_1 \ldots \lambda_s$. A priory the tensor fields have no symmetries with respect to the first index μ. The index a numerates the generators L_a of the Lie algebra L_G of a compact Lie group G with totally antisymmetric structure constants f_{abc}.

The tensor fields (1) can be considered as the components of a composite gauge field $\mathcal{A}_\mu(x,e)$ which depends on additional translationally invariant space-like unite vector:[11,15–17]

$$e_\lambda e^\lambda = -1. \tag{2}$$

A similar vector variable, in addition to the space–time coordinate x, was introduced earlier by Yakawa,[12] Fierz,[13] Wigner,[14] Ginzburg and Tamm[34,36] and others.[39] The variable e^λ is also reminiscent to the Grassmann variable θ in supersymmetric theories where the superfield $\Psi(x,\theta)$ depends on two variables x and θ.[18,19] We shall consider all tensor gauge fields (1) as the components appearing in the expansion over the above mentioned vector variable:[11]

$$\mathcal{A}_\mu(x,e) = \sum_{s=0}^{\infty} \frac{1}{s!} A^a_{\mu\lambda_1\ldots\lambda_s}(x) \, L_a e^{\lambda_1} \ldots e^{\lambda_s}. \tag{3}$$

The gauge field $A^a_{\mu\lambda_1\ldots\lambda_s}$ carries indices $a, \lambda_1, \ldots, \lambda_s$ which are labelling the generators $L_a^{\lambda_1\ldots\lambda_s} = L_a e^{\lambda_1} \ldots e^{\lambda_s}$ of extended current algebra $L_\mathcal{G}$ associated with the Lie algebra L_G.[11,47] The algebra $L_\mathcal{G}$ has infinitely many generators $L_a^{\lambda_1\ldots\lambda_s}$ and is given by the commutator[11,47,48]

$$[L_a^{\lambda_1\ldots\lambda_k}, L_b^{\lambda_{k+1}\ldots\lambda_s}] = if_{abc} L_c^{\lambda_1\ldots\lambda_s}, \quad s = 0, 1, 2 \ldots. \tag{4}$$

The generators $L_a^{\lambda_1\ldots\lambda_s}$ commute to themselves forming an infinite series of commutators of current algebra $L_\mathcal{G}$ which cannot be truncated, so that the index s runs from zero to infinity. Because the generators $L_a^{\lambda_1\ldots\lambda_s}$ are space–time tensors, the full algebra should include the Poincaré generators P^μ, $M^{\mu\nu}$ as well. This naturally leads to the extension $L_G(\mathcal{P})$ of the Poincaré algebra $L_\mathcal{P}$:[47–49]

$$[P^\mu, P^\nu] = 0,$$

$$[M^{\mu\nu}, P^\lambda] = \eta^{\nu\lambda} P^\mu - \eta^{\mu\lambda} P^\nu,$$

$$[M^{\mu\nu}, M^{\lambda\rho}] = \eta^{\mu\rho} M^{\nu\lambda} - \eta^{\mu\lambda} M^{\nu\rho} + \eta^{\nu\lambda} M^{\mu\rho} - \eta^{\nu\rho} M^{\mu\lambda},$$

$$[P^\mu, L_a^{\lambda_1\ldots\lambda_s}] = 0,$$

$$[M^{\mu\nu}, L_a^{\lambda_1\ldots\lambda_s}] = \eta^{\nu\lambda_1} L_a^{\mu\lambda_2\ldots\lambda_s} - \eta^{\mu\lambda_1} L_a^{\nu\lambda_2\ldots\lambda_s} + \cdots$$
$$+ \eta^{\nu\lambda_s} L_a^{\mu\lambda_1\ldots\lambda_{s-1}} - \eta^{\mu\lambda_s} L_a^{\nu\lambda_1\ldots\lambda_{s-1}},$$

$$[L_a^{\lambda_1\ldots\lambda_k}, L_b^{\lambda_{k+1}\ldots\lambda_s}] = if_{abc} L_c^{\lambda_1\ldots\lambda_s}. \tag{5}$$

We have here an extension of the Poincaré algebra by generators $L_a^{\lambda_1\ldots\lambda_s}$ which carry the *internal charges and spins*. The algebra $L_G(\mathcal{P})$ incorporates the Poincaré algebra $L_\mathcal{P}$ and an internal algebra L_G in a nontrivial way, which is different from the direct product.

There is no conflict with the Coleman–Mandula theorem[20,21] because the theorem applies to the symmetries that act on S-matrix elements and not on all the other symmetries that occur in quantum field theory. The above symmetry group (5) is the symmetry which acts on the gauge field $\mathcal{A}_\mu(x,e)$ and is not the symmetry of the S-matrix. The theorem assumes among other things that the vacuum is nondegenerate and that there are no massless particles in the spectrum. As we shall see, the spectrum of the extended Yang–Mills theory is massless.

In order to define the gauge field $\mathcal{A}_\mu(x,e)$ in (3) and find out its helicity content one should specify the representation of the generators $L_a^{\lambda_1...\lambda_s}$ in algebra (5). In the next section we shall describe the so called transversal representation, which is used to define the tensor gauge fields in the decomposition (3).

3. Transversal Representation of Algebra $L_G(\mathcal{P})$

The important property of the algebra (5) is its invariance with respect to the following "gauge" transformations:[47-49]

$$L_a^{\lambda_1...\lambda_s} \to L_a^{\lambda_1...\lambda_s} + \sum_1 P^{\lambda_1} L_a^{\lambda_2...\lambda_s} + \sum_2 P^{\lambda_1} P^{\lambda_2} L_a^{\lambda_3...\lambda_s} + \cdots + P^{\lambda_1} \ldots P^{\lambda_s} L_a$$

$$M^{\mu\nu} \to M^{\mu\nu}, \quad P^\lambda \to P^\lambda, \tag{6}$$

where the sums $\sum_1, \sum_2, \ldots$ are over all inequivalent index permutations. The above transformations contain polynomials of the momentum operator P^λ and are reminiscent of the gauge field transformations. This is "off-shell" symmetry because the invariant operator P^2 can have any value. As a result, to any given representation of $L_a^{\lambda_1...\lambda_s}$, $s = 1, 2, \ldots$ one can add the longitudinal terms, as it follows from the transformation (6). All representations are therefore defined modulo longitudinal terms, and we can identify these generators as "gauge generators".

The second general property of the extended algebra is that each gauge generator $L_a^{\lambda_1...\lambda_s}$ cannot be realized as an irreducible representation of the Poincaré algebra of a definite helicity, i.e. to be a *symmetric and traceless tensor*. The reason is that the commutator of two symmetric traceless generators in (5) is not any more a traceless tensor. Therefore the generators $L_a^{\lambda_1...\lambda_s}$ realize a reducible representation of the Poincaré algebra and each of them carries a spectrum of helicities, which we shall describe below.

The algebra $L_G(\mathcal{P})$ has representation in terms of differential operators of the following general form:

$$P^\mu = k^\mu,$$

$$M^{\mu\nu} = i\left(k^\mu \frac{\partial}{\partial k_\nu} - k^\nu \frac{\partial}{\partial k_\mu}\right) + i\left(e^\mu \frac{\partial}{\partial e_\nu} - e^\nu \frac{\partial}{\partial e_\mu}\right),$$

$$L_a^{\lambda_1...\lambda_s} = e^{\lambda_1} \ldots e^{\lambda_s} \otimes L_a, \tag{7}$$

where e^λ is a translationally invariant space-like unite vector (2). The vector space of a representation is parametrized by the momentum k^μ and translationally invariant vector variables e^λ:

$$\Psi(k^\mu, e^\lambda). \tag{8}$$

The irreducible representations can be obtained from (7) by imposing invariant constraints on the vector space of functions (8) of the following form:[12-14,22]

$$k^2 = 0, \quad k^\mu e_\mu = 0, \quad e^2 = -1. \tag{9}$$

These equations have a unique solution [14]

$$e^\mu = \chi k^\mu + e_1^\mu \cos\varphi + e_2^\mu \sin\varphi, \tag{10}$$

where $e_1^\mu = (0, 1, 0, 0)$, $e_2^\mu = (0, 0, 1, 0)$ when $k^\mu = \omega(1, 0, 0, 1)$. The χ and φ remain as independent variables on the cylinder $\varphi \in S^1, \chi \in R^1$. The invariant subspace of functions (8) now reduces to the following form:

$$\Psi(k^\mu, e^\nu)\, \delta(k^2)\, \delta(k \cdot e)\, \delta(e^2 + 1) = \Phi(k^\mu, \varphi, \chi). \tag{11}$$

If we take into account (10) the generators $L_a^{\lambda_1 \ldots \lambda_s} = e^{\lambda_1} \ldots e^{\lambda_s} \otimes L^a$ take the following form:

$$L_a^{\perp\, \lambda_1 \ldots \lambda_s} = \prod_{n=1}^{s} (\chi k^{\lambda_n} + e_1^{\lambda_n} \cos\varphi + e_2^{\lambda_n} \sin\varphi) \otimes L_a. \tag{12}$$

This is a purely transversal representation because of (9):

$$k_{\lambda_1} L_a^{\perp\, \lambda_1 \ldots \lambda_s} = 0, \quad s = 1, 2, \ldots. \tag{13}$$

The generators $L_a^{\perp\, \lambda_1 \ldots \lambda_s}$ carry helicities in the following range:

$$h = (s, s-2, \ldots, -s+2, -s), \tag{14}$$

in total $s+1$ states. Indeed, this can be deduced from the explicit representation (12) by using helicity polarization vectors $e_\pm^\lambda = (e_1^\lambda \mp i e_2^\lambda)/2$:

$$L_a^{\perp\, \lambda_1 \ldots \lambda_s} = \prod_{n=1}^{s} (\chi k^{\lambda_n} + e^{i\varphi} e_+^{\lambda_n} + e^{-i\varphi} e_-^{\lambda_n}) \oplus L_a. \tag{15}$$

Performing the multiplication in (15) and collecting the terms of a given power of momentum we shall get the following expression:

$$L_a^{\perp\, \mu_1 \ldots \mu_s} = \prod_{n=1}^{s} (e^{i\varphi} e_+^{\mu_n} + e^{-i\varphi} e_-^{\mu_n}) \oplus L_a \tag{16}$$

$$+ \sum_{1}^{s} \chi k^{\mu_1} \prod_{n=1}^{s-1} (e^{i\varphi} e_+^{\mu_n} + e^{-i\varphi} e_-^{\mu_n}) \oplus L_a + \cdots + \chi k^{\mu_1} \ldots \chi k^{\mu_s} \oplus L_a,$$

where the first term $\prod_{n=1}^{s}(e^{i\varphi}e_{+}^{\mu_n} + e^{-i\varphi}e_{-}^{\mu_n})$ represents the *helicity generators* $(L_a^{+\cdots+}, \ldots, L_a^{-\cdots-})$, while their helicity spectrum is described by the formula (14). The rest of the terms are purely longitudinal and proportional to the increasing powers of momentum k. The last formula also illustrates the realization of the transformation (6), that is, the helicity generators $(L_a^{+\cdots+}, \ldots, L_a^{-\cdots-})$ are defined modulo longitudinal terms proportional to $k^{\lambda_1} \ldots k^{\lambda_n}$, $n = 1, \ldots, s$.

The very fact that the representation of the generators $L_a^{\perp \lambda_1 \ldots \lambda_s}$ is transversal plays an important role in the definition of the gauge field $\mathcal{A}_\mu(x, e)$ in (3). Indeed, substituting the transversal representation (16) of the generators $L_a^{\perp \lambda_1 \ldots \lambda_s}$ into the expansion (3) and collecting the terms in front of the helicity generators $(L_a^{+\cdots+}, \ldots, L_a^{-\cdots-})$ we shall get

$$\mathcal{A}_\mu(x, e) = \sum_{s=0}^{\infty} \frac{1}{s!} (\tilde{A}^a_{\mu \lambda_1 \ldots \lambda_s} e_+^{\lambda_1} \ldots e_+^{\lambda_s} \oplus L_a + \cdots + \tilde{A}^a_{\mu \lambda_1 \ldots \lambda_s} e_-^{\lambda_1} \ldots e_-^{\lambda_s} \oplus L_a)$$

$$= \sum_{s=0}^{\infty} \frac{1}{s!} (\tilde{A}^a_{\mu+\cdots+} L_a^{+\cdots+} + \cdots + \tilde{A}^a_{\mu-\cdots-} L_a^{-\cdots-}), \tag{17}$$

where s is the number of negative indices. This formula represents the projection $\tilde{A}^a_{\mu \lambda_1 \ldots \lambda_s}$ of the components of the non-Abelian tensor gauge field $A^a_{\mu \lambda_1 \ldots \lambda_s}$ into the plane transversal to the momentum. The projection contains only positive definite space-like components of the helicities:[47–49]

$$h = \pm(s+1), \quad \frac{\pm(s-1)}{\pm(s-1)}, \frac{\pm(s-3)}{\pm(s-3)}, \ldots, \tag{18}$$

where the lower helicity states have double degeneracy. The analysis of the kinetic terms of the Lagrangian and of the corresponding equation of motions, which will be considered in the next section, confirms that indeed the propagating degrees of freedom are described by helicities (18).

In order to define the gauge invariant Lagrangian one should know the Killing metric of the algebra $L_G(\mathcal{P})$. The explicit transversal representation of the $L_G(\mathcal{P})$ generators given above (12), (15) and (16) allows to calculate the corresponding Killing metric:[47,48,50]

$$L_G: \quad \langle L_a; L_b \rangle = \delta_{ab}, \tag{19}$$

$$L_\mathcal{P}: \quad \langle P^\mu; P^\nu \rangle = 0,$$

$$\langle M_{\mu\nu}; P_\lambda \rangle = 0, \tag{20}$$

$$\langle M^{\mu\nu}; M^{\lambda\rho} \rangle = \eta^{\mu\lambda}\eta^{\nu\rho} - \eta^{\mu\rho}\eta^{\nu\lambda}$$

$$L_G(\mathcal{P}): \quad \langle P^\mu; L_a^{\perp\, \lambda_1...\lambda_s} \rangle = 0,$$

$$\langle M^{\mu\nu}; L_a^{\perp\, \lambda_1...\lambda_s} \rangle = 0, \qquad (21)$$

$$\langle L_a; L_b^{\perp\, \lambda_1} \rangle = 0,$$

$$\langle L_a^{\perp\, \lambda_1}; L_b^{\perp\, \lambda_2} \rangle = \delta_{ab}\, \bar{\eta}^{\lambda_1\lambda_2},$$

$$\langle L_a; L_b^{\perp\, \lambda_1\lambda_2} \rangle = \delta_{ab}\, \bar{\eta}^{\lambda_1\lambda_2},$$

$$\langle L_a^{\perp\, \lambda_1}; L_b^{\perp\, \lambda_2\lambda_3} \rangle = 0, \qquad (22)$$

....

$$\langle L_a^{\perp\, \lambda_1...\lambda_n}; L_b^{\perp\, \lambda_{n+1}....\lambda_{2s+1}} \rangle = 0, \quad s = 0,1,2,3,...$$

$$\langle L_a^{\perp\, \lambda_1...\lambda_n}; L_b^{\perp\, \lambda_{n+1}....\lambda_{2s}} \rangle = \delta_{ab}\, s!\, (\bar{\eta}^{\lambda_1\lambda_2}\bar{\eta}^{\lambda_3\lambda_4} \ldots \bar{\eta}^{\lambda_{2s-1}\lambda_{2s}} + \text{perm}),$$

where $\bar{\eta}^{\lambda_1\lambda_2}$ is the projector into the two-dimensional plane transversal to the momentum k^μ:[30]

$$\bar{\eta}^{\lambda_1\lambda_2} = \frac{k^{\lambda_1}\bar{k}^{\lambda_2} + \bar{k}^{\lambda_1}k^{\lambda_2}}{k\bar{k}} - \eta^{\lambda_1\lambda_2}, \quad k_{\lambda_1}\bar{\eta}^{\lambda_1\lambda_2} = k_{\lambda_2}\bar{\eta}^{\lambda_1\lambda_2} = 0, \qquad (23)$$

and $\bar{k}^\mu = \omega(1,0,0,-1)$. It follows then that the transversality conditions (13) are fulfilled:

$$k_{\lambda_i} \langle L_a^{\perp\, \lambda_1...\lambda_n}; L_b^{\perp\, \lambda_{n+1}...\lambda_{2s}} \rangle = 0, \quad i = 1,2,\ldots,2s. \qquad (24)$$

The Killing metric on the internal L_G and on the Poincaré $L_\mathcal{P}$ subalgebras (19), (20) are well known. The important conclusion which follows from the above result is that the Poincaré generators $P^\mu, M^{\mu\nu}$ are orthogonal to the gauge generators $L_a^{\lambda_1...\lambda_s}$ (21). The last formulas (22) represent the Killing metric on the $L_\mathcal{G}$ current algebra (4), (5) and will be used in the definition of the Lagrangian in the next section. It should be stressed that the metric (22) is defined modulo longitudinal terms. This is because under the "gauge" transformation of the generators (6) the metric will receive terms which are polynomial in momentum. The provided metric (22) is written in a particular gauge. This peculiar property of the metric is mirrored in the definition of the Lagrangian which can be written in different gauges. The spectrum of the propagating modes does not depend on the gauges chosen, as one can get convinced by inspecting the expression (17).

Notice that the reducible representation (7), without any of the constraints (9), should also be considered, as well as the representation in which only the last constrain in (9) is imposed. In that cases the transversality of the representation (24) will be lost, but instead one arrives to the homogeneous Killing metric in (22) $\bar{\eta}^{\lambda_1\lambda_2} \to \eta^{\lambda_1\lambda_2}$ and the longitudinal terms which can be gauged away.

With this Killing metric in hands one can define the Lagrangian of the theory.

4. The Lagrangian

The gauge transformation of the field $\mathcal{A}_\mu(x,e)$ is defined as[1,11,47]

$$\mathcal{A}'_\mu(x,e) = U(\xi)\mathcal{A}_\mu(x,e)U^{-1}(\xi) - \frac{i}{g}\partial_\mu U(\xi)\, U^{-1}(\xi), \tag{25}$$

where the group parameter $\xi(x,e)$

$$U(\xi) = e^{i\xi(x,e)}$$

has the decomposition[11,47]

$$\xi(x,e) = \sum_s \frac{1}{s!}\, \xi^a_{\lambda_1\ldots\lambda_s}(x)\, L_a e^{\lambda_1}\ldots e^{\lambda_s}$$

and $\xi^a_{\lambda_1\ldots\lambda_s}(x)$ are totally symmetric gauge parameters. Using the commutator of the covariant derivatives $\nabla^{ab}_\mu = (\partial_\mu - ig\mathcal{A}_\mu(x,e))^{ab}$

$$[\nabla_\mu, \nabla_\nu]^{ab} = gf^{acb}\mathcal{G}^c_{\mu\nu}, \tag{26}$$

we can define the extended field strength tensor

$$\mathcal{G}_{\mu\nu}(x,e) = \partial_\mu \mathcal{A}_\nu(x,e) - \partial_\nu \mathcal{A}_\mu(x,e) - ig[\mathcal{A}_\mu(x,e)\, \mathcal{A}_\nu(x,e)], \tag{27}$$

which transforms homogeneously:

$$\mathcal{G}'_{\mu\nu}(x,e)) = U(\xi)\mathcal{G}_{\mu\nu}(x,e)U^{-1}(\xi). \tag{28}$$

It is useful to have an explicit expression for the transformation law of the field components:[9–11]

$$\delta A^a_\mu = (\delta^{ab}\partial_\mu + gf^{acb}A^c_\mu)\xi^b, \tag{29}$$
$$\delta A^a_{\mu\nu} = (\delta^{ab}\partial_\mu + gf^{acb}A^c_\mu)\xi^b_\nu + gf^{acb}A^c_{\mu\nu}\xi^b,$$
$$\delta A^a_{\mu\nu\lambda} = (\delta^{ab}\partial_\mu + gf^{acb}A^c_\mu)\xi^b_{\nu\lambda} + gf^{acb}(A^c_{\mu\nu}\xi^b_\lambda + A^c_{\mu\lambda}\xi^b_\nu + A^c_{\mu\nu\lambda}\xi^b),$$

$\ldots\ldots\, .\,\ldots\ldots\ldots\ldots\ldots\ldots\ldots$

These extended gauge transformations generate a closed algebraic structure. The component field strengths tensors take the following form:[9–11]

$$G^a_{\mu\nu} = \partial_\mu A^a_\nu - \partial_\nu A^a_\mu + gf^{abc} A^b_\mu A^c_\nu, \tag{30}$$
$$G^a_{\mu\nu,\lambda} = \partial_\mu A^a_{\nu\lambda} - \partial_\nu A^a_{\mu\lambda} + gf^{abc}(A^b_\mu A^c_{\nu\lambda} + A^b_{\mu\lambda} A^c_\nu),$$
$$G^a_{\mu\nu,\lambda\rho} = \partial_\mu A^a_{\nu\lambda\rho} - \partial_\nu A^a_{\mu\lambda\rho} + gf^{abc}(A^b_\mu A^c_{\nu\lambda\rho} + A^b_{\mu\lambda} A^c_{\nu\rho} + A^b_{\mu\rho} A^c_{\nu\lambda} + A^b_{\mu\lambda\rho} A^c_\nu),$$

$\ldots\ldots\, .\,\ldots\ldots\ldots\ldots\ldots\ldots\ldots$

and transform homogeneously with respect to the transformations (29):

$$\delta G^a_{\mu\nu} = g f^{abc} G^b_{\mu\nu} \xi^c, \tag{31}$$

$$\delta G^a_{\mu\nu,\lambda} = g f^{abc} (G^b_{\mu\nu,\lambda} \xi^c + G^b_{\mu\nu} \xi^c_\lambda),$$

$$\delta G^a_{\mu\nu,\lambda\rho} = g f^{abc} (G^b_{\mu\nu,\lambda\rho} \xi^c + G^b_{\mu\nu,\lambda} \xi^c_\rho + G^b_{\mu\nu,\rho} \xi^c_\lambda + G^b_{\mu\nu} \xi^c_{\lambda\rho}),$$

$$\ldots \ldots \ldots \ldots \ldots \ldots \ldots \ldots \ldots \ldots \ldots$$

The field strength tensors are antisymmetric in their first two indices and are totally symmetric with respect to the rest of the indices. The symmetry properties of the field strength $G^a_{\mu\nu,\lambda_1\ldots\lambda_s}$ remain invariant in the course of these transformations.

The first gauge invariant density is given by the expression[9–11]

$$\mathcal{L}(x) = \langle \mathcal{L}(x,e) \rangle = -\frac{1}{4} \langle \mathcal{G}^a_{\mu\nu}(x,e) \mathcal{G}^{a\mu\nu}(x,e) \rangle, \tag{32}$$

where the trace of the generators is given in (22). One can get convinced that the variation of the (32) with respect to the gauge transformations (25) and (28) vanishes:

$$\delta \mathcal{L}(x,e) = -\frac{1}{2} \mathcal{G}^a_{\mu\nu}(x,e) \, g f^{abc} \, \mathcal{G}^{b\mu\nu}(x,e) \, \xi^c(x,e) = 0.$$

The invariant density (32) allows to extract *gauge invariant, totally symmetric, tensor densities* $\mathcal{L}_{\lambda_1\ldots\lambda_s}(x)$ by using expansion with respect to the vector variable e^λ:

$$\mathcal{L}(x,e) = \sum_{s=0}^{\infty} \frac{1}{s!} \mathcal{L}_{\lambda_1\ldots\lambda_s}(x) \, e^{\lambda_1} \ldots e^{\lambda_s}. \tag{33}$$

In particular, the expansion term which is quadratic in powers of e^λ is

$$\mathcal{L}_{\lambda_1\lambda_2} = -\frac{1}{4} G^a_{\mu\nu,\lambda_1} G^a_{\mu\nu,\lambda_2} - \frac{1}{4} G^a_{\mu\nu} G^a_{\mu\nu,\lambda_1\lambda_2}. \tag{34}$$

The gauge invariant density thus can be represented in the following form:[9–11]

$$\mathcal{L}(x) = \langle \mathcal{L}(x,e) \rangle = \sum_{s=0}^{\infty} \frac{1}{s!} \mathcal{L}_{\lambda_1\ldots\lambda_s}(x) \langle e^{\lambda_1} \ldots e^{\lambda_s} \rangle \tag{35}$$

and the density for the lower-rank tensor fields is

$$\mathcal{L}_2 = -\frac{1}{4} G^a_{\mu\nu,\lambda} G^a_{\mu\nu,\lambda} - \frac{1}{4} G^a_{\mu\nu} G^a_{\mu\nu,\lambda\lambda}.$$

Let us consider the second gauge invariant density of the form[9–11]

$$\mathcal{L}'(x) = \langle \mathcal{L}'(x,e) \rangle = \frac{1}{4} \langle \mathcal{G}^a_{\mu\rho_1}(x,e) e^{\rho_1} \, \mathcal{G}^{a\mu}_{\rho_2}(x,e) e^{\rho_2} \rangle'. \tag{36}$$

It is gauge invariant because its variation is also equal to zero:

$$\delta \mathcal{L}'(x,e) = \frac{1}{4} g f^{acb} \, \mathcal{G}^c_{\mu\rho_1}(x,e) e^{\rho_1} \, \xi^b(x,e) \mathcal{G}^{a\mu}{}_{\rho_2}(x,e) e^{\rho_2}$$
$$+ \frac{1}{4} \mathcal{G}^a_{\mu\rho_1}(x,e) e^{\rho_1} \, g f^{acb} \, \mathcal{G}^{c\mu}{}_{\rho_2}(x,e) e^{\rho_2} \, \xi^b(x,e) = 0. \quad (37)$$

The Lagrangian density (36) generates the second series of *gauge invariant tensor densities* $(\mathcal{L}'_{\rho_1\rho_2})_{\lambda_1...\lambda_s}(x)$ when we expand it in powers of the vector variable e^λ:

$$\mathcal{L}'(x) = \langle \mathcal{L}'(x,e) \rangle = \sum_{s=0}^{\infty} \frac{1}{s!} (\mathcal{L}'_{\rho_1\rho_2})_{\lambda_1...\lambda_s}(x) \, \langle e^{\rho_1} e^{\rho_2} e^{\lambda_1} \ldots e^{\lambda_s} \rangle'. \quad (38)$$

The term quartic in variable e^λ after contraction of the vector variables takes the following form:

$$\mathcal{L}'_2 = \frac{1}{4} G^a_{\mu\nu,\lambda} G^a_{\mu\lambda,\nu} + \frac{1}{4} G^a_{\mu\nu,\nu} G^a_{\mu\lambda,\lambda} + \frac{1}{2} G^a_{\mu\nu} G^a_{\mu\lambda,\nu\lambda}. \quad (39)$$

One can get convinced that it is gauge invariant under the transformation (29) and (31). The total Lagrangian density is a sum of two invariants (32) and (36):

$$L = \mathcal{L} + \mathcal{L}' = -\frac{1}{4} \langle \mathcal{G}^a_{\mu\nu}(x,e) \mathcal{G}^{a\mu\nu}(x,e) \rangle + \frac{1}{4} \langle \mathcal{G}^a_{\mu\rho_1}(x,e) e^{\rho_1} \mathcal{G}^{a\mu}{}_{\rho_2}(x,e) e^{\rho_2} \rangle'. \quad (40)$$

The Lagrangian for the lower-rank tensor gauge fields has the following form:

$$\mathcal{L} = \mathcal{L}_1 + \mathcal{L}_2 + \mathcal{L}'_2 + \cdots = -\frac{1}{4} G^a_{\mu\nu} G^a_{\mu\nu} \quad (41)$$
$$-\frac{1}{4} G^a_{\mu\nu,\lambda} G^a_{\mu\nu,\lambda} - \frac{1}{4} G^a_{\mu\nu} G^a_{\mu\nu,\lambda\lambda}$$
$$+ \frac{1}{4} G^a_{\mu\nu,\lambda} G^a_{\mu\lambda,\nu} + \frac{1}{4} G^a_{\mu\nu,\nu} G^a_{\mu\lambda,\lambda} + \frac{1}{2} G^a_{\mu\nu} G^a_{\mu\lambda,\nu\lambda} + \cdots.$$

The above Lagrangian defines the kinetic operators for the rank-1 A^a_μ and rank-2 $A^a_{\mu\lambda_1}$ fields, as well as trilinear and quartic interactions with the *dimensionless coupling constant g* (see Figs. 1 and 2).

As we found in Refs. 9–11, the corresponding free field equations coincide with the equations introduces in the classical works[25–27] and describe the propagation of the *helicity-two and zero* $h = \pm 2, 0$ *massless charged tensor gauge bosons*, and there are no propagating negative norm states. This is in agreement with the spectrum presented in (18). The next term in expansion of the Lagrangian density has the

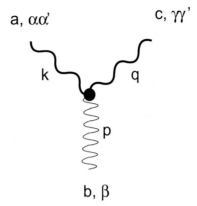

Fig. 1. The interaction vertex for the vector gauge boson V and two tensor gauge bosons T — the VTT vertex — $\mathcal{V}^{abc}_{\alpha\acute{\alpha}\beta\gamma\acute{\gamma}}(k,p,q)$ in non-Abelian tensor gauge field theory.[11] Vector gauge bosons are conventionally drawn as thin wave lines, tensor gauge bosons are thick wave lines. The Lorentz indices $\alpha\acute{\alpha}$ and momentum k belong to the first tensor gauge boson, the $\gamma\acute{\gamma}$ and momentum q belong to the second tensor gauge boson, and Lorentz index β and momentum p belong to the vector gauge boson.

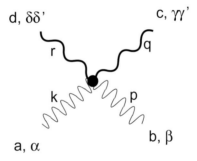

Fig. 2. The quartic vertex with two vector gauge bosons and two tensor gauge bosons — the VVTT vertex — $\mathcal{V}^{abcd}_{\alpha\beta\gamma\acute{\gamma}\delta\acute{\delta}}(k,p,q,r)$ in non-Abelian tensor gauge field theory,[11] Vector gauge bosons are conventionally drawn as thin wave lines, tensor gauge bosons are thick wave lines. The Lorentz indices $\gamma\acute{\gamma}$ and momentum q belong to the first tensor gauge boson, $\delta\acute{\delta}$ and momentum r belong to the second tensor gauge boson, the index α and momentum k belong to the first vector gauge boson and Lorentz index β and momentum p belong to the second vector gauge boson.

following form:[28,29]

$$\mathcal{L}_3 + \mathcal{L}'_3 = -\frac{1}{4}G^a_{\mu\nu,\lambda\rho}G^a_{\mu\nu,\lambda\rho} - \frac{1}{8}G^a_{\mu\nu,\lambda\lambda}G^a_{\mu\nu,\rho\rho} - \frac{1}{2}G^a_{\mu\nu,\lambda}G^a_{\mu\nu,\lambda\rho\rho} - \frac{1}{8}G^a_{\mu\nu}G^a_{\mu\nu,\lambda\lambda\rho\rho}$$

$$+\frac{1}{3}G^a_{\mu\nu,\lambda\rho}G^a_{\mu\lambda,\nu\rho} + \frac{1}{3}G^a_{\mu\nu,\nu\lambda}G^a_{\mu\rho,\rho\lambda} + \frac{1}{3}G^a_{\mu\nu,\nu\lambda}G^a_{\mu\lambda,\rho\rho}$$

$$+\frac{1}{3}G^a_{\mu\nu,\lambda}G^a_{\mu\lambda,\nu\rho\rho} + \frac{2}{3}G^a_{\mu\nu,\lambda}G^a_{\mu\rho,\nu\lambda\rho} + \frac{1}{3}G^a_{\mu\nu,\nu}G^a_{\mu\lambda,\lambda\rho\rho} + \frac{1}{3}G^a_{\mu\nu}G^a_{\mu\lambda,\nu\lambda\rho\rho}$$

$$(42)$$

and the corresponding free field equations for the tensor gauge field $A_{\mu\lambda_1\lambda_2}$ in four-dimensional space–time describe the *propagation of helicity-three and one* $h = \pm 3, \pm 1, \pm 1$ *massless charged gauge bosons* in agreement with the spectrum (18). There are no propagating negative norm states. The comparison of these equations with the Schwinger–Fronsdal equations [30–33] can be found in Ref. 44.

Considering the free field equation for the general rank-$(s+1)$ tensor gauge field one can find that the quadratic part of the Lagrangian has the following form:[29]

$$\mathcal{L}_{s+1} + \mathcal{L}'_{s+1}|_{\text{quadratic}} = \frac{1}{2} A^a_{\alpha\lambda_1...\lambda_s} \mathcal{H}^{\alpha\lambda_1...\lambda_s \gamma\lambda_{s+1}...\lambda_{2s}} A^a_{\gamma\lambda_{s+1}...\lambda_{2s}} \quad (43)$$

and is invariant with respect to the group of gauge transformations

$$\delta A^a_{\alpha\lambda_1...\lambda_s} = \partial_\alpha \xi^a_{\lambda_1...\lambda_s}, \quad \tilde{\delta} A^a_{\alpha\lambda_1...\lambda_s} = \partial_{\lambda_1} \zeta^a_{\lambda_2...\lambda_s \alpha} + \cdots + \partial_{\lambda_s} \zeta^a_{\lambda_1...\lambda_{s-1} \alpha}, \quad (44)$$

which should fulfil the following constraints:

$$\partial_\rho \zeta^a_{\rho\lambda_1...\lambda_{s-1}} - \frac{1}{s-2} (\partial_{\lambda_1} \zeta^a_{\lambda_2...\lambda_{s-1}\rho\rho} + \cdots + \partial_{\lambda_{s-1}} \zeta^a_{\lambda_1...\lambda_{s-2}\rho\rho}) = 0,$$

$$\partial_{\lambda_1} \zeta^a_{\lambda_2...\lambda_{s-1}\rho\rho} - \partial_{\lambda_2} \zeta^a_{\lambda_1...\lambda_{s-1}\rho\rho} = 0. \quad (45)$$

In momentum representation the kinetic operator has the following general form:

$$\mathcal{H}_{\alpha\lambda_1...\lambda_s \gamma\lambda_{s+1}...\lambda_{2s}}$$

$$= +\frac{1}{s!} \left(\sum_P \eta_{\lambda_{i_1}\lambda_{i_2}} \cdots \eta_{\lambda_{i_{2s-1}}\lambda_{i_{2s}}} \right) (-k^2 \eta_{\alpha\gamma} + k_\alpha k_\gamma)$$

$$+ \frac{1}{(s+1)!} \left(\sum_P \eta_{\alpha\lambda_{i_1}} \eta_{\lambda_{i_2}\lambda_{i_3}} \cdots \eta_{\lambda_{i_{2s-2}}\lambda_{i_{2s-1}}} \eta_{\gamma\lambda_{i_{2s}}} \right) k^2$$

$$- \frac{1}{(s+1)!} \left(\sum_P \eta_{\rho\lambda_{i_1}} \eta_{\lambda_{i_2}\lambda_{i_3}} \cdots \eta_{\lambda_{i_{2s-2}}\lambda_{i_{2s-1}}} \eta_{\gamma\lambda_{i_{2s}}} \right) k_\alpha k_\rho \quad (46)$$

$$- \frac{1}{(s+1)!} \left(\sum_P \eta_{\rho\lambda_{i_1}} \eta_{\lambda_{i_2}\lambda_{i_3}} \cdots \eta_{\lambda_{i_{2s-2}}\lambda_{i_{2s-1}}} \eta_{\alpha\lambda_{i_{2s}}} \right) k_\rho k_\gamma$$

$$+ \frac{1}{(s+1)!} \eta_{\alpha\gamma} \left(\sum_P \eta_{\rho\lambda_{i_1}} \eta_{\lambda_{i_2}\lambda_{i_3}} \cdots \eta_{\lambda_{i_{2s-2}}\lambda_{i_{2s-1}}} \eta_{\sigma\lambda_{i_{2s}}} \right) k_\rho k_\sigma,$$

where the sum $\sum_P$ runs over all non-equal permutations of λ_i's. The solution of the free field equation for the rank-$(s+1)$ field[29]

$$\mathcal{H}^{\alpha\lambda_1...\lambda_s \gamma\lambda_{s+1}...\lambda_{2s}} A_{\gamma\lambda_{s+1}...\lambda_{2s}} = 0 \quad (47)$$

describes the propagation of the helicities:

$$h = \pm(s+1), \quad \frac{\pm(s-1)}{\pm(s-1)}, \frac{\pm(s-3)}{\pm(s-3)}, \ldots \quad (48)$$

It is convenient to represent the spectrum (48) of tensor gauge bosons in the form which combines the helicity spectrum of all bosons. It is unbounded and has the following form:[47]

$$\begin{array}{cccccc}
\pm 1 \\
\pm 2, & 0 \\
\pm 3, & \pm 1, & \pm 1 \\
\pm 4, & \pm 2, & \pm 2, & 0 \\
\pm 5, & \pm 3, & \pm 3, & \pm 1, & \pm 1 \\
\pm 6, & \pm 4, & \pm 4, & \pm 2, & \pm 2, & 0 \\
& & \cdots\cdots\cdots\cdots\cdots
\end{array} \quad (49)$$

In summary, we defined the composite gauge field (3) which takes a value in the transversal representation (12), (15), (16) of the extended Poincaré algebra $L_G(\mathcal{P})$. We constructed the invariant Lagrangian (40), (32), (36) which contains infinity many tensor gauge fields (1) and found their helicity content (48), (49).

The theory has unexpected symmetry with respect to the duality transformation of the gauge fields.[86,87] The complementary gauge transformation $\tilde{\delta}$ is defined as:

$$\tilde{\delta}A^a_\mu = (\delta^{ab}\partial_\mu + gf^{acb}A^c_\mu)\eta^b,$$
$$\tilde{\delta}A^a_{\mu\lambda_1} = (\delta^{ab}\partial_{\lambda_1} + gf^{acb}A^c_{\lambda_1})\eta^b_\mu + gf^{acb}A^c_{\mu\lambda_1}\eta^b, \quad (50)$$
$$\tilde{\delta}A^a_{\mu\lambda_1\lambda_2} = (\delta^{ab}\partial_{\lambda_1} + gf^{acb}A^c_{\lambda_1})\eta^b_{\mu\lambda_2} + (\delta^{ab}\partial_{\lambda_2} + gf^{acb}A^c_{\lambda_2})\eta^b_{\mu\lambda_1}$$
$$+ gf^{acb}(A^c_{\mu\lambda_1}\eta^b_{\lambda_2} + A^c_{\mu\lambda_2}\eta^b_{\lambda_1} + A^c_{\lambda_1\lambda_2}\eta^b_\mu + A^c_{\lambda_2\lambda_1}\eta^b_\mu + A^c_{\mu\lambda_1\lambda_2}\eta^b),$$
$$\cdots\cdots\cdots\cdots\cdots$$

The transformations δ in (29) and $\tilde{\delta}$ in (50) do not coincide and are *complementary* to each other in the following sense: in δ the derivatives of the gauge parameters $\{\xi\}$ are over the first index μ, while in $\tilde{\delta}$ the derivatives of the gauge parameters $\{\eta\}$ are over the rest of the totally symmetric indices $\lambda_1 \ldots \lambda_s$. One can construct the new field strength tensors $\tilde{G}^a_{\mu\nu,\lambda_1\ldots\lambda_s}$ which are transforming homogeneously with respect to the $\tilde{\delta}$ transformations and then to construct the corresponding gauge invariant Lagrangian $\tilde{L}(A)$.[86,87] The relation between these two Lagrangians was found in the form of duality transformation:[86,87]

$$\tilde{A}_{\mu\lambda_1} = A_{\lambda_1\mu},$$
$$\tilde{A}_{\mu\lambda_1\lambda_2} = \tfrac{1}{2}(A_{\lambda_1\mu\lambda_2} + A_{\lambda_2\mu\lambda_1}) - \tfrac{1}{2}A_{\mu\lambda_1\lambda_2}, \quad (51)$$
$$\tilde{A}_{\mu\lambda_1\lambda_2\lambda_3} = \tfrac{1}{3}(A_{\lambda_1\mu\lambda_2\lambda_3} + A_{\lambda_2\mu\lambda_1\lambda_3} + A_{\lambda_3\mu\lambda_1\lambda_2}) - \tfrac{2}{3}A_{\mu\lambda_1\lambda_2\lambda_3},$$
$$\cdots\cdots\cdots\cdots\cdots$$

which maps the Lagrangian $L(\tilde{A})$ into the Lagrangian $\tilde{L}(A)$. This takes place because $G_{\mu\nu,\lambda_1...\lambda_s}(\tilde{A}) = \tilde{G}_{\mu\nu,\lambda_1...\lambda_s}(A)$ and therefore $L(\tilde{A}) = \tilde{L}(A)$.

The Lagrangian (40) defines not only a free propagation of tensor gauge bosons, but also their interactions. The interaction diagrams for the lower-rank bosons are presented in Figs. 1 and 2. The high-rank bosons also interact through the triple and quartic interaction vertices. It is therefore important to calculate and study the scattering amplitudes, the quantum loop corrections and their high energy behavior. By using the diagram technique it is possible to calculate the scattering amplitude, but the difficulties lie in the evaluation and contraction of high-rank tensors structures appearing in the diagram approach. In the next section we shall use alternative approach based on spinor representation of amplitudes developed recently in Refs. 54–71.

5. Scattering Amplitudes and Splitting Functions

A scattering amplitude for the massless particles of momenta p_i and polarization tensors ε_i ($i = 1, \ldots, n$), which are described by irreducible massless representations of the Poincaré group, can be represented in the following form:

$$M_n = M_n(p_1, \varepsilon_1; p_2, \varepsilon_2; \ldots; p_n, \varepsilon_n).$$

It is more convenient to represent the momenta p_i and polarization tensors ε_i in terms of spinors. In that case the scattering amplitude M_n can be considered as a function of spinors λ_i, $\tilde{\lambda}_i$ and helicities h_i:[54–71]

$$M_n = M_n(\lambda_1, \tilde{\lambda}_1, h_1; \ldots; \lambda_n, \tilde{\lambda}_n, h_n). \tag{52}$$

The advantage of the spinor representation is that introducing a complex deformation of the particles momenta one can derive a general form for the three-particle interaction vertices:[45,46,53,66,72]

$$M_3(1^{h_1}, 2^{h_2}, 3^{h_3}).$$

The dimensionality of the three-point vertex $M_3(1^{h_1}, 2^{h_2}, 3^{h_3})$ is

$$[\text{mass}]^{D=+(h_1\,\mid\,h_2+h_3)}.$$

In the generalized Yang–Mills theory,[9–11,47] which we described in the previous sections, all interaction vertices between high-spin particles have *dimensionless couplings constants*, which means that the helicities of the interacting particles in the vertex are constrained by the relation

$$D = \pm(h_1 + h_2 + h_3) = 1.$$

Therefore the interaction vertex between massless tensor-bosons, the TTT-vertex, has the following general form:[53,72]

$$M_3 = gf^{abc}\langle 1,2\rangle^{-2h_1-2h_2-1}\langle 2,3\rangle^{2h_1+1}\langle 3,1\rangle^{2h_2+1}, \quad h_3 = -1-h_1-h_2,$$

$$M_3 = gf^{abc}[1,2]^{2h_1+2h_2-1}[2,3]^{-2h_1+1}[3,1]^{-2h_2+1}, \quad h_3 = 1-h_1-h_2, \quad (53)$$

where f^{abc} are the structure constants of the internal gauge group G. In particular, considering the interaction between a boson of helicity $h_1 = \pm 1$ and a tensor-boson of helicity $h_2 = \pm s$, the VTT-vertex, one can find from (53) that

$$h_3 = \pm|s-2|, \pm s, \pm|s+2| \qquad (54)$$

and the corresponding vector–tensor–tensor interaction vertices VTT have the following form:

$$M_3^{a_1a_2a_3}(1^{-s}, 2^{-1}, 3^{+s}) = g\, f^{a_1a_2a_3} \frac{\langle 1,2\rangle^4}{\langle 1,2\rangle\langle 2,3\rangle\langle 3,1\rangle} \left(\frac{\langle 1,2\rangle}{\langle 2,3\rangle}\right)^{2s-2},$$

$$M_3^{a_1a_2a_3}(1^{-s}, 2^{+1}, 3^{s-2}) = g\, f^{a_1a_2a_3} \frac{\langle 1,3\rangle^4}{\langle 1,2\rangle\langle 2,3\rangle\langle 3,1\rangle} \left(\frac{\langle 1,2\rangle}{\langle 2,3\rangle}\right)^{2s-2}. \quad (55)$$

These are the vertices which reduce to the standard triple YM vertex when $s = 1$. Using these vertices one can compute the scattering amplitudes of vector and tensor bosons. The color-ordered scattering amplitudes involving two tensor-bosons of helicities $h = \pm s$, one negative helicity vector-boson and $(n-3)$ vector-bosons of positive helicity were found in Ref. 72:

$$\hat{M}_n(1^+, ..i^-, ...k^{+s}, ..j^{-s}, ..n^+) = ig^{n-2}(2\pi)^4\delta^{(4)}(P^{a\dot b})\frac{\langle ij\rangle^4}{\prod_{l=1}^n\langle ll+1\rangle}\left(\frac{\langle ij\rangle}{\langle ik\rangle}\right)^{2s-2}, \quad (56)$$

where n is the total number of particles and the dots stand for any number of positive helicity vector-bosons, i is the position of the negative-helicity vector, while k and j are the positions of the tensors with helicities $+s$ and $-s$ respectively. The expression (56) reduces to the famous Parke–Taylor formula[59] when $s = 1$. In particular, the five-particle amplitude takes the following form:

$$\hat{M}_5(1^+, 2^-, 3^+, 4^{+s}, 5^{-s}) = ig^3(2\pi)^4\delta^{(4)}(P^{a\dot b})\frac{\langle 25\rangle^4}{\prod_{i=1}^5\langle ii+1\rangle}\left(\frac{\langle 25\rangle}{\langle 24\rangle}\right)^{2s-2}, \quad (57)$$

where $P^{a\dot b} = \sum_{m=1}^n \lambda_m^a \tilde{\lambda}_m^{\dot b}$ is the total momentum. Notice that the scattering amplitudes (56) and (57) have large validity area: in the limit $s \to 1/2$ they reduce to the tree level gluon scattering amplitudes into a quark pair and into a pair of scalars as $s \to 0$.

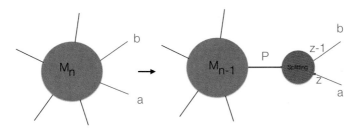

Fig. 3. The scattering amplitudes (56) and (57) can be used to extract splitting amplitude of a vector boson into two tensor bosons. Considering the amplitude in the limit when the tensor bosons become collinear, $k_a \parallel k_b$, that is, $k_a = zk_P$, $k_b = (1-z)k_P$, $k_P^2 \to 0$ and z describes the longitudinal momentum sharing, one can deduce that the corresponding behavior of spinors is $\lambda_a = \sqrt{z}\lambda_P$, $\lambda_b = \sqrt{1-z}\lambda_P$, and that the factorization form of the amplitude is (59).

The scattering amplitudes (56) and (57) can be used to extract splitting amplitudes of vector and tensor bosons.[73] The collinear behavior of the tree amplitudes has the following factorized form:[58–60,70,71]

$$M_n^{\text{tree}}(\ldots, a^{\lambda_a}, b^{\lambda_b}, \ldots) \xrightarrow{a \parallel b} \sum_{\lambda=\pm 1} \text{Split}_{-\lambda}^{\text{tree}}(a^{\lambda_a}, b^{\lambda_b}) \times M_{n-1}^{\text{tree}}(\ldots, P^{\lambda}, \ldots), \quad (58)$$

where $\text{Split}_{-\lambda}^{\text{tree}}(a^{\lambda_a}, b^{\lambda_b})$ denotes the splitting amplitude and the intermediate state P has momentum $k_P = k_a + k_b$ and helicity λ. Considering the amplitude (57) in the limit when the particles 4 and 5 become collinear, $k_4 \parallel k_5$, that is, $k_4 = zk_P$, $k_5 = (1-z)k_P$, $k_P^2 \to 0$ and z describes the longitudinal momentum sharing, one can deduce that the corresponding behavior of spinors is $\lambda_4 = \sqrt{z}\lambda_P$, $\lambda_5 = \sqrt{1-z}\lambda_P$, and that the amplitude (57) takes the following factorization form[73] (see Fig. 3):

$$M_5(1^+, 2^-, 3^+, 4^{+s}, 5^{-s}) = A_4(1^+, 2^-, 3^+, P^-) \times \text{Split}_+(a^{+s}, b^{-s}), \quad (59)$$

where

$$\text{Split}_+(a^{+s}, b^{-s}) = \left(\frac{1-z}{z}\right)^{s-1} \frac{(1-z)^2}{\sqrt{z(1-z)}} \frac{1}{\langle a, b \rangle}. \quad (60)$$

In a similar way one can deduce that

$$\text{Split}_+(a^{-s}, b^{+s}) = \left(\frac{z}{1-z}\right)^{s-1} \frac{z^2}{\sqrt{z(1-z)}} \frac{1}{\langle a, b \rangle}. \quad (61)$$

Considering different collinear limits $k_1 \parallel k_5$ and $k_3 \parallel k_4$ one can get[73]

$$\text{Split}_{+s}(a^+, b^{-s}) = \frac{(1-z)^{s+1}}{\sqrt{z(1-z)}} \frac{1}{\langle a, b \rangle}, \quad \text{Split}_{+s}(a^{-s}, b^+) = \frac{z^{s+1}}{\sqrt{z(1-z)}} \frac{1}{\langle a, b \rangle} \quad (62)$$

and

$$\text{Split}_{-s}(a^{+s}, b^+) = \frac{z^{-s+1}}{\sqrt{z(1-z)}} \frac{1}{\langle a, b \rangle}, \quad \text{Split}_{-s}(a^+, b^{+s}) = \frac{(1-z)^{-s+1}}{\sqrt{z(1-z)}} \frac{1}{\langle a, b \rangle}. \quad (63)$$

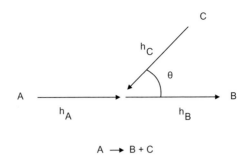

Fig. 4. The decay of a gluon of helicity h_A into the tensorgluons of helicities h_B and h_C. The arrows show the directions of the helicities. The corresponding splitting probability is defined as P_{BA}.

The set of splitting amplitudes (60)–(63) $V \to TT$, $T \to VT$ and $T \to TV$ reduces to the full set of gluon splitting amplitudes[58–60,70,71] when $s = 1$.

Since the collinear limits of the scattering amplitudes are responsible for parton evolution[75] we can extract from the above expressions the Altarelli–Parisi splitting probabilities for tensor-bosons. Indeed, the residue of the collinear pole in the square (of the factorized amplitude (58)) gives Altarelli–Parisi splitting probability $P(z)$:

$$P(z) = C_2(G) \sum_{h_P, h_a, h_b} |\text{Split}_{-h_P}(a^{h_a}, b^{h_b})|^2 \, s_{ab}, \qquad (64)$$

where $s_{ab} = 2k_a \cdot k_b = \langle a, b \rangle [a, b]$. The invariant operator C_2 for the representation R is defined by the equation $t^a t^a = C_2(R)\,\mathbf{1}$ and $\text{tr}(t^a t^b) = T(R)\delta^{ab}$. Substituting the splitting amplitudes (60)–(63) into (64) we are getting (see Fig. 4)

$$P_{TV}(z) = C_2(G)\left[\frac{z^4}{z(1-z)}\left(\frac{z}{1-z}\right)^{2s-2} + \frac{(1-z)^4}{z(1-z)}\left(\frac{1-z}{z}\right)^{2s-2}\right],$$

$$P_{VT}(z) = C_2(G)\left[\frac{1}{z(1-z)}\left(\frac{1}{1-z}\right)^{2s-2} + \frac{(1-z)^4}{z(1-z)}(1-z)^{2s-2}\right], \qquad (65)$$

$$P_{TT}(z) = C_2(G)\left[\frac{z^4}{z(1-z)}z^{2s-2} + \frac{1}{z(1-z)}\left(\frac{1}{z}\right)^{2s-2}\right].$$

The momentum conservation in the vertices clearly fulfils because these functions satisfy the relations

$$P_{TV}(z) = P_{TV}(1-z), \quad P_{VT}(z) = P_{TT}(1-z), \quad z < 1. \qquad (66)$$

In the leading order the kernel $P_{TV}(z)$ has a meaning of variation per unit transfer momentum of the probability density of finding a tensor-boson inside the vector-boson, $P_{VT}(z)$ — of finding a vector inside the tensor and $P_{TT}(z)$ — of finding a tensor inside the tensor. For completeness we shall present also quark and vector-boson kernels:[75–79]

$$P_{qq}(z) = C_2(R)\frac{1+z^2}{1-z},$$

$$P_{Vq}(z) = C_2(R)\frac{1+(1-z)^2}{z}, \qquad (67)$$

$$P_{qV}(z) = T(R)[z^2 + (1-z)^2],$$

$$P_{VV}(z) = C_2(G)\left[\frac{1}{z(1-z)} + \frac{z^4}{z(1-z)} + \frac{(1-z)^4}{z(1-z)}\right],$$

where $C_2(G) = N, C_2(R) = \frac{N^2-1}{2N}, T(R) = \frac{1}{2}$ for the SU(N) groups.

Having in hand the new set of splitting probabilities for tensor-bosons (65) we can consider a possible generalization of quantum chromodynamics.[52] In so generalized theory in addition to the quarks and gluons there should be tensorgluons. We can hypothesize that a possible emission of tensorgluons by gluons, as it is shown in Figs.1 and 3 should produce a non-zero density of tensorgluons inside the proton in additional to the quark and gluon densities. Our next goal is to derive DGLAP equations[75–82] which will take into account these new emission processes.

6. Generalization of DGLAP Equation. Calculation of Callan-Symanzik Beta Function

In this section we shall consider a possibility that inside the proton and, more generally, inside hadrons there are additional partons — tensorgluons, which can carry a part of the proton momentum.[51–53] Tensorgluons have zero electric charge, like gluons, but have a larger spin. Inside the proton a nonzero density of the tensorgluons can be generated by the emission of tensorgluons by gluons.[9–11,47] The last mechanism is typical for non-Abelian tensor gauge theories, in which there exists a gluon–tensor–tensor vertex of order g (see Fig. 2).[9–11,47] Therefore a number of gluons changes not only because a quark may radiate a gluon or because a gluon may split into a quark–antiquark pair or into two gluons,[75,83,84] but also because a gluon can split into two tensorgluons.[9–11,47,72,73] The process of gluon splitting into tensorgluons suggests that part of the proton momentum which was carried by neutral partons can be shared between vector and tensorgluons. Our aim is to calculate the scattering amplitudes and splitting function in QCD generalized in this way.

It is well known that the deep inelastic structure functions can be expressed in terms of quark distribution densities. If $q^i(x)$ is the density of quarks of type i (summed over colors) inside a proton target with fraction x of the proton longitudinal momentum in the infinite momentum frame[74] then the scaling structure functions can be represented in the following form:

$$2F_1(x) = F_2(x)/x = \sum_i Q_i^2[q^i(x) + \bar{q}^i(x)]. \qquad (68)$$

Fig. 5. We are considering a possibility that inside the proton and, more generally, inside hadrons there are additional partons — tensorgluons, which can carry a part of the proton momentum. Tensorgluons have zero electric charge, like gluons, but have a larger spin. Inside the proton a nonzero density of the tensorgluons can be generated by the emission of tensorgluons by gluons (see Figs. 1 and 3), thus we shall introduce the corresponding density $T(x,t)$ of tensorgluons inside the proton.

The scaling behavior of the structure functions is broken and the results can be formulated by assigning a well determined Q^2 dependence to the parton densities. This can be achieved by introducing integro-differential equations which describe the Q^2 dependence of quark $q^i(x,t)$ and gluon densities $G(x,t)$, where $t = \ln(Q^2/Q_0^2)$.[75–82]

Let us see what will happen if one supposes that there are additional partons — tensorsgluons — inside the proton. In accordance with our hypothesis there is an additional emission of tensorgluons in the proton, therefore one should introduce the corresponding density $T(x,t)$ of tensorgluons (summed over colors) inside the proton in the P_∞ frame.[74] We can derive integro-differential equations that describe the Q^2 dependence of parton densities in this general case:[52]

$$\frac{dq^i(x,t)}{dt} = \frac{\alpha(t)}{2\pi} \int_x^1 \frac{dy}{y} \left[\sum_{j=1}^{2n_f} q^j(y,t)\, P_{q^i q^j}\left(\frac{x}{y}\right) + G(y,t)\, P_{q^i G}\left(\frac{x}{y}\right) \right],$$

$$\frac{dG(x,t)}{dt} = \frac{\alpha(t)}{2\pi} \int_x^1 \frac{dy}{y} \left[\sum_{j=1}^{2n_f} q^j(y,t)\, P_{Gq^j}\left(\frac{x}{y}\right) + G(y,t)\, P_{GG}\left(\frac{x}{y}\right) \right.$$

$$\left. + T(y,t)\, P_{GT}\left(\frac{x}{y}\right) \right], \qquad (69)$$

$$\frac{dT(x,t)}{dt} = \frac{\alpha(t)}{2\pi} \int_x^1 \frac{dy}{y} \left[G(y,t) P_{TG}\left(\frac{x}{y}\right) + T(y,t)\, P_{TT}\left(\frac{x}{y}\right) \right].$$

The $\alpha(t)$ is the running coupling constant ($\alpha = g^2/4\pi$). In the leading logarithmic approximation $\alpha(t)$ is of the form

$$\frac{\alpha}{\alpha(t)} = 1 + b\,\alpha\,t, \tag{70}$$

where $\alpha = \alpha(0)$ and b is the one-loop Callan–Symanzik coefficient, which, as we shall see below, receives an additional contribution from the tensorgluon loop. Here the indices i and j run over quarks and antiquarks of all flavors. The number of quarks of a given fraction of momentum changes when a quark looses momentum by radiating a gluon, or a gluon inside the proton may produce a quark–antiquark pair.[75] Similarly the number of gluons changes because a quark may radiate a gluon or because a gluon may split into a quark–antiquark pair or into two gluons or *into two tensorgluons*. This last possibility is realized, because, as we have seen, in non-Abelian tensor gauge theories there is a triple vertex VTT (55) of a gluon and two tensorgluons of order g.[9-11,47] This interaction should be taken into consideration, and we added the term $T(y,t)\,P_{GT}(\frac{x}{y})$ in the second equation (69). The density of tensorgluons $T(x,t)$ changes when a gluon splits into two tensorgluons or when a tensorgluon radiates a gluon. This evolution is described by the last equation (69).

In order to guarantee that the total momentum of the proton, that is, of all partons is unchanged, one should impose the following constraint:

$$\frac{d}{dt}\int_0^1 dz\,z\left[\sum_{i=1}^{2n_f} q^i(z,t) + G(z,t) + T(z,t)\right] = 0. \tag{71}$$

Using the evolution equations (69) one can express the derivatives of the densities in (71) in terms of kernels and to see that the following momentum sum rules should be fulfilled:

$$\int_0^1 dz\,z[P_{qq}(z) + P_{Gq}(z)] = 0,$$

$$\int_0^1 dz\,z[2n_f P_{qG}(z) + P_{GG}(z) + P_{TG}(z)] = 0, \tag{72}$$

$$\int_0^1 dz\,z[P_{GT}(z) + P_{TT}(z)] = 0.$$

Before analyzing these momentum sum rules let us first inspect the behavior of the gluon-tensorgluon kernels (65) at the end points $z = 0, 1$. As one can see, they are singular at the boundary values similarly to the case of the standard kernels (67). Though there is a difference here: the singularities are of higher order compared to the standard case.[75] Therefore one should define the regularization procedure for the singular factors $(1-z)^{-2s+1}$ and z^{-2s+1} reinterpreting them as the distributions

$(1-z)_+^{-2s+1}$ and z_+^{-2s+1}, similarly to the Altarelli–Parisi regularisation.[75] We shall define them in the following way:

$$\int_0^1 dz \frac{f(z)}{(1-z)_+^{2s-1}} = \int_0^1 dz \frac{f(z) - \sum_{k=0}^{2s-2} \frac{(-1)^k}{k!} f^{(k)}(1)(1-z)^k}{(1-z)^{2s-1}},$$

$$\int_0^1 dz \frac{f(z)}{z_+^{2s-1}} = \int_0^1 dz \frac{f(z) - \sum_{k=0}^{2s-2} \frac{1}{k!} f^{(k)}(0) z^k}{z^{2s-1}}, \quad (73)$$

$$\int_0^1 dz \frac{f(z)}{z_+(1-z)_+} = \int_0^1 dz \frac{f(z) - (1-z)f(0) - zf(1)}{z(1-z)},$$

where $f(z)$ is any test function which is sufficiently regular at the end points and, as one can see, the defined substraction guarantees the convergence of the integrals. Using the same arguments as in the standard case[75] we should add the delta function terms into the definition of the diagonal kernels so that they will completely determine the behavior of $P_{qq}(z)$, $P_{GG}(z)$ and $P_{TT}(z)$ functions. The first equation in the momentum sum rule (72) remains unchanged because there is no tensorgluon contribution into the quark evolution. The second equation in the momentum sum rule (72) will take the following form:

$$\int_0^1 dz z [2n_f P_{qG}(z) + P_{GG}(z) + P_{TG}(z) + b_G \delta(z-1)]$$

$$= \int_0^1 dzz \left[2n_f T(R)[z^2 + (1-z)^2] + C_2(G) \left[\frac{1}{z(1-z)} + \frac{z^4}{z(1-z)} + \frac{(1-z)^4}{z(1-z)} \right] \right.$$

$$\left. + C_2(G) \left[\frac{z^4}{z(1-z)} \left(\frac{z}{1-z} \right)^{2s-2} + \frac{(1-z)^4}{z(1-z)} \left(\frac{1-z}{z} \right)^{2s-2} \right] \right] + b_G$$

$$= \frac{2}{3} n_f T(R) - \frac{11}{6} C_2(G) - \frac{12s^2 - 1}{6} C_2(G) + b = 0. \quad (74)$$

From this result we can extract an additional contribution to the one-loop Callan–Symanzik beta function arising from the tensorgluon loop. Indeed, the first beta-function coefficient enters into this expression because the momentum sum rule (72) implicitly comprises unitarity, thus the one-loop effects.[75] In (74) we have three terms which come from gluon and quark loops:

$$b_1 = \frac{11}{6} C_2(G) - \frac{2n_f}{3} T(R), \quad (75)$$

and from the tensorboson loop of spin s:

$$b_T = \frac{12s^2 - 1}{6} C_2(G), \quad s = 1, 2, 3, 4, \ldots. \tag{76}$$

It is a very interesting result because at $s = 1$ we are rediscovering the asymptotic freedom result.[83–85] For larger spins the tensorgluon contribution into the Callan-Symanzik beta function has the same signature as the standard gluons, which means that tensorgluons "accelerate" the asymptotic freedom (70) of the strong interaction coupling constant $\alpha(t)$. The contribution is increasing quadratically with the spin of the tensorgluons, that is, at large transfer momentum the strong coupling constant tends to zero faster compared to the standard case:

$$\alpha(t) = \frac{\alpha}{1 + b\,\alpha\,t}, \tag{77}$$

where

$$b = \frac{(12s^2 - 1)C_2(G) - 4n_f T(R)}{12\pi}, \quad s = 1, 2, \ldots. \tag{78}$$

Surprisingly, a similar result based on the parametrization of the charge renormalization taken in the form $b = (-1)^{2s}(A + Bs^2)$ was conjectured by Curtright.[89] Here A represents an orbital contribution and Bs^2 — the anomalous magnetic moment contribution.[90–92] The unknown coefficients A and B were found by comparing the suggested parametrization with the known results for $s = 0, 1/2$ and 1.

It is also possible to consider a straightforward generalization of the result obtained for the effective action in Yang–Mills theory long ago[90–93] to the higher spin gauge bosons. With the spectrum of the tensorgluons in the external chromomagnetic field $\lambda = (2n + 1 + 2s)gH + k_\parallel^2$ one can perform a summation of the modes and get an exact result for the one-loop effective action similarly to:[90,93]

$$\epsilon = \frac{H^2}{2} + \frac{(gH)^2}{4\pi} b \left[\ln \frac{gH}{\mu^2} - \frac{1}{2} \right], \tag{79}$$

where

$$b = -\frac{2C_2(G)}{\pi} \zeta\left(-1, \frac{2s+1}{2}\right) = \frac{12s^2 - 1}{12\pi} C_2(G), \tag{80}$$

and $\zeta(-1, q) = -\frac{1}{2}(q^2 - q + \frac{1}{6})$ is the generalized zeta function.[b] Because the coefficient in front of the logarithm defines the beta function,[90,91] one can see that (80) is in agreement with the result (76).

It is also natural to ask what will happen if one takes into consideration the contribution of tensorgluons of all spins into the beta function.[c] One can suggest

---

[b]The generalized zeta function is defined as $\zeta(p, q) = \sum_{k=0}^{\infty} \frac{1}{(k+q)^p} = \frac{1}{\Gamma(p)} \int_0^\infty dt\, t^{-1+p} \frac{e^{-qt}}{1-e^{-t}}$.
[c]I would like to thank John Iliopoulos and Constantin Bachas for raising this question.

two scenarios. In the first one the high spin gluons, let us say, of $s \geq 3$, will get large mass and therefore they can be ignored at a given energy scale. In the second case, when all of them remain massless, then one can suggest the Riemann zeta function regularization, similar to the Brink–Nielsen regularisation.[94] The summation over the spectrum in (49) gives:[53]

$$b_{11} = C_2(G) \left[\sum_{s=1}^{\infty} \frac{(12s^2 - 1)}{12\pi} + \sum_{s=0}^{\infty} \frac{(12s^2 - 1)}{12\pi} \right.$$
$$\left. + \sum_{s=1}^{\infty} \frac{(12s^2 - 1)}{12\pi} + \sum_{s=0}^{\infty} \frac{(12s^2 - 1)}{12\pi} + \cdots \right]$$
$$= C_2(G) \left[\frac{1}{\pi}\zeta(-2) - \frac{1}{12\pi}\zeta(0) - \frac{1}{12\pi} + \frac{1}{\pi}\zeta(-2) - \frac{1}{12\pi}\zeta(0) + \cdots \right]$$
$$= C_2(G) \left[\frac{1}{24\pi} - \frac{1}{12\pi} + \frac{1}{24\pi} + \cdots \right] = 0, \tag{81}$$

where $\zeta(-2) = 0$, $\zeta(0) = -1/2$, leading to the theory which is *conformally invariant* at very high energies. The above summation requires explicit regularization and further justification.

7. Unification of Coupling Constants of Standard Model

It is interesting to know how the contribution of tensorgluons changes the high energy behavior of the coupling constants of the Standard Model.[96,97] The coupling constants are evolving in accordance with the formulae

$$\frac{1}{\alpha_i(M)} = \frac{1}{\alpha_i(\mu)} + 2b_i \ln\frac{M}{\mu}, \quad i = 1, 2, 3, \tag{82}$$

where we shall consider only the contribution of the lower $s = 2$ tensorbosons:

$$2b = \frac{58 C_2(G) - 4 n_f T(R)}{6\pi}. \tag{83}$$

For the $SU(3)_c \times SU(2)_L \times U(1)$ group with its coupling constants α_3, α_2 and α_1 and six quarks $n_f = 6$ and $SU(5)$ unification group we will get

$$2b_3 = \frac{1}{2\pi}54, \quad 2b_2 = \frac{1}{2\pi}\frac{104}{3}, \quad 2b_1 = -\frac{1}{2\pi}4,$$

so that solving the system of equations (82) one can get

$$\ln\frac{M}{\mu} = \frac{\pi}{58}\left(\frac{1}{\alpha_{el}(\mu)} - \frac{8}{3}\frac{1}{\alpha_s(\mu)} \right), \tag{84}$$

where $\alpha_{el}(\mu)$ and $\alpha_s(\mu)$ are the electromagnetic and strong coupling constants at scale μ. If one takes $\alpha_{el}(M_Z) = 1/128$ and $\alpha_s(M_Z) = 1/10$ one can get that coupling constants have equal strength at energies of order

$$M \sim 4 \times 10^4 \, \text{GeV} = 40 \, \text{TeV},$$

which is much smaller than the scale $M \sim 10^{14}$ GeV in the absence of the tensorgluons contribution. The value of the weak angle[96,97] remains intact:

$$\sin^2 \theta_W = \frac{1}{6} + \frac{5}{9} \frac{\alpha_{el}(M_Z)}{\alpha_s(M_Z)}, \tag{85}$$

as well as the coupling constant at the unification scale remains of the same order $\bar{\alpha}(M) = 0,01$.

8. Conclusion

In the present article we describe a possible extension the Yang–Mills gauge principle[1] which includes tensor gauge fields. In this extension of the Yang–Mills theory the vector gauge boson becomes a member of a bigger family of gauge bosons of arbitrary large integer spins.

The proposed extension of Yang–Mills theory is essentially based on the existence of the enlarged Poincaré algebra and on an appropriate transversal representations of that algebra. The invariant Lagrangian is expressed in terms of new higher-rank field strength tensors. The Lagrangian does not contain higher derivatives of tensor gauge fields and all interactions take place through three- and four-particle exchanges with a dimensionless coupling constant (see Figs. 1 and 2).

We calculated the scattering amplitudes of non-Abelian tensor gauge bosons at tree level, as well as their one-loop contribution into the Callan–Symanzik beta function. This contribution is negative and corresponds to the asymptotically free theory. The proposed extension may lead to a natural inclusion of the standard theory of fundamental forces into a larger theory in which vector gauge bosons, leptons and quarks represent a low-spin subgroup.

In the line with the above development we considered a possible extension of QCD. In so extended QCD inside the proton and, more generally, inside hadrons there should be additional partons — tensorgluons, which can carry a part of the proton momentum. Among all parton distributions, the gluon density $G(x,t)$ is one of the least constrained functions since it does not couple directly to the photon in deep-inelastic scattering measurements of the proton F_2 structure function. Therefore it is only indirectly constrained by scaling violations and by the momentum sum rule which resulted in the fact that only half of the proton momentum is carried by charged constituents — the quarks — and that the other part is ascribed to the neutral constituents.

As it was suggested, the process of gluon splitting leads to the emission of tensorgluons and therefore a part of the proton momentum which is carried by the neutral constituents can be shared between gluons and tensorgluons. The density of neutral partons in the proton is therefore given by the sum of two functions: $G(x,t)+T(x,t)$, where $T(x,t)$ is the density of the tensorgluons. To disentangle these contributions and to decide which piece of the neutral partons is the contribution of gluons and which one is of the tensorgluons one should measure the helicities of the neutral components, which seems to be a difficult task.

The gluon density can be directly constrained by jet production.[95] In the suggested model the situation is such that the standard quarks cannot radiate tensorgluons (such a vertex is absent in the model[9-11,47]), therefore only gluons are radiated by quarks. A radiated gluon then can split into a pair of tensorgluons without obscuring the structure of the observed three-jet final states. Thus it seems that there is no obvious contradiction with the existing experimental data. Our hypotheses may be wrong, but the uniqueness and simplicity of suggested extension seems to be the reasons for serious consideration.

This extension of QCD influences the unification scale at which the coupling constants of the Standard Model merge. In the last section we observed that the unification scale at which standard coupling constants are merging is shifted to lower energies telling us that it may be that a new physics is round the corner. Whether all these phenomena are consistent with experiment is an open question.

Acknowledgments

I would like to thank the organizers of the Conference on 60 Years of Yang–Mills Gauge Field Theories for their kind hospitality in the Institute of Advanced Studies of the Nanyang Technological University in Singapore and Prof. Kok-Khoo Phua and Prof. Yong Min Cho for invitation. This work was supported in part by the General Secretariat for Research and Technology of Greece and the European Regional Development Fund (NSRF 2007-15 ACTION,KRIPIS).

References

1. C. N. Yang and R. L. Mills, Conservation of isotopic spin and isotopic gauge invariance, *Phys. Rev.* **96**, 191 (1954).
2. S. S. Chern, *Topics in Differential Geometry*, Ch. III "Theory of Connections" (The Institute for Advanced Study, Princeton, 1951).
3. H. Weyl, Electron and gravitation, *Z. Phys.* **56**, 330 (1929).
4. H. Weyl, *Space, Time, Matter* (Dover, New York, 1952).
5. E. Cartan, Sur les variétés á connexion affine et la théorie de la relativité généralisée, *Annales Sci. Ecole Norm. Sup.* **40**, 325 (1923); **41**, 1 (1924); **42**, 17 (1925).
6. E. Cartan, *La méthode de repére mobile, la théorie des groupes continus, et les espaces généralisés* (Hermann, Paris, 1935).
7. R. Utiyama, Invariant theoretical interpretation of interaction, *Phys. Rev.* **101**, 1597 (1956).

8. T. W. B. Kibble, Lorentz invariance and the gravitational field, *J. Math. Phys.* **2**, 212 (1961).
9. G. Savvidy, Non-Abelian tensor gauge fields: Generalization of Yang-Mills theory, *Phys. Lett. B* **625**, 341 (2005).
10. G. Savvidy, Non-Abelian tensor gauge fields. I, *Int. J. Mod. Phys. A* **21**, 4931 (2006).
11. G. Savvidy, Non-Abelian tensor gauge fields. II, *Int. J. Mod. Phys. A* **21**, 4959 (2006).
12. H. Yukawa, Quantum theory of non-local fields. Part I. Free fields, *Phys. Rev.* **77**, 219 (1950).
13. M. Fierz, Non-local fields, *Phys. Rev.* **78**, 184 (1950).
14. E. Wigner, Invariant quantum mechanical equations of motion, in *Theoretical Physics*, ed. A. Salam (International Atomic Energy, Vienna, 1963), p. 59.
15. G. K. Savvidy, Conformal invariant tensionless strings, *Phys. Lett. B* **552**, 72 (2003).
16. G. K. Savvidy, Tensionless strings: Physical Fock space and higher spin fields, *Int. J. Mod. Phys. A* **19**, (2004) 3171.
17. G. Savvidy, Tensionless strings, correspondence with SO(D,D) sigma model, *Phys. Lett. B* **615**, 285 (2005).
18. Yu. A. Golfand and E. P. Likhtman, Extension of the algebra of Poincare group generators and violation of p invariance, *JETP Lett.* **13**, 323 (1971).
19. A. Salam and J. A. Strathdee, Feynman rules for superfields, *Nucl. Phys. B* **86**, 142 (1975).
20. S. R. Coleman and J. Mandula, All possible symmetries of the S matrix, *Phys. Rev.* **159**, 1251 (1967).
21. R. Haag, J. T. Lopuszanski and M. Sohnius, All possible generators of supersymmetries of the S matrix, *Nucl. Phys. B* **88**, 257 (1975).
22. E. Wigner, On unitary representations of the inhomogeneous Lorentz group. *Ann. Math.* **40**, 149 (1939).
23. E. Majorana, Teoria relativistica di particelle con momento intrinseco arbitrario, *Nuovo Cimento* **9**, 335 (1932).
24. P. A. M. Dirac, Relativistic wave equations, *Proc. Roy. Soc. A* **155**, 447 (1936); Unitary representation of the Lorentz group, *Proc. Roy. Soc. A* **183**, 284 (1944).
25. M. Fierz, Über die relativistische theorie kräftefreier teilchen mit beliebigem spin, *Helv. Phys. Acta.* **12**, 3 (1939).
26. M. Fierz and W. Pauli, On relativistic wave equations for particles of arbitrary spin in an electromagnetic field, *Proc. Roy. Soc. A* **173**, 211 (1939).
27. P. Minkowski, Versuch einer konsistenten Theorie eines spin-2 mesons, *Helv. Phys. Acta.* **32**, 477 (1966).
28. G. Savvidy, Particle spectrum of non-Abelian tensor gauge fields, *Mod. Phys. Lett. A* **25**, 1137 (2010). [arXiv:0909.3859 [hep-th]].

29. G. Savvidy, Solution of free field equations in non-Abelian tensor gauge field theory, *Phys. Lett. B* **682**, 143 (2009).
30. J. Schwinger, *Particles, Sourses, and Fields* (Addison-Wesley, Reading, MA, 1970).
31. L. P. S. Singh and C. R. Hagen, Lagrangian formulation for arbitrary spin. I. The boson case, *Phys. Rev. D* **9**, 898 (1974).
32. C. Fronsdal, Massless fields with integer spin, *Phys. Rev. D* **18**, 3624 (1978).
33. S. Weinberg, Feynman rules for any spin, *Phys. Rev.* **133**, B1318 (1964).
34. V. L. Ginzburg and I. E. Tamm, To the theory of spin, *ZETP* **17**, 227 (1947).
35. P. Ramond, Dual theory for free fermions, *Phys. Rev. D* **3**, 2415 (1971).
36. V. L. Ginzburg and V. I. Manko, Relativistic wave equations with inner degrees of freedom and partons, *Sov. J. Part. Nucl.* **7**, 1 (1976).
37. Y. Nambu, Relativistic groups and infinite-component fields, in *Elementary Particle Theory. Proceedings of the Nobel Symposium, Lerum, Sweden, Stockholm 1968*, ed. N. Svartholm, pp. 105–117.
38. E. S. Fradkin and M. A. Vasiliev, *Dokl. Acad. Nauk.* **29**, 1100 (1986); *Ann. of Phys.* **177**, 63 (1987).
39. M. A. Vasiliev *et al.* Nonlinear higher spin theories in various dimensions, arXiv:hep-th/0503128.
40. F. A. Berends, G. J. H Burgers and H. Van Dam, On the theoretical problems in constructing interactions involving higher-spin massless particles, *Nucl. Phys. B* **260**, 295 (1985).
41. A. Sagnotti, E. Sezgin and P. Sundell, On higher spins with a strong Sp(2,R) condition, arXiv:hep-th/0501156.
42. R. R. Metsaev, Cubic interaction vertices of massive and massless higher spin fields, *Nucl. Phys. B* **759**, 147 (2006).
43. R. Manvelyan, K. Mkrtchyan and W. Ruhl, General trilinear interaction for arbitrary even higher spin gauge fields, *Nucl. Phys. B* **836**, 204 (2010).
44. S. Guttenberg and G. Savvidy, Schwinger-Fronsdal theory of Abelian tensor gauge fields, *SIGMA* **4**, 061 (2008) [arXiv:0804.0522 [hep-th]].
45. A. K. Bengtsson, I. Bengtsson and L. Brink, Cubic interaction terms for arbitrary spin, *Nucl. Phys. B* **227**, 31 (1983).
46. A. K. Bengtsson, I. Bengtsson and L. Brink, Cubic interaction terms for arbitrarily extended supermultiplets, *Nucl. Phys. B* **227**, 41 (1983).
47. G. Savvidy, Extension of the Poincaré group and non-abelian tensor gauge fields, *Int. J. Mod. Phys. A* **25**, 5765 (2010) [arXiv:1006.3005 [hep-th]].
48. G. Savvidy, Non-Abelian tensor gauge fields, *Proc. Steklov Inst. Math.* **272**, 201 (2011) [arXiv:1004.4456 [hep-th]].
49. I. Antoniadis, L. Brink and G. Savvidy, Extensions of the Poincare group, *J. Math. Phys.* **52**, 072303 (2011) [arXiv:1103.2456 [hep-th]].
50. G. Savvidy, Invariant scalar product on extended Poincar algebra, *J. Phys. A* **47**, 5 (2014), 055204 [arXiv:1308.2695 [hep-th]].

51. G. Savvidy, Asymptotic freedom of non-Abelian tensor gauge fields, *Phys. Lett. B* **732**, 150 (2014).
52. G. Savvidy, Proton structure and tensor gluons, *J. Phys. A* **47**, 35 (2014), 355401 [arXiv:1310.0856 [hep-th]].
53. G. Savvidy, Tensor gluons and proton structure, *Theor. Math. Phys.* **182**, 1 (2015), 114 [*Teor. Mat. Fiz.* **182**, 1 (2014), 140] [arXiv:1406.5334 [hep-ph]].
54. F. A. Berends, R. Kleiss, P. De Causmaecker, R. Gastmans and T. T. Wu, Single bremsstrahlung processes in gauge theories, *Phys. Lett. B* **103**, 124 (1981).
55. R. Kleiss and W. J. Stirling, Spinor techniques for calculating P Anti-P $\to$ W^{+-}/Z^0 + Jets, *Nucl. Phys. B* **262**, 235 (1985).
56. Z. Xu, D. H. Zhang and L. Chang, Helicity amplitudes for multiple bremsstrahlung in massless nonabelian gauge theories, *Nucl. Phys. B* **291**, 392 (1987).
57. J. F. Gunion and Z. Kunszt, Improved analytic techniques for tree graph calculations and the G G Q anti-Q lepton anti-lepton subprocess, *Phys. Lett. B* **161**, 333 (1985).
58. L. J. Dixon, Calculating scattering amplitudes efficiently, arXiv:hep-ph/9601359.
59. S. J. Parke and T. R. Taylor, An amplitude for n gluon scattering, *Phys. Rev. Lett.* **56**, 2459 (1986).
60. F. A. Berends and W. T. Giele, Recursive calculations for processes with n gluons, *Nucl. Phys. B* **306**, 759 (1988).
61. E. Witten, Perturbative gauge theory as a string theory in twistor space, *Commun. Math. Phys.* **252**, 189 (2004) [arXiv:hep-th/0312171].
62. F. Cachazo, P. Svrcek and E. Witten, Gauge theory amplitudes in twistor space and holomorphic anomaly, *JHEP* **0410**, 077 (2004). [hep-th/0409245].
63. F. Cachazo, Holomorphic anomaly of unitarity cuts and one-loop gauge theory amplitudes, hep-th/0410077.
64. R. Britto, F. Cachazo and B. Feng, New recursion relations for tree amplitudes of gluons, *Nucl. Phys. B* **715**, 499 (2005) [arXiv:hep-th/0412308].
65. R. Britto, F. Cachazo, B. Feng and E. Witten, Direct proof of tree-level recursion relation in Yang-Mills theory, *Phys. Rev. Lett.* **94**, 181602 (2005) [arXiv:hep-th/0501052].
66. P. Benincasa and F. Cachazo, Consistency conditions on the S-matrix of massless particles, arXiv:0705.4305 [hep-th].
67. F. Cachazo, P. Svrcek and E. Witten, MHV vertices and tree amplitudes in gauge theory, *JHEP* **0409**, 006 (2004) [arXiv:hep-th/0403047].
68. G. Georgiou, E. W. N. Glover and V. V. Khoze, Non-MHV tree amplitudes in gauge theory, *JHEP* **0407**, 048 (2004) [arXiv:hep-th/0407027].
69. N. Arkani-Hamed and J. Kaplan, On tree amplitudes in gauge theory and gravity, *JHEP* **0804**, 076 (2008) [arXiv:0801.2385 [hep-th]].

70. F. A. Berends and W. T. Giele, Multiple soft gluon radiation in parton processes, *Nucl. Phys. B* **313**, 595 (1989).
71. M. L. Mangano and S. J. Parke, Quark–gluon amplitudes in the dual expansion, *Nucl. Phys. B* **299**, 673 (1988).
72. G. Georgiou and G. Savvidy, Production of non-Abelian tensor gauge bosons. Tree amplitudes and BCFW recursion relation, *Int. J. Mod. Phys. A* **26**, 2537 (2011) [arXiv:1007.3756 [hep-th]].
73. I. Antoniadis and G. Savvidy, Conformal invariance of tensor boson tree amplitudes, *Mod. Phys. Lett. A* **27**, 1250103 (2012) [arXiv:1107.4997 [hep-th]].
74. J. D. Bjorken and E. A. Paschos, Inelastic electron proton and gamma proton scattering, and the structure of the nucleon, *Phys. Rev.* **185**, 1975 (1969).
75. G. Altarelli and G. Parisi, Asymptotic freedom in parton language, *Nucl. Phys. B* **126**, 298 (1977).
76. Y. L. Dokshitzer, Calculation of the structure functions for deep inelastic scattering and e+ e- annihilation by perturbation theory in quantum chromodynamics, *Sov. Phys. JETP* **46**, 641 (1977); [*Zh. Eksp. Teor. Fiz.* **73**, 1216 (1977)].
77. V. N. Gribov and L. N. Lipatov, Deep inelastic e p scattering in perturbation theory, *Sov. J. Nucl. Phys.* **15**, 438 (1972); [*Yad. Fiz.* **15**, 781 (1972)].
78. V. N. Gribov and L. N. Lipatov, e+ e- pair annihilation and deep inelastic e p scattering in perturbation theory, *Sov. J. Nucl. Phys.* **15**, 675 (1972); [*Yad. Fiz.* **15**, 1218 (1972)].
79. L. N. Lipatov, The parton model and perturbation theory, *Sov. J. Nucl. Phys.* **20**, 94 (1975); [*Yad. Fiz.* **20**, 181 (1974)].
80. V. S. Fadin, E. A. Kuraev and L. N. Lipatov, On the Pomeranchuk singularity in asymptotically free theories, *Phys. Lett. B* **60**, 50 (1975).
81. E. A. Kuraev, L. N. Lipatov and V. S. Fadin, The Pomeranchuk singularity in nonabelian gauge theories, *Sov. Phys. JETP* **45**, 199 (1977); [*Zh. Eksp. Teor. Fiz.* **72**, 377 (1977)].
82. I. I. Balitsky and L. N. Lipatov, The Pomeranchuk singularity in quantum chromodynamics, *Sov. J. Nucl. Phys.* **28**, 822 (1978); [*Yad. Fiz.* **28**, 1597 (1978)].
83. D. J. Gross and F. Wilczek, Asymptotically free gauge theories. 1, *Phys. Rev. D* **8**, 3633 (1973).
84. D. J. Gross and F. Wilczek, Asymptotically free gauge theories. 2., *Phys. Rev. D* **9**, 980 (1974).
85. H. D. Politzer, Reliable perturbative results for strong interactions?, *Phys. Rev. Lett.* **30**, 1346 (1973).
86. J. K. Barrett and G. Savvidy, A dual lagrangian for non-Abelian tensor gauge fields, *Phys. Lett. B* **652**, 141 (2007).
87. S. Guttenberg and G. Savvidy, Duality transformation of non-Abelian tensor gauge fields, *Mod. Phys. Lett. A* **23**, 999 (2008) [arXiv:0801.2459 [hep-th]].
88. J. Fang and C. Fronsdal. Deformation of gauge groups. Gravitation, *J. Math. Phys.* **20**, 2264 (1979).

89. T. L. Curtright, Charge renormalization and high spin fields, *Phys. Lett. B* **102**, 17 (1981).
90. G. K. Savvidy, Infrared instability of the vacuum state of gauge theories and asymptotic freedom, *Phys. Lett. B* **71**, 133 (1977).
91. S. G. Matinyan and G. K. Savvidy, Vacuum polarization induced by the intense gauge field, *Nucl. Phys. B* **134**, 539 (1978).
92. I. A. Batalin, S. G. Matinyan and G. K. Savvidy, Vacuum polarization by a source-free gauge field, *Sov. J. Nucl. Phys.* **26**, 214 (1977) [*Yad. Fiz.* **26**, 407 (1977)].
93. D. Kay, Supersymmetric savvidy state for SU(2), SU(3) and SU(4), *Phys. Rev. D* **28**, 1562 (1983).
94. L. Brink and H. B. Nielsen, A physical interpretation of the Jacobi imaginary transformation and the critical dimension in dual models, *Phys. Lett. B* **43**, 319 (1973).
95. J. R. Ellis, M. K. Gaillard and G. G. Ross, Search for gluons in e+ e- annihilation, *Nucl. Phys. B* **111**, 253 (1976).
96. H. Georgi and S. L. Glashow, Unity of all elementary particle forces, *Phys. Rev. Lett.* **32**, 438 (1974).
97. H. Georgi, H. R. Quinn and S. Weinberg, Hierarchy of interactions in unified gauge theories, *Phys. Rev. Lett.* **33**, 451 (1974).

Some Thoughts About Yang–Mills Theory

A. Zee

University of California, Santa Barbara, CA 93106, USA

President Guaning Su mentioned that he was very much inspired by the news of parity violation in 1957 and its impact on the Chinese community worldwide. I like to add that I remember very clearly that my father came home one day saying that he heard that two Chinese had overthrown Einstein. I was quite young at that time and knew only vaguely who Einstein was, but still I felt that this was very important and that I should perhaps study physics later.

Now since this is a celebratory occasion, I would like to show a few photos of Professor C. N. Yang from my private collection, which have never been shown before in public. But instead of bunching them at the beginning of the talk or the end of the talk, I have inserted them randomly.

I will start by letting Professor Yang speak for himself. He wrote that he first had this idea as a student and that he kept coming back to it year after year.

> "Most such ideas are eventually discarded or shelved. But some persist and may become obsessions. Occasionally an obsession does finally turn out to be something good."

So I always tell students to have obsessions.

Here is a picture from my collection and in case you do not recognize who these people are, they are — from right to left and in alphabetical order — Chan, Tsou, Yang and Zee. All four are in the room. But time has done its work.

From right: Chan, Tsou, Yang and Zee.

Let me begin talking about the unreasonable effectiveness of notation in theoretical physics, to paraphrase Eugene Wigner. We all know that good notation is very important; for example, the Einstein repeated summation convention, the Dirac bra and ket, etc. At the most pedestrian level, just hiding indices is a good thing to do. Differential forms do that for us.

$$A^a_\mu \to A \equiv A^a_\mu T^a dx^\mu \quad \text{a matrix 1-form}$$

A good notation would often be backed by good mathematics. In 1954, we wrote the electromagnetic potential as A_μ, and of course Yang's and Mills's great contribution is to add an index a. But now we can hide the two indices by introducing a matrix 1-form.

The challenge for Yang and Mills was that, given the transformation property of A, how would you construct the field strength?

$$A \to UAU^\dagger + UdU^\dagger: \quad \text{How to construct F?}$$

In the modern language, you ask, given the 1-form A, how do you construct a 2-form that transforms correctly? The answer is almost immediate, because there are only two 2-forms — one is dA, and the other A^2. You can only add them.

$$F = dA + A^2$$

Now we all know that the Chern–Simons theory in higher dimensions has played an enormous role in modern physics: modern physics and differential geometry. I refer in particular to a paper (1984) that I wrote with the late Bruno Zumino and Yong-Shi Wu (who is also here in the audience).

$$n^{\text{th}} \text{ Chern character} : \Omega_{2n}(A) \equiv \operatorname{tr} F^n = \operatorname{Str} F^n$$

$$d\Omega_{2n}(A) = 0 \text{ and Poincaré lemma imply } \Omega_{2n}(A) = d\omega_{2n-1}^0(A)$$

The nth Chern character Ω is just a trace of the Yang–Mills field strength to the nth power. It is necessary to introduce here a Str which stands for a symmetric trace. Since $d\Omega = 0$, together with the Poincaré lemma, this immediately imply that $\Omega = d\omega$. So the question is: how would you solve this equation for ω. And that was of course solved by Chern and Simons.

Perhaps in a more modern way of doing things or at least the physicist's way of doing things (I learned this from Yong-Shi Wu), you would make an arbitrary variation of both sides and later specialize to $A \to tA$, which t is a real variable. Integrating over t, you obtain this integral representation for the Chern–Simons form.

$$\omega_{2n-1}^0(A) = n \int_0^1 dt \, t^{n-1} \operatorname{Str}(A, (dA + tA^2)^{n-1}).$$

The familiar case is just for $n = 2$.

$$\text{The familiar Chern–Simons form} : \omega_3^0(A) = \operatorname{tr}\left(AdA + \frac{2}{3}A^3\right)$$

But obviously for large n, this is going to lead to rather involved expressions for the non-Abelian anomalies that would be very difficult to derive using the indices notation used in the paper of Yang and Mills.

I said I would insert some pictures randomly that I found at home. Too bad Kerson Huang is not around. I am sorry I don't remember who the person on the left is, but (referring to the photo) the other four are Li, Zee, Huang and Yang.

From left: ?, Li, Zee, Huang, Yang.

One of the lessons you should draw from this picture is that Professor Yang is truly an exceptional physicist, because you can see that he was the only one who was not required to hold a cup in his hand.

My next mention of Yang–Mills theory is a paper that Frank Wilczek and I wrote in 1984, in which we pointed out that the non-Abelian gauge structure will arise very naturally [F. Wilczek and A. Zee, Appearance of gauge structure in simple dynamical systems, *Phys. Rev. Lett.* **52**, 2111 (1984)]. In fact, any system with states evolving adiabatically would exhibit this gauge structure exactly as Yang and Mills said it would 30 years earlier in 1954.

As of May of this year, this paper has quite a few citations, 73 in 2014 alone. But the significant point is not the number of citations, but the variety of fields represented. Almost none of these citations are in particle physics. And this shows how important Yang–Mills theory is, not just for particle physics, but for all areas in physics. I like to emphasize the universality of non-Abelian gauge structure.

Appearance of gauge structure in simple dynamical systems

☐ Search within citing articles

Light-induced gauge fields for ultracold atoms.
N Goldman, G Juzeliūnas, P Ohberg... - Reports on progress in ..., 2014 - europepmc.org
Gauge fields are central in our modern understanding of physics at all scales. At the highest energy scales known, the microscopic universe is governed by particles interacting with each other through the exchange of gauge bosons. At the largest length scales, our ...
Cited by 84 Related articles All 8 versions Cite Save More

Measuring Z 2 topological invariants in optical lattices using interferometry
F Grusdt, D Abanin, E Demler - Physical Review A, 2014 - APS
Abstract We propose an interferometric method to measure Z 2 topological invariants of time-reversal invariant topological insulators realized with optical lattices in two and three dimensions. We suggest two schemes which both rely on a combination of Bloch ...
Cited by 12 Related articles All 7 versions Cite Save

Wilson-loop characterization of inversion-symmetric topological insulators
A Alexandradinata, X Dai, BA Bernevig - Physical Review B, 2014 - APS
Abstract The ground state of translationally invariant insulators comprises bands which can assume topologically distinct structures. There are few known examples where this distinction is enforced by a point-group symmetry alone. In this paper we show that 1D ...
Cited by 6 Related articles All 6 versions Cite Save

Universal non-adiabatic holonomic gates in quantum dots and single-molecule magnets
VA Mousolou, CM Canali, E Sjöqvist - New Journal of Physics, 2014 - iopscience.iop.org
Abstract Geometric manipulation of a quantum system offers a method for fast, universal and robust quantum information processing. Here, we propose a scheme for universal all-geometric quantum computation using non-adiabatic quantum holonomies. We propose ...
Cited by 5 Related articles All 9 versions Cite Save

Experimental realization of universal geometric quantum gates with solid-state spins
C Zu, WB Wang, L He, WG Zhang, CY Dai, F Wang... - Nature, 2014 - nature.com
Experimental realization of a universal set of quantum logic gates is the central requirement for the implementation of a quantum computer. In an 'all-geometric' approach to quantum computation 1, 2, the quantum gates are implemented using Berry phases 3 and their non ...
Cited by 8 Related articles All 8 versions Cite Save

Classical chiral kinetic theory and anomalies in even space-time dimensions
V Dwivedi, M Stone - Journal of Physics A: Mathematical and ..., 2014 - iopscience.iop.org
Abstract We propose a classical action for the motion of massless Weyl fermions in a background gauge field in (2N+ 1)+ 1 space-time dimensions. We use this action to derive the collisionless Boltzmann equation for a gas of such particles, and show how classical ...
Cited by 4 Related articles All 5 versions Cite Save

Factorized three-body S-matrix restrained by the Yang–Baxter equation and quantum entanglements
LW Yu, Q Zhao, ML Ge - Annals of Physics, 2014 - Elsevier
Abstract This paper investigates the physical effects of the Yang–Baxter equation (YBE) to quantum entanglements through the 3-body S-matrix in entangling parameter space. The explicit form of 3-body S-matrix View the MathML source
Cited by 4 Related articles All 4 versions Cite Save

Of course I don't expect you to read all these words on the slide, but I want to point out the different areas of physics. Look at the terms here (referring to the slide): ultracold atoms, optical lattices, inversion-symmetric topological insulators, quantum dots and single-molecule magnets, geometric quantum gates, Yang–Baxter equation, and so on. So in all these different areas, non-Abelian gauge structure comes in.

I would like to also mention that three years after I wrote that paper with Wilczek, I had the idea that Yang–Mills structure could also be seen in nuclear quadrupole resonance experiments.

Non-Abelian gauge structure in nuclear quadrupole resonance

A. Zee

Institute for Theoretical Physics, University of California, Santa Barbara, California 93106
(Received 26 October 1987)

We elucidate the non-Abelian gauge structure associated with nuclear quadrupole resonance. The Abelian part of this structure has been experimentally observed. The phases to be observed in various non-Abelian experiments are computed.

The extract of the paper said the Abelian phase has been observed, but the phases to be observed in various non-Abelian experiments are computed. Later, the experiments were done and of course the result was exactly as predicted by Yang and Mills in 1954. The proposed experiments are exactly the analogs of what Yang and Mills had in mind.

Here I will show the Yang and Mills paper. The physical motivation — and I would like to ask Professor Yang afterwards, whether they actually had this in mind — was that they said that (reading and slightly summarizing the slide) the differentiation between a neutron and a proton is purely arbitrary. As usually conceived, it is subject to one limitation, which is that once you choose what to call a proton and what a neutron at one location in space–time, then you are not free to make any choices at other space–time points.

Yang and Mills said that this was not consistent with the localized field concept and then they explored the possibility of imposing this requirement.

This paragraph if you like, was precisely realized in the nuclear quadrupole resonance experiment that I described.

> isotopic spin is of no physical significance. The differentiation between a neutron and a proton is then a purely arbitrary process. As usually conceived, however, this arbitrariness is subject to the following limitation: once one chooses what to call a proton, what a neutron, at one space-time point, one is then not free to make any choices at other space-time points.
>
> It seems that this is not consistent with the localized field concept that underlies the usual physical theories. In the present paper we wish to explore the possibility of requiring all interactions to be invariant under *independent* rotations of the isotopic spin at all space-time points, so that the relative orientation of the isotopic spin at two space-time points becomes a physically meaningless quantity (the electromagnetic field being neglected).

Since I googled the paper of Yang–Mills, I could not resist also — since it required only one click — to look up the citation history.

In the year 1954 to 1958, there were a total of nine citations. Interestingly, two of them are by my friend Sid Bludman. In case you don't know, he was actually the first one to write down $SU(2) \times U(1)$, not Glashow, not Schwinger. But unfortunately, his paper was wrong — it contained a group theoretical error. So, too bad. One citation is by Lee and Yang, but remarkably, four citations are by Japanese authors. Of course I have very limited data here, but it has been suggested to me that perhaps Japanese physicists are more receptive to new ideas than anyone else. Besides Lee and Yang, no Chinese physicists bothered to work on this paper at all.

Of course, in the next four years, 1958 to 1962, there were 60 citations, and the rest is what they call history, which I will not go into.

Talking about history, I might say that recently Glashow came to visit Santa Barbara (for the first time in his career actually). I was chatting with him about that period — it was much before my time — and he mentioned that I should look at this paper he wrote with Gell-Mann in 1961, in which they talked about generalized Yang–Mills theory associated with the simple Lie algebra. (They had a very strict interpretation that the Yang–Mills theory was associated with $SU(2)$ and they were going to generalize it to any Lie algebra.)

Just to show the young people, the students in the back, what the level of mathematical sophistication was at that time: in this paper they have to explain what the Lie algebra was, what $SU(3)$ was, what $SU(4)$ was and so on. Nobody knew these things. I actually read this paper, probably the first one in many years. Glashow and Gell-Mann (note the order of names) felt it necessary to list the algebras $SU(2)$, $SU(3)$, $Sp(2)$, G_2 and so on. And you see that they end with $SO(9)$

and then notice this sentence (reading from slide) "It is hard to imagine that any higher Lie algebras will be of physical interest." But as almost everybody here knows, many years later, Georgi and Glashow were to grand unify using $SO(10)$, the very next group on this list. But Glashow and Gell-Mann just stopped with $SO(9)$.

Glashow and Gell-Mann, *Ann. Phys.* (1961)

In the generalized Yang–Mills theory associated with the simple Lie algebra, we have gone over to a new particle representation. Instead of the n real fields

The simple Lie algebras of smallest dimension are those of the groups $SU(2)$, with $n = 3$; $SU(3)$, with $n = 8$; $Sp(2)$, with $n = 10$; G_2, with $n = 14$; $SU(4)$, with $n = 15$; $Sp(3)$, with $n = 21$; $O(7)$, with $n = 21$; $SU(5)$, with $n = 24$; $O(8)$, with $n = 28$; $SU(6)$, with $n = 35$; $Sp(4)$, with $n = 36$; and $O(9)$, with $n = 36$; It is hard to imagine that any higher Lie algebras will be of physical interest.

Let us now come back to the physical picture that Yang and Mills had about moving a proton and a neutron around.

I now show the concluding pages of my nuclear quadrupole resonance paper. I said that Yang and Mills spoke of the degeneracy of the proton and neutron and they imagined transporting a proton from one point in the universe to another. That a proton at one point can be interpreted as a neutron at another in an isospin invariant world requires the introduction of a non-Abelian gauge potential. My final sentence was that we find it amusing that this discussion can now be realized analogously in the laboratory.

So, the idea of the experiment is to find a nucleus with two states — one we call the proton and one the neutron — and we make an excursion. But we are not making the excursion in space–time, we are making the excursion in parameter space. When we come back, we should find a non-Abelian gauge structure.

It so happens that this paragraph I copied was next to the acknowledgment so I also included the acknowledgment here. I already said that I learned a great deal from Professor Wu and that he is here. I see that I thanked various people like Andy Strominger, and also Frank Wilczek for encouraging me to publish this paper — I was not going to publish this paper, I thought that it was almost a trivial remark that the original idea of Yang and Mills could be realized experimentally. It is also of interest to see that I benefitted from conservations with Pines and Tycko, who eventually did the proposed experiment (and of course the theory and experiment agree completely).

er. That a proton at one point can be interpreted as a neutron at another in an isospin invariant world necessitates the introduction of a non-Abelian gauge potential. We find it amusing that this discussion can now be realized analogously in the laboratory.

We are grateful to Y. S. Wu for helpful discussions. We would also like to thank M. Goodman and A. Strominger for useful conversations and F. Wilczek for encouraging us to publish this paper. We have also benefited from conversations with A. Pines and R. Tycko.

Professor Yang has always talked about the connection between gauge concept and geometry. And of course in my career I have come across this many times. We can talk about non-Abelian gauge theory in particle physics, but it is well-known, so on and so forth: standard model et cetera.

Here is a perhaps unusual connection, a work that Xiao-Gang Wen and I did in 1992 after he left Santa Barbara for MIT. We noticed something that was already known since the 1930s. I am sure Professor Yang could correct me on this, but I think it was noticed by Fierz early on that on a plane, the number of magnetic flux quanta going through the plane and the number of electrons moving around are simply related by definition by the filling factor. In contrast, on a closed topological manifold such as a sphere, this relation is shifted by a number. And if you are doing this on a Riemannian manifold with some other value of the genus, this shift would be a different number, independent of N_Φ and N_e. Of course, for an actual microscopic sample in a thermodynamic limit, this is irrelevant: both N_Φ and N_e are essentially infinite.

Shift and Spin Vector: New Topological Quantum Numbers for the Hall Fluids

X. G. Wen

Department of Physics, Massachusetts Institute of Technology, Cambridge, Massachusetts 02139-4307

A. Zee

Institute for Theoretical Physics, University of California, Santa Barbara, California 93106-4030
(Received 16 March 1992)

motivated by the following fact. On a closed topological 2-manifold, such as a sphere [4], over which the Hall fluid is defined, the number of electrons N_e and the number of magnetic flux quanta N_ϕ going through the manifold are not simply related by $\nu N_\phi = N_e$ (where ν is by definition the filling factor) as is the case on the plane. Rather there is a shift $\mathcal{S}$ in the relation between N_e and N_ϕ:

$$N_\phi = \nu^{-1} N_e - \mathcal{S}. \tag{2}$$

This suggests that the quantum Hall fluid state actually couples to the curvature of space. In order to include such a coupling, we recognize that on a curved manifold, there is another 1-form besides the electromagnetic gauge potential 1-form: the connection 1-form which gives rise to the curvature. We couple a conserved current to it, and then we go through two lines of arithmetic, literally two lines of trivial arithmetics. Diagonalizing a 2×2 matrix, we obtain a formula for the shift.

Now you may think that this is just a theoretical, mathematical game that is of no relevance, but in fact this paper has had some influence on the condensed matter literature. In fact just recently, there is a paper by Fradkin and several Russians. This (referring to lower portion of slide) was published in *Physical Review Letters* this year [E. Fradkin *et al.*, *Phys. Rev. Lett.* **114**, 016805 (2015)].

> We consider the geometric part of the effective action for the fractional quantum Hall effect (FQHE). It is shown that accounting for the framing anomaly of the quantum Chern-Simons theory is essential to obtain the correct gravitational linear response functions. In the lowest order in gradients, the linear response generating functional includes Chern-Simons, Wen-Zee, and gravitational Chern-Simons terms. The latter

Fradkin was in Santa Barbara and he told me about it. If he didn't tell me about it, I wouldn't know since I had long left this field. So people are worried about framing anomaly and such things, I understand tomorrow we would hear more about framing. The framing anomaly of the quantum Chern–Simons theory — this is from their paper, not from our paper — "is essentially to obtain the correct gravitational linear response functions." In the lowest order in gradients, the linear response must include "Chern–Simons, Wen–Zee and gravitational Chern–Simons terms".

It is amusing that many years later, this is still part of the condensed matter physics community's collective consciousness. I was told recently that there is by now a substantial literature on the shift.

I would like to conclude by saying that Yang–Mills theory still holds many mysteries. I personally believe that we are only at the beginning of understanding Yang–Mills theory. There is much yet to be understood. I will merely mention two examples.

1. Einstein Gravity = Yang–Mills Squared?

There is increasing evidence that the Einstein theory of gravity is just Yang–Mills theory squared. I put a question mark here because this is of course not proven. Now again, the relevant person, Henry Tye, is in the audience. This intriguing possibility was first suggested by Kawai, Lewellen and Tye in string theory. But just in the last six months, there have been tremendous progress and excitement. If you are interested in this, you should look at the recent paper by Monteiro, O'Connell, and White for example. In particular, O'Connell was in Santa Barbara.

2. Beyond the Higgs Mechanism

Could we have fermion masses without giving mass to the gauge bosons?

The second example I would like to mention involves my recent work with Yoni BenTov. I think it is time for particle physics to go beyond the Higgs mechanism. The Higgs mechanism is something that we have been talking about for the last 40 to 50 years and it is a completely weak coupling phenomenon. Our condensed matter colleagues have gone far beyond. They have discovered many new fascinating phenomena.

One question that we raised is whether we could have fermion masses without giving mass to the gauge bosons. I will tell you why this is important. If this is the case, then we may be able to solve the family problem with an $SO(18)$ gauge theory which was proposed by Gell-Mann, Ramond and Slansky, and by Wilczek and me, and also by Fuijimoto (whom I do not know), back in 1980.

The observation from basic group theory is that while the tensor representations do not decompose repetitively, the spinors do decompose repetitively. I see David Gross nodding his head. Spinors decompose into a bunch of spinors — this is very suggestive of the family structure.

Why does nature repeat itself three times? The spinors repeat themselves.

The simplest model with enough room for three families is $SO(18)$. But $SO(18)$, unfortunately, contains right-hand fermion also — and this goes again back to Lee and Yang, to parity violation — because it is vector-like, and again David is nodding in his head so I know I am not talking nonsense.

I want to mention a recent work inspired by Kitaev and Wen and many others. Kitaev has proposed some condensed matter models in which it is possible to give mass to fermion, without giving mass to gauge bosons. And so BenTov — who is my former student, and now a postdoc at Caltech working for Kitaev — and I put out a paper [BenTov and Zee, arXiv:1505.04312 (2015)]. If this idea turns out to be correct, then we can in fact explain the family structure.

I would like to close with a couple of photos. My next photo in particular, is truly historic, one I took almost forty years ago. And I hope that — because I had never shown this in public — perhaps even Professor Yang is seeing it for the first time. Unfortunately Kerson Huang is not here. I should explain that this photo was engineered and arranged by him; my role as a young guy was simply to follow orders. Kerson told me to bring my camera and showed up in his hotel room, promptly at a time he specified, when there would be a historic meeting. I was instructed to quickly take some pictures. (I am not in this picture because I was taking the photos.)

So here it is. In case you don't recognize the people, they are, from left to right, Yang, Huang, Huang, Lee, and Lee. Not all of them are with us today, unfortunately.

From left: Yang, Huang, Huang, Lee and Lee.

For the next picture, I would like to fast-forward 40 years. The occasion was Professor Yang's ninetieth birthday, almost 40 years after the previous picture. The people in this are Zee, Yang, Yang and Zee. (Notice that I do not have to specify whether this is from left to right, or from right to left.)

Zee, Yang, Yang and Zee.

So here we are at a birthday for a theory — Happy 60! If this were a birthday for a person, I think it would be dishonest to say that we will all be here for another 60 years as the Jewish saying goes. (When people turn 60, in Jewish culture you always say you wish for another 60.) But Yang–Mills theory will for sure endure far beyond another 60 years, and quite likely, if I am optimistic, for at least another 600 if not much more.

Thank you very much!

Acknowledgment

I would like to thank Lim Wen Yee (Germanie) and Chee-Hok Lim for transcribing my talk & text.

Gauging Quantum Groups: Yang–Baxter Joining Yang–Mills

Yong-Shi Wu

Key State Laboratory of Surface Physics, Department of Physics,
Center for Field Theory and Particle Physics,
Fudan University, Shanghai 200433, China
Collaborative Innovation Center of Advanced Microstructure,
Fudan University, Shanghai 200433, China
Department of Physics and Astronomy, University of Utah,
Salt Lake City, UT 84112, USA
wu7731@gmail.com

This review is an expansion of my talk at the conference on Sixty Years of Yang–Mills Theory. I review and explain the line of thoughts that lead to a recent joint work with Hu and Geer [Hu *et al.*, arXiv:1502.03433] on the construction, exact solutions and ubiquitous properties of a class of quantum group gauge models on a honey-comb lattice. Conceptually the construction achieves a synthesis of the ideas of Yang–Baxter equations with those of Yang–Mills theory. Physically the models describe topological anyonic states in 2D systems.

Keywords: Yang–Mills; Yang–Baxter; gauge theory; anyons.

1. Introduction

The first two talks I attended of Prof. Yang were in 1972, when he visited Peking (or Beijing) University in China. One was on the integral formalism for the gauge fields; another was on the Bethe Ansatz solutions for one-dimensional solvable many-body problems. At that time I had just left the job in a plasma laboratory assigned to me in 1965 when I graduated, and returned to research in theoretical physics which I had liked since my college years. The talks came to me just in time, when I was searching for research directions. As it turned out, the two talks inspired me in my career as a theoretical physicist for many years.

On the one hand, the fact that Prof. Yang gave at the same time the two talks in different areas of theoretical physics, in particle theory and in many-body theory respectively, deeply impressed me: These two areas, though seemingly divergent, must have some fundamental and profound connections. This promoted me to begin paying attention to progress in condensed matter theory while working on particle theory. In retrospect, this certainly has affected my later career in a profound way.

On the other hand, the two topics of the talks also attracted my interests, first in gauge theory in particle physics (particularly in topological aspects) and later in lower-dimensional systems in condensed matter physics (particularly in exactly solvable models), for years to come. Next I will recall how I achieved doing research in the two fields under constant inspiration from Prof. Yang. Also I will report on a recent work of mine with Hu et al.,[1] which synthesizes the essential ideas contained in the mentioned two talks of Yang in early 1970s.

More concretely, I am going to talk about the formulation of a class of solvable discrete models, in which a *quantum group* is *gauged*. These models are exactly solved, and their full dyon excitaton spectrum is organized by representations of the *quantum double* associated with the (gauged) quantum group. Physically this class of models can be used to describe topological states of *anyon matter* in two-dimensional systems. Mathematically, as a nontrivial extension of the existing string-net models, our models are believed to provide a viable framework for studying *electric-magnetic duality* in topological anyonic matter.

2. Yang–Mills Gauge Theory and Integral Formalism

The original paper of Yang and Mills[2] was published in 1954. Besides the formulation of the non-Abelian $SU(2)$ gauge theory, this paper proposed, the first time in the physics literature, the Gauge Principle as a *unifying* principle for fundamental interactions (except gravity). When combined with the Einstein's relativity principle, the gauge principle uniquely fixes the interactions between (spin-$\frac{1}{2}$) fermionic matter particles, such as leptons and quarks, as well as self-interactions of gauge bosons, if the symmetry of the matter particles is known. As Yang later summarizes, *symmetry dictates interactions*.

2.1. *Differential formalism*

In Yang–Mills theory, the interaction between a fermion and a gauge boson is usually formulated through the minimal coupling prescription; namely one substitutes the ordinary derivative by the gauge-covariant derivative: (e is the coupling constant)

$$\partial_\mu \psi(x) \to D_\mu \psi(x) = \left(\partial_\mu - ie \sum_a A_\mu^a T_a \right) \psi(x), \qquad (1)$$

where $\psi(x)$ is the fermion field, while A_μ^a is the vector potential of the gauge field. (Here μ is the index of spacetime coordinates, and T_a are generators of the Lie algebra of the gauge group in the representation of $\psi(x)$). The field strength can be defined through the commutator: $[D_\mu, D_\nu] = -ieF_{\mu\nu}$, which is gauge covariant. This is the origianl differential formalism for gauge fields.

In this way, the gauge principle predicted the existence of spin-1 gauge bosons A_μ^a as the quanta of the force fields that mediate the fundamental forces (except gravity). It also fixes the self-interactions between gauge bosons through the gauge invariant Yang–Mills action density: $-(1/4)F^a_{\mu\nu}F^{a\mu\nu}$. This principle has been realized and well tested in the Standard Model for particle physics, with gauge group $U(1) \times SU(2) \times SU(3)$, to describe the electromagnetic, weak and strong interactions.

2.2. *Integral formalism*

In a talk at Beijing in 1972, Yang proposed the following integral definition for a gauge field: Consider the space of all (continuous) paths $P(b, a)$ from point a to point b in space/spacetime for all pairs a, b. A gauge field is a group-valued functional of paths, $g[P; b, a] \in G$, satisfying the properties:

(i) Composition: If point c is a point on path P between a and b, which divides $P(b, a)$ into two paths $P_1(c, a)$ and $P_2(b, c)$, then

$$g[P; b, a] = g[P_2; b, c] \cdot g[P_1; c, a]. \tag{2}$$

Here the right-hand side is the group multiplication.

(ii) Gauge transformation: Under a gauge transformation $g(x)$, which is a group-valued function of points, the functional $g[P; b, a]$ transforms as

$$g'[P; b, a] = g(b) \cdot g[P; b, a] \cdot g(a)^{-1}. \tag{3}$$

A big advantage of the integral definition is its simplicity and generality. It can be applied to pure gauge field theories, without the pre-existence of a matter field. In a continuous space/spacetime, one may consider infinitesimal open paths $P(x + dx, x)$. By requiring in addition that

$$g[P; x + dx, x] = \mathbf{1} + A^a_\mu(x)T_a dx^\mu, \tag{4}$$

one recovers the gauge potential field $A^a_\mu(x)$ in the differential approach. When the integral definition applies to discretized space/spacetime, a lattice or more generally a graph, the above additional requirement is not necessary. So to gauge a finite group, one just assigns a group element $g(\ell)$ to each link ℓ of the lattice/graph, and a gauge transformation assigned a group element $g(v)$ to each vertex v. This is the starting point of usual lattice gauge theory.

The beauty of the formalism impressed me so much, I immediately changed my research from the relativistic quark model, popular in China then, to gauge theory.

3. Yang–Baxter Equations

Another talk of Prof. Yang in 1972 at Beijing was on the Bethe Ansatz of exactly solvable one-dimensional many-body models. My interest and background at that time were mainly in nuclear and particle physics; I did not really appreciate this talk. After the talk I heard private comments, saying that these things were of only mathematical interest and would have no connection with real physics. Despite the comments, somehow I believed in the importance of the works of Yang in this area in late 1960s, since they dealt with fundamental problems in many-body systems. I was confident that if one-dimensional materials can be made in laboratory, the exact solutions must be relevant, providing fundamental insights into the 1D systems.

But I did not have chance to learn Yang's classical work on Bethe Ansatz solutions until early 1981. Upon his invitation, I visited for six months the Institute for Theoretical Physics (now CN Yang Institute for Theoretical Physics) at Stony Brook, New York. I joined a small study group on Bethe Ansatz, meeting in Yang's office. I also had private discussions with him. I began to appreciate the beauty and depth of the exactly solvable many-body models and to become fascinated with the jewels Yang had found in 1D new physics. In late 1980s these were widely recognized as fundamental insights into strongly-correlated effects in lower dimensions.

3.1. *Bethe Ansatz for one-dimensional many-body problem*

In his original paper,[4] Yang introduced operators R_{ij}, which acts on the ith and jth particles, in the connecting equations of his generalized Bethe Ansatz. For the consistency of these equations, he identified a property key to his success:

$$R_{12}(\lambda) R_{23}(\lambda + \mu) R_{12}(\mu) = R_{23}(\mu) R_{12}(\lambda + \mu) R_{23}(\lambda). \qquad (5)$$

Here R_{ij} depend on a real (spectral) parameter λ. These equations are called the quantum Yang–Baxter equations (QYBE), because later they also appeared in Baxter's paper,[5] again as the key property of his method. Frankly speaking, I did not realize the profound significance of the QYBE until three years later.

3.2. *Braid group and anyons*

I went to University of Washington in 1982, where the particle theory group and Prof. Thouless's group shared a very big room. I got interested in the integer and fractional quantum Hall effect in 2D electron gas, and particularly in the possibility that quasiparticle excitations in the fractional quantum Hall systems may obey exotic fractional statistics. It has been well-known that the symmetric (or permutation) group underlies the usual Bose–Einstein and Fermi–Dirac statistics. So I was curious and puzzled about whether there could be a group that underlies possible exotic quantum statistics in two dimensions.

In spring 1984, I realized that exchanging two particles in a plane is actually a *path-dependent* process. Then using path integral approach, I developed a general theory[6] of quantum statistics for identical particles. In higher than two dimensions, there is no new story; but at $d = 2$, the symmetric group is replaced by the *braid group*, because the particle world lines in $(2+1)$-dimensional spacetime braid naturally.

Like the symmetric group S_N, the braid group can be generated by operators σ_i ($1 \leq i < N$): σ_i exchanges the ith and $(i+1)$th particles along a counter-clockwise loop that does *not enclose* any of other particles. The symmetric group is generated by products of these generators subject to following three constraints:

(a) Orientation Independence: $\sigma_i^2 = 1$.
(b) Commutativity of σ_i and σ_j ($|i-j| > 1$): $\sigma_i \sigma_j = \sigma_j \sigma_i$.
(c) Braiding Relation (or YBE): $\sigma_i \sigma_{i+1} \sigma_i = \sigma_{i+1} \sigma_i \sigma_{i+1}$.

If condition (a) is not required, we get the braid group B_N. After sending Prof. Yang my preprint, I received a phone call from him, reminding me that the Braiding Relation (c) is a special case of the QYBE (5): $\sigma_i = R_{i,i+1}(\lambda = 0)$. An immediate consequence is the prediction of non-Abelian anyons as YBE matrix solutions.

The YBE involve three anyons and are crucial, as necessary and sufficient conditions, for non-Abelian anyons, since the braiding matrix for two anyons can always be diagonalized. Soon systematic ways to find YBE solutions were invented by mathematical physicists.

3.3. *Quantum groups*

The term "quantum groups" was coined by V. G. Drinfel'd for a new mathematical structure invented by him[7] and Jimbo[8] independently. Having been working on integrable statistical models, they both emphasized the inspiration from QYBE.

Essentially quantum groups (or equivalently, quantized universal enveloping algebras), are deformation of Lie groups or Lie algebras: They are some specific generalization (in the form of Hopf algebra) of a Lie group G (or Lie algebra) with an additional complex parameter q, such that when $q \to 1$, the quantum group becomes an ordinary Lie group G. The situation is similar to the relationship between quantum and classical mechanics: Namely quantum mechanics may be said to be a deformation of classical mechanics, with the deformation parameter $q = \exp(\hbar)$ ($\hbar$ the Planck constant). In the limit $\hbar \to 0$, or $q \to 1$, quantum mechanics becomes classical mechanics.

The most interesting case in both physics and mathematics is when q is a root of unity: $q = \exp(i2\pi/r)$, where r is a non-negative integer. For example, for $G = SU(2)$, one may take $r = k+2$ ($k > 0$ integer), and call k the level and denote the quantum group as $SU(2)_k$. Explicitly, this quantum group is formed by linear combinations of arbitrary products of the generators E^+, E^- and K, subject to

$$[E^+, E^-] = \frac{K - K^{-1}}{q - q^{-1}}, \quad K E^{\pm} K^{-1} = q^{\pm 2} E^{\pm}. \tag{6}$$

As $q \to 1$ one recovers $SU(2)$, with $E^{\pm} = J^{\pm}$ and $K = \exp(\hbar J_3)$.

Why this particular generalization? The key property of this particular generalization is that it has a *good representation theory* similar/parallel to that of a Lie group/algebra. For example, Lie group $SU(2)$ has irreducible representations labeled by spin s ($s = 0, 1/2, 1, 3/2, \ldots$). And the tensor product of two representations s_1 and s_2 is a direct sum of irreducible representations of spin s

with $s = |s_1 - s_2|, |s_1 - s_2| + 1, \ldots, s_1 + s_2$ (the so-called triangle rule). Interestingly, the quantum group $SU(2)_k$ has k irreducible representations, labeled also by $2s = 0, 1, \ldots, k$, which are closed under the tensor product decomposition (or fusion), with a similar triangle rules plus the modulo-k constraint (here we only consider the irreducible representations with nonzero quantum dimensions). Therefore, $SU(2)_k$ has only a finite number, $k + 1$, of irreducible representations, like a finite group. When $k \to \infty$, $SU(2)_k$ becomes Lie group $SU(2)$.

How is the quantum group related to Yang–Baxter equations? The point is the exchange symmetry of the Clebsch–Gordan decomposition. The states in the fusion of two irreducible representations are symmetric, up to a sign, under exchange for either finite or Lie groups. But this is *not* true for quantum groups: Namely the Clebsch–Gordan states in the decomposition of tensoring two irreducible representations will suffer a braiding under exchange, which satisfies the YBE.

In quantum theory, symmetry described by finite or Lie groups plays a central role to constrain the Hamiltonians of the theory, while the irreducible representations are used to describe quantum numbers of excitations, or particle species. So one naturally expect that quantum groups and their representation theory should play a similar and important role in future physics.

4. Gauging Quantum Groups

In last decade, we have seen the tip of an iceberg of new physics for topological states of quantum matter. These states may be realized in solid state materials, cold atoms or photonic crystals; and they may be useful for fault-tolerant quantum computation. A recent hot topic is possible anyonic matter in either one or two dimensions, such as anyon condensation. Since anyons in two-dimensional systems are described as quantum group representations, one naturally expect that the *effective* theory of a large class of topological anyonic matter could be described by a quantum group gauge theory on a lattice/graph (topological states may be described by discrete models as in combinatoric topology).

Next, I will report a collaborative work of mine[1] on construction and solutions of such models. The concepts of Yang–Mills gauge theory and of Yang–Baxter equations jointly played a central role in this development.

Space limitation does not allow me to get much into the details. But I will talk about the motivations, explain the essence of the model construction and summarize its ubiquitous properties.

4.1. *Duality between group elements and representations*

Kitaev's Toric Model,[9] in its simplest form, is a Z_2 lattice gauge theory, with elementary dyon excitations behaving like anyons. Dynamic gauge degrees of freedom are described by group elements on each link. A quantum group has infinitely many elements. If one assigns an element to each link, the Hilbert space associated with a single link would become infinite-dimensional. Horrifying divergence will appear even in a discrete setting. To gauge quantum groups, a new way is needed to describe the dynamic link degrees of freedom.

Turaev–Viro's topological quantum field theory[10] (TQFT) first hinted about a new possibility in a discrete setting by generalizing $6j$-symbols in group representation theory. The String-Net Model of Levin and Wen[11] provided a Hamiltonian formulation of TV's TQFT, good for the degenerate ground states. Our models finally complete the formulation of a quantum group gauge theory on a lattice/graph, with a full dyon excitation spectrum closed under fusion.

This new way may be motivated by the well-known Fourier transformation, interpreted as a duality between group elements and irreducible representations. The Fourier transform for a periodic function $f(\theta)$,

$$f(\theta) = \sum_{n=-\infty}^{+\infty} f_n \exp(in\theta), \qquad (7)$$

gives us a function f_n of the integer n. We can view n as a label for irreducible representations of Lie group $U(1)$, whose elements are labeled by the angular variable θ. In this way, θ and n are dual variables for group elements and irreducible representations, respectively.

This duality holds also for a finite group G: We have the following known transform between group elements, $g \in G$, and irreducible representations, ρ_j:

$$|g\rangle = \frac{1}{|G|} \sum_j \sqrt{d_j} [(\rho_j)_{ab}(g)]^* |j; ab\rangle. \qquad (8)$$

Here a, b are the matrix indices of the irreducible representation. When G is the symmetry group of a theory, in accordance with the well-known Eckart–Wigner theorem, one may pay attention only to the representation j's, neglecting the matrix indices a, b.

So the basis $\{g\}$ (with $g \in G$) of the Hilbert space on a single link can be replaced by the dual basis $\{j\}$ (with j irreducible representation of G). Motivated by this duality, which is true also to quantum groups, one may try to assign to each link an irreducible representation j of a finite or quantum group G, to formulate dynamic gauge degrees of freedom. For the case of quantum groups, they are known to have a *finite* number of irreducible representations with nonvanishing quantum dimensions; the Hilbert space on each link is thus finite-dimensional.

This was the starting point of Levin–Wen's string-net model. In their terminology, j is called string-type. A configuration of j's on a trivalent graph is shown in Fig. 1, which is viewed as a basis vector in the Hilbert space of the system.

4.2. *Fusion category as input data*

Thus, for the input data of a lattice gauge theory one can use the *class* (*not a set*) of all representations of G, generated from the irreducible representations by

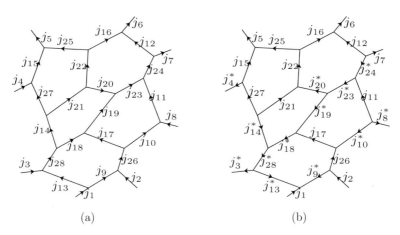

Fig. 1. A configuration of string types on a directed trivalent graph. The configuration (b) is treated the same as (a), with some of the direction of some edges reversed and the corresponding labels j changed to conjugated j^*.

tensor products. Mathematically this class with the operations of tensor product and direct sum constitutes a *fusion category* satisfying certain axioms, which formulate the conditions for the reconstruction of G in terms of its representations. The reconstruction theorem is the conceptual basis for our statement that assigning irreducible representations on links generates field configurations in a gauge theory.

Instead, Levin and Wen[11] derived the data for the input category (of representations) from a set of conditions that require the state or wave function of the model be *topologically invariant* or, in their term, be renormalization group fixed point.

Here I do not get into the details, but just mention that the input data include three ingredients: (i) (quantum) dimension d_j of the irreducible representation j; (ii) fusion rule multiplicity, a set of non-negative integers $\{N^i_{jk}\}$ in the fusion rules:

$$j \otimes k = \oplus_i N^i_{jk} i \qquad (9)$$

(i.e. N^i_{jk} being the number of times i appears in decomposition of $j \otimes k$), and (iii) the $6j$-symbols G^{ijm}_{kln}, similar to those in angular momentum theory in quantum mechanics. The latter is the unitary transformation connecting the two sides of the associativity of the tensor product:

$$(i \otimes j) \otimes k = i \otimes (j \otimes k). \qquad (10)$$

Assuming the multiplicity-free case, i.e. $N^i_{jk} \equiv \delta_{ijk} = 0, 1$, the unitary matrix elements depend on 6 j's, namely i, j, k, l, m and n: Symbolically, if ignoring arrows,

$$\begin{array}{c}\text{\Large (diagram)}\end{array} \sim \sum_n G^{ijm}_{kln} \begin{array}{c}\text{\Large (diagram)}\end{array} \qquad (11)$$

The detailed convention of notations is presented in Ref. 1.

Algebraically, one may define $6j$-symbols as solutions to the following equations: (assuming all $d_j > 0$)

$$G^{ijm}_{kln} = G^{mij}_{nk^*l^*} = G^{klm^*}_{ijn^*} = \overline{G^{j^*i^*m^*}_{l^*k^*n}},$$

$$\sum_n d_n G^{mlq}_{kp^*n} G^{jip}_{mns^*} G^{js^*n}_{lkr^*} = G^{jip}_{q^*kr^*} G^{riq^*}_{mls^*},$$

$$\sum_n d_n G^{mlq}_{kp^*n} G^{l^*m^*i^*}_{pk^*n} = \frac{\delta_{iq}}{d_i} \delta_{mlq} \delta_{k^*ip}.$$

(12)

The Levin–Wen Hamiltonian[11] is constructed in terms of the above input data. Similar to the Kitaev Hamiltonian,[9] it is a sum of two terms, consisting of commuting projectors acting on links of a vertex or on links of a plaquette.

It turns out that the Levin–Wen models are good to describe properties of topological phases in the ground states and with fluxon excitations at plaquettes. But they are *not appropriate* to describe the *charge* excitations at vertices. The reason is simple: The Levin–Wen Hilbert space has no label for charge excitations at vertices. So the fluxon spectrum of the model does not close under fusion and, therefore, is not the full spectrum of a gauge theory.

Reference 1 was motivated to resolve this problem.

4.3. *Extension of Hilbert space and Hamiltonian*

To incorporate explicitly states with charge excitations, we need extra string-type labels at the vertices of a trivalent graph. To maintain the simplicity of trivalent graphs, we introduce a tail labeled by irreducible representations j at one link connected to each vertex in the trivalent graph. The number of tails is equal to the number of vertices. We require that all vertices after adding tails always satisfy the fusion rule ($\delta_{ijk} = 1$), to avoid double counting of charges. In this way, all the vertices are neutral, with only the tails representing charge excitations at one nearby vertex. For example, we may arrange the underlying graph on a honeycomb lattice to be one shown in Fig. 2(b).

Obviously if one assigns the trivial representation $j = 0$ to all tails, the original Levin–Wen Hilbert space is recovered.

In Ref. 1, a simple and exactly solvable Hamiltonian in the extended Hilbert space is constructed as follows:

$$H = -\sum_v \mathcal{Q}_v - \sum_p \mathcal{B}_p,$$

(13)

where the sum run over vertices v and plaquettes p of the trivalent graph before being decorated with tails [see Fig. 2(a)]. For example, $\mathcal{Q}_v$ acting on vertex V is defined as

$$\mathcal{Q}_v \left| \begin{array}{c} i\ j \\ k_1\ q_1 \\ l \\ k_2\ q_2 \end{array} \right\rangle = \delta_{q_1,0} \left| \begin{array}{c} i\ j \\ k_1\ q_1 \\ l \\ k_2\ q_2 \end{array} \right\rangle.$$

(14)

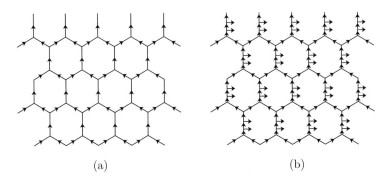

Fig. 2. Extension of the Hilbert space by one extra tail for each vertex.

When $q_1 = 0$, $\mathcal{Q}_v = 1$ expresses the Gauss's law (local neutrality). $\mathcal{B}_p$ acts on plaquette p; its expression is given in Ref. 1. $\mathcal{B}_p = 1$ implies there is no fluxon at p (flat connection).

The main property of $\mathcal{Q}_v$ and $\mathcal{B}_p$ is that they are mutually-commuting projection operators: (1) $[\mathcal{Q}_v, \mathcal{Q}_{v'}] = 0 = [\mathcal{B}_p, \mathcal{B}_{p'}]$, $[\mathcal{Q}_v, \mathcal{B}_p] = 0$; (2) and $(\mathcal{Q}_v)^2 = \mathcal{Q}_v$ and $(\mathcal{B}_p)^2 = \mathcal{B}_p$. Thus, the Hamiltonian is exactly solvable. The elementary excitations are given by common eigenvectors of all these projections. The ground states satisfy $\mathcal{Q}_v = \mathcal{B}_p = 1$ for all v and p, while the excited states violate these constraints for some plaquettes or vertices.

We have been able to exactly solved the models and get the full spectrum of excitations. Energy levels are easy, but the difficult part is the complete set of good quantum numbers labeling all energy eigenstates.

4.4. R-matrix and Yang–Baxter equations

Irreducible representations of quantum groups are known to be equipped with R matrices, which describe the effects of braiding two irreducible representations (or anyon species) on their fusion. Algebraically the R matrix is a phase factor R^a_{bc} depending on three irreducible representations a, b and c (b, c being fused into a) that satisfies the hexagon equations:

$$\sum_g d_g G^{cad^*}_{be^*g} R^e_{gc} G^{abg^*}_{ce^*f} = R^d_{ac} G^{acd^*}_{be^*f} R^f_{bc}, \qquad (15)$$

$$\sum_g d_g G^{e^*bd}_{cag} R^e_{ad} G^{e^*ag}_{bcf} = R^d_{ac} G^{e^*bd}_{acf} R^f_{ab}. \qquad (16)$$

These are the Yang–Baxter equations in terms of quantum group representations.

With the help of the R-matrix, we have been able to construct three local observables: charge (at a vertex), fluxon (at a plaquette) and twist (for a dyon, a composite of charge and fluxon), which are commuting with the Hamiltonian. We also have shown that the appropriate algebraic structure that characterizes the full dyon excitation spectrum is Drinfel'd's quantum double $D(G)$ (G being the gauged quantum group), which closes under fusion. For the details, we refer to our paper.[1]

4.5. *Example: The double Fibonacci model*

The quantum group $SU(2)_3$ (with level $k = 3$) is a deformation of $SU(2)$ with the q-deformation parameter $q = -\exp(\pi i/5)$. It has four irreducible representations labeled by $j \equiv 2s = 0, 1, 2, 3$. The $j = 0$ and $j = 2$ are closed under fusion:

$$0 \otimes j = j \, (j = 0, 2), \quad 2 \otimes 2 = 0 \otimes 2. \tag{17}$$

Let $\phi = (1 + \sqrt{5})/2$ be the golden ratio. The quantum dimensions of $j = 0, 2$ are $d_0 = 1$ and $d_2 = \phi$, respectively. The fusion rule coefficients are

$$\delta_{000} = \delta_{022} = \delta_{222} = 1, \delta_{002} = 0 \tag{18}$$

and the nonzero $6j$-symbols are given by

$$\begin{aligned} G^{000}_{000} &= 1, \quad G^{022}_{022} = G^{022}_{222} = 1/\phi, \\ G^{000}_{222} &= 1/\sqrt{\phi}, \quad G^{222}_{222} = -1/\phi^2. \end{aligned} \tag{19}$$

The other nonzero symmetrized $6j$-symbols are obtained through the tetrahedral symmetry. The nontrivial R matrices are $R^0_{22} = \exp(-4\pi i/5)$ and $R^2_{22} = \exp(3\pi i/5)$.

The elementary dyon excitations are irreducible representations of the quantum double $D(SU(2)_k)$. There are four quantum double labels: $J = \{j_1 \overline{j_2}\}$ ($j_1, j_2 = 0, 2$). The modular matrix S matrix of the quantum double is given by

$$S = \begin{pmatrix} 1 & \phi & \phi & \phi^2 \\ \phi & -1 & \phi^2 & -\phi \\ \phi & \phi^2 & -1 & -\phi \\ \phi^2 & -\phi & -\phi & 1 \end{pmatrix}. \tag{20}$$

For the quantum double labels, the four pairs $0\bar{0}$, $0\bar{2}$, $2\bar{0}$ and $2\bar{2}$, the twists are $1, \exp(4\pi i/5), \exp(-4\pi i/5)$ and 1, respectively.

5. Summary

I summarize the main achievements of our extended string-net models as follows:

- We have shown the gauge invariance and topological invariance of the model. When the input data are from a quantum group, our model achieves a *synthesis* of the ideas of Yang–Bater and Yang–Mills. Physically it is an *effective* (rather than microscopic) theory at large distances, describing a topological phase/order of *anyonic* matter in two dimensions. The anyon species are associated with the irreducible representations of the input quantum group.

- We have exactly solved the lattice quantum-group gauge theory, including ground states and localized excitations. These excitations are generally dyons, i.e. composite excitations of both charges at vertices and fluxons at plaquettes. The algebra of quantum numbers of excitations is *closed* under fusion, a property non-existent in previous string-net models.
- The excited dyon species is characterized by its charge, fluxon type and twist, forming representations of the quantum double of the input quantum group. It has a naturally defined braiding property, so Yang–Baxter equations also play a vital role in organizing the excitations.
- The full excitation spectrum with input data from a finite group exhibits the electric-magnetic duality to the Kitaev's Toric Code.[9] The unextended models has this duality only for the ground state sector. We conjecture that when gauging a quantum group, our models also have a full electric-magnetic duality for excited states as well.

Acknowledgments

The author thanks the organizers for the invitation. He also thanks Y. Hu and N. Geer for the pleasant collaboration in this research.

References

1. Y. Hu, N. Geer and Y. S. Wu, arXiv:1502.03433, http://arxiv.org/pdf/1502.03433.pdf.
2. C. N. Yang and R. L. Mills, *Phys. Rev.* **96**, 191 (1954).
3. C. N. Yang, *Phys. Rev. Lett.* **33**, 445 (1974); Erratum-*ibid.* **35**, 1748 (1975).
4. C. N. Yang, *Phys. Rev. Lett.* **19**, 1312 (1967).
5. R. J. Baxter, *Ann. Phys.* **70**, 193 (1972).
6. Y.-S. Wu, *Phys. Rev. Lett.* **52**, 2103 (1984).
7. V. G. Drinfel'd, *Sov. Math. Doklady* **32**, 254 (1985); reprinted in M. Jimbo, *Yang–Baxter Equation in Integrable Systems* (World Scientific; Singapore, 1990), p. 264.
8. M. Jimbo, *Lett. Math. Phys.* **10**, 63 (1985).
9. A. Kitaev, *Ann. Phys.* **303**, 2 (2003); arXiv:quant-ph/9707021.
10. V. Turaev and O. V. Viro, *Topology* **31**, 865 (1992).
11. M. Levin and X. G. Wen, *Phys. Rev. B* **71**, 045110 (2005).

The Framed Standard Model (I) — A Physics Case for Framing the Yang–Mills Theory?*

Chan Hong-Mo

Rutherford Appleton Laboratory, Chilton, Didcot, Oxon, OX11 0QX, UK
h.m.chan@stfc.ac.uk

Tsou Sheung Tsun

Mathematical Institute, University of Oxford, Radcliffe Observatory Quarter,
Woodstock Road, Oxford, OX2 6GG, UK
tsou@maths.ox.ac.uk

Introducing, in the underlying gauge theory of the Standard Model, the frame vectors in internal space as field variables (framons), in addition to the usual gauge boson and matter fermions fields, one obtains:

- the standard Higgs scalar as the framon in the electroweak sector;
- a global $\widetilde{su}(3)$ symmetry dual to colour to play the role of fermion generations.

Renormalization via framon loops changes the orientation in generation space of the vacuum, hence also of the mass matrices of leptons and quarks, thus making them rotate with changing scale μ. From previous work, it is known already that a rotating mass matrix will lead automatically to:

- CKM mixing and neutrino oscillations,
- hierarchical masses for quarks and leptons,
- a solution to the strong-CP problem transforming the theta-angle into a Kobayashi–Maskawa phase.

Here in the framed standard model (FSM), the renormalization group equation has some special properties which explain the main qualitative features seen in experiment both for mixing matrices of quarks and leptons, and for their mass spectrum. Quantitative results will be given in Paper II. The present paper ends with some tentative predictions on Higgs decay, and with some speculations on the origin of dark matter.

When the Yang–Mills theory was discovered 60 years ago, its significance was immediately recognized, although it was unclear at that stage in what physics context it should be applied. Now, 60 years later, the theory has found itself enthroned as the theoretical basis of, among other things, the standard model of particle physics, namely of all known physical phenomena apart from gravity. And this standard model can justly claim to be the most successful theory ever, given the

---

*Invited talk given by CHM at the Conference on 60 Years of Yang–Mills Gauge Theories, IAS, NTU, Singapore, 25–28 May 2015.

range of phenomena it covers and the resilience it has shown in surviving the many detailed experimental tests to which it has been subjected.

However, some things are still missing from this beautiful picture, at least to the fastidious theoretical mind, notably an understanding of the origin of:

- The Higgs boson needed to break the electroweak symmetry,
- Three generations of quarks and leptons needed to fit experiment,

neither of which is part of the original Yang–Mills structure or has any other theoretical explanation. The lack of understanding of the second, especially, is practically significant, since the masses and mixing matrices of quarks and leptons fall into a bizarre hierarchical pattern, and they account for about two-thirds of the standard model's twenty-odd empirical parameters. There are besides some ominous clouds on the horizon, such as the unresolved strong CP-problem, the missing right-handed neutrino, and the mysterious dominance of dark matter in the universe, plus, of course, the wanting link to gravity already mentioned.

To address these short-comings of the standard model, if indeed short-comings they are, one will obviously have to enrich its starting assumptions in some way. One direction, the most popular one, is to keep the Yang–Mills structure as it is, but enrich the superstructure by enlarging the gauge symmetry beyond the standard $su(3) \times su(2) \times u(1)$ (GUT, Supersymmetry), or the dimension of space–time beyond the standard $3 + 1$ (Kaluza–Klein), or both (superstring, branes). These extensions often open up grand vistas of things to come but are less successful in answering some of the detailed questions of immediate interest. For example, instead of reducing the number of parameters in the standard model by explaining the known values of some of them, supersymmetric models usually end up with a hundred parameters or more. For this reason, one thought, it might be worth trying a different path.

Suppose one keeps the same gauge symmetry $su(3) \times su(2) \times u(1)$ and the same four-dimensional space–time as for the standard model, but enrich instead the underlying structure by requiring the Yang–Mills theory be "framed." By "framing" a gauge theory here, we mean the introduction also as dynamical variables of the frame vectors in the internal symmetry space in addition to the usual gauge potential A_μ and the fermionic matter fields ψ. Such frame vector (or "framon") fields are analogues of the vierbeins e^a_μ in the theory of gravity in which they are often used in place of the metric $g_{\mu\nu}$ as dynamical variables. By taking them as dynamical variables too in the particle theory makes it closer in spirit to the gravity theory, and may eventually facilitate the union of the two. Our immediate aim, however, is not to attempt this union but, while staying within particle physics itself, to address the shortcomings of the standard model listed above.

What then will framing give us in particle theory? Like the vierbeins e^a_μ in gravity, the framons we need for the particle theory may be regarded as column vectors of a matrix relating the local gauge frame to a global reference frame, carrying hence both a local symmetry index (analogous to μ in e^a_μ) and a global frame index (analogous to a in e^a_μ). They thus transform both under the local gauge transformations of $G = su(3) \times su(2) \times u(1)$ and under the global transformations of the reference frame, say $\tilde{G} = \widetilde{su}(3) \times \widetilde{su}(2) \times \tilde{u}(1)$, but are just scalars under the Lorentz transformations in space–time. Two immediate consequences of framing then result:

(A) In the electroweak sector, the framon is an $su(2)$ doublet but Lorentz scalar field, exactly as needed for the Higgs field to break the electroweak symmetry;

(B) Since physics is independent of the choice of the reference frame, the framed theory has to be invariant not only under the local symmetry G, but also under its dual, the global symmetry $\tilde{G}$. Of this the 1st component $\widetilde{su}(3)$, a 3-fold symmetry, can function as fermion generation, while the 2nd component $\widetilde{su}(2)$ can act as up-down flavour, and the 3rd component $\tilde{u}(1)$ as $(B-L)$,

giving thus both the Higgs field and fermion generation each an hitherto lacking geometric significance.

To push these advantages further, there is some ambiguity first to settle as to the exact form that the framons should take. Minimality considerations in the number of scalar fields to be introduced suggest then that the framons in FSM belong to the representation $(\mathbf{3+2}) \times \mathbf{1}$ in G but to $\tilde{\mathbf{3}} \times \tilde{\mathbf{2}} \times \tilde{\mathbf{1}}$ in $\tilde{G}$ and that some of its components may be taken as dependent on others[1,2] leaving just:

- a "weak framon" of the form:

$$\boldsymbol{\alpha} \otimes \boldsymbol{\phi}, \tag{1}$$

where $\boldsymbol{\alpha}$ is a triplet in $\widetilde{su}(3)$, which may be taken without loss of generality[2] as a real unit vector in generation space, but is constant in space–time, while $\boldsymbol{\phi}$ is an $su(2)$ doublet but Lorentz scalar field over space–time which has the same properties as, and may thus be identified with, the standard Higgs field;

- the "strong framon":

$$\boldsymbol{\beta} \otimes \boldsymbol{\phi}^{\tilde{a}}, \quad \tilde{a} = \tilde{1}, \tilde{2}, \tilde{3}, \tag{2}$$

where $\boldsymbol{\beta}$ is a doublet of unit length in $\widetilde{su}(2)$ space but constant in space–time, while $\boldsymbol{\phi}^{\tilde{a}}$ are three colour $su(3)$ triplet Lorentz scalar fields over space–time, which when taken as column vectors give a matrix Φ transforming by $su(3)$ transformation from the left but by $\widetilde{su}(3)$ transformations from the right.

As usual the mass matrices at tree-level of quarks and leptons are to be obtained from the Yukawa couplings of the fermions to the Higgs scalar field (weak framon) by replacing it with its vacuum expectation value. But now since the weak framon

(1) carries a factor $\boldsymbol{\alpha}$, the fermion mass matrices will also carry this factor. Then by a simple relabeling of the right-handed singlet fields,

(C) The mass matrices of all quarks and leptons can be rewritten conveniently in the following form:
$$m = m_T \boldsymbol{\alpha}^\dagger \boldsymbol{\alpha}, \qquad (3)$$

where $\boldsymbol{\alpha}$, coming from the framon, is "universal," i.e. the same for up-type quark (U), down-type quark (D), charges leptons (L) and neutrinos (N), and only the coefficient m_T depends on the fermions species.

Now such a mass matrix has long been coveted by phenomenologists[3,4] as a starting approximation, since it gives only one massive generation for each species, which may be interpreted as embryonic mass hierarchy, and zero mixing between up- and down-states, which is not a bad approximation, at least for quarks.

The question now, of course, is what happens above the tree-level. This is the point at which the FSM first shows its power, beyond what can be done by just phenomenology. Because of the double invariance under both $G = su(3) \times su(2) \times u(1)$ and $\tilde{G} = \widetilde{su}(3) \times \widetilde{su}(2) \times \tilde{u}(1)$, the action for framons is much restricted in form, which allows some radiative corrections to be calculated. In particular, since the strong framon in (2) above carries both the local colour $su(3)$ and global dual colour $\widetilde{su}(3)$ (or generation) indices, renormalization by strong framon loops will change the orientation of the vacuum in generation space. This change will depend in general on the renormalization scale μ, thus inducing a μ-dependent $\widetilde{su}(3)$ transformation (rotation) on the vector $\boldsymbol{\alpha}$ which appears in the fermion mass matrix (3).

Now, we have studied the consequences of a rotating rank-1 mass matrix (R2M2) for some years and it has been shown, e.g. in Ref. 5, that a mass matrix of the form (3), with $\boldsymbol{\alpha}$ rotating with changing scale, will automatically give rise to the following effects:

(D1) Mixing between the up and down states, i.e. a nontrivial CKM matrix between up and down quarks, and a PMNS matrix between the charged leptons and neutrinos leading to neutrino oscillations. [This can be easily seen in, for example, the CKM matrix element V_{tb} which is the dot product between the state vectors **t** and **b** of respectively t and b in generations space. For (3), **t** is the value of $\boldsymbol{\alpha}$ at the scale $\mu = m_t$ while **b** is the value of $\boldsymbol{\alpha}$ at $\mu = m_b$, and since $m_t > m_b$ and $\boldsymbol{\alpha}$ rotates, it follows that **t** and **b** are not aligned and $V_{tb} \neq 1$, or that there is mixing.]

(D2) Fermion mass hierarchy in each species, with the mass in the heaviest generation in (3) of each species (e.g. t) "leaking" to the lower ones (e.g. c and u), giving each a small but nonzero mass. [This can be seen as follows. The state vector **t** for t is the vector $\boldsymbol{\alpha}$ at $\mu = m_t$, and the state vector **c** is a vector orthogonal to **t**, and having a zero eigenvalue for (3) at $\mu = m_t$. But this is not the mass m_c for c, which is to be measured at $\mu = m_c$, where $\boldsymbol{\alpha}$ will have rotated already to a different direction with nonzero component in $\mathbf{c}$, and hence $m_c \neq 0$.]

(D3) A solution to the strong-CP problem by transforming away a nonzero theta-angle in the QCD action by turning it, via rotation, into a nonzero Kobayashi–Maskawa phase in, and giving CP-violation to, the CKM matrix. [At every μ, the mass matrix (3) has two zero eigenvalues, so that a chiral transformation can be performed to eliminate the theta-angle from the QCD action without making the mass matrix complex. The effects of this chiral transformation, however, is transmitted by rotation to other μ values and make the CKM matrix complex leading to a KM phase.]

Any R2M2 scheme with a rotating rank-1 mass matrix will give (D1)–(D3) but the details will depend on the rotation trajectory of $\boldsymbol{\alpha}$, i.e how it actually changes with scale μ. Let us see now what FSM has to say about this trajectory. Recall that $\boldsymbol{\alpha}$ is itself a global quantity with no gauge interactions, and therefore not subject directly to radiative corrections. But it is coupled to the strong vacuum, and if that rotates with scale μ, $\boldsymbol{\alpha}$ will rotate also. Now, information of how the strong vacuum rotates can be obtained by studying the renormalization of any quantity which depends on the strong vacuum. We have, mainly for historical reasons, focussed on the Yukawa coupling of the strong framon, and obtained therefrom the renormalization group equations; hence also the equations governing the rotation of $\boldsymbol{\alpha}$. The implications of the rotation equation can be divided conveniently into two bits:

- The shape of the curve Γ traced out by $\boldsymbol{\alpha}$ on the unit sphere in generation space,
- The variable speed with respect to scale μ at which this curve Γ is traced.

The shape of the curve Γ turns out to be a consequence just of symmetry residual in the problem and depends only on a single integration constant, say a. This is shown in Fig. 1, where it is seen that Γ bends sharply near $\theta - 0$, $\phi = \pi$, thus giving it there a considerable local value for the geodesic curvature κ_g, especially for a small value of a. But κ_g needs not be large elsewhere, and indeed changes sign further along the curve. [Notice that for a Γ on the unit sphere, the torsion $\tau_g = 0$ and the normal curvature $\kappa_n = 1$, and only the geodesic curvature κ_g is variable.]

The shape of Γ in Fig. 1 then immediately implies:

(E1) The corner elements, V_{ub}, V_{td} of the CKM matrix for quarks, and similarly U_{e3}, $U_{\tau 1}$ of the PMNS matrix for leptons, both due to twist in Γ, are much smaller than the other off-diagonal elements because $\tau_g = 0$.

(E2) The elements V_{us}, V_{cd} in the CKM matrix for quarks, due to sideways bending of Γ (i.e. governed by κ_g) can be much larger than the elements V_{cb}, V_{ts} governed by $\kappa_n = 1$, although the corresponding elements in the PMNS matrix can all be of similar magnitudes.

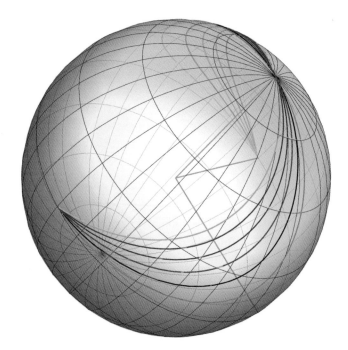

Fig. 1. The curve Γ traced out by the rotating $\boldsymbol{\alpha}$ on the unit sphere in generation space, for various values of the integration constant a.

and also the following, though less obviously, from the already noted fact that κ_g eventually changes sign:

(E3) $m_u < m_d$, despite that for the two heavier generations, $m_t \gg m_b$, $m_c \gg m_s$.

The points (E1) and (E2) are seen to be borne out by experiment which give the approximate CKM and PMNS matrices as:

$$V_{\text{CKM}} = \begin{pmatrix} 0.97428 & 0.2253 & 0.00347 \\ 0.2252 & 0.97345 & 0.0410 \\ 0.00862 & 0.0403 & 0.999152 \end{pmatrix},$$

$$U_{\text{PMNS}} = \begin{pmatrix} 0.82 & 0.55 & 0.17 \\ 0.50 & 0.52 & 0.70 \\ 0.30 & 0.66 & 0.70 \end{pmatrix}. \tag{4}$$

The last (E3) is of course a crucial empirical fact, without which the proton would be unstable, and we ourselves would not be here, but a fact which is, at first sight, theoretically very hard to understand.

Apart from other things, the equation governing the speed at which Γ is traced shows that $\boldsymbol{\alpha}$ has a fixed point at $\mu = \infty$, so that its rotation starts slowly as μ

lowers from ∞, but will accelerate with decreasing μ. This then immediately implies the following results:

(F1) Since lower generation masses according to (D)(ii) comes from "leakage" through rotation and will increase with rotation speed, it follows from the fact that $m_t > m_b > m_\tau$ that:

$$m_c/m_t < m_s/m_b < m_\mu/m_\tau. \tag{5}$$

This agrees with experiment which obtained the mass ratios as respectively 0.0074, 0.0227, 0.0595.

(F2) Since mixing, according to (D)(i), also comes from rotation and will increase with rotation speed, it follows from the fact that quarks are generally heavier than leptons that the off-diagonal mixing elements in the CKM matrix for quarks will generally be larger than the correspondent elements in the PMNS matrix for leptons. That this also agrees with experiment can be seen in (4).

(F3) Since the rotation is generally slow at quark mass scales, one can make a small angle approximation, which allows one to estimate the amount of CP-violation in the CKM matrix obtained via rotation as per (D3) from any given value the theta-angle in the QCD action, giving a Jarlskog invariant as:[5]

$$|J| \sim 7 \times \sin(\theta/2)10^{-5}, \tag{6}$$

which, for θ of order unity, is of the same order of magnitude as the experimentally measured value of $J \sim 2.95 \times 10^{-5}$.

Since, however, one has already derived the renormalization group equations governing the rotation for $\boldsymbol{\alpha}$, there is no need at all to stop at just this qualitative level. True, the rotation equations still depend at present on a number of parameters, but if one is willing to supply some experimental information to determine these parameters, then one can proceed to evaluate essentially all the masses and mixing angles which are taken as inputs from experiment in the usual formulation of the standard model. What the FSM does essentially is to replace 17 independent parameters of the standard model by 7.

(G) By fitting the 7 adjustable parameters of the rotation equations in FSM to 6 pieces of data, we have then calculated the values of 17 independent parameters of the standard model. Not all the 17 quantities have been measured experimentally, but of those 12 that have been measured, the agreement is good.[6]

I shall leave you to judge for yourselves the significance of this result which you will hear from my collaborator Tsou Sheung Tsun in the next talk (Paper II). But it does seem that, with the freedom still left in theory, there will not be much difficulty in reproducing the data as are known at present.

However, what proves a theory, of course, is not its ability to reproduce known results, but its predictions which can be tested against and then confirmed by experiment. Here, one is unfortunately hampered by not knowing how to calculate in general with a theory where the mass matrix rotates. To deduce the results reported, we had some patchy rules for calculating single particle properties like masses and mixing angles, but we have not worked out logically the rules for more general calculations, so any predictions we can make at this stage are only tentative.

Two tentative predictions, however, stand out, both concerned with the Higgs boson. The rotating mass matrix for quarks and leptons were deduced, from a Yukawa coupling. So, if α in the mass matrix rotates, it would seem that it should do the same in the Yukawa coupling too, and so affect the decay of the Higgs boson to fermion–antifermion pairs. An analysis along these lines then suggests that:

(H1?) Branching ratios of the Higgs boson to the second generation fermions such as $H \to \mu^+\mu^-$, $c\bar{c}$ would be suppressed compared to the standard model predictions.

(H2?) There can be flavour-violating decays, such as $H \to \tau\mu$ with branching ratios at the 10^{-4} level.

Experimentally, present sensitivity for these decays are still a couple of orders of magnitude higher than is required for these effects to be detected.

The other tentative prediction involve the mixing between the weak framon (Higgs boson) and the strong. Given that the FSM has to be invariant under the local gauge symmetry G and also under its dual the global symmetry $\tilde{G}$, the form of the interaction potential between the framons can be worked out, which has already been used above in the calculations reported. It is then an easy step to evaluate the mass matrix of the framons to tree-level, and this shows:

(H3?) There is mixing between the Higgs boson with several of its strong analogues. This mixing depends on a couple of yet unknown parameters and so its details are still unknown, but could make the Higgs' decays depart from the standard model predictions.

The predictions (H1?)–(H3?) all come from the weak framon (1). What is likely to be even more interesting, however, is how the strong framons of (2) will manifest themselves. After all, they are the truely new ingredients added by the FSM to the standard model. We recall that it is strong framons loops which gave rise, via renormalization, to rotation in the mass matrix of quarks and leptons, and hence to their mixing and mass hierarchy. So these important effects should already be regarded, in the FSM context, as indirect manifestations of the strong framon. Thus, rather, the question that one is actually asking now is whether strong framons could manifest themselves more directly in some other experimentally detectable physical phenomena. The answer to this, however, is not obvious and is at present perforce speculative, the strong framon being an entirely new type of field with special properties, including probably some unusual soft (nonperturbative) colour physics. But the question is nevertheless sufficiently intriguing, with potentially far-reaching consequences, to deserve some speculations, to which let us then, for a little while, indulge.

We recall that the strong framon carries colour, and so, because of colour confinement, cannot exist as a particle in free space, only inside hadronic matter. But it can combine with other constituents of hadronic matter with the opposite colour to form a colour neutral state, which then appear as a hadron in free space. For example, a framon can combine with an antiframon to form bosons, which in the lowest s-wave state are colour analogues of the standard Higgs boson in electroweak $su(2)$ and mix with it, as mentioned in (H3?) above.

There are altogether nine such states, and they are likely to be the lowest batch of the new hadrons which contain a strong framon as constituent. As also mentioned, their mass matrix at tree-level has already been worked out from the known framon self-interaction potential, and it is straightforward to extend the calculation to their mutual couplings, which has now also been completed. These show that they can readily decay into one another, but the lowest among them, called, for historical reason, H_-, H_4, H_5 are, of course, stable against such decays. These H_K's are electrically neutral, but have for constituents the strong framons (2) with charge $-1/3$, which would allow, among them, H_-, made from a framon and an antiframon with the same generation $\widetilde{su}(3)$ index, to decay into photons. But this does not apply to H_4 and H_5 which are made from a framon and antiframon carrying different $\widetilde{su}(3)$ indices. Further, since even their framon constituents carry no weak charge, these last H_f's (f = four or five) also cannot decay weakly, and would thus seem to be stable altogether.

Now, if these H_f states do exist and are stable, an obvious question would be why they have not been seen. Presumably, like quarks and gluons, strong framons would be present in the sea of hadronic matter inside a proton. And since they are both coloured and charged, they too, like quarks, can be "knocked out" by a hard kick from a gluon or a photon. A quark so "knocked out" will pick up and combine with an antiquark in the sea and emerge as a meson. Cannot then a "knocked out" framon pick up and combine with an antiframon from the sea and emerge as a H_f? If so, why do we not see it?

There is a difference, however, between a quark and a framon in that the latter has an imaginary mass, for like the Higgs scalar field, the quadratic term in the self-interaction potential has a negative coefficient. One can interpret this as meaning that the framon in hadronic matter, unlike the quark, has only a finite lifetime. It has thus only a limited time to seek out a partner from the sea to form an H_f, and in this it may not succeed if the time is short. Hence, again unlike the quark, a "knocked out" framon may not succeed to emerge from the host proton at all. Whether this happens will depend on its lifetime and the amount of hadronic matter that it can traverse inside the proton during its lifetime. Let us say here, for the sake of argument, that the conditions are such that this will not happen, so that no H_f's can be produced in ordinary hadronic reactions, and so explain their nonobservation so far in experiment.

However, that no H_f's are produced in ordinary hadron collisions in present experiments need not mean that the same is true under other circumstances. For example, in the primordial universe (or even now in, say, the galactic centre) both temperatures and densities are much higher than can be found under present experimental conditions in our laboratories. This may allow then these H_f to be formed, and once formed, being stable according to our previous argument, they would still be around with us today.

The question then leaps out whether they may be candidates for dark matter. A preliminary investigation does indeed indicate that they may have rather little interaction with ordinary matter, and also with themselves. That the strong framon is short-lived suggests, by an argument similar to that given above for the non-production of H_f's in present laboratory experiments, that they do not have the usual strong interactions of ordinary hadrons (although they are formally hadrons, being colour neutral bound states by colour confinement of coloured constituents). Besides, they are charged neutral, both electrically and in colour. And being relatively light (though not light enough to be hot dark matter) it may not be impossible for them to satisfy the otherwise very stringent bounds already set by recent experiments such as Xenon 100 and LUX. Hence, perhaps:

(I??) New constituents of Dark Matter(?)

However, if indeed relatively light, it will take a lot of them to make up a sufficient mass so as to matter in the dark matter problem. Is there then any reason why they should be produced in the early universe in such an abundance, say, as compared to luminous baryonic matter? Amusingly, there is, or at least may be, a possible reason. At some stage in its development after the Big Bang, the universe is presumably just a large blob of hot, dense hadronic matter with, among other things, quarks and framons swimming around. As it cools and expands further, the quarks and framons inside would be frantically seeking partners to hitch up as colour neutral bound states so as to survive into the next epoch as hadrons. For the framon to survive as an H_f, all it needs is to find an antiframon of opposite colour. For the quark to survive as a nucleon, however, it will have to find two others of the right colour at the same time, which looks an altogether tougher proposition. Hence, the much greater abundance of H_f than nucleons in our world today. Indeed, it would seem a very lucky chance that enough luminous baryonic matter managed to survive, or else most of the things we know would not be here.

As matters stand, of course, all this discussion about the H_f's as dark matter is merely an exercise in imagination. But, since some of the parameters in FSM have already been determined in the work to be reported by Tsou in the next talk, there may be a chance that some of these imaginings can actually be investigated, and either substantiated or repudiated in the not too distant future.

If there is any truth in these speculations, however, it would indeed seem an extraordinary stroke of good luck that enough luminous baryonic matter is left around from the early universe for us humans to come into existence. Then we are even luckier than we think, today, to be able to come together here to celebrate the 60th anniversary of the Yang–Mills theory. For this we have to thank, first Professor Yang for giving us the cause to celebrate, and secondly Professor Phua and the other organisers for giving us the opportunity to enjoy this celebration.

The work summarized in this talk has almost all been done in collaboration with Jose Bordes, and in part in collaboration with Mike Baker. We have benefitted also from discussions with, and constant interest and encouragement from, James Bjorken.

References

1. H.-M. Chan and S. T. Tsou, *Int. J. Mod. Phys. A* **27**, 1230002 (2012), arXiv:1111.3832.
2. M. J. Baker, J. Bordes, H.-M. Chan and S. T. Tsou, *Int. J. Mod. Phys. A* **27**, 1250087 (2012), arXiv:1111.5591.
3. H. Fritsch, *Nucl. Phys. B* **155**, 189 (1978).
4. H. Harari, H. Haut and J. Weyers, *Phys. Lett. B* **78**, 459 (1978).
5. M. J. Baker, J. Bordes, H.-M. Chan and S. T. Tsou, *Int. J. Mod. Phys. A* **26**, 2087 (2011), arXiv:1103.5615.
6. J. Bordes, H.-M. Chan and S. T. Tsou, *Int. J. Mod. Phys. A* **30**, 1550051 (2015), arXiv:1410.8022.

The Framed Standard Model (II) —
A First Test Against Experiment*

Chan Hong-Mo

Rutherford Appleton Laboratory, Chilton, Didcot, Oxon, OX11 0QX, UK
h.m.chan@stfc.ac.uk

Tsou Sheung Tsun

Mathematical Institute, University of Oxford, Radcliffe Observatory Quarter,
Woodstock Road, Oxford, OX2 6GG, UK
tsou@maths.ox.ac.uk

Apart from the qualitative features described in Paper I (Ref. 1), the renormalization group equation derived for the rotation of the fermion mass matrices are amenable to quantitative study. The equation depends on a coupling and a fudge factor and, on integration, on 3 integration constants. Its application to data analysis, however, requires the input from experiment of the heaviest generation masses $m_t, m_b, m_\tau, m_{\nu_3}$ all of which are known, except for m_{ν_3}. Together then with the theta-angle in the QCD action, there are in all 7 real unknown parameters. Determining these 7 parameters by fitting to the experimental values of the masses m_c, m_μ, m_e, the CKM elements $|V_{us}|, |V_{ub}|$, and the neutrino oscillation angle $\sin^2 \theta_{13}$, one can then calculate and compare with experiment the following 12 other quantities $m_s, m_u/m_d, |V_{ud}|, |V_{cs}|, |V_{tb}|, |V_{cd}|, |V_{cb}|, |V_{ts}|, |V_{td}|$, $J, \sin^2 2\theta_{12}, \sin^2 2\theta_{23}$, and the results all agree reasonably well with data, often to within the stringent experimental error now achieved. Counting the predictions not yet measured by experiment, this means that 17 independent parameters of the standard model are now replaced by 7 in the FSM.

In this talk I shall endeavour to quantify the conclusions derived from the framed standard model as described in Paper I (Ref. 1), to put actual numbers on what were estimates or inequalities, and to show that indeed the theory is capable of giving a reasonable overall fit to all the data on quark and lepton masses and mixing.

We shall start with a very brief summary of the framed standard model (FSM),[2-4] pointing out only those of its salient features which we shall refer to. In FSM, frame vectors form part of the geometry of gauge theory, and by promoting these into fields which we call framons, we have built into our system scalar fields which can play the role of the Higgs fields. Moreover they entail a doubling of the

---

*Invited talk given by TST at the Conference on 60 Years of Yang–Mills Gauge Theories, IAS, NTU, Singapore, 25–28 May 2015.

gauge symmetry

$$SU(3) \times SU(2) \times U(1) \times \widetilde{SU(3)} \times \widetilde{SU(2)} \times \widetilde{U(1)}.$$

This results in a tree-level mass matrix of rank 1[5,6]

$$m = m_T \boldsymbol{\alpha}^\dagger \boldsymbol{\alpha}$$

with no mixing at tree-level. Since the mass matrix is scale-dependent, under renormalization this will lead to nonzero lower generation masses and nonzero mixing,[6] according to the following formulae. Denote the state vectors (in generation space) of the t, c, u quarks respectively by $\mathbf{t}$, $\mathbf{c}$, and $\mathbf{u}$ (in the absence of a strong CP phase which does not affect the masses). Then these are obtained by

$$\begin{aligned}\mathbf{t} &= \boldsymbol{\alpha}(\mu = m_t), \\ \mathbf{c} &= \mathbf{u} \times \mathbf{t}, \\ \mathbf{u} &= \frac{\boldsymbol{\alpha}(\mu = m_t) \times \boldsymbol{\alpha}(\mu = m_c)}{|\boldsymbol{\alpha}(\mu = m_t) \times \boldsymbol{\alpha}(\mu = m_c)|}.\end{aligned} \quad (1)$$

Using these vectors, the lower generation masses are determined by

$$\begin{aligned} m_t &= m_U, \\ m_c &= m_U |\boldsymbol{\alpha}(\mu = m_c) \cdot \mathbf{c}|^2, \\ m_u &= m_U |\boldsymbol{\alpha}(\mu = m_u) \cdot \mathbf{u}|^2. \end{aligned} \quad (2)$$

When we take into account the strong CP phase θ_{CP}, the state vectors become complex and are given by

$$\begin{aligned} \tilde{\mathbf{t}} &= \boldsymbol{\alpha}(\mu = m_t), \\ \tilde{\mathbf{c}} &= \cos\omega_U \boldsymbol{\tau}(\mu = m_t) - \sin\omega_U \boldsymbol{\nu}(\mu = m_t)e^{-i\theta_{CP}/2}, \\ \tilde{\mathbf{u}} &= \sin\omega_U \boldsymbol{\tau}(\mu = m_t) + \cos\omega_U \boldsymbol{\nu}(\mu = m_t)e^{-i\theta_{CP}/2}. \end{aligned} \quad (3)$$

The same procedure applies to the down quark triad.

The direction cosines between these two triads then give the CKM mixing matrix:

$$V_{\text{CKM}} = \begin{pmatrix} \tilde{\mathbf{u}} \cdot \tilde{\mathbf{d}} & \tilde{\mathbf{u}} \cdot \tilde{\mathbf{s}} & \tilde{\mathbf{u}} \cdot \tilde{\mathbf{b}} \\ \tilde{\mathbf{c}} \cdot \tilde{\mathbf{d}} & \tilde{\mathbf{c}} \cdot \tilde{\mathbf{s}} & \tilde{\mathbf{c}} \cdot \tilde{\mathbf{b}} \\ \tilde{\mathbf{t}} \cdot \tilde{\mathbf{d}} & \tilde{\mathbf{t}} \cdot \tilde{\mathbf{s}} & \tilde{\mathbf{t}} \cdot \tilde{\mathbf{b}} \end{pmatrix}$$

with a complex phase corresponding to the Kobayashi–Maskawa phase giving CP violation. Thus we see explicitly how the QCD θ angle is transformed via rotation into the Kobayashi–Maskawa phase.[7]

The case of the leptons is similar, except that we are free not to consider a CP violating phase in the PMNS matrix (as not yet known experimentally).

With these formulae in hand our task is to confront them with actual data, and our main object of interest is the rotating vector $\boldsymbol{\alpha}$. By definition it is a unit vector, and under renormalization it rotates on the unit sphere, tracing out a trajectory say Γ. To study this we compute to 1-loop the relevant Feynman diagrams below which are self-energy diagrams involving the exchange of framons.

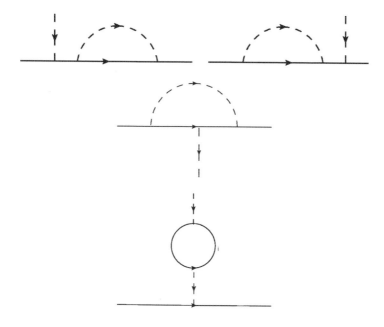

Recalling that the framon potential is given by:[3,4]

$$V[\boldsymbol{\alpha}, \boldsymbol{\phi}, \Phi] = -\mu_W |\phi|^2 + \lambda_W (|\phi|^2)^2 - \mu_S \sum_{\tilde{a}} |\phi^{\tilde{a}}|^2$$

$$+ \lambda_S \left(\sum_{\tilde{a}} |\phi^{\tilde{a}}|^2 \right)^2 + \kappa_S \sum_{\tilde{a},\tilde{b}} |\phi^{\tilde{a}*} \cdot \phi^{\tilde{b}}|^2$$

$$+ \nu_1 |\phi|^2 \sum_{\tilde{a}} |\phi^{\tilde{a}}|^2 - \nu_2 |\phi|^2 \left| \sum_{\tilde{a}} \alpha^{\tilde{a}} \phi^{\tilde{a}} \right|^2 \quad (4)$$

we deduce the renormalization equation for $\boldsymbol{\alpha}$.

If we write $\boldsymbol{\alpha}$ in spherical polar coordinates as usual

$$\boldsymbol{\alpha} = \begin{pmatrix} \sin\theta \cos\psi \\ \sin\theta \sin\phi \\ \cos\theta \end{pmatrix} \quad (5)$$

and introduce the parameter

$$R = \frac{\zeta_W^2 \nu_2}{2\kappa_S \zeta_S^2}, \quad (6)$$

we obtain RGE for the parameters of $\boldsymbol{\alpha}$ as follows:

$$\dot{R} = -\frac{3\rho_S^2}{16\pi^2} \frac{R(1-R)(1+2R)}{D} \left(4 + \frac{R}{2+R} - \frac{3R\cos^2\theta}{2+R}\right), \tag{7}$$

$$\dot{\theta} = -\frac{3\rho_S^2}{32\pi^2} \frac{R\cos\theta\sin\theta}{D} \left(12 - \frac{6R^2}{2+R} - \frac{3k(1-R)(1+2R)}{2+R}\right) \tag{8}$$

and

$$\cos\theta \tan\phi = a \text{ (constant)}, \tag{9}$$

where

$$D = R(1+2R) - 3R\cos^2\theta + k(1-R)(1+2R). \tag{10}$$

Here a dot denotes differentiation with respect to $t = \log \mu^2$.

The trajectory Γ traced out by $\boldsymbol{\alpha}$ on the sphere as we vary the scale μ depends on two functions (of scale) which we may call the shape function and the speed function. The shape function is a consequence of symmetry and depends only on one real parameter a (9); so it is simple to deal with, and has been discussed in some detail in Ref. 1 (see Fig. 1). The speed function, on the other hand, is much less precisely predicted, since first the RGE is only to 1-loop, and secondly it depends on three parameters ρ_S and two integration constants. Even more seriously it depends on some unknown and perhaps uncalculable effects represented by the function $k(\mu)$.

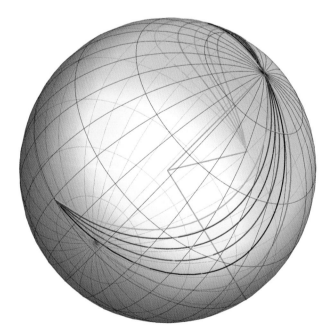

Fig. 1. The curve Γ traced out by the vector $\boldsymbol{\alpha}$ on the unit sphere in generation space for various values of the integration constant a, decreasing in magnitude from $a = -0.6$ in green to $a = -0.1$ in orange.

Clearly one can do little phenomenologically with an unknown function. With some justification we replace it by a constant k.

Before we actually confront our theory with the fermion masses and mixing data, let us see what kind of data we are faced with. There are three points to note.

- There is a large amount of data.
- They have vastly different percentage errors.
- The masses range over 13 orders of magnitude.

To be concrete, let me quote the data as given by the PDG[8] in the summer of 2014, when we did the fit to be reported below.[4]

The quark masses:

$$m_t = 173.07 \pm 0.52 \pm 0.72 \text{ GeV},$$
$$m_c = 1.275 \pm 0.025 \text{ GeV},$$
$$m_u = 2.3^{+0.7}_{-0.5} \text{ MeV (at 2 GeV)},$$
$$m_b = 4.18 \pm 0.03 \text{ GeV},$$
$$m_s = 0.095 \pm 0.005 \text{ GeV (at 2 GeV)},$$
$$m_d = 4.8^{+0.5}_{-0.3} \text{ MeV (at 2 GeV)}.$$

The lepton masses:

$$m_\tau = 1776.82 \pm 0.16 \text{ MeV},$$
$$m_\mu = 105.6583715 \pm 0.0000035 \text{ MeV},$$
$$m_e = 0.510998928 \pm 0.000000011 \text{ MeV}$$

with the squared mass differences for the neutrinos:

$$\left(m^{ph}_{\nu_3}\right)^2 - \left(m^{ph}_{\nu_2}\right)^2 = \left(2.23^{+0.12}_{-0.08}\right) \times 10^{-3} \text{ eV}^2,$$
$$\left(m^{ph}_{\nu_2}\right)^2 - \left(m^{ph}_{\nu_1}\right)^2 = (7.5 \pm 0.20) \times 10^{-5} \text{ eV}^2.$$

The quark CKM matrix:

$$\begin{pmatrix} |V_{ud}| & |V_{us}| & |V_{ub}| \\ |V_{cd}| & |V_{cs}| & |V_{cb}| \\ |V_{td}| & |V_{ts}| & |V_{tb}| \end{pmatrix},$$

$$\begin{pmatrix} 0.97427 \pm 0.00015 & 0.22534 \pm 0.00065 & 0.00351^{+0.00015}_{-0.00014} \\ 0.22520 \pm 0.00065 & 0.97344 \pm 0.00016 & 0.0412^{+0.0011}_{-0.0005} \\ 0.00867^{+0.00029}_{-0.00031} & 0.0404^{+0.0011}_{-0.0004} & 0.999146^{+0.000021}_{-0.000046} \end{pmatrix},$$

$$|J| = \left(2.96^{+0.20}_{-0.16}\right) \times 10^{-5}.$$

Table 1. The heaviest fermion from each type as input.

	Expt (September 2014)	Input value
m_t	$173.07 \pm 0.52 \pm 0.72$ GeV	173.5 GeV
m_b	4.18 ± 0.03 GeV	4.18 GeV
m_τ	1776.82 ± 0.16 MeV	1.777 GeV

The neutrino oscillation angles:

$$\sin^2 2\theta_{13} = 0.095 \pm 0.010,$$
$$\sin^2 2\theta_{12} = 0.857 \pm 0.024,$$
$$\sin^2 2\theta_{23} > 0.95.$$

The last two angles are also known as the solar angle and the atmospheric angle, respectively. I think we could appropriately call the first the Daya Bay angle.

We are now ready to vary the parameters of the theory so as to produce a trajectory Γ for $\boldsymbol{\alpha}$, which will fit best the masses and mixing data via the formulae (2) and (3).

Remembering that all the fermions lie on the same trajectory Γ, we fix the positions for each type (U quark, D quark, charged lepton, neutrino) by inputting the heaviest generation masses, as shown in Table 1.

The neutrino masses are assumed to be generated by some see-saw mechanism, so we put in some assumed value of m_{ν_3} for the Dirac mass of the heaviest neutrino to fit the data. The value affects only the lepton sector.

Next we shall do a parameter count in detail. The theory has 7 adjustable parameters (4),

$$a, \rho_S, k, R_I, \theta_I, m_{\nu_3}, \theta_{CP}. \tag{11}$$

We shall choose experimental data to fix these, using the following criteria for the choice:

- that they are sufficient to determine the 7 parameters adequately,
- that they have been measured in experiment to reasonable accuracy,
- that they are sufficiently sensitive to the values of the parameters,
- that they are strategically placed in $t = \ln \mu^2$ over the interesting range,

and we end up with the following choice:

- the masses m_c, m_μ, m_e,
- the elements $|V_{us}|, |V_{ub}|$ of the CKM matrix for quarks,
- neutrino oscillation angle $\sin^2 2\theta_{13}$.

Because of the special role played by the Cabibbo angle $|V_{us}|$ with respect to the geodesic curvature of the trajectory Γ, it fixes by itself already 2 of the parameters above (11).

Table 2. The input experimental values compared with calculated values.

	Expt (June 2014)	FSM Calc	Agree to
INPUT			
m_c	1.275 ± 0.025 GeV	1.275 GeV	$< 1\sigma$
m_μ	0.10566 GeV	0.1054 GeV	0.2%
m_e	0.511 MeV	0.513 MeV	0.4%
$|V_{us}|$	0.22534 ± 0.00065	0.22493	$< 1\sigma$
$|V_{ub}|$	$0.00351^{+0.00015}_{-0.00014}$	0.00346	$< 1\sigma$
$\sin^2 2\theta_{13}$	0.095 ± 0.010	0.101	$< 1\sigma$

We now demand that, by varing the 7 parameters (11), the calculations give us back the 6 inputted data within the desired accuracy: either within experimental errors or within half a percent, as shown in Table 2.

Note that the functional form for the trajectory for α having already been prescribed by the RGE (7), (8), (9), it is not at all obvious that the 6 targeted quantities can be so fitted with the given 7 parameters. That it can indeed be done to the accuracy stipulated constitutes already quite a nontrivial test.

This test done, we can proceed to calculate the following 23 quantities of the standard model:[a]

- 8 lower generation masses,
- the absolute values of all 9 CKM elements,
- the Jarlskog invariant J,
- 3 neutrino oscillation angles,
- m_{ν_3},
- θ_{CP}.

Of these, 17 ($= 23 - 6$) are independent in SM

- 8 lower generation masses,
- 4 CKM parameters,
- 3 neutrino oscillation angles,
- m_{ν_3},
- θ_{CP}

and of which 12 (=17 − 5) can be compared to experiment (the remaining five being not yet measured):

- m_c, m_s, m_μ, m_e, m_u/m_d,
- 4 CKM parameters,
- 3 neutrino oscillation angles.

---

[a]By the standard model here, we mean that in which the now established fact, that neutrinos have masses and oscillate, is incorporated. This means it will have to carry the Dirac masses of the neutrinos also as parameters. Further, we count θ_{CP} also as a parameter of the standard model although it is often arbitrarily put to zero.

Table 3. The calculated output values using inputs in Table 2.

	Expt (June 2014)	FSM Calc	Agree to
OUTPUT			
m_s	0.095 ± 0.005 GeV (at 2 GeV)	0.169 GeV (at m_s)	QCD running
m_u/m_d	0.38–0.58	0.56	$< 1\sigma$
$|V_{ud}|$	0.97427 ± 0.00015	0.97437	$< 1\sigma$
$|V_{cs}|$	0.97344 ± 0.00016	0.97350	$< 1\sigma$
$|V_{tb}|$	$0.999146^{+0.000021}_{-0.000046}$	0.99907	1.65σ
$|V_{cd}|$	0.22520 ± 0.00065	0.22462	$< 1\sigma$
$|V_{cb}|$	$0.0412^{+0.0011}_{-0.0005}$	0.0429	1.55σ
$|V_{ts}|$	$0.0404^{+0.0011}_{-0.0004}$	0.0413	$< 1\sigma$
$|V_{td}|$	$0.00867^{+0.00029}_{-0.00031}$	0.01223	41%
$|J|$	$(2.96^{+0.20}_{-0.16}) \times 10^{-5}$	2.35×10^{-5}	20%
$\sin^2 2\theta_{12}$	0.857 ± 0.024	0.841	$< 1\sigma$
$\sin^2 2\theta_{23}$	> 0.95	0.89	$> 6\%$

However, we need to check 18 $(= 23 - 5)$ experimental values to ensure that we have good accuracy, although these are not all independent in the standard model. For example, although the CKM matrix is unitary and has only 4 independent parameters, ensuring that only these 4 fall within error does not imply that the remaining elements are within error too. These 18 quantities we checked are listed in Tables 2 and 3.

We note that of the 12 output quantities shown in Table 3, 6 are within experimental error or else (m_μ, m_e) within 0.5 percent of the accurate measured values, while two are within $\sim 1.5\sigma$. Of the remaining 4, one (m_s) can only be roughly compared with experiment, because of QCD running, and it does so compare quite reasonably. The other 3: $|V_{td}|, J, \sin^2 2\theta_{23}$, are all outside the stringent experimental errors, but still not outrageously so. Besides, $|V_{td}|$ and J both being small and therefore delicate to reproduce, obtaining them with the right order of magnitude as they are here is already no mean task.

The fit gives in addition the following values for the 5 other standard model parameters which, not being measured, cannot be checked against experiment at present:

$$\theta_{CP} = 1.78, \quad m_u(\mu = m_u) = 0.22 \text{ MeV} \quad [\text{or } m_d(\mu = m_d) = 0.39 \text{ MeV}],$$
$$m_{\nu_3} = 29.5 \text{ MeV}, \quad m_{\nu_2} = 16.8 \text{ MeV}, \quad m_{\nu_1} = 1.4 \text{ MeV}.$$

Figures 2 and 3 show the actual trajectory of $\boldsymbol{\alpha}$ corresponding to the fit above. There are many interesting features, in accordance with the qualitative expectations described in Ref. 1. Here we would like to comment on one particular aspect.

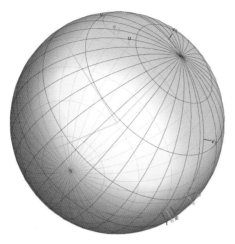

Fig. 2. The trajectory for $\boldsymbol{\alpha}$ on the unit sphere in generation space obtained from the parameter values obtained as described, showing the locations on the trajectory where the various quarks and leptons are placed: high scales in front.

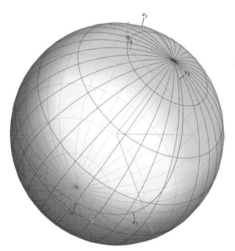

Fig. 3. The trajectory for $\boldsymbol{\alpha}$ on the unit sphere in generation space obtained from the parameter values obtained as described, showing the locations on the trajectory where the various quarks and leptons are placed: low scales in front.

As already mentioned in Ref. 1, because the change in sign of the geodesic curvature of Γ, we have, generically as a consequence of symmetry and not only for this fit, that $m_u < m_d$, despite the fact that $m_t \gg m_b$, $m_c \gg m_s$. Thus we are able to reproduce this crucial empirical fact, without which the proton would be unstable and we ourselves would not exist. To understand this result a bit further, we note that the geodesic curvature changes sign around the scale of order MeV, where masses of the lowest generation quarks occur. Now according to (2) above,

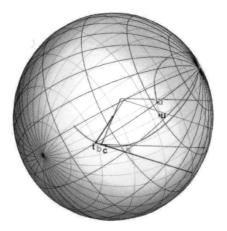

Fig. 4. Figure illustrating the reason why $m_u < m_d$ in Table 3.

the mass of the u and d quarks in FSM are to be given respectively by solution of the equations:

$$|\langle \mathbf{u}|\boldsymbol{\alpha}(\mu)\rangle|^2 = \mu, \quad |\langle \mathbf{d}|\boldsymbol{\alpha}(\mu)\rangle|^2 = \mu, \tag{12}$$

where $\mathbf{u}$ the state vector of u is of course orthogonal to $\mathbf{t}$ and $\mathbf{c}$, the state vectors of t and c. Similarly for the triad $\mathbf{b}$, $\mathbf{s}$, $\mathbf{d}$. The masses of u (d) being only of order MeV, this means that one has an approximate solution for m_u (m_d) whenever the vector $\boldsymbol{\alpha}$ crosses the $\mathbf{tc}$-plane ($\mathbf{bs}$-plane). Given the ordering of the masses of t, b and that, as noted before, $m_c/m_t < m_s/m_b$, the picture is as shown in Fig. 4. It is thus clear that in the MeV region where the geodesic curvature has the opposite sign to that in the high scale region, the vector $\boldsymbol{\alpha}$ must cross the $\mathbf{bs}$-plane before (i.e. at a higher scale than) the $\mathbf{tc}$-plane. In other words, m_d must be larger than m_u, as experiment wants.

It is interesting to note that a scale of a few MeV (at which our geodesic curvature changes sign) occurs also, but for a different reason, in another rotating mass scheme[9] quite similar to ours.

In summary, we can say that

- with 7 adjustable parameters,
- can calculate 23 quantities,
- of which 18 are measurable
  - 10 within errors
  - 2 within 0.5%
  - 2 within $\sim 1.5\sigma$
  - 3 within order of magnitude (or better)
  - 1 with QCD running (cannot calculate at present),
- 17 independent in SM,
- 12 both measurable and independent,
- bonus point: $m_d > m_u$ generically.

In this short talk I did not present the framed standard model in full, but only a small part of it: a fit to data. We did not explore all parameter space; our purpose was just to show that it is possible to obtain a decent (or to our biased eyes, a good) fit. This fit fixes for us a number of parameters in the theory, which we shall use to further explore consequences of FSM.

The work reported was done in collaboration with Jose Bordes.

References

1. H.-M. Chan and S. T. Tsou, invited talk (by CHM) at the *Conference on 60 Years of Yang–Mills Gauge Field Theories*, 25–28 May 2015, Singapore, to be published in the Proceedings, eds. L. Brink and K. K. Phua, arXiv:1505.0547.
2. H.-M. Chan and S. T. Tsou, *Int. J. Mod. Phys. A* **27**, 1230002 (2012), arXiv:1111.3832.
3. M. J. Baker, J. Bordes, H.-M. Chan and S. T. Tsou, *Int. J. Mod. Phys. A* **27**, 1250087 (2012), arXiv:1111.5591.
4. J. Bordes, H.-M. Chan and S. T. Tsou, *Int. J. Mod. Phys. A* **30**, 1550051 (2015), arXiv:1410.8022.
5. H. Fritsch, *Nucl. Phys. B* **155**, 189 (1978).
6. M. J. Baker, J. Bordes, H.-M. Chan and S. T. Tsou, *Int. J. Mod. Phys. A* **26**, 2087 (2011), arXiv:1103.5615.
7. J. Bordes, H.-M. Chan and S. T. Tsou, *Int. J. Mod. Phys. A* **25**, 5897 (2010), arXiv:1002.3542 [hep-ph].
8. PDG website: http://durpdg.dur.ac.uk/lbl/.
9. J. Bjorken, private communication; see also the website: bjphysicsnotes.com.

On the Study of the Higgs Properties at a Muon Collider

Mario Greco

Dipartimento di Matematica e Fisica, Università di Roma Tre
and
INFN, Sezione di Roma Tre,
Via della Vasca Navale 84, I-00146 Rome, Italy
mario.greco@roma3.infn.it

The discovery of the Higgs particle at 125 GeV is demanding a detailed knowledge of the properties of this fundamental component of the Standard Model. To that aim various proposals of electron and muon colliders have been put forward for precision studies of the partial widths of the various decay channels. It is shown that in the case of a Higgs factory through a muon collider, sizeable radiative effects — of order of 50% — must be carefully taken into account for a precise measurement of the leptonic and total widths of the Higgs particle. Similar effects do not apply in the case of Higgs production in electron–positron colliders.

1. Introduction

The announcement of the discovery of a new scalar particle by the LHC experiments ATLAS[1] and CMS[2] has immediately triggered the question of the real nature of this particle. The determination of the spin–parity quantum numbers and the couplings to other Standard Model (SM) particles strongly suggest it to be the Higgs boson, i.e. the particle responsible for the electroweak symmetry breaking. From the available data however, it cannot be concluded yet that we have found the SM Higgs boson and not one of the scalars postulated within the possible extensions of the SM.

Therefore, a detailed study of this particle is required and various types of Higgs factories have been proposed, which can precisely determine the properties of the Higgs boson, as an important step in the future of high energy physics. In particular, the Higgs width's measurement is an essential ingredient to determine the partial width and the coupling constants to the fermions and bosons, and to this aim it is well known that lepton colliders are most appropriate for precision measurements, as happened in the past for LEP and SLC. Indeed the natural width of a 125 GeV SM Higgs boson is about 4 MeV, which is far from the precision which can be achieved at LHC.

Then circular electron–positron colliders are receiving again considerable attention. Design studies have been launched at CERN with the Future Circular Colliders (FCC), formerly called TLEP,[3] of which an e^+–e^- collider hosted in a tunnel of about 100 km is a potential first step (FCC-ee), and in China with the Circular Electron Positron Collider (CEPC).[4] Both projects can deliver very high luminosity ($L > 10^{34}$) from the Z peak to HZ threshold. The Higgs bosons are produced mainly through Higgs-strahlung, at center-of-mass energy of about 250 GeV, and therefore the Higgs boson is produced in association with a Z boson. Then the Higgs signal can be tagged with the Z boson decay, especially if the Z boson decays into a pair of leptons. The HZZ coupling can be inferred in this way, but it seems to us that the Higgs width is not achieved so easily.

On the other hand the idea of a muon collider with $L > 10^{32}$ seems much more appealing.[5–7] The collider ring is much smaller, with $R \sim 50$ m, and here the direct H^0 cross section is greatly enhanced with respect to e^+-e^- since the s-channel coupling is proportional to the square of the initial lepton mass. In analogy to the case of the Z^0 at LEP/SLC, the production of a single H^0 scalar in the s-state offers unique conditions of experimentation because the H^0 has a very narrow width and most of its decay channels can be directly compared to the SM predictions with a very high accuracy. The drawback of this project however is that a very powerful "muon cooling" is needed.[a] A practical realization of a demonstrator of a muon cooled Higgs factory has been recently suggested by C. Rubbia.[7]

In the present note we show that in the case of a muon collider very important QED radiative corrections change drastically the lowest order Higgs production results, and must be taken in full account for an accurate description of the Higgs total and partial widths, as well as the various H^0 couplings. Such type of effects do not apply in the case of electron positron colliders, when the Higgs bosons are produced mainly through Higgs-strahlung in association with a Z boson. The origin of these radiative effects is well known and has been discussed in very great detail in the case of collisions of electrons and positrons with production of narrow resonances in the s-channel like the J/Psi[8] and the Z boson.[9] Namely a correction factor $\propto (\Gamma/M)^{(4\alpha/\pi)\log(2E/m)}$ modifies the lowest order cross section, where M and Γ are the mass and width of the s-channel resonance, $W = 2E$ is the total initial energy and m is the initial lepton mass. Physically this is understood by saying that the width provides a natural cutoff in damping the energy loss for radiation in the initial state. To be more specific, defining

$$\beta_i = \frac{4\alpha}{\pi}\left[\log\frac{W}{m_i} - \frac{1}{2}\right], \tag{1}$$

where m_i is the initial lepton mass,

$$y = W - M,$$

$$\tan\delta_R(W) = \frac{1}{2}\Gamma/(-y)$$

---

[a]For a detailed discussion and a complete set of references, see for example, Ref. 7.

then the infrared factor $C^{\text{res}}_{\text{infra}}$ due to the soft radiation emitted from the initial charged leptons is given by[8]

$$C^{\text{res}}_{\text{infra}} = \left(\frac{y^2 + (\Gamma/2)^2}{(M/2)^2}\right)^{\beta_i/2} \left[1 + \beta_i \frac{y}{\Gamma/2}\delta_R\right], \quad (2)$$

so that the observed resonant cross section can be written as

$$\sigma^c = C^{\text{res}}_{\text{infra}}\sigma_{\text{res}}(1 + C^{\text{res}}_F). \quad (3)$$

In the above equation σ_{res} is the Born resonant cross section (of Breit–Wigner form) and C^{res}_F is a finite standard correction of order α which we will neglect in the following. In the case of Higgs production at a muon collider with $W = 2E = 125\,\text{GeV}$ the factor $\beta_i = 0.061$ and at the resonance ($y = 0$) the factor $C^{\text{res}}_{\text{infra}} = (\Gamma/M)^{\beta_i} = 0.53$, assuming the Higgs width $\Gamma = 4\,\text{MeV}$, which gives a substantial reduction of the Born cross section and therefore can mimic a smaller initial (and/or final) partial decay width of the Higgs.

As it is well known, since the produced resonance is quite narrow, one has to integrate over the machine resolution, which is assumed to be

$$G(W' - W) = \frac{1}{\sqrt{2\pi}\sigma}e^{-(W'-W)^2/(2\sigma^2)}, \quad (4)$$

where σ is the machine dispersion, such that $(\Delta W)_{\text{FWHM}} = 2.3548\sigma$. Then the experimentally observed cross section into a final state $|f\rangle$ is given by

$$\tilde{\sigma}(W) = \int G(W' - W)dW'\sigma(W'), \quad (5)$$

where

$$\sigma(W') = \frac{4\pi}{W'^2}\frac{\Gamma_i\Gamma_f}{\Gamma^2}\sin^2\delta_R(W')\left\{\frac{\Gamma}{W'\sin\delta_R(W')}\right\}^{\beta_i}$$
$$\times(1 - \beta_i\delta_R\cot\delta_R)(1 + C^{\text{res}}_F). \quad (6)$$

By integrating the last equation, a useful formula for the observed cross section at the peak is[8]

$$\tilde{\sigma}(M) = \frac{2\pi^2\Gamma_i\Gamma_f}{\sqrt{2\pi}\sigma M^2\Gamma}\left(\frac{\Gamma}{M}\right)^{\beta_i}e^{\left(\frac{\Gamma}{2\sqrt{2}\sigma}\right)^2}$$
$$\times\left\{\text{erfc}\left(\frac{\Gamma}{2\sqrt{2}\sigma}\right) + \frac{1}{2}\beta_i E_1\left(\frac{\Gamma^2}{8\sigma^2}\right)\right\}(1 + C^{\text{res}}_F), \quad (7)$$

where the second term in the square bracket represents the contribution from the radiative tail.

Numerically, and neglecting the contribution from C^{res}_F, the overall radiative correction factor C to the Born cross section at the peak is $C = 0.47, 0.37, 0.30, 0.25$ for $\sigma = 1\,\text{MeV}, 2\,\text{MeV}, 3\,\text{MeV}, 4\,\text{MeV}$, respectively. This result shows once again the importance of the radiative effects for a precision measurement of the Higgs couplings.

Our analytical approach, with an accuracy of order of a few %, gives a good estimate of the radiation effects expected for the Higgs signal around the resonance. On the other hand, the background final states, as b quark–antiquark pairs and W pairs, coming from the direct Z production and decay, estimated for example in Ref. 10, are not substantially affected from the initial state radiation. Of course a more precise numerical evaluation of the ratio Signal/Background is in order, when the muon collider will be finalized.

To conclude, we have shown that in the case of a Higgs factory through a muon collider, sizeable radiative effects — of order of 50% — must be carefully taken into account for a precise measurement of the leptonic and total widths of the Higgs particle. Similar effects do not apply in the case of Higgs production in electron–positron colliders.

References

1. G. Aad *et al.* [ATLAS Collaboration], *Phys. Lett. B* **716**, 1 (2012) [arXiv:1207.7214 [hep-ex]].
2. S. Chatrchyan *et al.* [CMS Collaboration], *Phys. Lett. B* **716**, 30 (2012) [arXiv:1207.7235 [hep-ex]].
3. M. Koratzinos, A. P. Blondel, R. Aleksan, O. Brunner, A. Butterworth, P. Janot, E. Jensen and J. Osborne *et al.*, arXiv:1305.6498 [physics.acc-ph].
4. D. Wang and J. Gao, Study on Beijing Higgs Factory (BHF) and Beijing Hadron Collider (BHC), IHEP-AC-LC-Note 2012-012
5. D. B. Cline, *Nucl. Instrum. Meth. A* **350**, 24 (1994).
6. V. D. Barger, M. S. Berger, J. F. Gunion and T. Han, *Phys. Rept.* **286**, 1 (1997) [hep-ph/9602415].
7. C. Rubbia, arXiv:1308.6612 [physics.acc-ph].
8. M. Greco, G. Pancheri-Srivastava and Y. Srivastava, *Nucl. Phys. B* **101**, 234 (1975).
9. M. Greco, G. Pancheri-Srivastava and Y. Srivastava, *Nucl. Phys. B* **171**, 118 (1980) [Erratum-*ibid.* **197**, 543 (1982)].
10. C. Han and Z. Liu, arXiv:1210.7803v2 [hep-ph].

Aharonov–Bohm Types of Phases in Maxwell and Yang–Mills Field Theories

Bruce H. J. McKellar

ARC Centre of Excellence for Particle Physics at the Terrascale (CoEPP),
School of Physics, University of Melbourne, Vic 3000, Australia
bhjmckellar@mac.com
www.coepp.org.au

I review the topological phases of the Aharonov–Bohm type associated with Maxwell and Yang–Mills fields.

Keywords: Aharonov–Bohm; topological phases; electromagnetic duality.

1. Introduction

The quantum phase of Aharonov and Bohm[1] demonstrated that the vector potential of electrodynamics was not just a mathematical construct giving a convenient description of the electromagnetic field, it had physical effects. This phase is produced when a charged particle traverses a closed path along which the magnetic field **B** vanishes, but the magnetic vector potential **A** does not, and is determined by the vector potential, independently of the magnetic field on the path of the test particle. Later the AB phase was shown to be a particular case of the Berry phase, and is topological in nature. The study of the AB phase has had many consequences for both Maxwell and Yang–Mills fields.

A number of phases, which in one way or another are dual to the AB phase, have been introduced. The first dual to the AB effect proposed was the Aharonov Casher (AC) effect — the phase acquired by a magnetic dipole traversing a closed path in an electric field. The potentials no longer play a central role, but the phase is still topological. The duality between the AB and the AC phases arises from the interchange of the source of the field and the nature of the text particle. The dual AB effect (magnetic monopole traversing a closed path outside an electric field) and dual AC effect (electric dipole traversing a closed path in an magnetic field) have also been studied.

The AB effect can limit the mass of the photon, and also limit electromagnetic Lorentz breaking terms.

The generalization of the AB phase to Yang–Mills fields led Wu and Yang[2] to the realization that fiber bundles were the mathematical structure underlying field theories. Wu and Yang also described the modification of the non-Abelian AB effect when the fields are massive, and proposed an experimental test, which has subsequently been studied.

In this paper I will be reviewing the AB phase and all of the above developments from it, some in more detail than others.

2. The Aharonov–Bohm Phase

The Aharonov–Bohm phase was predicted by Ehrenberg and Siday[3] in 1949, and rediscovered, and more thoroughly studied, by Aharonov and Bohm[1] in 1959.[a] This phase is produced when a charged particle traverses a closed path along which the magnetic field **B** vanishes, but the magnetic vector potential **A** does not, as illustrated in Fig. 1.

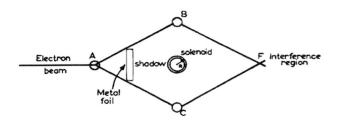

Fig. 1. The Aharonov–Bohm concept, after Ref. 1.

The first measurements were made by Chambers[4] in 1960 and by G. Möllenstedt and W. Bayh[5] in 1962. The charged particles were electrons in an electron microscope with a biprism-like beam splitter, and the magnetic fields were generated by magnetic whiskers, or by meticulously wound micro-scale solenoids. Because of finite size of the magnetic field source, there were concerns that stray magnetic field could have been be present at the path of the particle. The concerns about stray fields producing a fake effect were laid to rest by the definitive experiment in 1986 of Tonomura et al.[6] They confined the magnetic field in a toroidal magent, completely eliminating the possibility of stray fields.

To review some details about the AB phase, I briefly review a proof of the effect, using the Dirac equation, simply because first-order equations are easier

---

[a]There is thus a case for referring to the phase as the Ehrenberg–Siday-Aharonov–Bohm phase, or even the Ehrenberg–Siday phase. We retain the usual designation, so that readers know what we are talking about.

to manipulate than second-order equations, and not because we are necessarily considering relativistic particles. Start with the free Dirac equation

$$(i\gamma^\mu \partial_\mu - m)\psi = 0 \tag{1}$$

and include the electromagnetic field through its potential A_μ, when the equation becomes

$$A_\mu : \quad (i\gamma^\mu [\partial_\mu + ieA_\mu] - m)\psi = 0. \tag{2}$$

I am using the usual particle physics units, in which $\hbar = c = 1$. Now make a phase transformation of the Dirac wave function

$$\psi \to \psi' = e^{i\chi}\psi. \tag{3}$$

The new wave function ψ' satisfies the free Dirac equation

$$(i\gamma^\mu \partial_\mu - m)\psi' = 0 \quad \text{if } A_\mu = e\partial_\mu \chi. \tag{4}$$

A vector potential of this form is a pure gauge field, and it is very well known that pure gauge fields have no physical effects. Naively one would not expect such a field to have any experimental consequences.

However consider a vector potential which is purely spatial, $A_0 = 0$, and introduce the phase

$$\chi(\boldsymbol{x}) = \int_P^{\boldsymbol{x}} \boldsymbol{A}(\boldsymbol{x}') \cdot d\boldsymbol{s}'. \tag{5}$$

This satisfies the second of Eq. (4), and thus the transformed ψ' satisfies the free field equation. How can a simple magnetic vector potential ever have physical effects? It can when the magnetic field related to $\boldsymbol{A}$ by $\boldsymbol{B} = \nabla \times \boldsymbol{A}$ is nonzero in some region, since the line integral defining the phase χ is then not uniquely defined. The potential is genuinely pure gauge only if there is no magnetic field anywhere.

When the magnetic field is excluded from the region in which the charged particle moves, Aharonov and Bohm demonstrated that the phase has physical consequences. The situation considered by Aharonov and Bohm was one in which the magnetic field, perhaps produced by a very long solenoid, is excluded from the region in which the charged particle moves, as shown in Fig. 1. Because of this exclusion, the $\boldsymbol{B} = 0$ region is not simply connected.

They then considered two paths $\mathcal{C}_1$ and $\mathcal{C}_2$ from an initial point P to a final point Q. These paths will have a phase difference

$$e(\chi_1 - \chi_2) = e\int_\mathcal{C} \boldsymbol{A}(\boldsymbol{x}) \cdot d\boldsymbol{s} = e\int_\mathcal{S} \boldsymbol{B} \cdot d\boldsymbol{S} = e\Phi_\mathcal{C}, \tag{6}$$

where
$$C = C_1 - C_2 = \partial S \tag{7}$$
and Φ_C is the magnetic flux through the closed curve C.

If we go back to SI units, the AB phase is $e\Phi/\hbar$.

3. Topological or Geometric Phase

Whether or not a phase is topological is often the subject of much discussion in the literature. To answer this about a particular phase it is necessary to have a clear answer to the question:

What is meant by topological phase, or geometric phase? [b]

My answer is that the phase is topological phase, or geometric phase when:

(1) the phase $\phi(\mathbf{x})$ is such that $\psi'(\mathbf{x}) = \exp\{i\phi(\mathbf{x})\}\psi(\mathbf{x})$ satisfies the free wave equation,
(2) the phase is independent of the path, except for the number of times it circles some interior excluded region, and
(3) the phase is determined by properties of the fields in the excluded region

The AB phase will be topological if the magnetic field $\mathbf{B}$ vanishes in the interference region — the region which contains all of the relevant closed paths C. Then the phase difference between two closed paths, C and C', each of which go from P to Q and back to P, will be

$$e(\chi - \chi') = e \int_{S'} \mathbf{B} \cdot d\mathbf{S} = 0, \tag{8}$$

where S' is the surface bounded by the closed curve $C - C'$. Moreover this condition guarantees that only the excluded regions can contribute to the magnetic flux Φ_C through the curve C.

Thus, when deciding whether or not an experiment is measuring a topological AB effect, it is important to ask the question: *Is the magnetic field zero in the interference region?* When the answer to this question is *YES* then the experiment measures a topological phase. The early experiments used magnetic whiskers or small solenoids, and, while they had good grounds for believing that stray magnetic fields did not contribute, it was not until the definitive experiment of Tonomura and his collaborators using a toroidal magnet, which was clad with a superconductor to prevent leakage of the magnetic field, and with gold to prevent the electrons from entering the magnetic field region, settled that question.

I want to emphasize that the paths for the AB phase do not have to be in the same plane. And, while the long straight magnet/solenoid and the effective two-dimensional geometry implied by Fig. 1 is convenient for picturing the physics, it is

---

[b] I regard the adjectives topological and geometric as synonymous and interchangeable.

by no means essential — witness the toroidal geometry of the definitive Tonomura experiment. The AB phase is genuinely a three-dimensional effect.

It is of course possible to do the calculations with the Schrödinger equation. It just takes a little more time and I will not give the details here.

4. The Many Dual Effects

There are many effects or phases which can be regarded as dual to the AB effect. The AB effect involves an electric charge passing around a confined magnetic field. Electromagnetic duality — interchanging electric and magnetic fields and electrical and magnetic charges — leads to the case of a magnetic monopole passing around a confined electric field. This dual AB effect would seem to be of little interest in the absence of magnetic monopoles, but it has been useful in several contexts.

If we regard the magnetic field in the AB effect as that of a line of magnetic dipoles, the AB effect comes from an electric charge moving around a line of magnetic dipoles, and another type of duality is possible. One can interchange the charge and the dipoles and consider a magnetic dipole moving around a line of electric charges. This is the duality introduced by Aharonov and Casher,[7] and gives rise to the Aharonov–Casher (AC) phase.

Then electromagnetic duality can be applied to the AC phase, and one considers an electric dipole moving around a line of magnetic monopoles, as was done by He and McKellar[8] and by Wilkens,[9] introducing the He–McKellar–Wilkens (HMW) phase. At first sight the physical realization of the HMW phase seems impossible, but it turns out that one does not need magnetic monopoles and the Toulouse group measured the HMW phase in 2012.[15,16] The various dual effects to the AB effect are illustrated in Fig. 2

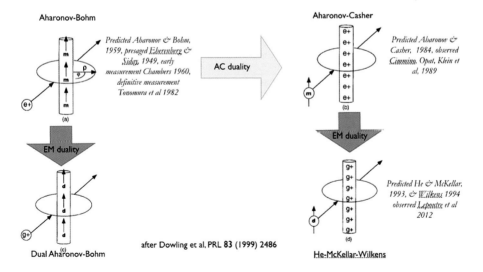

Fig. 2. The dual Aharonov–Bohm effects, after Ref. 19.

4.1. The AC dual of the AB effect

Continue to work with the Dirac equation, now for a particle with zero charge but a nonzero magnetic moment. The neutron is a prime example, but neutral beams of atoms or molecules have also been used in experiments. The Dirac equation is now

$$\left(i\gamma^\mu \partial_\mu - \frac{1}{2}\mu\sigma^{\mu\nu} F^{\mu\nu} - m\right)\psi = 0. \tag{9}$$

It is immediately clear that we must impose special conditions if we are to transform this equation into the desired form

$$\left(i\gamma^\mu[\partial_\mu + i\mu S_\mu] - m\right)\psi = 0, \tag{10}$$

with some *effective vector potential* S_μ, to have any chance of having a topological phase. The physical picture which Aharonov and Casher[7] used to derive their phase gives us hints about what conditions may be required. As illustrated in Fig. 3, they imagined the magnetic flux of the AB effect as generated by a line of magnetic dipoles. The AB phase was rederived from the interaction between an electric charge and a magnetic dipole, integrating over a line of magnetic dipoles. They then switched the integration to integrate over a line of charges.

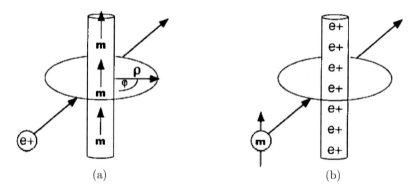

Fig. 3. Aharonov–Casher duality, after Ref. 19.

The geometry of a magnetic dipole going around a line of charges suggests making the line of charges infinitely long, and thus imposing translational symmetry, say along the z axis on the equation. Our analysis follows Hagen[20] and He and McKellar.[21] This effectively reduces the geometry to $2+1$ dimensions, so consider Eq. (9) in 2+1 dimensions, with the following conventions for the (2+1)-dimensional metric $g_{\mu\nu}$ and the antisymmetric tensor $\epsilon_{\mu\nu\alpha}$:

$$g_{\mu\nu} = \text{diag}(1,-1,-1) \quad \text{and} \quad \epsilon_{012} = +1. \tag{11}$$

Only three Dirac matrices are now needed. These may be a suitably chosen set of Pauli matrices, and 2-spinors instead of 4-spinors are used.

There are two inequivalent representations of the Dirac matrices in 2+1 dimensions which generate different Clifford algebras. They may be distinguished by the value of s in the relationship $\sigma_{12} = s\gamma^0$. The parameter s takes the values ± 1

which correspond to spin up and spin down in the "hidden" third spatial dimension. The Clifford Algebra in $2+1$ dimensions has just four basis operators, the unit operator and the three two-dimensional Dirac matrices, and the defining equation for it is

$$\gamma^\mu \gamma^\nu = g^{\mu\nu} + is\epsilon^{\mu\nu\lambda}\gamma_\lambda. \tag{12}$$

The interaction term in the Dirac equation is then proportional to

$$\sigma^{\mu\nu}\psi F_{\mu\nu} = -F^{\mu\nu}s\epsilon_{\mu\nu\lambda}\gamma^\lambda\psi, \quad \text{with } F^{\mu\nu} = \begin{pmatrix} 0 & -E^1 & -E^2 \\ E^1 & 0 & -B^3 \\ E^2 & B^3 & 0 \end{pmatrix}, \tag{13}$$

where E^i and B^i are the electric and magnetic fields, respectively. The indices "1" and "2" indicate the coordinates on the x–y plane along the x and y directions. The index "3" indicates that the magnetic field in this configuration is normal to the x–y plane, in the notional z direction.

Now it is clear that we have a chance of generating a topological phase, because the interaction term is of the form in Eq. (10), with the "effective vector potential" S_μ as the dual of the field strength tensor

$$S_\mu = (1/2)\epsilon_{\mu\alpha\beta}F^{\alpha\beta}. \tag{14}$$

In the AC configuration, the magnetic field vanishes and E_1, E_2 are constant in time. Then $S_\mu = (0, E_2, -E_1) = (0, \mathbf{E} \times \mathbf{k})$, where $\mathbf{k}$ is a unit vector in the z direction, i.e. the direction of the magnetic moment.

Making a transformation $\psi' = \exp[-is\mu_m \int^r \mathbf{S}(\mathbf{r}') \cdot d\mathbf{s}']\psi$, one finds that ψ' satisfies the free Dirac equation.

The reduction of the problem to $2+1$ dimensions was a critical part of this result, because the magnetic moment — electromagnetic field term in the Dirac equation can be converted into a γ^μ interaction with the vector dual of the em field. It then has the same structure as the AB effect, and we now have a topological phase as long as

(1) curl$\mathbf{S} = 0$ in the interference region,
(2) curl$\mathbf{S} \neq 0$ in the excluded region.

As curl$\mathbf{S} =$ curl$\mathbf{E} \times \mathbf{k} = \mathbf{k}(\text{div }\mathbf{E}) - (\mathbf{k}\cdot\nabla)\mathbf{E}$, the simplest way for the excluded region to generate a nonvanishing contribution to the phase is for it to contain some charges, giving rise to div $\mathbf{E} \neq 0$. To preserve the $(2+1)$-dimensional geometry, the charges should be extended uniformly and infinitely in the z direction. The simplest such charge configuration, that chosen by Aharonov and Casher is a line of charges on the z axis, with a linear electric charge density λ_e.

The phase, relative to the configuration without any field, developed in the wave function when the particle travels along a closed path which encircles the line of charge is

$$\chi_{AC} = s\mu_m \oint \mathbf{S} \cdot d\mathbf{r} = -s\mu_m \int_S (\nabla \cdot \mathbf{E})\mathbf{k} \cdot d\mathbf{S} = -s\mu_m \lambda_e, \tag{15}$$

which is just the result of Aharonov and Casher.

This demonstration of the existence of the AC phase depends critically on the fact that the magnetic dipole interaction can be written as an interaction between the electromagnetic field strength $F_{\mu\nu}$ and the dual of the Dirac matrix $\epsilon^{\mu\nu\lambda}\gamma_\lambda$. The dual operator is then transferred to act on the electromagnetic field strength tensor, producing the effective vector potential interacting with the Dirac current. This is the origin of the AC topological phase representation of the magnetic dipole interaction in this $(2+1)$-dimensional configuration. Working in $2+1$ dimensions is the key to making this into a topological or geometric phase, because in $2+1$ dimensions the dual of a tensor is a vector, and vice versa. Then the whole argument leading to the AB phase goes straight through. I remark that, with Xiao-Gang He[22,23] I have shown that this conversion of the dipole interaction to an effective vector potential — Dirac current interaction can be generalized to arbitrary spins, even spin zero, using the fact that such particles satisfy the Bargmann–Wigner equation.

We have chosen to work with the Dirac equation because that is the most elegant structure. While the Dirac equation is not central to the derivation of the topological phase, the $(2+1)$-dimensional structure is. It is then necessary to decide what is vital to the proof as we return to the real world, and what can be relaxed.

Key features of the derivation are that the magnetic moment $\boldsymbol{\mu}_m = \mu_m \boldsymbol{k}$ is normal to the plane defined by the electric field and the direction of motion of the dipole, and that the magnetic moment, or more precisely the component of the magnetic moment along the normal to that plane, is constant. These are the conditions which appear in alternative derivations. The need to go to two spatial dimensions to obtain the topological phase shift was emphasized by Hagen,[20] working in both the Dirac equation and Schrödinger equation contexts. In the latter case the Hamiltonian

$$H_{\rm NR} = \frac{1}{2m}\boldsymbol{\sigma}\cdot(\mathbf{p}-i\mu\mathbf{E})\boldsymbol{\sigma}\cdot(\mathbf{p}-i\mu\mathbf{E}) \qquad (16)$$

is reduced to

$$H_{\rm NR} = \frac{1}{2m}\sum_{j=1}^{2}\left[\boldsymbol{p}_j - i(\boldsymbol{E}\times\boldsymbol{\mu})_j\right]^2 + \frac{1}{2m}\left\{[\boldsymbol{p}_3 - i(\boldsymbol{E}\times\boldsymbol{\mu})_3]^2 - \mu^2\boldsymbol{E}^2\right\}. \qquad (17)$$

The first term produces the topological phase, and only it survives if

- We restrict the problem to two spatial dimensions, OR
- $E_3 = 0$ and $\partial_3\mathbf{E} = 0$, which achieves the same end.

The magnetic moment being normal to the plane determined by the electric field and the path of the particle is a necessary condition for the topological nature of the AC phase, as has been stressed by Sangster et al.[24]

In the AB effect there is no field and no force on the charge in the interference region. In the AC effect there is clearly an electric field, but is there a force? The interaction is that of a magnetic dipole with the effective magnetic field, $\mathbf{v} \times \mathbf{E}$, so the force on the dipole is

$$\mathbf{F} = -\nabla\{\boldsymbol{\mu} \cdot (\mathbf{v} \times \mathbf{E})\}. \tag{18}$$

When $\mathbf{v} \times \mathbf{E}$ is constant in space and time there is no force on the dipole. That is the case for the AC configuration. Again we see that the equivalence of the AB and AC effects requires restrictions on the geometry in the latter case, and that these restrictions are just those required to obtain a topological phase.

One should also consider whether or not the dipole will experience a torque. The torque on the dipole is

$$\boldsymbol{T} = \boldsymbol{\mu} \times (\boldsymbol{v} \times \boldsymbol{E}), \tag{19}$$

which vanishes since the magnetic moment is parallel to $\boldsymbol{v} \times \boldsymbol{E}$.

Because one can regard the AC effect as a consequence of the motional magnetic field $\boldsymbol{B}_{\text{eff}} = \boldsymbol{v} \times \boldsymbol{E}$ seen by the moving magnetic dipole it is important to emphasize that the AC phase is independent of the velocity of the dipole. This velocity independence is a necessary, but not a sufficient, condition for the phase to be a topological phase.

The original observation of the AC phase, due to Cimmino et al.,[26] used cold neutrons and followed the original AC geometry. While the precision was not high it was certainly measuring a topological phase.

The most precise measurement of the AC phase is the experiment of Sangster et al.[24] who used a completely different geometry. They pass coherent beams of magnetic dipoles of different orientation through the same electric field, and observe a phase difference between the beams. They are able to replace the charged wire by a capacitor, as shown in Fig. 4. As they point out themselves, the magnetic moment in this case is not constant in magnitude or direction, and so the phase they measure is not topological, as they themselves acknowledge. The same geometry was used in a number of subsequent measurements of the phase using atoms.

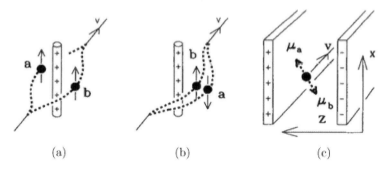

Fig. 4. Sangster et al. proposed modifying the original AC geometry (a), in which particles of the same spin pass on opposite sides of the charged wire, to the geometry (b), in which particles of opposite spin pass on the same side of the wire. They realise (b) with the geometry (c), using a capacitor to generate the electric field, after Ref. 24.

Recently there has been a measurement of the AC phase with atoms which is topological, by the Toulouse group.[27] They use Li atoms in a separated arm atom interferometer — just the apparatus they built to measure the HMW phase. In fact both the HMW and AC phases appear in their experiment and an important part of each experiment was to find ways of getting one with no perturbation of the other one.

An interesting variant of the AC geometry was proposed by Casella.[25] Now the spin is in the plane determined by the path, and the electric field is normal to the spin and the plane of the path, as shown in Fig. 5.

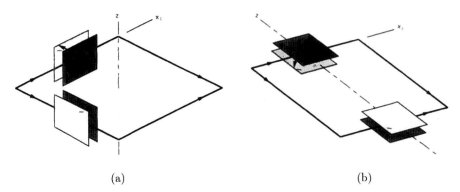

(a) (b)

Fig. 5. Casella proposed converting the AC geometry of figure (a) in which the spin is quantized normal to the pane of the path, and the electric field is in that plane, to the geometry (b) with the spin quantized in the pane of the path, and the electric field normal to that plane, and in opposite directs on opposite sections of the path, after Ref. 25.

Is the phase topological in this case? At first sight one would say no, and Casella does not explicitly claim it to be so. Indeed the $(2+1)$-dimensional reduction is not possible. However, the phase is

$$\chi_c = -\oint_C (\boldsymbol{\mu} \times \boldsymbol{E}) \cdot d\boldsymbol{s}, \qquad (20)$$

and

(1) in the region of the path $\text{curl}(\boldsymbol{\mu} \times \boldsymbol{E}) = \boldsymbol{0}$, so the phase is path independent;
(2) in the plane of the circuit, interior to the circuit, there is a region in which $\text{curl}(\boldsymbol{\mu} \times \boldsymbol{E}) \neq \boldsymbol{0}$, not because $\text{div}\,\boldsymbol{E} \neq 0$, but because $(\boldsymbol{\mu} \cdot \boldsymbol{\nabla})(\boldsymbol{E} \cdot d\boldsymbol{S}) \neq 0$. The phase depends on the fields in the excluded region, even though there are no charges in that region;[c]
(3) the phase is independent of the velocity of the dipole.

The AC phase for the Casella geometry satisfies the last two of our conditions for a topological phase, but not the first, and it is velocity independent. We conclude

---

[c]A simple estimate shows that the line integral and the surface integral indeed give the same phases.

that a measurement of the AC phase using the Casella geometry does not measure a topological phase.

4.2. *The HMW dual of the AC effect*

Just as a magnetic dipole moving through in an electric field which has the appropriate geometry may acquire a topological quantum phase, the AC phase, an electric dipole in a magnetic field which has the appropriate geometry may also acquire a topological quantum phase. This electromagnetic dual phenomenon was pointed out by He and McKellar[8] in 1993, and independently by Wilkens[9] in 1994. Dowling *et al.* in 1999 named it the He–McKellar–Wilkens (HMW) phase. This duality concept is illustrated in Fig. 6.

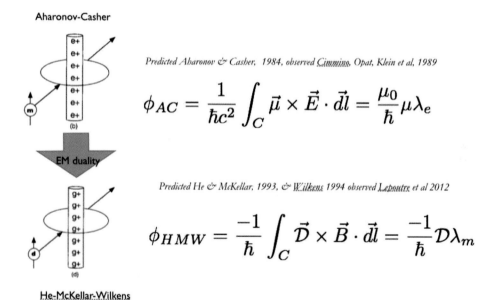

Fig. 6. The electromagnetic dual of the Aharonov–Casher phase is the He–McKellar–Wilkens phase. Modified from Ref. 19.

The derivation of the quantum topological phase acquired by a magnetic dipole carries over with the obvious changes to the derivation of the quantum topological phase acquired by an electric dipole. Nevertheless I quickly review the derivation based on the Dirac equation.

For a neutral spin half particle with an electric dipole moment μ_e the Dirac equation is

$$\left(i\gamma^\mu \partial_\mu + \frac{1}{2}\mu_e \sigma^{\mu\nu} \gamma_5 F^{\mu\nu} - m\right)\psi = 0. \tag{21}$$

Using the relationship that

$$-iF_{\mu\nu}\sigma^{\mu\nu}\gamma_5 = \tilde{F}_{\mu\nu}\sigma^{\mu\nu}, \quad \text{where } \tilde{F}_{\mu\nu} \equiv \frac{1}{2}\epsilon_{\mu\nu\alpha\beta}F^{\alpha\beta} \tag{22}$$

is the $(3+1)$-dimensional dual of the electromagnetic field tensor, in which the electric and magnetic fields are interchanged with respect to their positions in $F^{\mu\nu}$, the Dirac equation (21) may be written as

$$\left(i\gamma^\mu\partial_\mu + i\frac{1}{2}\mu_e\sigma^{\mu\nu}\tilde{F}^{\mu\nu} - m\right)\psi = 0, \tag{23}$$

which is just Eq. (9) with $-\tilde{F}^{\mu\nu}$ replacing $F^{\mu\nu}$.

As before, the $(2+1)$-dimensional Dirac equation can be rewritten as

$$\left(i\gamma^\mu[\partial_\mu + i\mu_e T_\mu] - m\right)\psi = 0. \tag{24}$$

The "effective vector potential" T_μ is the $(2+1)$-dimensional dual of the $(3+1)$-dimensional dual $\tilde{F}^{\alpha\beta}$ of the electromagnetic field strength tensor $F^{\mu\nu}$,

$$T_\mu = (1/2)\epsilon_{\mu\alpha\beta}\tilde{F}^{\alpha\beta}. \tag{25}$$

In the HMW configuration, the electric field vanishes and B_1, B_2 are constant in time. Then $T_\mu = (0, \mathbf{T}) = (0, T_1, T_2) = (0, B_2, -B_1) = (0, \mathbf{B} \times \mathbf{k})$, where $\mathbf{k}$ is a unit vector in the z direction, i.e. the direction of the electric moment.

Making a transformation

$$\psi' = \exp\left[-is\mu_e \int^r \mathbf{T}(\mathbf{r}') \cdot d\mathbf{s}'\right]\psi \tag{26}$$

in Eq. (24), one finds that ψ' satisfies the free Dirac equation

$$(i\gamma^\mu\partial_\mu - m)\psi' = 0. \tag{27}$$

You see that the reduction of the problem to $2+1$ dimensions was a critical step towards this result, because the electric moment–electromagnetic field term in the Dirac equation can be converted into a γ^μ interaction with the vector dual of the tensor electromagnetic field. In $3+1$ dimensions the dual of the tensor electromagnetic field is a tensor and the dual of the $\sigma_{\mu\nu}$ tensor is also a tensor. We do not have the ability to transform the Dirac equation for an electric moment interacting with the dual electromagnetic field into the vector current interaction with an effective vector potential field, but we need to do that to be able to make the phase transformation to convert the wave function to one satisfying the free field equation.

However in $2+1$ dimensions we can make the phase transformation of Eq. (26) to recover the free Dirac equation and we now have a topological phase as long as

(1) curl $\mathbf{T} = 0$ in the interference region,
(2) curl $\mathbf{T} \neq 0$ in the excluded region.

As curl T = curl $B \times k = k(\text{div } B) - (k \cdot \nabla)B$, the simplest way for the excluded region to generate a nonvanishing contribution to the phase is for it to contain some magnetic charges, giving rise to div $\mathbf{B} \neq 0$. To preserve the $(2+1)$-dimensional geometry, the magnetic charges should be extended uniformly and infinitely in the z direction. The simplest such charge configuration, that chosen by He and McKellar, is a line of magnetic monopoles on the z axis, with a linear magnetic monopole charge density λ_m. In this configuration, $\partial_z \mathbf{B} = 0$, so $(\mathbf{k} \cdot \nabla)B = 0$.

Then the phase developed in the wave function when the particle travels along a closed path $\mathcal{C} = \partial \mathcal{S}$ which encircles the line of magnetic charge with a linear monopole density λ_m once is

$$\chi_{\text{HMW}} = s\mu_e \oint_\mathcal{C} T \cdot dr = -s\mu_e \int_\mathcal{S} (\nabla \cdot B) k \cdot d\mathcal{S} = -s\mu_e \lambda_m, \qquad (28)$$

as found by He and McKellar.

It is clear that, since no magnetic monopoles have yet been found this manifestation of the HMW phase is not capable of experimental observation.

The HMW system with magnetic monopoles as the source of the magnetic field, no electric field, electric dipoles and no electric charges is the precise electromagnetic dual of the AC system, which has electric charges as the source of the electric field, no magnetic field, magnetic dipoles and no magnetic monopoles. It is important to emphasize that the calculations are mathematically consistent.

To see this in detail look at the electromagnetic duality transformation.

The concept of electromagnetic duality is described by Jackson.[28] Consider Maxwell's equations and the Lorentz force equation, extended to include magnetic monopoles:

$$\nabla \cdot \mathbf{D} = \rho_e, \quad \nabla \times \mathbf{H} = \frac{\partial \mathbf{D}}{dt} + \mathbf{J}_e,$$

$$\nabla \cdot \mathbf{B} = \rho_m, \quad \nabla \times \mathbf{E} = \frac{\partial \mathbf{B}}{dt} + \mathbf{J}_m, \qquad (29)$$

$$\mathbf{F} = q_e(\mathbf{E} + \mathbf{v} \times \mathbf{B}) + q_m(\mathbf{H} - \mathbf{v} \times \mathbf{D}).$$

The source particles moving with velocity $\mathbf{v}$ carry both an electric and a magnetic charge. Normal Maxwellian electrodynamics is the case that $q_m = 0$, $\rho_m = 0$ and $\mathbf{J}_m = 0$. The duality transformation

$$\mathbf{E} = \mathbf{E}' \cos\xi + Z_0 \mathbf{H}' \sin\xi, \quad Z_0 \mathbf{D} = Z_0 \mathbf{D}' \cos\xi + \mathbf{B}' \sin\xi,$$

$$Z_0 \mathbf{H} = -\mathbf{E}' \sin\xi + \mathbf{B}' \cos\xi, \quad \mathbf{B} = -Z_0 \mathbf{D}' \sin\xi + \mathbf{B}' \cos\xi, \qquad (30)$$

$$Z_0 q_e = Z_0 q_e' \cos\xi + q_m' \sin\xi, \quad q_m = -Z_0 q_e' \sin\xi + q_m' \cos\xi,$$

retains the form of the equations, transforming them to

$$\nabla \cdot \mathbf{D}' = \rho'_e, \quad \nabla \times \mathbf{H}' = \frac{\partial \mathbf{D}'}{dt} + \mathbf{J}'_e,$$

$$\nabla \cdot \mathbf{B}' = \rho'_m, \quad \nabla \times \mathbf{E}' = \frac{\partial \mathbf{B}'}{dt} + \mathbf{J}'_m, \quad \text{and} \qquad (31)$$

$$\mathbf{F} = q'_e(\mathbf{E}' + \mathbf{v} \times \mathbf{B}') + q'_m(\mathbf{H}' - \mathbf{v} \times \mathbf{D}').$$

The choice $\xi = \pi/2$ transforms electric charges into magnetic monopoles, magnetic dipole moments into electric dipole moments, magnetic fields into electric fields and electric fields into magnetic fields. This is just the transformation we need to transform the AC effect into the HMW effect. The equations of normal Maxwellian electrodynamics are transformed to

$$\nabla \cdot \mathbf{D}' = 0, \quad \nabla \times \mathbf{H}' = \frac{\partial \mathbf{D}'}{dt},$$

$$\nabla \cdot \mathbf{B}' = \rho'_m, \quad \nabla \times \mathbf{E}' = \frac{\partial \mathbf{B}'}{dt} + \mathbf{J}'_m, \quad \text{and} \quad \mathbf{F} = q'_m(\mathbf{H}' - \mathbf{v} \times \mathbf{D}'). \qquad (32)$$

Equations (32) lead to the HWM phase and show that the calculations are mathematically consistent.

As an amusing aside note that, if all particles have the same ratio of electric charge to magnetic charge, then the general equations (29) can be converted by duality transformations to either the usual Maxwell equations or the magnetic monopole form of (32). In this sense our decision to describe the world in terms of electric charges and currents is purely an historical happenstance.

The alternative, independent, derivation by Wilkens[9] relied on the effective electric field[d] $\mathbf{E}_R = \mathbf{v} \times \mathbf{B}$ felt by the moving dipole in the magnetic field, but Wilkens also suggested a line of magnetic dipoles as the source of the magnetic field, and proposed possible approximate realizations of this concept.

4.2.1. An induced electric dipole

Wei, Han and Wei[10] pointed out that a practical realization of the HWM effect would require an electric field to induce an electric dipole in a neutral atom. That is indeed how the HMW phase was measured. One may think it would be possible to avoid the electric field by using a molecule with an intrinsic electric dipole moment. However a beam of polarized molecules would rapidly depolarize in the absence of an electric field to maintain the alignment of the dipole. In either case a strong electric field is necessary for the realization of the HMW phase. It is an important consequence of Wei, Han and Wei that the electric field changes the geometry, and allows a realization of the HMW effect <u>without</u> a region in which div $\mathbf{B} \neq 0$.

---

[d]This effective electric field felt by a charge moving in a magnetic field was introduced by Röntgen, and is called the Röntgen field. I therefore use the subscript R for it.

Now the electric field felt by the moving atom is the sum of the applied field $\mathbf{E}$ and the Röntgen field $\mathbf{v} \times \mathbf{B}$, and so the induced dipole is $\mathbf{d} = \alpha(\mathbf{E} + \mathbf{v} \times \mathbf{B})$, where α is the electric polarizability of the atom.

The Lagrangian is

$$\mathcal{L} = \frac{1}{2}m\mathbf{v}^2 + \frac{1}{2}\alpha(\mathbf{E} + \mathbf{v} \times \mathbf{B})^2 \,. \tag{33}$$

Working nonrelativistically the Schrödinger equation becomes

$$\frac{1}{2m}(-i\nabla - \alpha(\mathbf{B} \times \mathbf{E}))^2 \psi = 0 \,, \tag{34}$$

after neglecting terms $\alpha \mathbf{E}^2$, $\alpha \mathbf{B}^2$, and for $\mathbf{v} \cdot \mathbf{B} = 0$.

Now it is clear that a phase factor

$$\exp\left(-i\alpha \int_P^\mathbf{r} \mathbf{B} \times \mathbf{E} \cdot \mathbf{ds}\right) \tag{35}$$

will convert the solution of the Schrödinger equation in the presence of the fields to the free Schrödinger equation, in the same approximation.

The phase

$$\chi_{\text{HMW}} = \alpha \int_\mathcal{C} \mathbf{B} \times \mathbf{E} \cdot \mathbf{ds} \tag{36}$$

is topological if

$$\text{curl}(\mathbf{B} \times \mathbf{E}) = \mathbf{B}\,\text{div}\,\mathbf{E} - \mathbf{E}\,\text{div}\,\mathbf{B} + (\mathbf{B} \cdot \nabla)\mathbf{E} - (\mathbf{E} \cdot \nabla)\mathbf{B} \tag{37}$$

vanishes in the interference region and is nonzero in the excluded region. Now electric charges, which give div $\mathbf{E} = \mathbf{0}$ can generate the HMW topological phase. Working with an induced dipole not only changes the topology, it also removes the need to have magnetic monopoles as sources of $\mathbf{B} \times \mathbf{E}$.

The condition $\mathbf{v} \perp \mathbf{E}$ does not appear in this derivation, although the condition $\mathbf{v} \perp \mathbf{B}$ does. However from the relativistic equivalent of the Lagrangian of an electrically polarizable particle moving in electric and magnetic fields, one finds that a term $-\alpha(\mathbf{v} \cdot \mathbf{E})^2$ is missing from the Lagrangian of Eq. (33). With this term in place the demonstration of the phase factor of Eq. (35) reduces the wave function in the presence of fields to that without fields requires both

$$\mathbf{v} \cdot \mathbf{E} = 0 \quad \underline{\text{and}} \quad \mathbf{v} \cdot \mathbf{B} = 0 \,. \tag{38}$$

The relativistic discussion of polarizable materials has it origin in the famous 1908 paper of Minkowski.[11] There are accessible accounts in Pauli[12] and Møller[13] and Becker and Sauter.[14] This subject is nowadays not often treated in courses on electromagnetism. For example, there is no relativistic discussion of polarizable materials in Jackson's *Electrodynamics*,[28] and so I go into a little of the detail here.

Minkowski's proposal is that the relativistic version of **D** and **H** is the tensor $G_{\mu\nu}$ obtained by replacing **E** and **B** in $F_{\mu\nu}$ by **D** and **H**. The Lagrangian density is then proportional to $-G_{\mu\nu}F^{\mu\nu}$. As

$$\mathbf{D} = \mathbf{E} + \mathbf{P}$$

and

$$\mathbf{H} = \mathbf{B} - \mathbf{M},$$

the relativistic description of the electric and magnetic moments is a tensor (which Becker and Sauter[14] call the moments tensor) $K_{\mu\nu}$ constructed from $F_{\mu\nu}$ by replacing **E** with **P** and **B** with $-\mathbf{M}$. Then $G_{\mu\nu} = F_{\mu\nu} + K_{\mu\nu}$ and the interaction Lagrangian involving the moments is then

$$\mathcal{L}_{\text{int}} = -\frac{1}{4}K_{\mu\nu}F^{\mu\nu}. \tag{39}$$

For now I will ignore intrinsic moments which are proportional to the spin of the particle, and only consider induced moments, which are proportional to the applied fields. We need the generalization of

$$\mathbf{P} = \alpha\mathbf{E}, \quad \text{and} \quad \mathbf{M} = \chi\mathbf{B}, \tag{40}$$

which hold in the rest frame of the material. α is the electric polarizability and χ is the magnetic susceptibility. Following Minkowski I write

$$u^{\mu}K_{\mu\nu} = \alpha u^{\mu}F_{\mu\nu} \quad \text{and} \quad u^{\mu}\tilde{K}_{\mu\nu} = \chi u^{\mu}\tilde{F}_{\mu\nu}, \tag{41}$$

which is identical to Eq. (40) in the rest frame, and is a tensor equation, so it is the correct generalization.

Equation (VI 58) in Ref. 13:

$$K_{\mu\nu} = u_{\mu}K_{\nu\lambda}u^{\lambda} - u_{\nu}K_{\mu\lambda}u^{\lambda} + \epsilon_{\mu\nu\kappa\lambda}\tilde{K}^{\kappa\sigma}u_{\sigma}u_{\lambda} \tag{42}$$

shows how to construct $K_{\mu\mu}$ from $u^{\mu}K_{\mu\nu}$ and $u^{\mu}\tilde{K}_{\mu\nu}$. The result is

$$K_{\mu\nu} = \alpha\{u_{\mu}F_{\nu\lambda}u^{\lambda} - u_{\nu}F_{\mu\lambda}u^{\lambda}\} + \chi\epsilon_{\mu\nu\kappa\lambda}\tilde{F}^{\kappa\sigma}u_{\sigma}u^{\lambda}. \tag{43}$$

Consider only the induced electric dipole moment, and set $\chi = 0$. Then with the auxiliary field 4-vector

$$F^{\mu} = F^{\mu\nu}u_{\nu} = \gamma(\mathbf{E}\cdot\mathbf{v}, (\mathbf{E} + \mathbf{v}\times\mathbf{B})), \tag{44}$$

$$K_{\mu\nu} = \alpha\{u_{\mu}F_{\nu} - u_{\nu}F_{\mu}\}, \tag{45}$$

the interaction Lagrangian is

$$\mathcal{L}_{\text{int}} = -\frac{1}{4}K^{\mu\nu}F_{\mu\nu} = -\frac{1}{2}\alpha F_{\mu}F^{\mu} \tag{46}$$

$$= \frac{1}{2}\alpha\gamma^2\{(\mathbf{E} + \mathbf{v}\times\mathbf{B})^2 - (\mathbf{E}\cdot\mathbf{v})^2\}. \tag{47}$$

Clearly, when $\mathbf{v} = 0$, $F^{\mu} = (0, \mathbf{E})$, which is just the electric field in this reference frame. In a frame in which $\mathbf{v} \neq 0$, one can apply the traditional definition of the electric field is a material body, to see that $q\mathbf{E}$ is the force on a charge q is a longitudinal crevasse cut into the matter, and qF^μ is the equivalent 4-force on the charge. qF^i is the force on the test charge in the crevasse in the moving frame, and qF^0 is the energy acquired by the particle.

To $O(v^2)$ the first term on the r.h.s. of Eq. (47) is the interaction Lagrangian of Wei, Han and Wei, but the $\alpha(\mathbf{E}\cdot\mathbf{v})^2$ term is missing from their Lagrangian. To get the HMW phase not only must $\mathbf{B}\cdot\mathbf{v}=0$, as they require, but also $\mathbf{E}\cdot\mathbf{v}=0$. This second geometrical constraint is missing in their analysis, but it is satisfied in their example and in the Toulouse experiment. We may have escaped the restriction to 2+1 dimensions in the nonrelativistic limit, but there are still geometric constraints.

In the AB effect there is no field and *a fortiori* no force on the charge in the interference region. In the HWM effect, as realized with an induced electric dipole, there are clearly both electric magnetic fields. Just as for the AC effect there is no force on the dipole. A possible force comes from the interaction of the induced electric dipole $\mathbf{d}=\alpha(\mathbf{E}+\mathbf{v}\times\mathbf{B})$ with the sum of the applied and the Róntgen electric fields, $\mathbf{E}+\mathbf{v}\times\mathbf{B}$, so the force on the dipole is

$$\mathbf{F} = -\nabla\{\mathbf{d}\cdot(\mathbf{E}+\mathbf{v}\times\mathbf{B})\} = \alpha\nabla(\mathbf{E}+\mathbf{v}\times\mathbf{B})^2. \qquad (48)$$

When $|\mathbf{E}+\mathbf{v}\times\mathbf{B}|$ is constant there is no force on the electric dipole. That is a severe constraint on the experimental realization of the HMW effect, requiring uniformity of both fields, and a constant velocity. There is already the constraint that $\mathbf{v}\cdot\mathbf{E}=\mathbf{v}\cdot\mathbf{B}=0$, and with the constancy of $\mathbf{E}$, $\mathbf{B}$ and $\mathbf{v}$ the system is highly constrained if a topological HMW phase is to be produced.

One should also consider whether or not the dipole will experience a torque. The torque will be

$$\boldsymbol{T} = \boldsymbol{d}\times(\boldsymbol{E}+\boldsymbol{v}\times\boldsymbol{B}) = \alpha(\boldsymbol{E}+\boldsymbol{v}\times\boldsymbol{B})\times(\boldsymbol{E}+\boldsymbol{v}\times\boldsymbol{B}) = 0 \qquad (49)$$

which vanishes for the induced dipole. The induced dipole is always parallel to the effective electric field, and thus experiences no torque.

To summarize: the HMW phase for an induced electric dipole, as given in Eq. (36) is a topological phase when

- $\mathbf{v}\perp\mathbf{B}$ and $\mathbf{v}\perp\mathbf{E}$,
- curl($\mathbf{B}\times\mathbf{E}$) vanishes in the interference region, and
- curl($\mathbf{B}\times\mathbf{E}$) is nonzero in the excluded region.

Moreover although the induced electric dipole is in electric and magnetic fields it experiences no torque, and, if the effective electric field ($\mathbf{E}+\mathbf{v}\times\mathbf{B}$) is constant in magnitude, it experiences no force.

4.2.2. *Experimental observations of the HMW phase*

In 2012, the Toulouse group led by Jacques Vigué succeeded in measuring the HMW phase[15–17] using an induced electric dipole moment in ^{7}Li ions, in a geometry

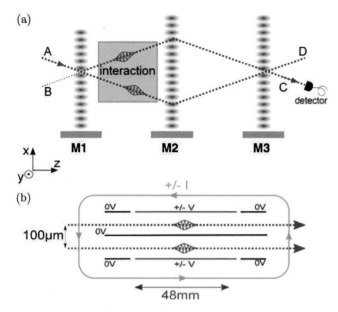

Fig. 7. The Toulouse experiment: (a) The atom interferometer, with two entrances A and B and two exits C and D (C is detected). An atomic beam (dotted lines) entering by A is diffracted by three quasiresonant laser standing waves produced by the mirrors Mi. The interaction region is placed where the distance between interferometer arms is largest, close to 100 μm. (b) The interaction region producing the electric and magnetic fields (not to scale — note the 100 μm vertical scale and the 48 mm horizontal scale). The interferometer arms (dotted lines) are separated by a septum, which is the common electrode of two plane capacitors producing opposite electric fields (high voltage electrodes labeled ± V; grounded electrodes labeled 0 V). Two rectangular coils (represented by a rectangle labeled ±I) produce the magnetic field, after Ref. 15.

which is a development of that proposed by Wei, Han and Wei. The experimental apparatus is summarized in Fig. 7.

The Toulouse group have taken great care to create uniform electric and magnetic fields, thus ensuring that there is no force on the atom. They analyzed very carefully the uniformity of their fields and the forces that may be felt by the induced dipole to confirm that they do not contribute to the observed HWM effect.[18] They have also verified that the measured phase is independent of the velocity of the atoms.

The electric dipole moment is induced by applying an electric field to ^{7}Li ions, and is in the plane of the path, not normal to it, and the magnetic field is normal to the plane of the path. However the dipole moment changes sign on the two sections of the path, and is not constant in direction. Were the central plate of the capacitor to shrink to a wire, the geometry would be just that of Wei, Han and Wei. The use of the double capacitor does not change the topology of the system but it both increases the magnitude of the possible the electric field, and increases the path over which the phase integral is performed. Both of these effects improve the observability of the phase.

The final result of this impressive experiment[16] is the observed phase for different ion velocities given in Table 1.

Table 1. Measured values of the HMW phase for different ion velocities, after Ref. 16. VI is given in VA.

Velocity in units ms^{-1}	Phase in units 10^{-6} rad VI
744 ± 18	1.41 ± 0.24
1062 ± 20	1.315 ± 0.071
1520 ± 38	1.270 ± 0.072

The phase is clearly independent of the velocity as it should be, and the weighted mean value is $\phi_{\text{HMW, obs}} = (1.29 \pm 0.10) \times 10^{-6}$ rad VI, to be compared to the calculated value $\phi_{\text{HMW, cal}} = (1.28 \pm 0.03) \times 10^{-6}$ rad VI. The agreement of the measured and calculated values is well within the errors, and there is no doubt the the HMW phase has been successfully observed.

5. Aharonov–Bohm Phases and Yang–Mills Fields

T. T. Wu and C. N. Yang[2] in 1975 provided the first discussion of AB phases in Yang–Mills fields. This was a very thorough investigation of the subject. The AB type of phase still exists for the Yang–Mills fields. The phase factor becomes

$$\Phi_{QP} = \mathcal{P}\left\{\exp\left(\int_Q^P -qA_\mu(x)dx^m u\right)\right\}. \qquad (50)$$

Now the "vector potential" $A_\mu(x)$ is a position dependent member of the group $\mathcal{G}$ of the field. It is clear that now $[A_\mu(x), A_{n\nu}(y)] \neq 0$ in general, i.e. the vector potentials at different points do not commute, and so the exponential defining the phase has to be path ordered as indicated by the symbol $\mathcal{P}$. The path joining Q to P is described by a curve $x^\mu = g^\mu(s)$ in terms of the parameter $s \in \{0, 1\}$. Path ordering of the exponential is then defined similarly to the familiar time ordering of field theory with the parameter s taking over the role of the time t.

They also proposed an experiment to test the isospin case, varying the AB experimental arrangement by replacing electrons with neutrons, and cylinder of magnetic field with a cylinder of neutron rich material, e.g. ^{238}U, as illustrated in Fig. 8(a). However it was recognized that the gauge field was massive and that this would create complications. Their comment is quoted in Fig. 8(b).

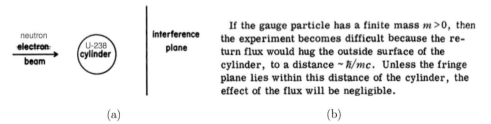

Fig. 8. (a) The Wu–Yang modification of the Aharonov–Bohm experiment for the non-Abelian isospin Yang–Mills field, after Ref. 2. (b) The Wu–Yang caveat concerning this experiment, from Ref. 2.

This observation did not quite close the investigations of Aharonov and Bohm with massive gauge fields. In simple cases the vector potential can be obtained by analytic solutions of the differential equation. Mindful of the need to keep the test particle close to the cylinder, when $m_g^{-1} \ll \ell$ where ℓ is the dimensions of the apparatus, Opat[29] discussed a variant of the Wu–Yang experiment in which the neutron beam passes through the rotating cylinder and showed that the existing experiment of Klein et al.[30] placed a limit of

$$m_g/\sqrt{\alpha_g} > 2.8 \text{ KeV}. \tag{51}$$

Unfortunately this is not a particularly interesting limit, and the experiment has not been improved.

5.1. *Using AB to measure the photon mass*

By considering the Maxwellian case where for $m_g^{-1} \gg \ell$ where ℓ is the dimensions of the apparatus it is possible to put a limit on the mass of the photon. This analysis was done by Boulware and Deser,[31] and they estimate that the Chambers-type experiments place limits of

$$m_\gamma^{-1} > 100 \text{ km}, \quad m_\gamma < 2 \times 10^{-11} \text{ eV}. \tag{52}$$

In contrast the current PDG limit is $m_\gamma < 10^{-18}$ eV, derived from observations of the solar wind.[32] The AB limits of Boulware and Deser are not competitive with this, but they are done on a table top. Spavieri et al.[33] have considered how the AB, AC and HMW phases can be used to improve the mass limits, and suggest that it should be possible to reach or improve on the limit set by Ryutov.

5.2. *AB experiment as a test of Lorentz invariance*

It was shown by Kobakhidze and myself[34] that the AB experiment could be used to test Lorentz invariance. A typical Lagrangian with Lorentz noninvariant terms, characterized by the parameters κ_F and κ_{AF}, is

$$\mathcal{L} = -\frac{1}{4} F_{\mu\nu} F^{\mu\nu} - j_\mu A^\mu$$
$$- \frac{1}{4}(\kappa_F)_{\mu\nu\rho\sigma} F^{\mu\nu} F^{\rho\sigma} + \frac{1}{2}(\kappa_{AF})^\alpha \epsilon_{\alpha\mu\nu\rho} A^\mu F^{\nu\rho}. \tag{53}$$

Because of constraints on the other parameters, we considered only the case

$$(\kappa_{DB})^{ij} = (\kappa_F)^{0ipq} \epsilon^{jpq} \quad \text{and introduce} \quad \kappa^i = \frac{1}{2}\epsilon^{ijk}(\kappa_{DB})^{jk}. \tag{54}$$

From optical cavities[35] one obtains the limit $\kappa^i \lesssim 10^{-11}$.

We found that nontopological AB type experiments, with an improvement of 1–2 orders of magnitude on present sensitivity, would improve this limit. Unfortunately the corrections to the classic topological AB experiments are of $O(\kappa^2)$ and are far too small to be useful.

6. Concluding Remarks

Just as Yang and Mills started a rich vein of physics, so did Aharonov and Bohm. Both are still being mined. The interaction between the two has not been fully explored. For example, the duality of electric and magnetic fields in electromagnetism, breaks down for Yang–Mills fields, and Wu and Yang emphasized. This area is ripe for further theoretical work.

Acknowledgments

I express my thanks to my collaborators, Xiao-Gang He, Archil Kobakhidze, and Tony Klein, for their contributions to my understanding of this subject.

I thank Professor K. K. Phua and the organizers for the opportunity to present the talk which led to this paper.

I acknowledge the support of this work by the Australian Research Council through their support of the ARC Centre of Excellence for Particle Physics at the Terrascale (CoEPP) at the School of Physics of the University of Melbourne.

References

1. Y. Aharonov and D. Bohm, *Phys. Rev.* **115**, 485 (1959).
2. T. T. Wu and C. N. Yang, *Phys. Rev. D* **12**, 3845 (1975), doi:10.1103/PhysRevD.12.3845.
3. W. Ehrenberg and R. E. Siday, *Proc. Phys. Soc. B* **62**, 8 (1949).
4. R. G. Chambers, *Phys. Rev. Lett.* **5**, 3 (1960).
5. G. Möllenstedt and W. Bayh, *Naturwissenschaften* **49**, 81 (1962).
6. A. Tonomura *et al*, *Phys. Rev. Lett.* **56**, 792 (1986).
7. Y. Aharonov and A. Casher, *Phys. Rev. Lett.* **53**, 319 (1984).
8. X.-G. He and B. H. J. McKellar, *Phys. Rev. A* **47**, 3424 (1993).
9. M. Wilkens, *Phys. Rev. Lett.* **72**, 5 (1994).
10. H. Wei, R. Han and X. Wei, *Phys. Rev. Lett.* **75**, 2071 (1995), doi:10.1103/PhysRevLett.75.2071.
11. H. Minkowski, *Nachr. Ges. Wiss. Göttingen* 53 (1908), This paper is available on http://de.wikisource.org/wiki/Die_Grundgleichungen_fr_die_elektromagnetischen_Vorgnge_in_bewegten_Krpern and Saha's 1920 English translation is available on http://en.wikisource.org/wiki/Author:Meghnad_Saha.
12. W. Pauli, *Theory of Relativity* (Dover Publications, New York, 1981), translated from the German by G. Field.
13. C. Møller, *The Theory of Relativity* (Oxford University Press, London, 1952).
14. R. Becker and F. Sauter, *Electromagnetic Fields and Interactions, Vol. 1, Electromagnetic Theory and Relativity* (Blackie, London and Glasgow, 1964).
15. S. Lepoutre, A. Gauguet, G. Trénec, M. Büchner and J. Vigué, *Phys. Rev. Lett.* **109**, 120404 (2012).

16. J. Gillot, S. Lepoutre, A. Gauguet, M. Büchner and J. Vigué, *Phys. Rev. Lett.* **111**, 030401 (2013).
17. S. Lepoutre, J. Gillot, A. Gauguet, M. Büchner and J. Vigué, *Phys. Rev. A* **88**, 043628 (2013).
18. S. Lepoutre, A. Gauguet, M. Büchner and J. Vigué, *Phys. Rev. A* **88**, 043627 (2013).
19. J. P. Dowling, C. P. Williams and J. D. Franson, *Phys. Rev. Lett.* **83**, 2486 (1999).
20. C. R. Hagen, *Phys. Rev. Lett.* **64**, 2347 (1990).
21. X.-G. He and B. H. J. McKellar, *Phys. Lett. B* **256**, 250 (1991).
22. X.-G. He and B. H. J. McKellar, *Phys. Rev. A* **64**, 022012 (2001).
23. X.-G. He and B. H. J. McKellar, *Phys. Lett. B* **559**, 263 (2003).
24. K. Sangster, E. A. Hinds, S. M. Barnett, E. Rijs and A. G. Sinclair, *Phys. Rev. A* **51**, 1776 (1995).
25. R. C. Casella, *Phys. Rev. Lett.* **65**, 2217 (1990).
26. A. Cimmino, G. I. Opat, A. G. Klein, H. Kaiser, S. A. Werner, M. Arif and R. Clothier, *Phys. Rev. Lett.* **63**, 380 (1989).
27. J. Gillot, S. Lepoutre, A. Gauguet, J. Vigué and M. Büchner, *Eur. Phys. J. D* **68**, 168 (2014).
28. J. D. Jackson, *Electrodynamics* (John Wiley and Sons, New York, 1998), pp. 273–275.
29. G. I. Opat, A neutron interferometric search for strongly interacting gauge vector bosons (1985) (unpublished University of Melbourne Preprint UM-P 85/7).
30. A. G. Klein, G. I. Opat, A. Cimmino, W. Treimer, A. Zeilinger and R. Gahler, *Phys. Rev. Lett.* **46**, 1551 (1981).
31. D. G. Boulware and S. Deser, *Phys. Rev. Lett.* **63**, 2319 (1989), doi:10.1103/PhysRevLett.63.2319.
32. D. D. Ryutov, *Plasma Phys. Control. Fusion* **49**, B429 (2007), doi:10.1088/0741-3335/49/12B/S40.
33. G. Spavieri, J. Quintero, G. T. Gillies and M. Rodriguez, *Eur. Phys. J. D* **61**, 531 (2011), doi:10.1140/epjd/e2011-10508-7.
34. A. Kobakhidze and B. H. J. McKellar, *Phys. Rev. D* **76**, 093004 (2007).
35. P. Wolf, S. Bize, A. Clairon, G. Santarelli, M. E. Tobar and A. N. Luiten, *Phys. Rev. D* **70**, 051902 (2004), doi:10.1103/PhysRevD.70.051902.

Yang–Mills for Historians and Philosophers

R. P. Crease

Department of Philosophy, Stony Brook University, Stony Brook, NY 11794, USA
robert.crease@stonybrook.edu

The phrase "Yang–Mills" can be used (1) to refer to the specific theory proposed by Yang and Mills in 1954; or (2) as shorthand for any non-Abelian gauge theory. The 1954 version, physically speaking, had a famous show-stopping defect in the form of what might be called the "Pauli snag," or the requirement that, in the Lagrangian for non-Abelian gauge theory the mass term for the gauge field has to be zero. How, then, was it possible for (1) to turn into (2)? What unfolding sequence of events made this transition possible, and what does this evolution say about the nature of theories in physics? The transition between (1) and (2) illustrates what historians and philosophers a century from now might still find instructive and stimulating about the development of Yang–Mills theory.

Keywords: Yang–Mills; history of science; philosophy of science.

1. Introduction

When historians and philosophers look back on Yang–Mills theory a hundred years from now, what will they still find instructive and stimulating? They will be fascinated, I predict, by the peculiar fact that the phrase "Yang–Mills" can be used in two ways: (1) to refer to the specific theory proposed by Yang and Mills that was clearly wrong when it was published in October 1, 1954 issue of the *Physical Review* and (2) as shorthand for any non-Abelian gauge theory of the sort that has become indispensable for high-energy physics.

The 1954 version of Yang–Mills theory, physically speaking, had a famous show-stopping defect in the form of what might be called the "Pauli snag," or the requirement that in the Lagrangian for non-Abelian gauge theory the mass term for the gauge field has to be zero. How, then, was it possible for (1) to turn into (2)? What unfolding sequence of events made this transition possible, and what does this evolution say about the nature of theories in physics? Three elements of this transition, I suspect, will fascinate our future historians and philosophers. One is the plot: how Yang–Mills came together as a scientific tool. Historians love a great story on an important subject with complex twists. A second involves the dramatic changes in the physics workplace in which that tool was used; what it was like to be a theoretical physicist changed as the transition from (1) to (2) took place, and how much of this change was due to the development of Yang–Mills theory itself. A third involves what the Yang–Mills story reveals about the nature of theory and theory-making. Some discoveries not only contribute to science but can also tell us about science. The genesis of Yang–Mills is one.

2. Plot

First, the plot of the story of how Yang–Mills (1) turned into (2). I can only provide a brief and incomplete summary here; my point is not to tell the full story, much of which has been told elsewhere,[1] but to show what will fascinate our future historians. Several key background elements set the stage. One is Maxwell's formulation of electromagnetism, which was gauge-invariant although the full significance of this in classical theory went unrealized; it was viewed as more a convenient technical feature than a deep principle. Another is Einstein's work. He, with Minkowski, *did* realize the gauge invariance of Maxwell's equations and enlarged the concept of invariance. As C. N. Yang has written in several places, "Einstein initiated the principle that *symmetry dictates interactions*." Hermann Weyl tried to relate the coordinate invariance of gravitation, the gauge invariance of electromagnetism and the gauge invariance of differential geometry. In the process, Weyl not only introduced the term "gauge" which has stuck eversince, but proposed that it be used as a principle in the very existence of electromagnetic interactions. Gauge invariance is to conservation of electric charge, Weyl wrote in 1918, as coordinate invariance is to conservation of energy and momentum.

Weyl and Einstein then carried out an extended debate on whether Weyl's ambitions in this respect were possible, with Einstein arguing that they were not. This debate was tremendously stimulating and rivals in significance, I think, the much better known debate between Einstein and Bohr over quantum mechanics, about which many books have been written. Weyl's ideas made it into Pauli's paper on quantum theory in the *Handbuch der Physik*, which was published in 1926, and continued to be explored by Pauli into the early 1950s. Non-Abelian gauge theory was also explored by several others, including Klein, as O'Raifeartaigh discusses in his book on *The Dawning of Gauge Theory*.[1]

Enter Yang. As a graduate student in the 1940s, he had read Pauli's *Handbuch* paper, and his attention, as he writes in his *Selected Papers*, was grabbed by the proposal that gauge invariance *determined* electromagnetic interactions. He wondered whether the principle of gauge invariance could be moved to new contexts, specifically non-Abelian ones required to address strong interactions. As a graduate student he tried to generalize gauge invariance by associating it with isotopic spin interactions, found it "led to a mess" and gave up.

In 1953, Yang went to Brookhaven National Laboratory for a year, where one of his officemates was Robert Mills, then in his last year of getting his Ph.D. at Columbia University (their third officemate was Burton Richter). Theorists with incomplete ideas are like people with songs in their heads that they cannot identify; they cannot stop trying to place the tune; at Brookhaven, Yang recruited Mills in his attempt to place the tune. The two wondered whether isotopic spin, associated with an SU(2) global symmetry group, might provide a conserved quantity that could be converted into a local gauge symmetry. Just as in electromagnetism the phase of the wave function can be shifted arbitrarily in space and time because the interaction with the electromagnetic field will cancel out the effect of the alteration, so Yang and Mills proposed to do the same for isotopic spin, hypothesizing the existence of a "B field" to counteract the change. Just as the *raison d'être* of the electromagnetic field is to ensure the symmetry of the electromagnetic interactions with respect to local variations of the wave function phase, so the B field would maintain the gauge symmetry of strong interactions with respect to the orientation of isotopic spin. The invariance would determine the interaction. Yang and Mills finished what Yang, in his *Selected Papers*,[2] described as the "formal aspect" of the work in February 1954.

The historians of a hundred years hence will have read Yang's *Selected Papers*, and therefore will know the dramatic story of what happened next. That story, in fact, provides the focal point for what I think will be the interest of historians of the future in the development of Yang–Mills theories. At Oppenheimer's invitation, Yang went to Princeton to present the work at the Institute for Advanced Study seminar. Who should be present at the seminar, spending the year at the Institute, but Pauli. Pauli, who had been working on the issue as deeply as anyone but Weyl had identified the show-stopping issue. The future historian will know that this cranky perfectionist interrupted Yang demanding to be told the mass of the B field. Yang said he did not know and resumed the presentation. Pauli cut him off again with the same demand, to which Yang responded that he and Mills had looked at the matter but reached "no definite conclusions." Pauli remarked, "That is not sufficient excuse" in such a hostile way that Yang, distressed and uncertain, sat down. An awkward silence ensued, with the seminar effectively at a halt. Oppenheimer then said, "We should let Frank proceed." He did, but with the rest of the presentation having an awkward flavor in the shadow of Pauli's unanswered, and obviously all-important, question.

Pauli's question was on the money. He was channeling the voice of the quantum field theory of the day, embedded in a particular view of what nature looked like. In the Lagrangian for gauge theories both Abelian and non-Abelian, the mass term has to be zero. In the Abelian case for QED, this in fact provides our understanding of the zero mass of the photon. But Yang and Mills were out to build a non-Abelian theory, applicable to hadrons, where the gauge particles had to be massive. In light of the Pauli snag, and this view of nature's fundamental units, this ambition just could not get off the ground. The Yang–Mills work was wrong. It did not fit nature.

When Yang returned to Brookhaven, he and Mills decided to publish their work anyway. Why, if it was wrong? Bookmark this question for the moment; I'll return to it below. In their paper, which appeared in *Physical Review* later that year, they wrestled with the nature of the B quantum in the final section, which Yang wrote

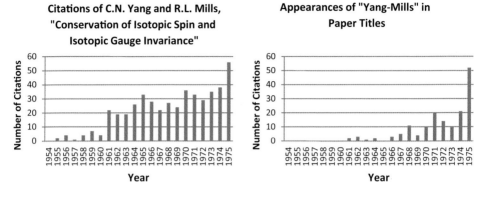

Fig. 1. Source: *Web of Science*.

was "more difficult to write than all earlier sections." In regard to its mass, the authors write, "we do not have a satisfactory answer."

Small wonder, then, that their work was initially regarded as only a mathematical curiosity. The Web of Science (Fig. 1) shows that the paper had only a handful of citations each year for the next few years; 1 citation in 1957, for instance, 4 in 1960. Another interesting metric to use is the times that "Yang–Mills" appears in the titles of papers, some of which may not even cite the work. According to the Web of Science, the first time this happens is 1961.

At this point the story effectively forks, splitting into two more or less parallel plot threads which recombine only about 1974, or two decades after the Yang–Mills theory first appeared. This is the interesting plot development that, I think, is sure to fascinate our historians.

One thread involves the weak interaction. Key events here include Glashow's 1961 paper, "Partial-Symmetries of Weak Interactions", which proposed an application of non-Abelian gauge theory to the electroweak interaction. He, too, encountered the Pauli snag, called it "the principal stumbling block in any pursuit of the analogy between hypothetical vector mesons and photons," but says "it is a stumbling block we must overlook." He also encountered the issue that the theory might not be renormalizable. Six years later, Steven Weinberg published "A Model of Leptons." Weinberg had been trying to make a Yang–Mills theory for various mesons, unsuccessfully. Then, one day while driving to his office at MIT, he suddenly realized that he might have been applying the right idea to the wrong problem: why not use the Yang–Mills mathematical apparatus for weak interactions and intermediate vector bosons, employing "spontaneous symmetry breaking" to give the gauge particles mass? Spontaneous symmetry breaking has its own plot line, proposed in 1964 by Englert and Brout, by Higgs, and by Hagen, Guralnik and Kibble, and its plot suggested the existence of a yet unseen, massive and spinless "Higgs boson", which was discovered experimentally almost half a century later. Weinberg brought these two plot lines together. "We do not usually expect non-Abelian gauge theories to be renormalizable if the vector-meson mass is not zero," Weinberg writes in the paper, but here the mesons in question "get their mass from the spontaneous breaking of their symmetry, not from a mass term put in at the beginning." But whether the theory *was* in fact renormalizable was not clear and as a result the paper was hardly cited for several years. (Salam, who had been following his own path with his long-time collaborator John Ward, wrote a similar proposal about the same time.)

Three years later, another key step was taken when Glashow, Iliopoulos and Maiani, in what was soon called the GIM mechanism, realized that adding an additional quark, bringing the total of quarks and leptons to four each, would solve many phenomenological and theoretical problems, reducing the divergences and eliminating neutral currents. In their first version of the paper they proclaimed that this theory might be renormalizable, a claim that was slapped down by a *Physical Review* referee. The authors then toned their claim down to the softer statement that a Yang–Mills theory "does not make the theory more divergent".[3] Though this was almost as unsupported as the original claim, *Physical Review* now accepted the paper, which was published in October 1970.

The success of all these attempts hung on their renormalizability. This path had been worked on by Faddeev and Popov and others, but finally answered in dramatic fashion at a conference in Amsterdam in August 1971, when Veltman introducing the work of 't Hooft, declared it made the theory "every bit as good as quantum electrodynamics." A non-Abelian gauge theory is renormalizable, 't Hooft showed, if and only if the mass term comes from spontaneous symmetry breaking. But 't Hooft's proofs, written in the language of Feynman diagrams, were difficult to understand; also, was the proposed theory true? Experimentally, confirmation shortly came from multiply confirmed observations of weak neutral currents. Theoretically, Ben Lee recast 't Hooft's work in the language of path integrals, which many found easier to understand and generalize.

In this weak interaction subplot, therefore, spontaneous symmetry breaking was the answer to the Pauli snag. With spontaneous symmetry breaking, you have a different world — one with more than one vacuum — where the Hamiltonian is invariant but the ground state is not. The world addressed by the new theory was a different one from the world toward which Pauli was directing his remark.

The strong interaction subplot was also winding. In this territory gauge theory had long seemed implausible. Until 1968, that is, when the results of deep inelastic electron–proton scattering at SLAC appeared to be describable in terms of scattering off point-like particles. Quarks were behaving not as mere mathematical entities but as if they played dynamical roles — free particles at short distances. This provoked theorists to look for a field theory that acted counter to those that were known, one in which the coupling gets weaker at short distances. After a dramatic scramble in which many theorists found key pieces and made close encounters, such an "asymptotically free" theory was found, with proofs (by Gross and Wilczek, and by Politzer) published back to back in the *Physical Review Letters* issue of November 15, 1973.

Here, too, the Pauli snag was overcome in a bizarre way. A local non-Abelian gauge theory turns out not to apply to the same conserved quantities as isospin does, but to a different world. Pauli's question was about things that today we'd call quarks and gluons. But the work on quantum chromodynamics showed that the relevant physical states are not quarks and gluons but color singlets. A non-Abelian gauge theory of the sort Yang and Mills were aiming to construct turned out to apply to entities different from the neutrons, protons and pions that had inspired their effort.

A historian knows, too, that a breakthrough does not necessarily become universally acknowledged and accepted at its publication. Its spread can be hindered by complicated mathematics and techniques that are difficult to generalize. Two key developments then ensured the spread of this theory: the work of Ken Wilson in 1974, which provided a nonperturbative formulation of gauge theory, making it easier to work with, and its further extension by Mike Creutz in 1979.

Not only do citations shoot up, but even more a sign of importance than citations is the rising use of the term "Yang–Mills" in the title of later papers, some of which, as mentioned, do not even bother to cite the original paper.

This, then, is a brief and incomplete sketch of what historian of a century from now will probably see as the rough plot outline of the way Yang–Mills went from (1) to (2), from a mathematical curiosity "known" not to apply to the world to an indispensable tool of theoretical physics. Again, my aim here is not to fill in all the plot twists, but to show that the story has two main subplots, each with tangled sub-subplots and many more figures and contributions than I have mentioned. Also fascinating is that the Pauli snag is resolved differently in each, with the solution being that the world turns out to be different from the one about which Pauli thought he was asking. But the end result was the creation of the loom on which modern gauge field theory is woven. The most remarkable aspect of the story, to paraphrase O'Raifeartaigh, is not that it took time, but that it came together at all.

3. Workplace

So that is the remarkable story of how the tool developed, which I said was the first aspect which will interest the historian of a hundred years hence. But historians are interested not only in the tools of science, but also the workplaces. I also suggested that another aspect will be the dramatic change in the workplace that took place in the time that tool developed. How did what it was like to be a physicist change during the period in which Yang–Mills theories developed, and how much of that was due to their development?

The early 1950s was the beginning of a new period for physics. The first accelerators to surpass 1 GeV were coming on line. No longer would experimenters have to climb mountains to hunt cosmic rays and strange particles; they could now create them conveniently and copiously in the safety of the laboratory.[4] The first of these was the Cosmotron at Brookhaven, but Berkeley's Bevatron was not far behind, and soon more appeared, in Argonne, Birmingham and Dubna — then, within a few years, a second generation based on alternating gradient focusing. The experimental workplace in particle physics was suddenly an exciting place.

With experimenters at these facilities discovering ever more particles, the early 1950s was also a thrilling time for theorists. What sense did it all make? What schemes could be devised to organize a particle equivalent of the periodic table? It was unclear which properties were the most important. It was a time of "educated theoretical guesses," as O'Raifeartaigh puts it. One notable feature of the theoretical workplace in the early 1950s was the segmentation of its theorists to different ethnicities, with those who worked on strong, weak, and electromagnetic interactions using different tools and speaking different mathematical languages. Electromagnetism had quantum electrodynamics (QED), which stemmed from the Dirac equation, an equation that Frank Wilczek has called "achingly beautiful." QED is a theory showing that particles of light (which we now call photons) arise naturally if you apply quantum mechanics to electromagnetism. The hope of quantum field theory was that you could treat other particles, such as electrons and protons, as quanta of generalizations of the electric field, shifting the basic structure of the theoretical framework from particles and waves to fields. QED, of course, includes the electron field as well as the photon field.

But while quantum electrodynamics scored a series of stunning successes in its first few years, by the 1950s many people regarded it as having lost its beauty. The wrinkles that spoiled this beauty were divergences that required using *ad hoc* procedures and introducing things by hand. Dirac himself called the theory "ugly." The theory did not flow from a single vision, he declared, in the way of truly beautiful things, but seemed stitched together in a way that made it seem fundamentally incomplete. Even though quantum field theory had some stunning successes — most notably its ability to predict with extraordinary accuracy the Lamb shift and the electron anomalous magnetic moment — the price one had to pay was having to "abandon logical deduction and replace it by working rules," Dirac said. "This is a very heavy price and no physicist should be content to pay it." Landau was a particular skeptic. So was Oppenheimer. Throughout the 1950s, Oppenheimer "struggled to maintain his disbelief" in QED, as Serber recalled. It had become a difficult language to use, with many people certain that it could not address all contemporary problems. It did not seem to apply to the strong interaction, which seemed amenable to other languages such as S-matrix theory. Even in 1956, when parity nonconservation was observed, it was not completely clear that there even might be a unified weak interaction.

At the beginning of the development of Yang–Mills theory, therefore, the theoretical workshop was a kind of Babel. At the other end — just 20 years after the Yang–Mills paper — the structural environment had changed. Yang–Mills theory had established field theory as the dominant theoretical language and unified the mathematical Babel. It also had some longer term structural effects: its development made possible the beginning of the time when looking at constraints at low energies could allow you to make predictions at extremely high energies.

The current form of such a theory is the "Standard Model" of electroweak and strong interactions, but that model by itself does not appear to be complete for energies much higher than the TeV range now being explored at the CERN Large Hadron Collider. Thus there *has* been a successful extrapolation of 2 to 3 orders of magnitude in energy, a remarkable achievement but still far from the end of the story. There are hopes of finding ways to complete the model, which have given rise to a new industry called BSM (beyond the Standard Model) physics. Many of these schemes involve generalizations of Yang–Mills theory, along with string theory. The latter started out from the old S-matrix theory, but by now has strong kinship with quantum field theory. So over time the landscape of high-energy or short-distance physics has been dramatically re-shaped by the introduction of Yang–Mills theory, which likely will remain a key part of future developments.

4. Theories

But there's yet a third aspect of the transition from (1) to (2) that is interesting to historians and philosophers, which has to do, not with how the tool or the workshop changed, but with what the episode reveals about theory-making. Pauli, after all, was not wrong; the theory proposed by Yang and Mills in 1954 did not and could not apply to the world. In terms of conventional philosophical notions about the nature of theory — that a theory is a hypothesis either to be confirmed and added to the store of knowledge, or disconfirmed and rejected — the Yang–Mills work was a nonstarter. How was it possible for such a clearly incorrect proposal to become such a seminal event in the history of physics? The fact that it did shows conventional philosophical notions are mistaken and points to a deeper story.

Let me explain by mentioning an analogous episode whose lesson is that discovery is sometimes an ambiguous and extended process. The episode, well-known to historians and philosophers of science, involves the discovery of oxygen. Who was the discoverer: Scheele, Priestley or Lavoisier? Scheele collected a gas that turns out to be oxygen by early 1774, but did not publish until after the work of the other two was well-known, so what he did played no role in the discovery. In late 1774, Priestley collected what turns out to be oxygen by heating red precipitate of mercury, but described it confusedly in obsolete terms as a previously known species of air, nitrous air or N_2O; he thought what he had was not new. Lavoisier, learning of Priestley's work and its confusions, gradually realized in 1776-7 that this gas isolated by Priestley and others was not an existing species but a distinct species itself. But he was only able to make this dawning realization because his assumptions had begun to change in what crystallized as the combustion theory — that burning bodies absorb something that's pure and a component of air. The discovery of oxygen therefore did not take place in an instant but was protracted. It required the work of many people, an awareness of confusions in that work, a willingness on the part of people like Lavoisier and others not to accept the inherited understanding of nature and to allow themselves to respond to what they were finding in unorthodox ways — all this ending up in a revision of inherited assumptions that would, only at its conclusion, provide a clear idea of what was being emitted by heated red precipitate of mercury. The story of the discovery of oxygen, in short, reveals that the conventional notion of discovery as a process of hypothesis and confirmation was but a historian's shorthand leaving out key deeper aspects of scientific practice. The lesson was that discovery is sometimes not adding something new to the world, but reinterpreting the world anew in a way that allows something to be seen in it for the first time.

To put it another way, the scientist's path is sometimes blocked by inherited assumptions about nature — inherited assumptions are unavoidable — in a way that reduced the effectiveness of the process of hypothesis and deduction. In his *Lectures on Physics*, Feynman famously compared the world to a giant chess game played by the gods, with scientists being observers who are only allowed to watch and who try to guess the rules. But the world always appears to the onlookers through a lens fashioned by assumptions inherited from previous generations, assumptions that even tell them what the basic pieces are. Understanding the game better therefore may require more than making guesses about moves, but also revising the lens through which the game appears.

The Yang–Mills story contains a similar lesson about theory. When did the transition from (1) to (2) — from what in conventional terms is a theory that does not apply to the world to one that does — take place? It cannot be pinned to any specific date between 1954 and, say, 1975, but required an interpretive shift in understanding nature, a shift to which the Yang–Mills proposal — which exploited what its authors considered a possibly fruitful analogy between strong and electromagnetic theory rooted in gauge invariance — itself contributed. This does not fit the shorthand "hypothesis-confirmation" view of theoretical progress. We can try to "save" this shorthand view by saying things like, "First the Yang–Mills theory was wrong and then it was right," or "The Yang–Mills theory was right; it was the world that was wrong," but stating the assumptions so baldly reveals the shallowness of the shorthand view. The shorthand view fails to capture the actual practice of what it is like to be a theorist seeking to discover the rules of the game, a practice in which it can be valuable to explore analogies for their own sake. The experienced practitioners Yang and Mills knew that, which gave them the confidence to publish work that did not fit nature even in the face of the vociferous criticism of the eminent Pauli.

In this way the Yang–Mills story, in which changes in the lens through which particles were viewed had to take place before the theory worked, points to a deeper understanding of theory-making. Not only that, but the really interesting issue —

which makes this episode even more suggestive than the one involving oxygen — is that Yang–Mills proposal preceded the re-interpretation that changed the lens. The Yang–Mills proposal, that is, helped to usher in the re-interpretation of nature that made the proposal applicable to the world.

5. Conclusion

The Yang–Mills story suggests that theory-making is not always a question of seeking something provable and applicable to the world, but also of articulating a sense of the world that is not yet articulated theoretically. It can summarize and organize some pre-existing sense of the world (via, say, exploring a key analogy) that is not yet explicitly stated before proof or evidence (afterwards, we can say it was this way all along). This ability to envision nature other than the way it shows itself before proof or evidence is not something arbitrary or undisciplined — it does not involve simple belief — but involves the exercise of expertise. The 1954 Yang–Mills theory was a momentous step forward because it summarized and organized much of what had been learned about particles and field theory, pointing towards and making explicit the problems that needed to be solved, in order to account for the present. It laid out what would make it possible for the resources of quantum field theory to apply. The Yang–Mills story thus points to a way to revise conventional shorthand notions of the practice of science.

Acknowledgment

Thanks to Alfred S. Goldhaber, Robert Shrock, George Sterman, and Peter van Nieuwenhuizen for help.

References

1. L. O'Raifeartaigh, *The Dawning of Gauge Theory* (Princeton Univ. Press, 1997).
2. C. N. Yang, *Selected Papers, 1945–1980, with Commentary* (W. H. Freeman, 1983).
3. R. P. Crease and C. C. Mann, *The Second Creation: Makers of the Revolution in 20th Century Physics* (Macmillan, 1986), reprinted (Rutgers Univ. Press, 1996).
4. R. P. Crease, *Making Physics: A Biography of Brookhaven National Laboratory, 1946–1972* (University of Chicago Press, 1999).

Gauge Concepts in Theoretical Applied Physics

Seng Ghee Tan*,† and Mansoor B. A. Jalil‡

*Data Storage Institute, Agency for Science,
Technology and Research (A*STAR) 2 Fusionopolis Way,
#08-01 DSI, Innovis, Singapore 138634

‡Department of Electrical Engineering,
National University of Singapore, 4 Engineering
Drive 3, Singapore 117576
† Tan_Seng_Ghee@dsi.a-star.edu.sg

Gauge concept evolves in the course of nearly one century from Faraday's rather obscure electrotonic state of matter to the physically significant Yang–Mills that underpin today's standard model. As gauge theories improve, links are established with modern observations, e.g. in the Aharonov–Bohm effect, the Pancharatnam–Berry's phase, superconductivity, and quantum Hall effects. In this century, emergent gauge theory is formulated in numerous fields of applied physics like topological insulators, spintronics, and graphene. We will show in this article the application of gauge theory in two particularly useful spin-based phenomena, namely the spin orbit spin torque and the spin Hall effect. These are important fields of study in the engineering community due to great commercial interest in the technology of magnetic memory (MRAM), and magnetic field sensors. Both spin orbit torque and spin Hall perform magnetic switching at low power and high speed. Furthermore, spin Hall is also a promising source of pure spin current, as well as a reliable form of detection mechanism for the magnetic state of a material.

1. Introduction

Gauge is a theory of general physics which had evolved[1] in the course of nearly one century from Faraday's electrotonic state of matter, Maxwell's gauge freedom, Weyl's gauge invariance, to the physically significant Yang and Mills' non-commutative physics in the standard model. In the modern context, it is related to the Aharonov–Bohm effect, and the Pancharatnam–Berry's phase which has found strong links to useful electrical and optical phenomena like electrical conductance, localization and optical polarization. In condensed matter physics, gauge theory is an important concept in superconductivity as well as the quantum Hall effects. As recently as this century (21st), gauge concepts emerged with clear physical significance in the technologies of electronics, spintronics and photonics. The scale on which these technologies are pursued range from device relevant sizes of meso-scale, nano-scale, molecular, atomic, as well as the scale of individual quantum bit in the case of quantum information. These technologies can be realized in solid state materials from metal, semiconductor, carbon (graphene), to insulators (topological).

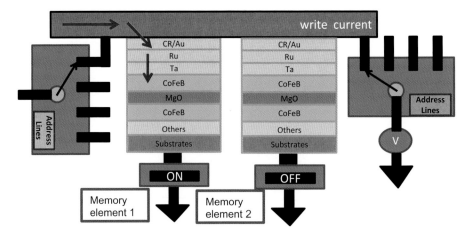

Fig. 1. The lateral view of an array of memory elements integrated in the device of magnetic memory.

In this paper, we will discuss the physics of spin orbit coupling (SOC) that exists in a wide range of material system in the context of the non-Abelian gauge concept.[2,3] We discuss its relation to the classical notion of forces and velocities,[4,5] and its measurement in technologically relevant parameters of conductivity and voltage. We will give a specific introduction to a physical phenomenon known as the "spin orbit torque", which is gaining popularity in the engineering physics community working on the magnetic non-volatile memory. Shown in Fig. 1 is a repeating part of an array formation that integrates many memory elements in one memory device.

We will also introduce the relevance of gauge theory to a rather well-known physical phenomenon known as the "spin Hall effect" (SHE), which can be understood as the spin version of the classical Hall effect. We briefly describe the treatment of SHE using the gauge theoretic approach, and show that gauge theory brings upon necessary corrections to SHE conductivity. Figure 2 is a summary of how the spin orbit gauge alongside a local frame transformation gauge that we call the Murakami–Fujita here can be related to the notion of velocity and force. This would lead to understanding of how these gauge potentials are further linked to physical phenomena in modern devices like the spin torque, spin Hall, anomalous Hall, magneto-resistance, and more. All of these phenomena have found applications in magnetic switching, microwave oscillation, magnetic field sensing and have been implemented in electronic or spintronic devices like MRAM, spin transistor, disk drive recording head.

2. Classical Notion of Forces and Velocities

Spin orbit torque and spin Hall effect, briefly introduce above, can be described in classical and intuitive languages, respectively, as some kind of force and velocity

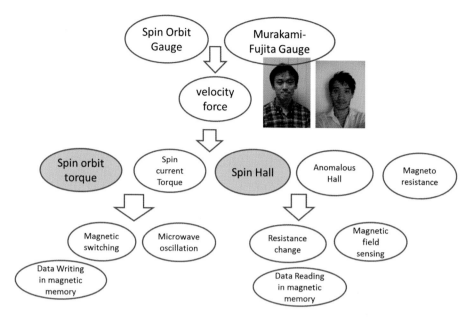

Fig. 2. Introduction to the idea of gauge fields and their relationship to physical phenomena as well as modern nano-scale devices.

associated with the spin of a charge carrier. Spin carrier refers simply to a charged particle (electron) with specific spin quantization axis defined by magnetic field, real or effective. As a prelude, we will briefly introduce the relationship of the gauge field to velocity and force, in which the spin orbit coupling and local magnetic moment are formulated as an emergent gauge field. This emergent field is then found to be irrevocably linked to the velocity and the force of a spin carrier.[4,5] A quick extrapolation leads to its linkage with the spin torque and the spin Hall. For simplicity, we will borrow a simple spin orbit system to illustrate this subtle relationship. A simple spin orbit system which is common in applied physics is the Rashba system that produces a linear form of SOC either in a semiconductor GaAs-based heterostructure[6,7] with inversion asymmetry at the interface, or in bulk material[8–10] with internal inversion asymmetry, e.g. bulk BiTeI, BiTeI/Bi2Te3 heterostructure. A linear SOC system is given by

$$H = \frac{p^2}{2m} + \lambda \sigma \cdot p \times z + eE \cdot r \qquad (1)$$

where one could identify a spin orbit gauge potential of $A = \frac{m\lambda}{e}\sigma \times z = \frac{m\lambda}{e}(\sigma_y, -\sigma_x, 0)$. One can then derive with the Heisenberg method via $f = \frac{dv}{dt} = [v, H]$, a quantity f_i^{Heis} that can be loosely described as a force. This quantity can be compared to another similar force-like quantity f_i^{YM} that one conjectures in parody of the classical Lorentz force picture of $f_i^{\text{Lorentz}} = ev_j B_k \varepsilon_{ijk}$ in classical Hall

Table 1. A summary of the derivation process for f_i^{YM} as inspired by the classical Lorentz force picture in classical Hall effect.

	Velocity	Spin Orbit Magnetic Field	Yang–Mills force
Classical	$v_j = \frac{p_j}{m} + A_j$	$B_k^{YM} = -\frac{ie}{\hbar}[A_i, A_j]\varepsilon_{ijk}$	$f_i^{YM} = -\frac{ie^2}{m\hbar}(p_j + eA_j)[A_i, A_j]$

effect. Both quantities are shown below as

$$f_i^{\text{Heis}} = -\frac{ie^2}{m\hbar}p_j[A_i, A_j]; \quad f_i^{YM} = -\frac{ie^2}{m\hbar}(p_j + eA_j)[A_i, A_j]. \quad (2)$$

These forces are spin forces in matrix form. Table 1 summarizes the process of deriving f_i^{YM} through comparison and relating velocity, force and magnetic field in a Lorentz manner.

While the Heisenberg spin force is prescribed by quantum mechanics, the Yang–Mills spin force is predominantly inspired by the Lorentz picture. Similarity between f_i^{Heis} and f_i^{YM} leads one to deduce that a spin Lorentz force description may be in order for f_i^{Heis} as well, i.e. f_i^{Heis} is a spin orbit version of the Lorentz force. In greater clarity, one uses the simple gauge $\boldsymbol{A} = \frac{m\lambda}{e}(\sigma_y, -\sigma_x, 0)$ to produce $\boldsymbol{f}^{\text{Heis}} = \frac{m\lambda^2}{\hbar}\langle\{\sigma_z, p_y\}i - \{\sigma_z, p_x\}j\rangle$ or $\boldsymbol{f}^{\text{Heis}} = \sigma_z \boldsymbol{p} \times \left(\frac{2m\lambda^2}{\hbar}\right)\boldsymbol{z}$. Since in a two-dimensional spin orbit system, spin current is given by $\boldsymbol{j}^z = \frac{\hbar}{4m}\{\boldsymbol{p} + e\boldsymbol{A}, \sigma_z\} = \frac{\hbar}{2m}\boldsymbol{p}\sigma_z$, one is led to the simple spin Lorentz relation of

$$\boldsymbol{f}^{\text{Heis}} = \boldsymbol{j}^z \times \left(\frac{2m\lambda}{\hbar}\right)^2 \boldsymbol{z} \quad (3)$$

which relates the spin current (current of a spin carrier) to the spin force. The above shows that the spin transverse current (a measurement of SHE) is related to the spin longitudinal force. Spin transverse current will lead to the accumulation of spin density on the lateral edges of the device.

3. Spin Orbit Spin Torque

The field of spin orbit torque owes its rise to the commercial need for memory devices of higher density that operate at higher speed, and lower power. Magnetic memory, alongside the optical phase-change memory are two promising options of non-volatile memory in the form of solid state device, that could replace the cumbersome, and bulkier hard disk drive. The progression of magnetic moment over time follows a trajectory that can be described by the Landau–Lifshitz–Gilbert (LLG) equation. The LLG tracks its departure from one stable energy state, and its evolution with constant strength over a spherical magnetic space, to its eventual unification with the other stable energy state. This process is known as magnetic switching, and the two stable energy states represent the binary electrical states for computer and memory use. The process of switching can be triggered by magnetic field, exchange force, and under the particular circumstance of SOC, the spin orbit torque.

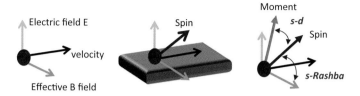

Fig. 3. A cartoon strip showing the spin orbit coupling effect on the moving electron and the local magnetic moment.

The RSOC is used to illustrate the generation and the operation of the spin orbit torque, which was first derived in the field-like form[11–13] and experimentally verified to display field-like behavior.[14] SOC is seen by a moving particle with momentum and charge, as an effective magnetic field. The spin of the particle aligns towards the effective magnetic field. As spin is also locked via s-d coupling to the magnetic moment, one imagines the magnetic moment being dragged along the spin trajectory towards an eventual unification with the spin and the effective magnetic field. Figure 3 provides a schematic illustration.

This is a loose physical illustration, but nonetheless a necessary mental process to inspire theoretical derivation along such thinking. The theoretical framework used here is one that involves the concept of spin orbit gauge field. In the presence of magnetic moment with smooth spatial variation, a local gauge transformation is performed. This transformation process aligns the reference spin axis (z) to the local moment, producing two gauge field terms in the Hamiltonian

$$H = \sum_{\mu}^{3} \frac{1}{2m} \left(p_\mu + e \left[\alpha U E_i \sigma_j \varepsilon_{ij\mu} U^\dagger - \frac{i\hbar}{e} U \partial_\mu U^\dagger \right] \right)^2 + \frac{eg\hbar}{4m} \sigma_z^r |M_z^r|. \quad (4)$$

The gauge field terms are the: (1) Spin orbit gauge, $A_\mu^{SO} = \alpha U E_i \sigma_j \varepsilon_{ij\mu} U^\dagger$ and (2) Chiral gauge of $A_\mu^{Ch} = \frac{-i\hbar}{e} U \partial_\mu U^\dagger$. In vacuum SOC, $\alpha = \frac{\hbar}{4mc^2} = 3.2 \times 10^{-22}$ s. In the Rashba SOC system, the material-dependent constant of $\langle \alpha E \rangle$ will be determined from the relation of $e\langle \alpha E \rangle = m/\hbar \alpha_R$ where theoretical and experimental values of α_R for various material systems can be found in literatures. The superscript r indicates that the M direction has been relabeled z as the original lab z is rotated to coincide with M. Under the adiabatic approximation, $A_\mu^{Ch} = \sigma_z a_\mu^{Ch}$, where a_μ^{Ch} is the top left diagonal element of the gauge field matrix A_μ^{Ch}. Explicitly,

$$a_\mu^{Ch} = -\frac{i\hbar}{e} \left(\frac{1 - \cos\theta}{2} \right) \frac{\partial \phi}{\partial \boldsymbol{n}} \cdot \frac{\partial \boldsymbol{n}}{\partial r_\mu} = a_{\text{mon}}^v(n) \partial_\mu n_v, \quad (5)$$

where it can be shown that a_{mon}^v is a magnetic monopole in the space of local M. The term a_μ^{Ch} has previously been associated with the adiabatic spin torque.[15] Of interest here is the spin orbit gauge potential $A_\mu^{SOC} = e\alpha U E_i \sigma_j \varepsilon_{ij\mu} U^\dagger$ which describes the modified momentum of an electron spin-aligned along the local moment under RSOC

$$\boldsymbol{A}^{\text{SOC}} = \alpha \sigma_r^z [(n_y E_z - n_z E_y)\boldsymbol{i} + (n_z E_x - n_x E_z)\boldsymbol{j} + (n_x E_y - n_y E_x)\boldsymbol{k}]. \quad (6)$$

In the adiabatic system where spin is constantly aligned to the local M, there is no probability of the spin assuming its other eigenstate. In electrodynamics, the emergent gauge potentials A_μ^{SO} and A_μ^{Ch} due to SOC and chiral local moment, produces an electromagnetic interaction between the current and the local moment. The interaction energy density is thus

$$E_{\text{INT}} = \langle \varphi_r | j_\mu A_\mu | \varphi_r \rangle$$

$$= \alpha \left[j_x \left(n_y E_z - n_z E_y + \frac{a_x^{\text{Ch}}}{\alpha} \right) + j_y \left(n_z E_x - n_x E_z + \frac{a_y^{\text{Ch}}}{\alpha} \right) \right.$$

$$\left. + j_z \left(n_x E_y - n_y E_x + \frac{a_z^{\text{Ch}}}{\alpha} \right) \right], \quad (7)$$

where $a_\mu = a_{\text{mon}}^v \partial_\mu n_v$, and j_μ is the charge current density, and $A_\mu = A_\mu^{\text{SO}} + A_\mu^{\text{Ch}}$. This gauge field derives its form from the continued spin locking to the local moment in the presence of SOC. The corresponding EOM will describe the dynamics of the local magnetic moment as a result of the above. Generally, the local moment will adjust its orientation in order to achieve minimum energy. Thus, one would expect the local moment dynamics to be governed by the energy gradient with respect to a change in the local moment orientation. The general expression for the effective field is

$$\boldsymbol{H}_{\text{ani}}^{\text{total}} = -\frac{1}{\mu_0} \frac{\delta E_{\text{int}}}{\delta \boldsymbol{M}} = \boldsymbol{H}^{\text{CH}} + \boldsymbol{H}^{\text{RSO}}. \quad (8)$$

The term $\boldsymbol{H}^{\text{CH}} = \frac{\hbar}{2M\mu_0} \frac{j_\lambda}{e} \nabla_\lambda \boldsymbol{n} \times \boldsymbol{n}$ is considered an effective anisotropy field due to the local chiral moment. The SOC effective anisotropy field is

$$\boldsymbol{H}^{\text{SOC}} = -\frac{\alpha(\boldsymbol{r})}{Me\hbar\mu_0} \boldsymbol{z} \times \boldsymbol{j} - \frac{m}{e\hbar\mu_0} \boldsymbol{z} \cdot (\boldsymbol{j} \times \boldsymbol{n}) \frac{\partial \alpha(\boldsymbol{r})}{\partial \boldsymbol{M}} - \frac{m}{e\hbar\mu_0} \nabla \cdot \frac{\partial \alpha(\boldsymbol{r})}{\partial \nabla \boldsymbol{M}} \boldsymbol{z} \cdot (\boldsymbol{j} \times \boldsymbol{n})$$

$$- \frac{m}{e\hbar\mu_0} \frac{\partial \alpha(\boldsymbol{r})}{\partial \nabla \boldsymbol{M}} \cdot \nabla (\boldsymbol{z} \cdot (\boldsymbol{n} \times \boldsymbol{j})). \quad (9)$$

In spatially homogenous magnetic system, i.e. where the Rashba constant is neither a function of $\boldsymbol{M}$ nor a function of $\nabla \boldsymbol{M}$, the total effective anisotropy field is therefore $\boldsymbol{H}^{\text{CH}}$ plus first term on the righthand side of $\boldsymbol{H}^{\text{RSO}}$, resulting in

$$\boldsymbol{H}_{\text{ani}}^{\text{total}} = \frac{1}{\mu_0 M} \left[j_x \left(\alpha \frac{\partial n_\mu}{\partial \boldsymbol{n}} E_v \varepsilon_{x\mu\nu} + \frac{\partial a_x^{\text{Ch}}}{\partial \boldsymbol{n}} \right) + j_y \left(\alpha \frac{\partial n_\mu}{\partial \boldsymbol{n}} E_v \varepsilon_{y\mu\nu} + \frac{\partial a_y^{\text{Ch}}}{\partial \boldsymbol{n}} \right) \right.$$
$$\left. + j_z \left(\alpha \frac{\partial n_\mu}{\partial \boldsymbol{n}} E_v \varepsilon_{z\mu\nu} + \frac{\partial a_z^{\text{Ch}}}{\partial \boldsymbol{n}} \right) \right] \quad (10)$$

where $\boldsymbol{n} = \boldsymbol{M}/M$ and $\mu_0 = 4\pi \times 10^{-7}$ TmA^{-1}. In the low-damping limit, the local moment will precess about the effective field, so that the general EOM can be written as $\frac{d\boldsymbol{M}}{dt} = \gamma \boldsymbol{M} \times \boldsymbol{H}$ where γ is the gyromagnetic ratio (in units of A^{-1}s^{-1}). One could now write the LLG as follows:

$$\frac{d\boldsymbol{M}}{dt} = \frac{\gamma}{\mu_0} \boldsymbol{n} \times j_\lambda \left(\alpha \frac{\partial n_\mu}{\partial \boldsymbol{n}} E_v \varepsilon_{\lambda\mu\nu} - \frac{i\hbar}{e} \boldsymbol{n} \times \partial_\lambda \boldsymbol{n} \right). \quad (11)$$

The modified LLG equation in a continuously magnetic medium, taking into account the effect of spin-polarized current injected externally and the SOC intrinsically present, is given by

$$\frac{d\boldsymbol{M}}{dt} = \gamma (\boldsymbol{M}_f \times \boldsymbol{H}) + \alpha_d \boldsymbol{M}_f \times (\boldsymbol{M}_f \times \boldsymbol{H}) + j_\lambda \left[a_C \boldsymbol{M}_f \times \left(\boldsymbol{M}_f \times \frac{\partial \boldsymbol{M}_f}{\partial \lambda} \right) \right.$$
$$\left. + b_C \left(\boldsymbol{M}_f \times \frac{\partial \boldsymbol{M}_f}{\partial \lambda} \right) + d_C \boldsymbol{M}_f \times \frac{\partial}{\partial \boldsymbol{M}_f} M_\mu E_v \varepsilon_{\lambda\mu\nu} \right], \quad (12)$$

where α_d is the damping constant, a_C, b_C are the usual spin torque constants in a magnetic system, e.g. domain wall or magnetic medium with chiral local moment texture, subscript "C" denotes a medium with continuous local spin variation. In ferromagnetic hetero-structure commonly used in magnetic memory (MRAM) where the inversion asymmetry E_z field is vertical to the two-dimensional plane that contains free-moving electron, and net electrical current flows along direction x, the spin orbit field simplifies to

$$H_y^{\text{SOC}} = \frac{j_x \langle \alpha E_z \rangle}{\mu_0 M}. \quad (13)$$

The spin orbit torque described above is now known as the field-like spin orbit torque. It has since been measured in many experiments,[16,17] but its field-like behavior was properly characterized by Kim et al.[14] Recently,[18] another type of spin orbit torque that is related to the Berry's phase of the system has been derived and this torque has a damping form.

$$\tau = \boldsymbol{M} \times \boldsymbol{H}_{\text{damping}}^{\text{SOC}},$$
$$\tau = \boldsymbol{M} \times (\boldsymbol{M} \times \boldsymbol{H}_{\text{field}}^{\text{SOC}}). \quad (14)$$

Both the field-like and the damping-like spin orbit torque have been experimentally proven. In magnetic conferences, spin orbit spin torque is an important session. It is heartening to know that both field and damping-like spin orbit torques have an elegant connection with the gauge theory. Whereas the field-like spin orbit torque is related to the diagonal components of the $A_\mu^{SOC} = e\alpha U E_i \sigma_j \varepsilon_{ij\mu} U^\dagger$, the damping-like spin orbit torque might be related to spin orbit gauge in the momentum space.

4. Spin Hall Effect (SHE)

The idea of a form of Hall effect that involves the spin carrier, known as the spin Hall effect (SHE), was first discussed by Dyakonov in 1971,[19] It was later brought up again for consideration in device applications.[20,21] Interestingly, modern development of gauge theory and Berry's phase has found subtle links to the classical SHE. Figure 4 shows an old photo of Dyakonov having a discussion with Perel, another scientist who had contributed significantly to the early development of theoretical semiconductor physics.

The spin Hall effect as described in the classical framework in the first section of this paper is related to the spin transverse velocity and the spin longitudinal force f_s. The spin transverse velocity is related to the spin longitudinal force in a classical Lorentz force framework, in which the spin orbit coupling provides an effective magnetic field, and the $SU(2)$ moment coupling constant $e\sigma_z$ to the velocity.

$$f_s = e\sigma_z v \times B. \tag{15}$$

The above suggests that one can derive the SHE in an intuitive manner by looking into the spin transverse velocity of a spin orbit system. In fact the classical notion of forces and velocities strongly indicates that one should look in the velocity. But it is the gauge notion that completes the thought process, leading to a complete and

Perel   Dyakonov

Fig. 4. (Courtesy of Dyakonov) A 1976 photo of Dyakonov and Perel in discussion, probably on spin Hall physics.

physically intuitive derivation of the SHE conductivity. Referring to past literatures, one notes that the SHE has indeed been discussed in the context of velocity and gauge in various ways as follows:

(1) Geometric & Gauge physicsc [22]
(2) Kinetic & Kubo formulation [23]
(3) Spin orbit gauge (Yang–Mills) [24]
(4) Numerous semi-classical treatments.

Here we show that with a locally gauge-transformed Hamiltonian, one can derive the spin transverse velocity and study the various SHE associated with it. In contrast to the gauge transformation in the spin orbit torque which was carried out in the real space, in the derivation of the SHE, transformation is performed in the momentum and time space.[25] The spin transverse velocities are:

$$\langle v_z^y \rangle = \langle v_z^y \rangle_{\text{KE}} + \langle v_z^y \rangle_{\text{YM}} + \langle v_z^y \rangle_{\text{MF}}.$$

$$\swarrow \qquad \downarrow \qquad \searrow$$

$$\text{kinetic} \qquad \text{Yang–Mills} \qquad \text{Murakami–Fujita} \qquad (16)$$

The first term on the RHS is the kinetic velocity. The second term arises directly from spin orbit coupling and is known as the Yang–Mills. The third term arises from local gauge transformation in momentum or time space, known as the Murakami–Fujita velocity. As details of this work can be found in Ref. 26, and the concept of gauge applied here is in the same spirit as that described in the spin orbit torque except that transformation is carried out in the time and momentum instead of the real space here, we will only present the final results.

The main advantage of considering SHE under the concept of gauge and its relation to force and velocity, is the physically intuitive picture it presents with respect to the physics of SHE. The contribution of the spin transverse velocity to SHE can be sensibly classified into individual components. Results show that by considering all velocity components, a sign reversal (see Table 2) of SHE conductivity is found in at least two semiconductor systems with RSOC, namely the Rashba 2DEG and the Rashba heavy hole systems.

Table 2. The gauge theoretic physics provides a full treatment of the SHE, showing results opposite in sign to previous treatments, summarized below.

	SHE Conductivity	New SHE Conductivity
Rashba 2DEG [23]	$-\dfrac{e}{8\pi}$	$+\dfrac{e}{8\pi}$
Rashba Heavy hole [27]	$-\dfrac{9e}{8\pi}$	$+\dfrac{9e}{8\pi}$

5. Conclusion

We have summarized two specific examples (spin orbit spin torque, spin Hall effect) of gauge theoretic methods expanding beyond the traditional realms of general and high energy physics. As a matter of fact, it is expanding beyond statistical mechanics and pure condensed matter physics to the world of applied condensed matter and modern device physics, which is increasingly defined by newly discovered materials, and ever decreasing sizes. From the material standpoint, gauge theory is applied beyond ferromagnetic metal, and semiconductor. In fact it has more versatile usage in new materials like graphene, silicene, and topological insulators. From the carrier standpoint, gauge theory is used in electronics, spintronics, and photonics, covering all aspects of applied physics in the 21st century.

Acknowledgments

M. B. A. Jalil acknowledges the financial support of MOE Tier II grant MOE2013-T2-2-125 (NUS Grant No. R-263-000-B10-112), and the National Research Foundation of Singapore under the CRP programs "Next Generation Spin Torque Memories: From Fundamental Physics to Applications" NRF-CRP12-2013-01 and "Non-Volatile Magnetic Logic and Momery Integrated Circuit Devices" NRF-CRP9-2011-01.

References

1. C. N. Yang, The conceptual origins of Maxwell's equations and gauge theory, *Physics Today* **67**, 45 (2014).
2. T. Fujita, M. B. A. Jalil, S. G. Tan and S. Murakami, Gauge field in spintronics, *J. Appl. Phys. [Appl. Phys. Rev.]* **110**, 121301 (2011).
3. S. G. Tan and M. B. A. Jalil, *Introduction to the Physics of Nanoelectronic*, (Woodhead Publishing Limited, Cambridge, U.K., 2012), Chapter 5.
4. S. G. Tan and M. B. A. Jalil, Spin Hall effect in a simple classical picture of spin forces, *J. Phys. Soc. Jpn.* **82**, 094714 (2013).
5. S.-Q. Shen, Spin transverse force on spin current in an electric field, *Phys. Rev. Lett.* **95**, 187203 (2005).
6. F. T. Vasko, Spin splitting in the spectrum of two-dimensional electrons due to the surface potential, *Pis'ma Zh. Eksp. Teor. Fiz.* **30**, 574 (1979) [*JETP Lett.*, 30, 541].
7. Y. A. Bychkov and E. I. Rashba, Properties of a 2D electron gas with lifted spectral degeneracy, *Pis'ma Zh. Eksp. Teor. Fiz.*, **39**, 66 (1984) [*JETP Lett.*, **39**, 78].
8. K. Ishizaka *et al.*, Giant Rashba-type spin splitting in bulk BiTeI, *Nature Materials* **10**, 521 (2011).
9. K. Tsutsui and S. Murakami, Spin-torque efficiency enhanced by Rashba spin splitting in three dimensions, *Phys. Rev. B* **86**, 115201 (2012).

10. J.-J. Zhou et al., Engineering topological surface states and giant rashba spin splitting in BiTeI/Bi2Te3 heterostructures, *Scientific Report* **4**, 3841 (2014).
11. S. G. Tan, M. B. A. Jalil and X.-J. Liu, Local spin dynamic arising from the non-perturbative SU(2) gauge field of the spin orbit effect, arXiv:0705.3502.
12. K. Obata and G. Tatara, Current-induced domain wall motion in Rashba spin-orbit system, *Phys. Rev. B* **77**, 214429 (2008).
13. S. G. Tan et al., Spin dynamics under local gauge fields in chiral spin–orbit coupling systems, *Ann. Phys. (NY)* **326**, 207 (2011).
14. J. Y. Kim et al., Layer thickness dependence of the current-induced effective field vector in Ta/CoFeB/MgO, *Nature Materials* **12**, 240 (2013).
15. Y. B. Bazaliy, B. A. Jones and S.-C. Zhang, Modification of the Landau-Lifshitz equation in the presence of a spin-polarized current in colossal- and giant-magnetoresistive materials, *Phys. Rev. B* **57** (1998) 3213(R).
16. I. M. Miron et al., Current-driven spin torque induced by the Rashba effect in a ferromagnetic metal layer, *Nature Materials* **9**, 230 (2010); X. Fan et al., Observation of the nonlocal spin-orbital effective field, *Nature Communications* **4**, 1799 (2013).
17. M. Jamali et al., Spin-orbit torques in Co/Pd multilayer nanowires, *Phys. Rev. Letts.* **111**, 246602 (2013); X. Qiu et al., Angular and temperature dependence of current induced spin-orbit effective fields in Ta/CoFeB/MgO nanowires, *Scientific Report* **4**, 4491 (2014).
18. H. Kurebayashi et al., An antidamping spin–orbit torque originating from the Berry curvature, *Nature Nanotechnology* **9**, 211 (2014).
19. M. I. Dyakonov and V. I. Perel, Possibility of orienting electron spins with current, *Pis'ma Zh. Eksp. Teor. Fiz.* **13**, 657 (1971) [*JETP Lett.* **13**, 467 (1971)].
20. J. E. Hirsch, Spin Hall effect, *Phys. Rev. Lett.* **83**, 1834 (1999).
21. Y. K. Kato et al., Observation of the spin Hall effect in semiconductors, *Science* **306**, 1910 (2004).
22. S. Murakami, N. Nagaosa and S. C. Zhang, Dissipationless quantum spin current at room temperature, *Science* **301**, 1348 (2003).
23. J. Sinova et al., Universal intrinsic spin Hall effect, *Phys. Rev. Lett.* **92**, 126603 (2004).
24. S. G. Tan et al., Transverse spin transverse separation in a two-dimensional electron-gas using an external magnetic field with a topological chirality, *Phys. Rev. B* **78**, 245321 (2008).
25. T. Fujita, M. B. A. Jalil, and S. G. Tan, Unified model of intrinsic spin-Hall effect in spintronic, optical, and graphene systems, *J. Phys. Soc. Jpn.* **78**, 104714 (2009).
26. S. G. Tan et al., Gauge physics of spin Hall effect, ArXiv 1504.04451.
27. B. A. Bernevig and S. C. Zhang, Intrinsic spin Hall effect in the two-dimensional hole gas, *Phys. Rev. Lett.* **95**, 016801 (2005).

Yang–Yang Equilibrium Statistical Mechanics: A Brilliant Method

Xi-Wen Guan[1,2,*] and Yang-Yang Chen[1]

[1]*State Key Laboratory of Magnetic Resonance and Atomic and Molecular Physics, Wuhan Institute of Physics and Mathematics, Chinese Academy of Sciences, Wuhan 430071, China*
[2]*Department of Theoretical Physics, Research School of Physics and Engineering, Australian National University, Canberra ACT 0200, Australia*

C. N. Yang and C. P. Yang in 1969 (*J. Math. Phys.* **10**, 1115 (1969)) for the first time proposed a rigorous approach to the thermodynamics of the one-dimensional system of bosons with a delta-function interaction. This paper was a breakthrough in exact statistical mechanics, after C. N. Yang (*Phys. Rev. Lett.* **19**, 1312 (1967)) published his seminal work on the discovery of the Yang–Baxter equation in 1967. Yang and Yang's brilliant method yields significant applications in a wide range of fields of physics. In this communication, we briefly introduce the method of the Yang–Yang equilibrium statistical mechanics and demonstrate a fundamental application of the Yang–Yang method for the study of thermodynamics of the Lieb–Liniger model with strong and weak interactions in a whole temperature regime. We also consider the equivalence between the Yang–Yang's thermodynamic Bethe ansatz equation and the thermodynamics of the ideal gas with the Haldane's generalized exclusion statistics.

1. The Yang–Yang Method

The Bethe ansatz (BA) should date back to 1931 when Hans Bethe introduced a particular form of wavefunction (Bethe ansatz) to obtain the energy eigenspectrum of the 1D Heisenberg spin chain.[3] For such exactly solved models, the energy eigenspectrum of the model Hamiltonian can be obtained exactly in terms of the BA equations, from which physical properties can be derived via mathematical analysis. The Lieb–Liniger Bose gas,[4] Yang–Gaudin model,[2,5] Hubbard model,[6] SU(N) interacting Fermi gases[7] are notable BA integrable models and have had tremendous impact in a variety of fields of physics. The BA approach has also found success in the realm of condensed matter physics, such as Kondo impurity problems,[8] BCS pairing models,[9] strongly correlated electron systems[10] and spin chain and ladder compounds,[11] quantum gases of cold atoms,[12,13] and many other problems in physics.[14–20]

In the context of exactly solvable models, the finite temperature problem for the Lieb–Liniger Bose gas was solved by C. N. Yang and C. P. Yang in 1969.[1] They showed that the thermodynamics can be determined from the minimization conditions of the Gibbs free energy subject to the BA equations. It turns out that the Yang–Yang method is an elegant way to analytically access not only the thermodynamics, but also correlation functions, quantum criticality and Luttinger liquid physics for a wide range of low-dimensional quantum many-body systems.

In order to carry out the Yang–Yang grand canonical method for the thermodynamics of the 1D Bose gas, we first introduce the Lieb–Liniger model. In 1963, Lieb and Liniger[4] applied the Bethe ansatz to the problem of one-dimensional bosons with a δ-function interaction. The quantum field theory Hamiltonian for describing N bosons with δ-function interaction reads

$$H = \int_0^L dx \left[\partial_x \Psi^\dagger(x) \partial_x \Psi(x) + c \Psi^\dagger(x) \Psi^\dagger(x) \Psi(x) \Psi(x) \right], \quad (1)$$

where c denotes the coupling constant, under periodic boundary conditions, with x in coordinate in length L. We let $\hbar = 2m = 1$ for convenience. The nonlinear Schrödinger (NS) equation is given by

$$i\partial_t \Psi = -\partial_x^2 \Psi + 2c \Psi^\dagger \Psi \Psi. \quad (2)$$

It is straightforward to show that solving an eigenvalvue problem of the quantum field theory Hamiltonian (1), or second quantization Hamiltonian for interacting bosons in one dimension, is equivalent to that of the quantum mechanical Hamiltonian

$$H = -\sum_{i=1}^N \frac{\partial^2}{\partial x_i^2} + 2c \sum_{1 \leq i \leq j \leq N} \delta(x_i - x_j), \quad (3)$$

where N is the number of particles in the system, c is the interaction strength between particles, and $\{x_i\}$ refers to the positions of the particles.

Lieb and Liniger[4] used a BA form of the wavefunction — superposition of all possible permutations of plane waves in a ring of size L, namely

$$\chi = \sum_{\mathcal{P}} A(\mathcal{P}) e^{i(k_{\mathcal{P}_1} x_1 + \cdots + k_{\mathcal{P}_N} x_N)}$$

where N is the number of particles and $\mathcal{P}_1, \ldots, \mathcal{P}_N$ stand for a permutation $\mathcal{P}$ of integers $1, 2, \ldots, N$. The $N!$ plane waves are N-fold products of individual exponential phase factors $e^{i k_i x_j}$. Here the N distinct wave numbers k_i are permuted among the N distinct coordinates x_j. The pseudo momenta k_i satisfies the following BA equations

$$e^{i k_i L} = -\prod_{j=1}^N \frac{k_i - k_j + ic}{k_i - k_j - ic}, \quad i = 1, 2, \ldots, N, \quad (4)$$

which are called the Lieb–Liniger equations.

However, the thermodynamics of this model was not solved until Yang and Yang's seminal paper[1] published in 1969. The first step for the Yang–Yang grand canonical method is to count degeneracies of a finite temperature macroscopic state in terms of the quantum numbers $I_1, I_2, \ldots$ associated BA equations for the Lieb–Linger gas. Such quantum numbers one-to-one determine the quasimomenta of the particles though the BA equations[1,4]

$$k_i L = 2\pi I_i - \sum_{j=1}^{N} 2\tan^{-1}\left(\frac{k_i - k_j}{c}\right), \tag{5}$$

$$i = 1, 2, \ldots, N,$$

where

$$I_i = \begin{cases} \text{integer}, & \text{if } N \text{ is odd}, \\ \text{half-odd integer}, & \text{if } N \text{ is even}. \end{cases} \tag{6}$$

The uniqueness of the solution $\{k_i\}$ to the BA (5) for a given set of real quantum numbers $\{I_i\}$ was proved in Refs. 1 and 4. Each set of the solution $\{k_i\}$ leads to the energy of the system through $E = \sum_{i=1}^{N} k_i^2$. For the ground state, all quasimomenta $\{k_i\}$ are bounded in an interval $(-Q, Q)$, namely the quantum numbers are given by $I_j = -\frac{N-1}{2} + j - 1$, $j = 1, \ldots, N$. Here Q is the cut-off.

In the thermodynamic limit, i.e. $N, L \to \infty$ and $n = N/L$ is a finite number, the BA roots can be treated as a density distribution for particles as well as holes (unoccupied BA vacancies), namely

$$L\rho(k)dk = \text{number of particles in } dk, \tag{7}$$

$$L\rho^h(k)dk = \text{number of holes in } dk. \tag{8}$$

We define a density distribution function $Ldh(k) = 2\pi L(\rho(k) + \rho^h(k))dk$, thus the BA equations (5) become

$$\rho(k) + \rho^h(k) = \frac{1}{2\pi} + \frac{c}{\pi} \int_{-\infty}^{\infty} \frac{\rho(k')dk'}{c^2 + (k-k')^2}, \tag{9}$$

where $\rho(k)$ and $\rho^h(k)$ are the particle and hole density distribution functions at finite temperatures, respectively. Due to the partitions (choices) of particles and holes in a small interval dk, the energy of a macroscopic state is degenerate. Hence, such degeneracies give a none zero entropy. The number of microscopic states associated with this energy is the total partitions in dk

$$dW = \frac{(L(\rho + \rho^h)dk)!}{(L\rho dk)!(L\rho^h dk)!}. \tag{10}$$

Using Sterling's approximation $\ln m! = m \ln m - m$ when $m \gg 1$, Yang and Yang for first time gave the expression of entropy per unit length

$$s = \int_{-\infty}^{\infty} \left[(\rho + \rho^h)\ln(\rho + \rho^h) - \rho \ln \rho - \rho^h \ln \rho^h\right] dk$$

$$= \int_{-\infty}^{\infty} \left[(\rho + \rho^h)\ln\left(1 + \frac{\rho}{\rho^h}\right) - \rho \ln\left(\frac{\rho}{\rho^h}\right)\right] dk. \tag{11}$$

This entropy measures disorder of the particle and hole distributions for a macroscopic state. We would like to emphasize that such subtle connection between the BA microscopic states and the macroscopic state of the system play a key role in the Yang–Yang brilliant method. It is also an essential part in the applications of this method to other systems of physical problems.

Maximizing the entropy is the next key step for the Yang–Yang method. The partition function is written in terms of ρ and ρ^h

$$\mathcal{Z} = \text{Tr}(e^{-H/T}) = \sum_{\rho,\rho^h} W(\rho, \rho^h) e^{-E(\rho,\rho^h)/T}, \tag{12}$$

where ρ and ρ^h satisfy the BA equation (9). Hence it gives

$$\mathcal{Z} = \sum_{\rho,\rho^h} e^{-(E(\rho,\rho^h) - S(\rho,\rho^h)T)/T}. \tag{13}$$

We define the Helmholtz free energy $F(\rho, \rho^h) = E(\rho, \rho^h) - S(\rho, \rho^h)T$. Then in the grand ensemble, the Gibbs free energy per unit length is given by $G/L = E/L - \mu n - Ts$ with the relation to the free energy $F = G + \mu N$. Here μ is the chemical potential. The true physical state is determined by minimizing G with respect to the density functions ρ and ρ^h, i.e. $\frac{\delta G}{L} = \frac{\delta E}{L} - \mu \delta n - T \delta s = 0$. Here the energy and the density per unit length of the system are given by

$$\frac{E}{L} = \int_{-\infty}^{\infty} k^2 \rho(k) dk, \quad n = \int_{-\infty}^{\infty} \rho(k) dk. \tag{14}$$

Taking infinitesimal variation of the entropy Eq. (11), we have

$$\delta s = \int_{-\infty}^{\infty} (\delta\rho + \delta\rho^h) \ln\left(1 + \frac{\rho}{\rho^h}\right) - \delta\rho \ln\left(\frac{\rho}{\rho^h}\right) dk. \tag{15}$$

Here the $\rho(k)$ and $\rho^h(k)$ are not independent, i.e.

$$\delta\rho(k) + \delta\rho^h(k) = \frac{c}{\pi} \int_{-\infty}^{\infty} \frac{\delta\rho(k') dk'}{c^2 + (k-k')^2}. \tag{16}$$

With the help of these relations, the minimization condition $\frac{\delta G}{L} = 0$ leads to the relation

$$k^2 - \mu + T \ln\left(\frac{\rho}{\rho^h}\right) - \frac{Tc}{\pi} \int_{-\infty}^{\infty} \frac{dq}{c^2 + (k-q)^2} \ln\left(1 + \frac{\rho}{\rho^h}\right) = 0.$$

Introducing the dressed energy of the system $\exp(\varepsilon(k)/T) = \rho^h/\rho$, then Yang and Yang obtained the so-called Yang–Yang thermodynamic Bethe ansatz (TBA) equation[1]

$$\varepsilon(k) = k^2 - \mu - \frac{Tc}{\pi}\int_{-\infty}^{\infty} \frac{dq}{c^2 + (k-q)^2} \ln\left(1 + e^{-\frac{\varepsilon(q)}{T}}\right) \tag{17}$$

which determines the thermodynamics of the system in a whole temperature regime.

The last step in Yang–Yang's approach is to derive the pressure. Using the BA equation again, Gibbs free energy $p = -\left(\frac{\partial G}{\partial L}\right)_{T,\mu,c}$ gives

$$p = \int_{-\infty}^{\infty} \{(\mu - k^2)\rho(k) + T[(\rho(k) + \rho^h(k))$$

$$\times \ln(\rho(k) + \rho^h(k)) - \rho(k)\ln\rho(k) - \rho^h(k)\ln\rho^h(k)]\}\,dk$$

$$= \int_{-\infty}^{\infty}\left\{\left[(\mu - k^2) - \ln\frac{\rho(k)}{\rho^h(k)} + T\ln\left(1 + \frac{\rho(k)}{\rho^h(k)}\right)\right]\rho(k)\right.$$

$$\left. + T\rho^h(k)\ln\left(1 + \frac{\rho(k)}{\rho^h(k)}\right)\right\}dk$$

$$= \int_{-\infty}^{\infty}\left\{T\rho(k)\ln\left(1 + \frac{\rho(k)}{\rho^h(k)}\right) - \frac{Tc}{\pi}\int\frac{dq}{c^2+(k-q)^2}\ln\left(1+\frac{\rho(q)}{\rho^h(q)}\right)\right.$$

$$\left. + T\rho^h(k)\ln\left(1 + \frac{\rho(k)}{\rho^h(k)}\right)\right\}dk$$

$$= T\int_{-\infty}^{\infty}\left\{\left[\rho(k) + \rho^h(k) - \frac{Tc}{\pi}\int\frac{\rho(q)dq}{c^2+(k-q)^2}\right]\ln\left(1+\frac{\rho(k)}{\rho^h(k)}\right)\right\}dk$$

$$= \frac{T}{2\pi}\int_{-\infty}^{\infty} \ln\left(1 + e^{-\varepsilon(k)/T}\right)dk. \tag{18}$$

Thus thermodynamic quantities can be derived from this pressure through the thermodynamic relations, for example, density, entropy, compressibility and specific heat are given by

$$n = \left(\frac{\partial p}{\partial \mu}\right)_{c,T}, \quad s = \left(\frac{\partial p}{\partial T}\right)_{c,\mu},$$

$$\kappa = \left(\frac{\partial^2 p}{\partial \mu^2}\right)_{c,T}, \quad c_V = T\left(\frac{\partial^2 p}{\partial T^2}\right)_{c,\mu}. \tag{19}$$

The Yang–Yang approach provide a powerful method for precisely treating thermodynamics of integrable models of interacting electrons, spins, fermionic and bosonic atoms. In particular, due to the extensive nature of the entropy, the Yang–Yang method has been successfully adapted to treat the thermodynamics for the models with internal degrees of freedom, see.[10,20] In the following section, we will demonstrate how one can obtain the exact thermodynamics of the Lie–Liniger model.

2. Thermodynamics of the 1D Bose Gas

So far, we have derived the Yang–Yang's TBA equation (17) which gives the dressed energy $\varepsilon(k)$ of the system in terms of chemical potential and temperature. It seems that there is no standard method to solve such nonlinear integral equation in mathematics. Here we first present the numerical solution of this TBA equation for a whole temperature regime, see Fig. 1. The TBA equation was solved analytically[21] in certain regimes, such as zero temperature regime, high temperature regime, strong

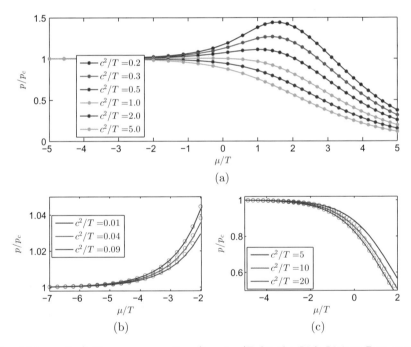

Fig. 1. (Color online) The pressure ratio p/p_c vs μ/T for the Lieb–Liniger Bose gas. Here $p_c = \frac{1}{2\sqrt{\pi}} T^{3/2} e^{\mu/T}$ is the pressure of the Boltzmann gas. (a) The pressure ratio for a variety of interaction strength from high temperature to low temperature. It shows round peaks for small values of c^2/T. (b) The pressure ratio at high temperatures. The solid lines show the numerical TBA result obtained from (17) which is in excellent agreement with the high temperature expansion result (37) (circles). (c) The solid lines show the numerical TBA result obtained from (17) which is in excellent agreement with the analytical result for the strong coupling regime (23) (circles).

coupling regime and weak coupling regime, see brief review.[22] We will mention few cases below.

Weak interaction. In this weak-coupling limit, the system behaves as free bosons. To show this, let us consider the kernel in this weak coupling limit, namely, $\lim_{c\to 0} \frac{c/\pi}{c^2+(k-q)^2} = \delta(k-q)$. The TBA equation (17) becomes

$$\varepsilon(k) = k^2 - \mu - T\ln\left(1 + e^{-\varepsilon(k)/T}\right). \qquad (20)$$

Using this relation and the BA equation (9), we obtain

$$2\pi\rho(k) = \exp(-\varepsilon(k)/T) = \frac{1}{e^{(k^2-\mu)/T} - 1},$$

which is precisely the distribution function for free bosons.

Strong interaction. In strong coupling limit, i.e. $c \to \infty$, the distribution function (9) shows a Fermi statistic $2\pi\rho(k) = 1/(e^{(k^2-\mu)/T} + 1)$. In this limit, the Yang–Yang equation (17) can be expanded in the powers of $1/c$ (up to the order of $O(1/c^4)$),[21]

$$\varepsilon(k) \approx \varepsilon^0(k) - \mu - \frac{2cp(T)}{c^2+k^2} - \frac{1}{2\sqrt{\pi}c^3}\frac{T^{\frac{5}{2}}}{\left(\frac{\hbar^2}{2m}\right)^{\frac{3}{2}}}f_{\frac{5}{2}}(A), \qquad (21)$$

where

$$A_0 = \mu + \frac{2p(T)}{c} - \frac{4\mu^{5/2}}{15\pi|c|^3} \qquad (22)$$

and $f_n(x) = \text{Li}_n(-e^{x/T})$. Here $\text{Li}_n(x) = \sum_{k=1}^{\infty}\frac{x^k}{k^n}$ denote the polylog function. Furthermore, with the help of the dressed energy (21), we can calculate the finite temperature pressure in terms of the polylog function, namely

$$p \approx -\sqrt{\frac{m}{2\pi\hbar^2}}T^{\frac{3}{2}}f_{\frac{3}{2}}(A)\left[1 + \frac{1}{2c^3\sqrt{\pi}}\left(\frac{T}{\frac{\hbar^2}{2m}}\right)^{\frac{3}{2}}f_{\frac{3}{2}}(A)\right] \qquad (23)$$

where

$$A = \mu + \frac{2p(T)}{c} + \frac{1}{2\sqrt{\pi}c^3}\frac{T^{\frac{5}{2}}}{\left(\frac{\hbar^2}{2m}\right)^{\frac{3}{2}}}f_{\frac{5}{2}}(A_0). \qquad (24)$$

This result (23) services as an accurate equation of state (EOS) for the Bose gas with a strong repulsion. For our convenience in the analysis of the critical behavior of this model, we introduce the interaction-rescaled EOS, i.e. $\tilde{p} \equiv p/(\frac{\hbar^2}{2m}c^3)$, which can be written in terms of the dimensionless temperature. From the EOS (23), we can analytically calculate the particle density, compressibility and the specific heat

$$n = -\frac{1}{2\sqrt{\pi}}T^{\frac{1}{2}}f_{\frac{1}{2}}\left\{1 - \frac{1}{\sqrt{\pi}c}T^{\frac{1}{2}}f_{\frac{1}{2}} + \frac{T}{\pi c^2}f_{\frac{1}{2}}^2 + \frac{1}{\sqrt{\pi}c^3}T^{\frac{3}{2}}\left[-\frac{1}{\pi}f_{\frac{1}{2}}^3 + \frac{3}{2}f_{\frac{3}{2}}\right]\right\}, \tag{25}$$

$$\kappa \approx -\frac{1}{2\sqrt{\pi}}T^{-\frac{1}{2}}f_{-\frac{1}{2}} + \frac{3}{2\pi c}f_{-\frac{1}{2}}f_{\frac{1}{2}} - \frac{3}{\pi^{3/2}c^2}T^{\frac{1}{2}}f_{-\frac{1}{2}}f_{\frac{1}{2}}^2 - \frac{1}{\pi c^3}Tf_{-\frac{1}{2}}f_{\frac{3}{2}}$$
$$+ \frac{5}{\pi^2 c^3}Tf_{-\frac{1}{2}}f_{\frac{1}{2}}^3 - \frac{3}{4\pi c^3}Tf_{\frac{1}{2}}^2, \tag{26}$$

$$\frac{c_V}{T} = -\frac{3}{8\sqrt{\pi}}T^{-\frac{1}{2}}f_{\frac{3}{2}} + \frac{1}{2\sqrt{\pi}}T^{-\frac{1}{2}}\frac{A}{T}f_{\frac{1}{2}} - \frac{1}{2\sqrt{\pi}}T^{-\frac{1}{2}}\left(\frac{A}{T}\right)^2 f_{-\frac{1}{2}} + O\left(\frac{1}{c}\right), \tag{27}$$

respectively. These results provide insight into the understanding of quantum anomalies at criticality.

Quantum critical. At zero temperature, the TBA equation (17) suggests two distinguishable regimes: a vacuum for the chemical potential is less than zero, i.e. $\mu < \mu_c$ with $\mu_c = 0$ for this model and a filled "Fermi sea" with finite density of bosons for $\mu > \mu_c$, see Fig. 2. For $\mu = \mu_c = 0$, the dressed energy $\epsilon(0) = 0$ that indicates a phase transition from vacuum into the Tomonaga–Luttinger liquid (TLL) at zero temperature. For $\mu < \mu_c$ and $0 < T < n^2\pi^2$, the vacuum will become populated with an exponentially small number of bosons for a small

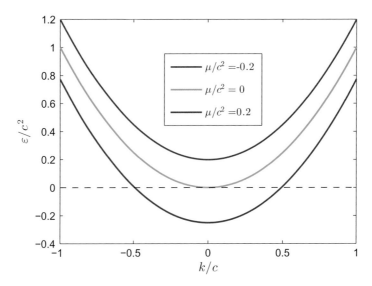

Fig. 2. (Color online) The dressed energy $\varepsilon(k)$ in quasimomentum space for different values of chemical potential μ. At zero temperature, for $\mu < 0$, there is no particle in the grand canonical ensemble, whereas for $\mu > 0$ there is a filled sea with two "Fermi points." At the critical point $\mu = 0$, the dressed energy $\varepsilon(0) = 0$ that leads to the vanishment of the liner dispersion near the two "Fermi points."

positive temperature, i.e. a classical gas phase persists due to the reason that the de Broglie wavelength is much smaller than the inter-particle mean spacing $1/n$ with the density $n \sim (1/\lambda)e^{-|\mu|/T}$. However, for $\mu > \mu_c$, the filled "Fermi sea" is regarded as the TLL where the conformal field theory describes power law behavior of correlation functions. Although a finite temperature quantum phase transition does not exist in the 1D Lieb–Liniger Bose gas, there exist a crossover temperature T^* below which the TLL with relativistic dispersions can be sustained. In this sense, the TLL phase persists for temperature $T < T^*$ where T^* is a crossover temperature which separates the TLL from the quantum critical regime. At finite temperatures, the correlation functions decay exponentially under the crossover temperature. If the temperature exceeds this crossover value T^*, the excitations involve free quasiparticles with non-relativistic dispersion, i.e. quadratic dispersion. This crossover temperature T^* significantly marks the breakdown of the linear dispersion of TLL, see Fig. 3. This crossover temperature was also manifested by the correlation length in the Lieb–Liniger model, more discussion was given in Ref. 23

In the TLL phase, the free energy derived from the TBA equations (40) does comprise the field theory form

$$F(T) \approx E_0 - \frac{\pi C(k_B T)^2}{6\hbar v_s}, \tag{28}$$

where the central charge $C = 1$ corresponding to a Gaussian field. The sound velocity v_s can be calculated from the ground state energy through $v_s = \sqrt{\frac{L}{mn}\frac{\partial^2 E}{\partial L^2}}$.

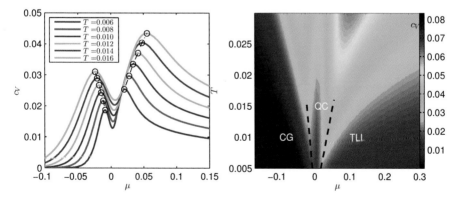

Fig. 3. (Color online) Left panel: specific heat c_V vs chemical potential μ for the 1D Bose gas with $c = 1$ at different temperatures. The left peaks mark the crossover temperatures T^* which separate the classical gas (CG) region from the quantum critical (QC) region, whereas the right peaks indicate the crossover temperatures T^* which distinguish the TLL phase from the QC region. Right panel: Contour plot of specific heat c_V in $T - \mu$ plane. It gives a finite temperature phase diagram in the $T - \mu$ plane. The two dashed lines are determined by the peak positions in the left panel.

For strong coupling regime, the ground state energy per length and the sound velocity are given by

$$E_0 \approx \frac{1}{3}n^3\pi^2\left(1 - \frac{4}{\gamma} + \frac{12}{\gamma^2} + \frac{32}{\gamma^3}\left(\frac{\pi^2}{15} - 1\right)\right),$$

$$v_s \approx \frac{\hbar\pi n}{m}\left(1 - \frac{4}{\gamma} + \frac{12}{\gamma^2} + \frac{16}{\gamma^3}\left(\frac{\pi^2}{3} - 2\right)\right). \tag{29}$$

The ground state energy per length and the sound velocity of the gas with weak repulsive interaction ($0 < c \ll 1$) are given by

$$E = \frac{\hbar^2}{2m}n^3\left(\gamma - \frac{4}{3\pi}\gamma^{3/2}\right),$$

$$v_s = \frac{\hbar n}{m}\sqrt{\gamma}\left(1 - \frac{\sqrt{\gamma}}{2\pi}\right)^{1/2}. \tag{30}$$

This implies that for temperatures below a crossover value T^*, the low-lying excitations have a linear relativistic dispersion relation $\omega(k) = v_s(k - k_F)$. The free fermion quantum critical region is expected for the temperature over the crossover value T^*.

On the other hand, near a quantum phase transition $\mu = \mu_c$ and $T > T^*$, thermal and quantum fluctuations destroy the forward scattering process, i.e. the back scattering process involves in the excitations. In this region, the density and the compressibility can be cast into universal scaling forms in quantum critical regime[21]

$$\tilde{n}(t, \tilde{\mu}) - \tilde{n}_0(t, \tilde{\mu}) \approx t^{\frac{d}{z}+1-\frac{1}{\nu z}}\mathcal{F}\left(\frac{\tilde{\mu} - \tilde{\mu}_c}{t^{\frac{1}{\nu z}}}\right), \tag{31}$$

$$\tilde{\kappa}(t, \tilde{\mu}) - \tilde{\kappa}_0(t, \tilde{\mu}) \approx t^{\frac{d}{z}+1-\frac{2}{\nu z}}\mathcal{Q}\left(\frac{\tilde{\mu} - \tilde{\mu}_c}{t^{\frac{1}{\nu z}}}\right), \tag{32}$$

where the dimensionless chemical potential $\tilde{\mu} = \mu/\epsilon_0$ as well as the dimensionless density and compressibility are defined as $\tilde{n} = n/c$ and $\tilde{\kappa} = \kappa\epsilon_0/c$ where $\epsilon_0 = c^2$. For such a type of phase transition, the critical behavior near the critical point is characterized by a divergent correlation length $\xi \sim |\mu - \mu_c|^{-\nu}$ and the energy gap Δ, which vanishes inversely proportional to the correlation length as $\Delta \sim \xi^{-z} \sim |\mu - \mu_c|^{z\nu}$. In these scaling forms, the dynamic exponent $z = 2$ and the correlation length exponent $\nu = 1/2$ with the scaling functions given by

$$\mathcal{F}(x) = -\frac{1}{2\sqrt{\pi}}\text{Li}_{\frac{1}{2}}(-e^x), \tag{33}$$

$$\mathcal{Q}(x) = -\frac{1}{2\sqrt{\pi}}\text{Li}_{-\frac{1}{2}}(-e^x) \tag{34}$$

for $T > |\mu - \mu_c|$. The background density and compressibility in the vacuum are zero, i.e., $n_0(t, \tilde{\mu}) = \kappa_0(t, \tilde{\mu}) = 0$. These scaling forms map out quantum criticality of the Lieb–Liniger gas with the critical exponents $z = 2$ and $\nu = 1/2$ in the critical regime.

Further more, near the critical chemical potential, the specific heat divided by the temperature $\tilde{c}_v \equiv c_v/T$ obeys the following scaling form:

$$\tilde{c}_v = -T^{-\frac{1}{2}}\left[\frac{3}{8\sqrt{\pi}}\mathrm{Li}_{\frac{3}{2}}\left(-e^{\frac{\mu}{T}}\right) - \frac{1}{2\sqrt{\pi}}\frac{\mu}{T}\mathrm{Li}_{\frac{1}{2}}\left(-e^{\frac{\mu}{T}}\right) + \frac{1}{2\sqrt{\pi}}\left(\frac{\mu}{T}\right)^2 \mathrm{Li}_{-\frac{1}{2}}\left(-e^{\frac{\mu}{T}}\right)\right]. \tag{35}$$

Figure 3 shows the specific heat evolves two round peaks near the critical point $\mu_c = 0$ for a fixed value of temperature. These peaks mark the crossover temperatures that distinguish the TLL and semi-classical gas phases from the quantum critical regime, see the finite temperature phase diagram (b) of Fig. 3.

High temperature. At high temperatures the Yang–Yang equation (17) gives rise to the Maxwell-Boltzmann statistic such that the particles are *distinguishable*. In the high temperature limits, it is very convenient to consider Viral expansions with the TBA equation (17). We take the following form in this limit:

$$\begin{aligned} e^{-\varepsilon(k)/T} &= \mathcal{Z}e^{-\frac{k^2}{T}} e^{\int_{-\infty}^{\infty} dq a(k-q) \ln\left(1+\mathcal{Z}e^{-\frac{q^2}{T}}\right)} \\ &\approx \mathcal{Z}e^{-\frac{k^2}{T}}\left[1 + \mathcal{Z}\int_{-\infty}^{\infty} dq a(k-q) e^{-\frac{q^2}{T}}\right]. \end{aligned} \tag{36}$$

Here $\mathcal{Z} = e^{\mu/T}$ is fugacity and the function $a(x) = \frac{1}{2\pi}\frac{2c}{c^2+x^2}$. Similarly, we take Taylor expansion in terms of the fugacity in the Eq. (18) and after a length algebra, we find the pressure up to the second Viral coefficient

$$p = p_0 + \frac{T^{\frac{3}{2}}}{\sqrt{2\pi}}\mathcal{Z}^2 p_2, \tag{37}$$

where $p_2 = -\frac{1}{2} + \int_{-\infty}^{\infty} dq' a(2q') e^{-\frac{2q'^2}{T}}$ gives the two-body interaction term. In the above equation, the $p_0 = -\frac{T}{2\pi}\int_{-\infty}^{\infty} dk \ln\left(1 - \mathcal{Z}e^{-\frac{k^2}{T}}\right)$ is the pressure of the free bosons. The result (37) gives the Maxwell–Boltzmann statistics in the limit of $T \to \infty$. This result contains subtle corrections from quantum statistics and dynamical interactions to the Boltzmann statistics. We summarize these different regimes in Table 1.

Generalized exclusion statistics. In the strong coupling limit $\gamma \to \infty$, Tonks–Girardeau gas is equivalent to the free gas with pure Fermi statistics. In fact, the 1D δ-function interacting bosons can map onto an ideal gas[24] with the generalized exclusion statistics (GES)[25] described by the statistical parameter α_{ij}. The GES parameter α_{ij} is defined through the linear relation $\Delta d_i/\Delta N_j = -\alpha_{ij}$,[25] i.e., the number of available single particle states of species i, denoted by d_i, depends on

Table 1. Different regimes in the 1D Lieb–Linger Bose gas. In this table $\gamma = c/n$ is the dimensionless interacting strength. BG stands for Bose gas, QC denotes quantum critical, TG is the initials of Tonks and Girardeau for denoting the strongly repulsive Bose gas. Here K is the Luttinger parameter with $K \approx 1 + 4/\gamma + 4/\gamma^2 + 16\pi^2/(3\gamma^3)$ and $K \approx \pi/\sqrt{\gamma}$ for strongly and weakly interactions, respectively.

γ/T	Weak interaction	Strong interaction
$T = 0, \mu > 0$	Bogoliubov BG	TG gas
$T < T^*, \mu > T$	TLL, $K \gg 1$	TLL, $K \sim 1$
$T_d > T > T^*, \mu < T$	nearly ideal BG	QC, $z = 1, \nu = 1/2$
$T < T^*, \mu < 0$	Quasiclassical gas	Quasiclassical gas
$T_d > T > T^*, \mu < 0$	QC, nearly TG gas	QC, $z = 1, \nu = 1/2$
$T > T_d$	Boltzmann gas	Boltzmann gas

the number of other species $\{N_j\}$ when one particle of species i is added. Here d_i is given by [24,26]

$$d_i(\{N_j\}) = G_i^0 - \sum_j \alpha_{ij} N_j. \tag{38}$$

Here $G_i^0 = d(\{0\})$ is the number of available single particle states with no particles present in the system. The GES parameter can be determined by the BA two-body scattering matrix

$$\alpha_{ij} := \alpha(k, k') = \delta(k, k') - \frac{1}{2\pi}\theta'(k - k') \tag{39}$$

through counting the available states in quasimomentum space. Here the function $\theta'(x) = 2c/(c^2 + x^2)$.

The equivalence between the 1D interacting bosons and ideal particles obeying GES was established on the equivalence between the TBA

$$\varepsilon(k) = \varepsilon^0(k) - \mu - \frac{T}{2\pi} \int_\infty^\infty dk' \theta'(k - k') \ln\left(1 + e^{-\frac{\varepsilon(k')}{T}}\right) \tag{40}$$

and the GES equation

$$(1 + w_i) \prod_j \left(\frac{w_j}{1 + w_j}\right)^{\alpha_{ji}} = e^{(\varepsilon_i - \mu_i)/K_B T} \tag{41}$$

with $w_i = e^{\varepsilon(k_i)/K_B T}$. Here the GES equation can be rewritten as [27]

$$\varepsilon(k_i) = \varepsilon_i - \mu_i - K_B T \sum_j (\delta_{ji} - \alpha_{ji}) \left(1 + e^{-\frac{\varepsilon(k_j)}{K_B T}}\right) \tag{42}$$

that is identical to the TBA equation (17). Here $\varepsilon(k)$ is the dressed energy measuring the energy above the Fermi surface, with $\varepsilon^0(k) = k^2$. In the above equations, the most probable distribution is given by $\sum_j (w_j \delta_{ij} + \beta_{ij}) n_j = 1$ with $n_i = N_i/G_i$ and $\beta_{ij} = \alpha_{ij} G_j / G_i$.[24] For $\alpha = 0$ the GES result (41) reduces to Bose statistics. For $\alpha_{ij} = 1$ it reduces to Fermi statistics. The above equivalence is valid for an arbitrary interaction strength.

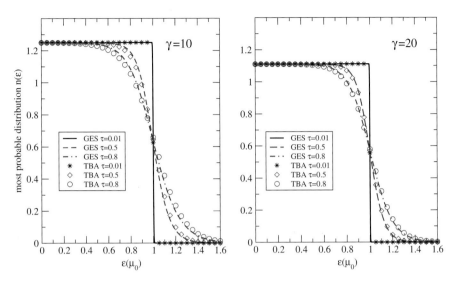

Fig. 4. Mapping the Tonks–Girardeau gas to the ideal gas with the nonmutual GES: the most probable distribution profiles $n(\varepsilon)$ for the values $\gamma = 10$ (left panel) and $\gamma = 20$ (right panel) at different values of the degeneracy temperature $\tau = K_B T/T_d$, where $T_d = \frac{\hbar^2}{2m} n^2$ is the quantum degeneracy temperature. At zero temperature $n(\varepsilon) = 1/\alpha$ leads to a Fermi surface at $\varepsilon = \mu_0$. The pure Fermi statistics can be obtained in the limit $\gamma \to \infty$. At the low temperatures a smooth distributions near the Fermi surface μ_0 are observed. The solid lines show the most probable GES distribution for ideal particles (44). The symbols stand for the corresponding distributions evaluated from the TBA equation (17) for interacting bosons. The figure from Ref. 11.

We note that this type of connection between TBA and GES is general for the 1D interacting systems, for example, for the interacting fermions.[28] This is mainly because the statistical interaction and dynamical interaction (δ-function interaction) are transmutable. It is obvious that in weak interaction regime, the quasimomentum k_i of one particle essentially depends on the ones of the all rest particles. This naturally gives a mutual statistics in the GES. Whereas, in the strong repulsive regime, the dynamical interaction drives the system into the nonmutual GES, i.e. the quasimomenta of the particle have a equal spacing in the quasimomentum space. In the strong coupling regime and at low temperatures, i.e. the hole density ρ_h in Eq. (9) is very small, then we have the relation between the particle density ρ and the hole density ρ_h is given by $2\pi (\alpha \rho + \rho_h) \approx 1$ that indicates a nonmutal GES with

$$\alpha \approx 1 - \frac{2}{\gamma}. \tag{43}$$

This statistical parameter determines the most probable distribution $n(\varepsilon) = 2\pi \rho$ as

$$n(\varepsilon) = \frac{w(\varepsilon)}{1 + (\alpha - 1)w(\varepsilon)}, \tag{44}$$

$$\alpha \ln(1 - w(\varepsilon)) - \ln w(\varepsilon) = \frac{\varepsilon - \mu}{K_B T}. \tag{45}$$

Figure 4 shows that the GES distribution (44) and (45) is consistent with the one obtained from the TBA equation (17) for the 1D interacting Bose gas. In this sense, the strong coupling Bose gas can map onto the ideal gas with a nonmutual statistics.

3. Discussion

Yang and Yang's seminal work[1] immediately triggered important applications in the thermodynamics of the Heisenberg–Ising chain[29] and a variety of 1D integrable models,[30–36] the Kondo problem,[37,38] etc. Recent developments of exactly solvable models largely involve the study of universal quantum many-body phenomena, such as quantum liquids, quantum correlations and quantum criticality in the context of the Yang–Yang's method, see Ref. 13. In particular, the exact result for the thermodynamics of interacting fermions in 1D provides precise understanding of the universal Bose–Einstein condensate–Bardeen–Cooper–Schreiffer crossover thermodynamics, spin-charge separation, Fulde–Ferrel–Larkin–Ovchinnikov pairing, quantum criticality, Tan's contact and universal dimensionless ratios.[39] Remarkably, the Yang–Yang thermodynamics of the 1D Bose gas has been tested in recent experiments.[40–48] Mathematical theory of this kind has now become testable in experiments.

Acknowledgments

XWG thanks professor C. N. Yang and professor Z.-Q. Ma for encouragements and helpful discussions. This work has been supported by the NNSFC under grant No. 11374331 and the key NNSFC grant No. 11534014.

References

1. C. N. Yang and C. P. Yang, *J. Math. Phys.* **10**, 1115 (1969).
2. C. N. Yang, *Phys. Rev. Lett.* **19**, 1312 (1967).
3. H. Bethe, *Z. Physik* **71**, 205 (1931).
4. E. H. Lieb and W. Liniger, *Phys. Rev.* **130**, 1605 (1963).
5. M. Gaudin, *Phys. Lett. A* **24**, 55 (1967).
6. E. H. Lieb and F. Y. Wu, *Phys. Rev. Lett.* **20**, 1445 (1968).
7. B. Sutherland, *Phys. Rev. Lett.* **20**, 98 (1968).
8. N. Andrei, K. Furuya, and J. H. Lowenstein, *Rev. Mod. Phys.* **55**, 331 (1983).
9. S. Dukelsky, S. Pittel, and G. Sierra, *Rev. Mod. Phys.* **76**, 643 (2004).
10. F. H. L. Essler, H. Frahm, F. Göhmann, A. Klümper and V. E. Korepin, *The One-Dimensional Hubbard Model* (Cambridge University Press, 2005).
11. M. T. Batchelor, X. W. Guan, N. Oelkers, and Z. Tsuboi, *Adv. Phys.* **56**, 465 (2007).
12. M. A. Cazalilla, R. Citro, T. Giamarchi, E. Orignac and M. Rigol, *Rev. Mod. Phys.* **83**, 1405 (2011).
13. X. W. Guan, M. T. Batchelor and C. Lee, *Rev. Mod. Phys.* **85**, 1633 (2013).

14. R. J. Baxter, *Exactly Solved Models in Statistical Mechanics*, (Academic Press, 1982).
15. C. N. Yang, *Selected Papers 1945–1980* (W. H. Freeman and Company, 1983).
16. C. N. Yang, *Selected Papers II* (World Scientific, 2013).
17. V. E. Korepin, A. G. Izergin, and N. M. Bogoliubov, *Quantum Inverse Scattering Method and Correlation Functions* (Cambridge University Press, 1993).
18. B. Sutherland, *Beautiful Models: 70 Years of Exactly Solved Quantum Many-Body Problems* (World Scientific, 2004).
19. M. Gaudin, *The Bethe Wavefunction* (Cambridge University Press, 2014).
20. M. Takahashi, *Thermodynamics of One-Dimensional Solvable Models* (Cambridge University Press, 1999).
21. X. W. Guan and M. T. Batchelor, *J. Phys. A: Math. Theor.* **44**, 102001 (2011).
22. Y.-Z. Jiang, Y.-Y. Chen and X.-W. Guan, *Chinese Phys. B* **24**, 050311 (2015).
23. A. Klumper, O. I. Patu, *Phys. Rev. A* **90**, 053626 (2014).
24. Y.-S. Wu, *Phys. Rev. Lett.* **73**, 922 (1994); D. Bernard and Y.-S. Wu, arXiv:cond-mat/9404025.
25. F. D. M. Haldane, *Phys. Rev. Lett.* **67**, 937 (1991).
26. S. B. Isakov, *Phys. Rev. Lett.* **73**, 2150 (1994).
27. M. T. Batchelor and X.-W. Guan, *Laser Phys. Lett.* **4**, 77 (2007).
28. X.-W. Guan, M. T. Batchelor, C. Lee, and M. Bortz, *Phys. Rev. B* **76**, 085120 (2007).
29. M. Gaudin, *Phys. Rev. Lett.* **26**, 1301 (1971).
30. M. Takahashi, *Prog. Theor. Phys.* **46**, 401 (1971).
31. M. Takahashi, *Prog. Theor. Phys.* **46**, 1388 (1971).
32. M. Takahashi, *Prog. Theor. Phys.* **47**, 69 (1972).
33. M. Takahashi, *Prog. Theor. Phys.* **50**, 1519 (1973).
34. M. Takahashi, *Prog. Theor. Phys.* **52**, 103 (1973).
35. P. Schlottmann, *J. Phys. Condens. Matter* **5**, 5869 (1993).
36. P. Schlottmann, *J. Phys. Condens. Matter* **6**, 1359 (1994).
37. V. M. Filyov, M. Tsvelick, and P. B. Wiegmann, *Phys. Lett. A* **81**, 175 (1981).
38. J. H. Lowenstein, *Surveys High Energy Phys.* **2**, 207 (1981).
39. Y.-C. Yu, Y.-Y. Chen, H.-Q. Lin, R. A. Roemer, X.-W. Guan, arXiv:1508.00763.
40. A. H. van Amerongen, J. J. P. van Es, P. Wicke, K.V. Kheruntsyan, and N. J. van Druten, *Phys. Rev. Lett.* **100**, 090402 (2008).
41. J. Armijo, T. Jacqmin, K.V. Kheruntsyan, and I. Bouchoule, *Phys. Rev. Lett.* **105**, 230402 (2010).
42. J. Armijo, T. Jacqmin, K.V. Kheruntsyan, and I. Bouchoule, *Phys. Rev. A* **83**, 021605(R) (2011).
43. J. Armijo, *Phys. Rev. Lett.* **108**, 225306 (2012).
44. T. Jacqmin, J. Armijo, T. Berrada, K.V. Kheruntsyan, and I. Bouchoule, *Phys. Rev. Lett.* **106**, 230405 (2011).
45. H.-P. Stimming, N. J. Mauser, J. Schmiedmayer, and I. E. Mazets, *Phys. Rev. Lett.* **105**, 015301 (2010).

46. P. Krüger, P., S. Hofferberth, I. E. Mazets, I. Lesanovsky, and J. Schmiedmayer, *Phys. Rev. Lett.* **105**, 265302 (2010).
47. Y. Sagi, M. Brook, I. Almog, and N. Davidson, *Phys. Rev. Lett.* **108**, 093002 (2012).
48. A. Vogler, R. Labouvie, F. Stubenrauch, G. Barontini, V. Guarrera, and H. Ott, *Phys. Rev. A* **88**, 031603 (2013).

Chern–Simons Theory, Vassiliev Invariants, Loop Quantum Gravity and Functional Integration Without Integration

Louis H. Kauffman

Mathematics Department, University of Illinois at Chicago,
Chicago, IL 60607-7045, USA
kauffman@uic.edu
www.math.uic.edu/~kauffman

This paper is an exposition of the relationship between Witten's Chern–Simons functional integral and the theory of Vassiliev invariants of knots and links in three-dimensional space. We conceptualize the functional integral in terms of equivalence classes of functionals of gauge fields and we do not use measure theory. This approach makes it possible to discuss the mathematics intrinsic to the functional integral rigorously and without functional integration. Applications to loop quantum gravity are discussed.

Keywords: Knot; link; Vassiliev invariant; Lie algebra; Chern–Simons form; functional integral; Kontsevich integral; loop quantum gravity; Kodama state.

1. Introduction

This paper is an introduction to how Vassiliev invariants in knot theory arise naturally in the context of Witten's functional integral. The relationship between Vassiliev invariants and Witten's integral has been known since Bar-Natan's thesis[6] where he discovered, through this connection, how to define Lie algebraic weight systems for these invariants.

This paper is written in a context of "integration without integration." The idea is as follows. Let $F(A)$, $G(A)$, $H(A)$ be functionals of a gauge field A that vanish rapidly as the amplitude of the field goes to infinity. We say that $F \sim G$ if $F - G = DH$ where D denotes a gauge functional derivative. We define $\int F(A)$ to be the equivalence class of $F(A)$. By definition, this integral satisfies integration by parts, and it is a useful conceptual substitute for a functional integral over all gauge fields (modulo gauge equivalence). We replace the usual notion of functional integral with such equivalence classes.

The paper is a sequel to Refs. 16 and 15. In these papers we show somewhat more about the relationship of Vassiliev invariants and the Witten functional integral. In particular, we show how the Kontsevich integrals (used to give rigorous definitions of these invariants) arise as Feynman integrals in the perturbative expansion of the Witten functional integral. See also the work of Labastida and Pérez[18] on this same subject. The result is an interpretation of the Kontsevich integrals in terms of the light-cone gauge and thereby extending the original work of Fröhlich and King.[9] The purpose of this paper is to give an exposition of the beginnings of these relationships, to introduce diagrammatic techniques that illuminate the connections, and to show how the integral can be fruitfully formulated in terms of certain equivalence classes of functionals of gauge fields.

The paper is divided into six sections beyond the introduction. Section 2 discusses Vassiliev invariants and invariants of rigid vertex graphs. Section 3 discusses the concept of replacing integrals by equivalence classes. Section 4 introduces the basic formalism and shows how the functional integral, regarded without integration, is related directly to knot invariants and particularly, Vassiliev invariants. Section 5 discusses the formalism of the perturbative expansion of the Witten integral. Section 6 is a sketch of the loop transform, useful in loop quantum gravity and ends with a quick discussion of the Kodama state with references to recent literature. Section 7 discusses how the Kontsevich integrals for Vassiliev invariants arise from the perturbation expansion.

2. Vassiliev Invariants and Invariants of Rigid Vertex Graphs

If $V(K)$ is a (Laurent polynomial valued, or more generally — commutative ring valued) invariant of knots, then it can be naturally extended to an invariant of rigid vertex graphs[11] by defining the invariant of graphs in terms of the knot invariant via an "unfolding" of the vertex. That is, we can regard the vertex as a "black box" and replace it by any tangle of our choice. Rigid vertex motions of the graph preserve the contents of the black box, and hence implicate ambient isotopies of the link obtained by replacing the black box by its contents. Invariants of knots and links that are evaluated on these replacements are then automatically rigid vertex invariants of the corresponding graphs. If we set up a collection of multiple replacements at the vertices with standard conventions for the insertions of the tangles, then a summation over all possible replacements can lead to a graph invariant with new coefficients corresponding to the different replacements. In this way each invariant of knots and links implicates a large collection of graph invariants. See Refs. 11 and 12.

The simplest tangle replacements for a 4-valent vertex are the two crossings, positive and negative, and the oriented smoothing. Let $V(K)$ be any invariant of knots and links. Extend V to the category of rigid vertex embeddings of 4-valent graphs by the formula

$$V(K_*) = aV(K_+) + bV(K_-) + cV(K_0),$$

where K_+ denotes a knot diagram K with a specific choice of positive crossing, K_- denotes a diagram identical to the first with the positive crossing replaced by a negative crossing and K_* denotes a diagram identical to the first with the positive crossing replaced by a graphical node.

This formula means that we define $V(G)$ for an embedded 4-valent graph G by taking the sum

$$V(G) = \sum_S a^{i_+(S)} b^{i_-(S)} c^{i_0(S)} V(S)$$

with the summation over all knots and links S obtained from G by replacing a node of G with either a crossing of positive or negative type, or with a smoothing of the crossing that replaces it by a planar embedding of non-touching segments (denoted 0). It is not hard to see that if $V(K)$ is an ambient isotopy invariant of knots, then, this extension is an rigid vertex isotopy invariant of graphs. In rigid vertex isotopy the cyclic order at the vertex is preserved, so that the vertex behaves like a rigid disk with flexible strings attached to it at specific points.

There is a rich class of graph invariants that can be studied in this manner. The Vassiliev invariants[5,7] constitute the important special case of these graph invariants where $a = +1$, $b = -1$ and $c = 0$. Thus $V(G)$ is a Vassiliev invariant if

$$V(K_*) = V(K_+) - V(K_-)$$

Call this formula the *exchange identity* for the Vassiliev invariant V. See Fig. 1. V is said to be of *finite type* k if $V(G) = 0$ whenever $|G| > k$ where $|G|$ denotes the number of (4-valent) nodes in the graph G. The notion of finite type is of extraordinary significance in studying these invariants. One reason for this is the following basic Lemma.

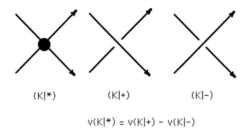

Fig. 1. Exchange identity for Vassiliev invariants.

Lemma. If a graph G has exactly k nodes, then the value of a Vassiliev invariant v_k of type k on G, $v_k(G)$, is independent of the embedding of G.

Proof. The different embeddings of G can be represented by link diagrams with some of the 4-valent vertices in the diagram corresponding to the nodes of G. It suffices to show that the value of $v_k(G)$ is unchanged under switching of a crossing. However, the exchange identity for v_k shows that this difference is equal to the evaluation of v_k on a graph with $k + 1$ nodes and hence is equal to zero. This completes the proof. □

The upshot of this Lemma is that Vassiliev invariants of type k are intimately involved with certain abstract evaluations of graphs with k nodes. In fact, there are restrictions (the four-term relations) on these evaluations demanded by the topology and it follows from results of Kontsevich[5] that such abstract evaluations actually determine the invariants. The knot invariants derived from classical Lie algebras are all built from Vassiliev invariants of finite type. All of this is directly related to Witten's functional integral.[29]

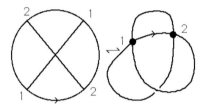

Fig. 2. Chord diagrams.

In the next few figures we illustrate some of these main points. In Fig. 2 we show how one associates a so-called chord diagram to represent the abstract graph associated with an embedded graph. The chord diagram is a circle with arcs connecting those points on the circle that are welded to form the corresponding graph. In Fig. 3 we illustrate how the four-term relation is a consequence of topological

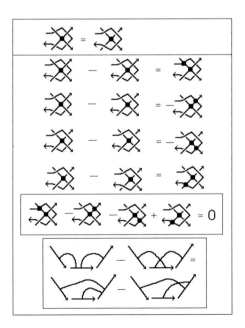

Fig. 3. The four-term relation from topology.

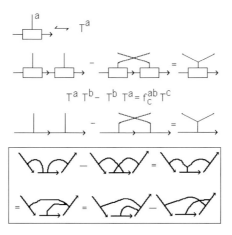

Fig. 4. The four-term relation from categorical Lie algebra.

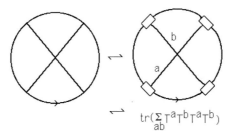

Fig. 5. Calculating Lie algebra weights.

invariance. In Fig. 4 we show how the four-term relation is a consequence of the abstract pattern of the commutator identity for a matrix Lie algebra. This shows that the four-term relation is directly related to a categorical generalization of Lie algebras. Figure 5 illustrates how the weights are assigned to the chord diagrams in the Lie algebra case — by inserting Lie algebra matrices into the circle and taking a trace of a sum of matrix products.

3. Integration Without Integration

Recall that if $Z = \int_{-\infty}^{\infty} e^{-x^2/2} dx$ then

$$Z^2 = \int_{-\infty}^{\infty} \int_{-\infty}^{\infty} e^{-(x^2+y^2)/2} dx\, dy$$

$$= \int_0^{2\pi} \int_0^{\infty} e^{-r^2/2} r dr\, d\theta$$

$$= 2\pi \int_0^{\infty} e^{-r^2/2} r dr = 2\pi.$$

Whence
$$Z = \sqrt{2\pi}.$$

Furthermore, if
$$Z(J) = \int_{-\infty}^{\infty} e^{-x^2/2+Jx} dx,$$
then
$$Z(J) = \int_{-\infty}^{\infty} e^{-(x-J)^2/2+J^2/2} dx$$
$$= e^{J^2/2} \int_{-\infty}^{\infty} e^{-(x-J)^2/2} dx$$
$$= e^{J^2/2} \int_{-\infty}^{\infty} e^{-x^2/2} dx$$
$$= e^{J^2/2} Z(0) = \sqrt{2\pi} e^{J^2/2}.$$

Now examine how much of this calculation could be done if we did not know about the existence of the integral, or if we did not know how to calculate explicitly the values of these integrals across the entire real line. Given that we believed in the existence of the integrals, and that we could use properties such as change of variable giving
$$\int_{-\infty}^{\infty} e^{-(x-J)^2/2} dx = \int_{-\infty}^{\infty} e^{-x^2/2} dx,$$
we could deduce the relative result stating that
$$Z(J) = \int_{-\infty}^{\infty} e^{-x^2/2+Jx} dx = e^{J^2/2} \int_{-\infty}^{\infty} e^{-x^2/2} dx.$$

From this we can deduce that
$$d^n Z(J)/dJ^n |_{J=0} = d^n/dJ^n \int_{-\infty}^{\infty} e^{-x^2/2+Jx} dx = \int_{-\infty}^{\infty} x^n e^{-x^2/2} dx.$$

Hence
$$\int_{-\infty}^{\infty} x^n e^{-x^2/2} dx = d^n (e^{J^2/2})/dJ^n |_{J=0} \int_{-\infty}^{\infty} e^{-x^2/2} dx.$$

But now, lets go a step further and imagine that we really have no theory of integration available. Then we are in the position of freshman calculus where one defines $\int f$ to be "any" function g such that $dg/dx = f$. One *defines* the integral in this form of elementary calculus to be the anti-derivative, and this takes care of the matter for a while! What are we really doing in freshman calculus? We are noting that for integration on an interval $[a,b]$, if two functions f and g satisfy $f - g = dh/dx$ for some differentiable function h, then we have that

$$\int_a^b (f-g) = \int_a^b dh/dx = h(b) - h(a).$$

If the function $h(x)$ vanishes as x goes to infinity, then we have that

$$\int_{-\infty}^{\infty} f dx = \int_{-\infty}^{\infty} g dx$$

when $f - g = dh/dx$. This suggests turning things upside down and *defining an equivalence relation on functions*

$$f \sim g$$

if

$$f - g = dh/dx$$

where $h(x)$ is a function vanishing at infinity. Then *we define the integral*

$$\int f(x)$$

to be the equivalence class of the function $f(x)$. This "integral" represents integration from minus infinity to plus infinity but it is defined only as an equivalence class of functions. An "actual" integral, like the Riemann, Lesbeque or Henstock integral is a well-defined real valued function that is constant on these equivalence classes.

We shall say that $f(x)$ is *rapidly vanishing at infinity* if $f(x)$ and all its derivatives are vanishing at infinity. For simplicity, we shall assume that all functions under consideration have convergent power series expansions so that $f(x+J) = f(x) + f'(x)J + f''(x)J^2/2! + \cdots$, and that they are rapidly vanishing at infinity. It then follows that

$$f(x+J) = f(x) + d(f(x)J + f'(x)J^2/2! + \cdots)/dx \sim f(x),$$

and hence we have that $\int f(x+J) = \int f(x)$, giving translation invariance when J is a constant.

We have shown the following Proposition.

Proposition. Let $f(x), g(x), h(x)$ be functions rapidly vanishing at infinity (with power series representations). Let $\int f$ denote the equivalence class of the function f where $f \sim g$ means that $f - g = Dh$ where $Dh = dh/dx$. Then this integral satisfies the following properties

(1) If $f \sim g$ then $\int f = \int g$.
(2) If k is a constant, then $\int (kf + g) = k \int f + \int g$.
(3) If J is a constant, then $\int f(x+J) = \int f(x)$.
(4) $\int Dh = 0$ where 0 denotes the equivalence class of the zero function. Hence $\int f(Dg) + \int (Df)g = \int D(fg) = 0$, so that integration by parts is valid with vanishing boundary conditions at infinity.

Note that $e^{-x^2/2}$ is rapidly vanishing at infinity. We now see that most of the calculations that we made about $e^{-x^2/2}$ were actually statements about the equivalence class of this function:

$$e^{-x^2/2 + Jx} = e^{-(x-J)^2/2 + J^2/2} = e^{J^2/2} e^{-(x-J)^2/2} \sim e^{J^2/2} e^{-x^2/2},$$

whence

$$\int e^{-x^2/2 + Jx} = e^{J^2/2} \int e^{-x^2/2}.$$

3.1. Functional derivatives

In order to generalize the ideas presented in this section to the context of functional integrals, we need to discuss the concept of functional derivatives. We are given a *functional* $F(\alpha(x))$ whose argument $\alpha(x)$ is a function of a variable x. We wish to define the *functional derivative* $\delta F(\alpha(x))/\delta\alpha(x_0)$ of $F(\alpha(x))$ with respect to $\alpha(x)$ at a given point x_0. The idea is to regard each $\alpha(x_0)$ as a separate variable, giving $F(\alpha(x))$ the appearance of a function of infinitely many variables. In order to formalize this notion one needs to use generalized functions (distributions) such as the Dirac delta function $\delta(x)$, a distribution with the property that $\int_a^b \delta(x_0) f(x) dx = f(x_0)$ for any integrable function $f(x)$ and point x_0 in the interval $[a,b]$. One defines the functional derivative by the formula

$$\delta F(\alpha(x))/\delta\alpha(x_0) = \lim_{\epsilon \to 0}[F(\alpha(x) + \delta(x_0)\epsilon) - F(\alpha(x))]/\epsilon.$$

Note that if

$$F(\alpha(x)) = \alpha(x)^2$$

then

$$\delta F(\alpha(x))/\delta\alpha(x_0) = \lim_{\epsilon \to 0}[(\alpha(x) + \delta(x_0)\epsilon)^2 - \alpha(x)^2]/\epsilon$$
$$= \lim_{\epsilon \to 0}[2\alpha(x)\delta(x_0)\epsilon + \delta(x_0)^2 \epsilon^2]/\epsilon$$
$$= 2\alpha(x)\delta(x_0).$$

While if

$$G(\alpha(x)) = \int_a^b \alpha(x)^2 dx$$

then

$$\delta G(\alpha(x))/\delta\alpha(x_0) = 2\alpha(x_0)$$

when $x_0 \in [a,b]$. More generally, if

$$G(\alpha(x)) = \int_a^b f(\alpha(x)) dx$$

for a differentiable function f, then

$$\delta G(\alpha(x))/\delta\alpha(x_0) = f'(\alpha(x_0)).$$

These examples show that the results of a functional differentiation can be either a distribution or a function, depending upon the context of the original functional.

In the case of a path integral of the type used in quantum mechanics, one wants to integrate a functional $F(p)$ over paths $p(t)$ with t in an interval $[0,1]$. The functional takes the form

$$F(p) = e^{(i/\hbar)\int_0^1 S(p(t))dt}$$

and the traditional Feynman path integral has the form

$$\int dP e^{(i/\hbar)\int_0^1 S(p(t))dt},$$

giving the amplitude for a particle to travel from $a = p(0)$ to $b = p(1)$, the integration proceeding over all paths with these initial and ending points.

Here the equivalence relation corresponding to the functional integral is $F \sim G$ if $F - G = DH$ where

$$DH = \delta H(p)/\delta p(t_0)$$

for some time t_0 and some $H(p)$. Again we need to specify the class of functionals and to say what it means for a functional to "vanish at infinity." Since we are integrating over all paths, we need a notion of size for a path. This can be defined by

$$||p|| = \left(\int_0^1 |p(t)|^2 dt\right)^{1/2}.$$

Note that for $F(p) = e^{(i/\hbar)\int_0^1 S(p(t))dt}$ we have

$$\delta F(p)/\delta p(t_0) = (i/\hbar)\left[\delta\int_0^1 S(p(t))dt/\delta p(t_0)\right]F(p)$$
$$= (i/\hbar)S'(p(t_0))F(p).$$

Here we see the fact that the integral can be dominated by contributions from paths where this variation is zero. Note that in order to estimate this stationary phase contribution to the functional integral, one needs more than just a definition of the integral as an equivalence class of functionals. Nevertheless, we shall see in the next section that these equivalence classes do give insight into the topology associated with Witten's integral.

4. Vassiliev Invariants and Witten's Functional Integral

In Ref. 29 Edward Witten proposed a formulation of a class of 3-manifold invariants as generalized Feynman integrals taking the form $Z(M)$ where

$$Z(M) = \int DA e^{(ik/4\pi)S(M,A)}.$$

Here M denotes a 3-manifold without boundary and A is a gauge field (also called a gauge potential or gauge connection) defined on M. The gauge field is a one-form on a trivial G-bundle over M with values in a representation of the Lie algebra of G. The group G corresponding to this Lie algebra is said to be the gauge group. In this integral the "action" $S(M, A)$ is taken to be the integral over M of the trace of the Chern–Simons three-form $A \wedge dA + (2/3) A \wedge A \wedge A$. (The product is the wedge product of differential forms.)

$Z(M)$ integrates over all gauge fields modulo gauge equivalence.

The formalism and internal logic of Witten's integral supports the existence of a large class of topological invariants of 3-manifolds and associated invariants of knots and links in these manifolds.

The invariants associated with this integral have been given rigorous combinatorial descriptions but questions and conjectures arising from the integral formulation are still outstanding. Specific conjectures about this integral take the form of just how it implicates invariants of links and 3-manifolds, and how these invariants behave in certain limits of the coupling constant k in the integral. Many conjectures of this sort can be verified through the combinatorial models. On the other hand, the really outstanding conjecture about the integral is that it exists! At the present time there is no measure theory or generalization of measure theory that supports it. Here is a formal structure of great beauty. It is also a structure whose consequences can be verified by a remarkable variety of alternative means.

In this section we will examine the formalism of Witten's approach via a generalization of our sketch of "integration without integration." In order to do this we need to consider functions $f(A)$ of gauge connections A and a notion of equivalence, $f \sim g$, taking the form $f - g = Dh$ where D is a gauge functional derivative. Since these notions need defining, we first discuss them in the context of the integrand of Witten's integral. Thus for a while, we shall speak of Witten's integral, but let it be known that this integral will soon be replaced by an equivalence class of functions just as happened in the last section!

The formalism of the Witten integral implicates invariants of knots and links corresponding to each classical Lie algebra. In order to see this, we need to introduce the Wilson loop. The Wilson loop is an exponentiated version of integrating the gauge field along a loop K in three space that we take to be an embedding (knot) or a curve with transversal self-intersections. For this discussion, the Wilson loop will be denoted by the notation

$$W_K(A) = \langle K|A\rangle$$

to denote the dependence on the loop K and the field A. It is usually indicated by the symbolism $\text{tr}(Pe^{\oint_K A})$. Thus

$$W_K(A) = \langle K|A\rangle = \text{tr}\left(Pe^{\oint_K A}\right).$$

Here the P denotes path ordered integration — we are integrating and exponentiating matrix valued functions, and so must keep track of the order of the operations. The symbol tr denotes the trace of the resulting matrix. This Wilson loop integration exists by normal means and will not be replaced by function classes.

With the help of the Wilson loop functional on knots and links, Witten writes down a functional integral for link invariants in a 3-manifold M:

$$Z(M,K) = \int DA e^{(ik/4\pi)S(M,A)} \text{tr}\left(Pe^{\oint_K A}\right)$$

$$= \int DA e^{(ik/4\pi)S} \langle K|A\rangle.$$

Here $S(M,A)$ is the Chern–Simons Lagrangian, as in the previous discussion. We abbreviate $S(M,A)$ as S and write $\langle K|A\rangle$ for the Wilson loop. Unless otherwise mentioned, the manifold M will be the three-dimensional sphere S^3.

An analysis of the formalism of this functional integral reveals quite a bit about its role in knot theory. This analysis depends upon key facts relating the curvature of the gauge field to both the Wilson loop and the Chern–Simons Lagrangian. The idea for using the curvature in this way is due to Lee Smolin[20] (see also Ref. 19). To this end, let us recall the local coordinate structure of the gauge field $A(x)$, where x is a point in three-space. We can write $A(x) = A_k^a(x) T_a dx^k$ where the index a ranges from 1 to m with the Lie algebra basis $\{T_1, T_2, T_3, \ldots, T_m\}$. The index k goes from 1 to 3. For each choice of a and k, $A_k^a(x)$ is a smooth function defined on three-space. In $A(x)$ we sum over the values of repeated indices. The Lie algebra generators T_a are matrices corresponding to a given representation of the Lie algebra of the gauge group G. We assume some properties of these matrices as follows:

1. $[T_a, T_b] = if^{abc}T_c$ where $[x,y] = xy - yx$, and f^{abc} (the matrix of structure constants) is totally antisymmetric. There is summation over repeated indices.
2. $\text{tr}(T_a T_b) = \delta_{ab}/2$ where δ_{ab} is the Kronecker delta ($\delta_{ab} = 1$ if $a = b$ and zero otherwise).

We also assume some facts about curvature. (The reader may enjoy comparing with the exposition in Ref. 13. But note the difference of conventions on the use of i in the Wilson loops and curvature definitions.) The first fact is the relation of Wilson loops and curvature for small loops:

Fact 1. The result of evaluating a Wilson loop about a very small planar circle around a point x is proportional to the area enclosed by this circle times the corresponding value of the curvature tensor of the gauge field evaluated at x. The curvature tensor is written

$$F^a_{rs}(x)T_a dx^r dy^s.$$

It is the local coordinate expression of $F = dA + A \wedge A$.

Application of Fact 1. Consider a given Wilson line $\langle K|S \rangle$. Ask how its value will change if it is deformed infinitesimally in the neighborhood of a point x on the line. Approximate the change according to Fact 1, and regard the point x as the place of curvature evaluation. Let $\delta\langle K|A\rangle$ denote the change in the value of the line. $\delta\langle K|A\rangle$ is given by the formula

$$\delta\langle K|A\rangle = dx^r\, dx^s\, F^{rs}_a(x) T_a \langle K|A\rangle.$$

This is the first order approximation to the change in the Wilson line.

In this formula it is understood that the Lie algebra matrices T_a are to be inserted into the Wilson line at the point x, and that we are summing over repeated indices. This means that each $T_a \langle K|A\rangle$ is a new Wilson line obtained from the original line $\langle K|A\rangle$ by leaving the form of the loop unchanged, but inserting the matrix T_a into that loop at the point x. In Fig. 6 we have illustrated this mode of insertion of Lie algebra into the Wilson loop. Here and in further illustrations in this section we use $W_K(A)$ to denote the Wilson loop. Note that in the diagrammatic version shown in Fig. 6 we have let small triangles with legs indicate dx^i. The legs correspond to indices just as in our work in the last section with Lie algebras and chord diagrams. The curvature tensor is indicated as a circle with three legs corresponding to the indices of F^{rs}_a.

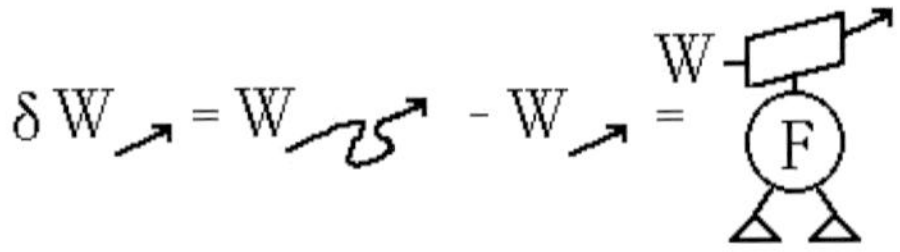

Fig. 6. Lie algebra and curvature tensor insertion into the Wilson loop.

Notation. In the diagrams in this section we have dropped mention of the factor of $(1/4\pi)$ that occurs in the integral. This convention saves space in the figures. In these figures L denotes the Chern–Simons Lagrangian.

Remark. In thinking about the Wilson line $\langle K|A\rangle = \text{tr}(Pe^{\oint_K A})$, it is helpful to recall Euler's formula for the exponential:
$$e^x = \lim_{n\to\infty}(1 + x/n)^n.$$
The Wilson line is the limit, over partitions of the loop K, of products of the matrices $(1 + A(x))$ where x runs over the partition. Thus we can write symbolically,
$$\langle K|A\rangle = \prod_{x\in K}(1 + A(x))$$
$$= \prod_{x\in K}(1 + A_k^a(x)T_a dx^k).$$
It is understood that a product of matrices around a closed loop connotes the trace of the product. The ordering is forced by the one-dimensional nature of the loop. Insertion of a given matrix into this product at a point on the loop is then a well-defined concept. If T is a given matrix then it is understood that $T\langle K|A\rangle$ denotes the insertion of T into some point of the loop. In the case above, it is understood from context in the formula that the insertion is to be performed at the point x indicated in the argument of the curvature.

Remark. The previous remark implies the following formula for the variation of the Wilson loop with respect to the gauge field:
$$\delta\langle K|A\rangle/\delta(A_k^a(x)) = dx^k T_a \langle K|A\rangle.$$
Varying the Wilson loop with respect to the gauge field results in the insertion of an infinitesimal Lie algebra element into the loop. Figure 7 gives a diagrammatic form for this formula. In that figure we use a capital D with up and down legs to denote the derivative $\delta/\delta(A_k^a(x))$. Insertions in the Wilson line are indicated directly by matrix boxes placed in a representative bit of line.

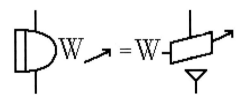

Fig. 7. Differentiating the Wilson line.

Proof.
$$\delta\langle K|A\rangle/\delta(A_k^a(x))$$
$$= \delta\prod_{y\in K}(1 + A_k^a(y)T_a dy^k)/\delta(A_k^a(x))$$

$$= \prod_{y<x\in K}(1+A_k^a(y)T_a dy^k)[T_a dx^k] \prod_{y>x\in K}(1+A_k^a(y)T_a dy^k)$$

$$= dx^k T_a \langle K|A\rangle.$$

Fact 2. The variation of the Chern–Simons Lagrangian S with respect to the gauge potential at a given point in three-space is related to the values of the curvature tensor at that point by the following formula:

$$F_{rs}^a(x) = \epsilon_{rst}\delta S/\delta(A_t^a(x)).$$

Here ϵ_{abc} is the epsilon symbol for three indices, i.e. it is $+1$ for positive permutations of 123 and -1 for negative permutations of 123 and zero if any two indices are repeated. A diagrammatic for this formula is shown in Fig. 8. □

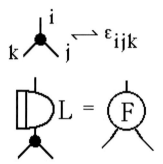

Fig. 8. Variational formula for curvature.

The Functional Equivalence Relation. With these facts at hand, we are prepared to define our equivalence relation on functions of gauge fields. Given a function $F(A)$ of a gauge field A, we let DF denote any gauge functional derivative of $f(A)$. That is

$$DF = \delta F(A)/\delta(A_k^a(x)).$$

Note that

$$D\langle K|A\rangle = \delta\langle K|A\rangle/\delta(A_k^a(x)) = dx^k T_a\langle K|A\rangle$$

with the insertion conventions as explained above. Then we say that functionals F and G are *integrally equivalent* ($F \sim G$) if there exists an H such that $DH = F - G$. We stipulate that all functionals in the discussion are rapidly vanishing at infinity, where this is taken to mean that $F(A)$ goes to zero as $||A||$ goes to infinity, and the same is true for all functional derivatives of F. Here the norm

$$||A|| = \Sigma_{i,a}\int_{R^3}(A_i^a)^2 d\text{ vol}$$

where d vol is the volume form on R^3 and it is assumed that all gauge fields have finite norm in this sense.

We then define the integral

$$Z(M,K) = \int DA e^{(ik/4\pi)S(M,A)} \operatorname{tr}(Pe^{\oint_K A}) = \int DA e^{(ik/4\pi)S} \langle K|A\rangle$$

to be the equivalence class of the functional

$$e^{(ik/4\pi)S(M,A)} \operatorname{tr}(Pe^{\oint_K A}).$$

We invite the reader to make this interpretation throughout the derivations that follow. It will then be apparent that much of what is usually taken for formal heuristics about the functional integral is actually a series of structural remarks about these equivalence classes. Of course, one needs to know that the equivalence classes are non-trivial to make a complete story. An existent integral would supply that key ingredient. It its absence, we can examine that structure that can be articulated at the level of the equivalence classes.

We are prepared to determine how the Witten integral behaves under a small deformation of the loop K.

Theorem.
1. Let $Z(K) = Z(S^3, K)$ and let $\delta Z(K)$ denote the change of $Z(K)$ under an infinitesimal change in the loop K. Then

$$\delta Z(K) = (4\pi i/k) \int dA e^{(ik/4\pi)S} [\mathrm{Vol}] T_a T_a \langle K|A\rangle$$

where $\mathrm{Vol} = \epsilon_{rst} dx^r dx^s dx^t$.

The sum is taken over repeated indices, and the insertion is taken of the matrices $T_a T_a$ at the chosen point x on the loop K that is regarded as the "center" of the deformation. The volume element $\mathrm{Vol} = \epsilon_{rst}\, dx_r\, dx_s\, dx_t$ is taken with regard to the infinitesimal directions of the loop deformation from this point on the original loop.
2. The same formula applies, with a different interpretation, to the case where x is a double point of transversal self-intersection of a loop K, and the deformation consists in shifting one of the crossing segments perpendicularly to the plane of intersection so that the self-intersection point disappears. In this case, one T_a is inserted into each of the transversal crossing segments so that $T_a T_a \langle K|A\rangle$ denotes a Wilson loop with a self-intersection at x and insertions of T_a at $x + \epsilon_1$ and $x + \epsilon_2$ where ϵ_1 and ϵ_2 denote small displacements along the two arcs of K that intersect at x. In this case, the volume form is nonzero, with two directions coming from the plane of movement of one arc, and the perpendicular direction is the direction of the other arc.

Proof.

$$\delta Z(K) = \int DA e^{(ik/4\pi)S} \delta\langle K|A\rangle$$

$$= \int DA e^{(ik/4\pi)S} dx^r dy^s F^a_{rs}(x) T_a \langle K|A\rangle$$

$$= \int DAe^{(ik/4\pi)S} dx^r dy^s \epsilon_{rst}(\delta S/\delta(A_t^a(x))) T_a \langle K|A \rangle$$

$$= (-4\pi i/k) \int DA(\delta e^{(ik/4\pi)S}/\delta(A_t^a(x))) \epsilon_{rst} dx^r dy^s T_a \langle K|A \rangle$$

$$= (4\pi i/k) \int DAe^{(ik/4\pi)S} \epsilon_{rst} dx^r dy^s (\delta T_a \langle K|A \rangle/\delta(A_t^a(x)))$$

(integration by parts and the boundary terms vanish)

$$= (4\pi i/k) \int DAe^{(ik/4\pi)S} [\text{Vol}] T_a T_a \langle K|A \rangle.$$

This completes the formalism of the proof. In the case of part 2, a change of interpretation occurs at the point in the argument when the Wilson line is differentiated. Differentiating a self-intersecting Wilson line at a point of self-intersection is equivalent to differentiating the corresponding product of matrices with respect to a variable that occurs at two points in the product (corresponding to the two places where the loop passes through the point). One of these derivatives gives rise to a term with volume form equal to zero, the other term is the one that is described in part 2. This completes the proof of the Theorem. □

The formalism of this proof is illustrated in Fig. 9.

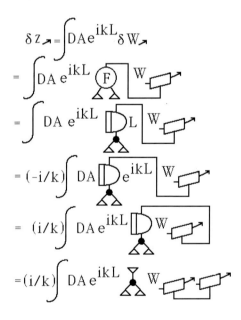

Fig. 9. Varying the functional integral by varying the line.

In the case of switching a crossing the key point is to write the crossing switch as a composition of first moving a segment to obtain a transversal intersection of the diagram with itself, and then to continue the motion to complete the switch. One then analyzes separately the case where x is a double point of transversal self-intersection of a loop K, and the deformation consists in shifting one of the crossing segments perpendicularly to the plane of intersection so that the self-intersection point disappears. In this case, one T_a is inserted into each of the transversal crossing segments so that $T^a T^a \langle K | A \rangle$ denotes a Wilson loop with a self-intersection at x and insertions of T^a at $x + \epsilon_1$ and $x + \epsilon_2$ as in part 2 of the Theorem above. The first insertion is in the moving line, due to curvature. The second insertion is the consequence of differentiating the self-touching Wilson line. Since this line can be regarded as a product, the differentiation occurs twice at the point of intersection, and it is the second direction that produces the non-vanishing volume form.

Up to the choice of our conventions for constants, the switching formula is, as shown (see Fig. 10).

$$Z(K_+) - Z(K_-) = (4\pi i/k) \int DA e^{(ik/4\pi)S} T_a T_a \langle K_{**} | A \rangle$$
$$= (4\pi i/k) Z(T^a T^a K_{**}),$$

where K_{**} denotes the result of replacing the crossing by a self-touching crossing. We distinguish this from adding a graphical node at this crossing by using the double star notation.

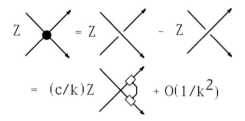

Fig. 10. The difference formula.

A key point is to notice that the Lie algebra insertion for this difference is exactly what is done (in chord diagrams) to make the weight systems for Vassiliev invariants (without the framing compensation). In order to extend the Heuristic at this point we need to assume the analog of a perturbative expansion for the integral. That is, we assume that there are invariants of regular isotopy of K, $Z_n(K)$ and that

$$e^{(ik/4\pi)S(A)} \langle K | A \rangle \sim \Sigma_{n=0}^{\infty} k^{-n} Z_n(K).$$

Note that since we have shown that the equivalence class of

$$e^{(ik/4\pi)S(A)} \langle K | A \rangle$$

is a regular isotopy invariant, it is not at all implausible to assume that there is a power series representative of this functional whose coefficients are numerical regular isotopy invariants. It is this assumption that allows one to make contact with numerical evaluations. The assumption of this power series representation corresponds to the formal perturbative expansion of the Witten integral. One obtains Vassiliev invariants as coefficients of the powers of $(1/k^n)$. Thus the formalism of the Witten functional integral takes one directly to these weight systems in the case of the classical Lie algebras. In this way the functional integral is central to the structure of the Vassiliev invariants.

5. Perturbative Expansion

Letting M^3 be a three-manifold and K a knot or link in M^3, we write

$$\psi(A) = e^{ikL(A)} W_K(A),$$

and replace A by $A/\sqrt{k}$ then we can write

$$\hat{\psi}(A) = e^{\frac{i}{4\pi}\int_{M^3} \operatorname{tr}(A \wedge dA)} e^{\frac{i}{6\pi\sqrt{k}}\int_{M^3} \operatorname{tr}(A \wedge A \wedge A)} W_K(A/\sqrt{k}).$$

It is the equivalence class of this functional of gauge fields that contains much topological information about knots and links in the three-manifold M^3. We can expand this functional by taking the explicit formula for the Wilson loop:

$$W_K(A/\sqrt{k}) = \operatorname{tr}\left(\prod_{x \in K}(1 + A(x)/\sqrt{k})\right),$$

$$\hat{\psi}(A) = e^{\frac{i}{4\pi}\int_{M^3} \operatorname{tr}(A \wedge dA)} e^{\frac{i}{6\pi}\int_{M^3} \operatorname{tr}\left(\frac{1}{\sqrt{k}} A \wedge A \wedge A\right)} \operatorname{tr}\left(\prod_{x \in K}(1 + A(x)/\sqrt{k})\right),$$

$$\operatorname{tr}\left(\prod_{x \in K}\left(1 + \frac{1}{\sqrt{k}} A(x)\right)\right) = \operatorname{tr}\left(1 + \frac{1}{\sqrt{k}}\int_K A + \frac{1}{k}\int_{K_1 < K_2} A(x_1)A(x_2) + \cdots\right)$$

where

$$\int_{K_1 < \cdots < K_n} A(x_1)A(x_2)\cdots A(x_n) = \int_{\underbrace{K \times \cdots \times K}_{n} = K^n} A(x_1)A(x_2)\cdots A(x_n),$$

$$\vec{x} = (x_1, x_2, \ldots, x_n) \in K^n \text{ with } x_1 < x_2 < \cdots < x_n.$$

This is an iterated integrals expression for the Wilson loop.

Our functional is transformed into a perturbative series in powers of $1/k$. The equivalence class of each term in the series (when M^3 is the three-sphere S^3) is formally a Vassiliev invariant as we have described in the previous section. A more intense look at the structure of these functionals can be accomplished by gauge-fixing as we show in the last section.

6. The Loop Transform and Loop Quantum Gravity

Suppose that $\psi(A)$ is a (complex-valued) function defined on gauge fields. Then we define formally the *loop transform* $\hat{\psi}(K)$, a function on embedded loops in three-dimensional space, by the formula

$$\hat{\psi}(K) = \int \psi(A) W_K(A).$$

Note that we could also write

$$\hat{\psi}(K) = \psi(A) W_K(A)$$

where it is understood that the right-hand side of the equation represents its integral equivalence class. Then we can look at it as a function of the loop K *and* as a function of the gauge field A. This changes one's point of view about the loop transform. We are really examining a hybrid function of both a possibly knotted loop K and a gauge field A. The important structure is the relationships that ensue in the integral equivalence class between varying A and varying K. Nevertheless, we shall continue to use integral signs to remind the reader that we are working with the integral equivalence classes of these functionals.

If Δ is a differential operator defined on $\psi(A)$, then we can use this integral transform to shift the effect of Δ to an operator on loops via integration by parts:

$$\widehat{\Delta\psi}(K) = \int \Delta\psi(A) W_K(A)$$

$$= -\int \psi(A) \Delta W_K(A).$$

When Δ is applied to the Wilson loop the result can be an understandable geometric or topological operation. In Figs. 11–13 we illustrate this situation with diagrammatically defined operators G and H.

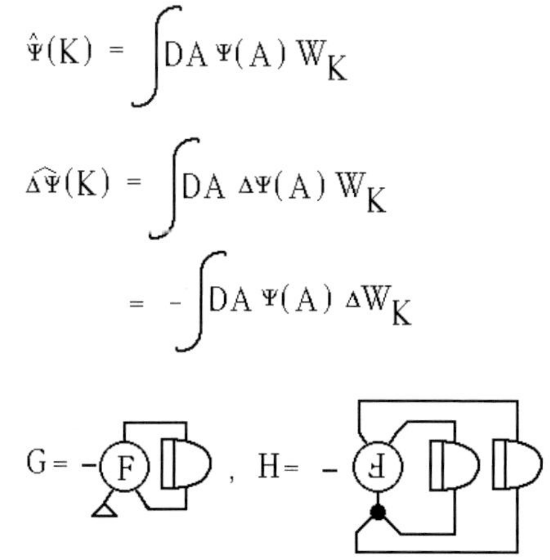

$$\hat{\Psi}(K) = \int DA \, \Psi(A) \, W_K$$

$$\widehat{\Delta\Psi}(K) = \int DA \, \Delta\Psi(A) \, W_K$$

$$= -\int DA \, \Psi(A) \, \Delta W_K$$

$$G = -\boxed{F}\,\,\,,\quad H = -\boxed{J}$$

Fig. 11. The loop transform and operators G and H.

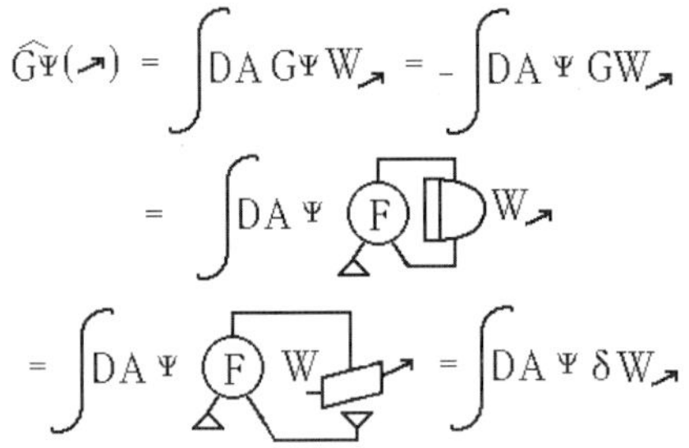

$$\widehat{G\Psi}(\nearrow) = \int DA\, G\Psi W_{\nearrow} = -\int DA\, \Psi\, GW_{\nearrow}$$

$$= \int DA\, \Psi\, \underset{\triangle}{(F)}\!\!\!\square W_{\nearrow}$$

$$= \int DA\, \Psi\, \underset{\triangle}{(F)}\, W_{\!\!\!\bot\!\!\nearrow} = \int DA\, \Psi\, \delta W_{\nearrow}$$

Fig. 12. The diffeomorphism constraint.

We see from Fig. 12 that

$$\widehat{G\psi}(K) = \delta\widehat{\psi}(K)$$

where this variation refers to the effect of varying K by a small loop. As we saw in this section, this means that if $\widehat{\psi}(K)$ is a topological invariant of knots and links, then $\widehat{G\psi}(K) = 0$ for all embedded loops K. This condition is a transform analogue of the equation $G\psi(A) = 0$. This equation is the differential analogue of an invariant of knots and links. It may happen that $\delta\widehat{\psi}(K)$ is not strictly zero, as in the case of our framed knot invariants. For example with

$$\psi(A) = e^{(ik/4\pi)\int \text{tr}(A\wedge dA + (2/3)A\wedge A\wedge A)}$$

we conclude that $\widehat{G\psi}(K)$ is zero for flat deformations (in the sense of this section) of the loop K, but can be nonzero in the presence of a twist or curl. In this sense the loop transform provides a subtle variation on the strict condition $G\psi(A) = 0$. This Chern–Simons functional $\psi(A)$ can be seen to be a state of loop quantum gravity.

In Ref. 2 and earlier publications by these authors, the loop transform is used to study a reformulation and quantization of Einstein gravity. The differential geometric gravity theory is reformulated in terms of a background gauge connection and in the quantization, the Hilbert space consists in functions $\psi(A)$ that are required to satisfy the constraints

$$G\psi = 0$$

and

$$H\psi = 0,$$

where H is the operator shown in Fig. 13. Thus we see that $\widehat{G}(K)$ can be partially zero in the sense of producing a framed knot invariant, and (from Fig. 13 and the antisymmetry of the epsilon) that $\widehat{H}(K)$ is zero for non-self-intersecting loops. This means that the loop transforms of G and H can be used to investigate a subtle variation of the original scheme for the quantization of gravity. The appearance of the Chern–Simons state

$$\psi(A) = e^{(ik/4\pi)\int \text{tr}(A\wedge dA + (2/3)A\wedge A\wedge A)}$$

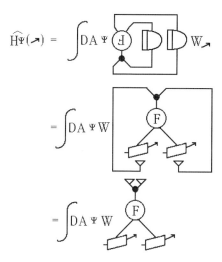

Fig. 13. The Hamiltonian constraint.

is quite remarkable in this theory, where it is commonly referred to as the *Kodama State*. See Refs. 21–28 for a number of references about this state, up to the present day. Many ways of weaving this relationship of knot theory and quantum gravity have been devised, from examining directly the Kodama state and its relationship with DeSitter space, to the evolution of spin networks and spin foams to handle the fundamental topological conditions in the theory.

7. Wilson Lines, Axial Gauge and the Kontsevich Integrals

In this section we follow the gauge fixing method used by Fröhlich and King.[9] Their paper was written before the advent of Vassiliev invariants, but contains, as we shall see, nearly the whole story about the Kontsevich integral. A similar approach to ours can be found in Ref. 18. In our case we have simplified the determination of the inverse operator for this formalism and we have given a few more details about the calculation of the correlation functions than is customary in physics literature. I hope that this approach makes this subject more accessible to mathematicians. A heuristic argument of this kind contains a great deal of valuable mathematics. It is clear that these matters will eventually be given a fully rigorous treatment. In fact, in the present case there is a rigorous treatment, due to Albevario and Sen-Gupta[1] of the functional integral *after* the light-cone gauge has been imposed.

Let (x^0, x^1, x^2) denote a point in three-dimensional space. Change to light-cone coordinates

$$x^+ = x^1 + x^2$$

and

$$x^- = x^1 - x^2.$$

Let t denote x^0.

Then the gauge connection can be written in the form

$$A(x) = A_+(x)dx^+ + A_-(x)dx^- + A_0(x)dt.$$

Let $CS(A)$ denote the Chern–Simons integral (over the three-dimensional sphere)

$$CS(A) = (1/4\pi) \int \mathrm{tr}(A \wedge dA + (2/3) A \wedge A \wedge A).$$

We define *axial gauge* to be the condition that $A_- = 0$. We shall now work with the functional integral of the previous section under the axial gauge restriction. In axial gauge we have that $A \wedge A \wedge A = 0$ and so

$$CS(A) = (1/4\pi) \int \mathrm{tr}(A \wedge dA).$$

Letting $\partial_\pm$ denote partial differentiation with respect to $x^\pm$, we get the following formula in axial gauge

$$A \wedge dA = (A_+ \partial_- A_0 - A_0 \partial_- A_+) dx^+ \wedge dx^- \wedge dt.$$

Thus, after integration by parts, we obtain the following formula for the Chern–Simons integral:

$$CS(A) = (1/2\pi) \int \mathrm{tr}(A_+ \partial_- A_0) dx^+ \wedge dx^- \wedge dt.$$

Letting ∂_i denote the partial derivative with respect to x_i, we have that

$$\partial_+ \partial_- = \partial_1^2 - \partial_2^2.$$

If we replace x^2 with ix^2 where $i^2 = -1$, then $\partial_+ \partial_-$ is replaced by

$$\partial_1^2 + \partial_2^2 = \nabla^2.$$

We now make this replacement so that the analysis can be expressed over the complex numbers.

Letting

$$z = x^1 + ix^2,$$

it is well known that

$$\nabla^2 \ln(z) = 2\pi \delta(z)$$

where $\delta(z)$ denotes the Dirac delta function and $\ln(z)$ is the natural logarithm of z. Thus we can write

$$(\partial_+ \partial_-)^{-1} = (1/2\pi) \ln(z).$$

Note that $\partial_+ = \partial_z = \partial/\partial z$ after the replacement of x^2 by ix^2. As a result we have that

$$(\partial_-)^{-1} = \partial_+(\partial_+\partial_-)^{-1} = \partial_+(1/2\pi)\ln(z) = 1/2\pi z.$$

Now that we know the inverse of the operator ∂_- we are in a position to treat the Chern–Simons integral as a quadratic form in the pattern

$$(-1/2)\langle A, LA \rangle = -iCS(A)$$

where the operator

$$L = \partial_-.$$

Since we know L^{-1}, we can express the functional integral as a Gaussian integral: We replace

$$Z(K) = \int DA e^{ikCS(A)} \, \mathrm{tr}(Pe^{\oint_K A})$$

by

$$Z(K) = \int DA e^{iCS(A)} \, \mathrm{tr}(Pe^{\oint_K A/\sqrt{k}})$$

by sending A to $(1/\sqrt{k})A$. We then replace this version by

$$Z(K) = \int DA e^{(-1/2)\langle A, LA \rangle} \, \mathrm{tr}(Pe^{\oint_K A/\sqrt{k}}).$$

In this last formulation we can use our knowledge of L^{-1} to determine the correlation functions and express $Z(K)$ perturbatively in powers of $(1/\sqrt{k})$.

Proposition. Letting

$$\langle \phi(A) \rangle = \int DA e^{(-1/2)\langle A, LA \rangle} \phi(A) \Big/ \int DA e^{(-1/2)\langle A, LA \rangle}$$

for any functional $\phi(A)$, we find that

$$\langle A_+^a(z,t) A_+^b(w,s) \rangle = 0,$$
$$\langle A_0^a(z,t) A_0^b(w,s) \rangle = 0,$$
$$\langle A_+^a(z,t) A_0^b(w,s) \rangle = \kappa \delta^{ab} \delta(t-s)/(z-w)$$

where κ is a constant.

Proof sketch. Let us recall how these correlation functions are obtained. The basic formalism for the Gaussian integration is in the pattern

$$\langle A(z)A(w) \rangle = \int DA e^{(-1/2)\langle A, LA \rangle} A(z)A(w) \Big/ \int DA e^{(-1/2)\langle A, LA \rangle}$$
$$= ((\partial/\partial J(z))(\partial/\partial J(w))|_{J=0}) e^{(1/2)\langle J, L^{-1} J \rangle}.$$

Letting $G * J(z) = \int dw G(z-w) J(w)$, we have that when

$$LG(z) = \delta(z)$$

($\delta(z)$ is a Dirac delta function of z.) then

$$LG * J(z) = \int dw LG(z-w) J(w) = \int dw \delta(z-w) J(w) = J(z).$$

Thus $G * J(z)$ can be identified with $L^{-1} J(z)$.

In our case

$$G(z) = 1/2\pi z$$

and

$$L^{-1} J(z) = G * J(z) = \int dw J(w)/(z-w).$$

Thus

$$\langle J(z), L^{-1} J(z) \rangle = \langle J(z), G * J(z) \rangle$$
$$= (1/2\pi) \int \text{tr}\left(J(z)\left(\int dw J(w)/(z-w)\right)\right) dz$$
$$= (1/2\pi) \iint dz dw \, \text{tr}(J(z) J(w))/(z-w).$$

The results on the correlation functions then follow directly from differentiating this expression. Note that the Kronecker delta on Lie algebra indices is a result of the corresponding Kronecker delta in the trace formula $\text{tr}(T_a T_b) = \delta_{ab}/2$ for products of Lie algebra generators. The Kronecker delta for the $x^0 = t, s$ coordinates is a consequence of the evaluation at J equal to zero. □

We are now prepared to give an explicit form to the perturbative expansion for

$$\langle K \rangle = Z(K) / \int DA e^{(-1/2)\langle A, LA \rangle}$$
$$= \int DA e^{(-1/2)\langle A, LA \rangle} \text{tr}\left(P e^{\oint_K A/\sqrt{k}}\right) / \int DA e^{(-1/2)\langle A, LA \rangle}$$
$$= \int DA e^{(-1/2)\langle A, LA \rangle} \text{tr}\left(\prod_{x \in K} (1 + (A/\sqrt{k}))\right) / \int DA e^{(-1/2)\langle A, LA \rangle}$$
$$= \sum_n (1/k^{n/2}) \oint_{K_1 < \cdots < K_n} \langle A(x_1) \cdots A(x_n) \rangle.$$

The latter summation can be rewritten (Wick expansion) into a sum over products of pair correlations, and we have already worked out the values of these. In the formula above we have written $K_1 < \cdots < K_n$ to denote the integration over variables $x_1, \ldots, x_n$ on K so that $x_1 < \cdots < x_n$ in the ordering induced on the loop K by choosing a basepoint on the loop. After the Wick expansion, we get

$$\langle K \rangle = \sum_m (1/k^m) \oint_{K_1 < \cdots < K_n} \sum_{P=\{x_i < x'_i | i=1,\ldots,m\}} \prod_i \langle A(x_i)A(x'_i) \rangle.$$

Now we know that

$$\langle A(x_i)A(x'_i) \rangle = \langle A^a_k(x_i) A^b_l(x'_i) \rangle T_a T_b dx^k dx^l.$$

Rewriting this in the complexified axial gauge coordinates, the only contribution is

$$\langle A^a_+(z,t) A^b_0(s,w) \rangle = \kappa \delta^{ab} \delta(t-s)/(z-w).$$

Thus

$$\langle A(x_i)A(x'_i) \rangle = \langle A^a_+(x_i) A^a_0(x'_i) \rangle T_a T_a dx^+ \wedge dt + \langle A^a_0(x_i) A^a_+(x'_i) \rangle T_a T_a dx^+ \wedge dt$$
$$= (dz - dz')/(z - z')[i/i']$$

where $[i/i']$ denotes the insertion of the Lie algebra elements $T_a T_a$ into the Wilson loop.

As a result, for each partition of the loop and choice of pairings $P = \{x_i < x'_i | i = 1, \ldots, m\}$ we get an evaluation D_P of the trace of these insertions into the loop. This is the value of the corresponding chord diagram in the weight systems for Vassiliev invariants. These chord diagram evaluations then figure in our formula as shown below:

$$\langle K \rangle = \sum_m (1/k^m) \sum_P D_P \oint_{K_1 < \cdots < K_n} \bigwedge_{i=1}^m (dz_i - dz'_i)/((z_i - z'_i)).$$

This is a Wilson loop ordering version of the Kontsevich integral. To see the usual form of the integral appear, we change from the time variable (parametrization) associated with the loop itself to time variables associated with a specific global direction of time in three-dimensional space that is perpendicular to the complex plane defined by the axial gauge coordinates. It is easy to see that this results in one change of sign for each segment of the knot diagram supporting a pair correlation where the segment is oriented (Wilson loop parameter) downward with respect to the global time direction. This results in the rewrite of our formula to

$$\langle K \rangle = \sum_m (1/k^m) \sum_P (-1)^{|P\downarrow|} D_P \int_{t_1 < \cdots < t_n} \bigwedge_{i=1}^m (dz_i - dz'_i)/((z_i - z'_i))$$

where $|P \downarrow|$ denotes the number of points (z_i, t_i) or (z'_i, t_i) in the pairings where the knot diagram is oriented downward with respect to global time. The integration around the Wilson loop has been replaced by integration in the vertical time direction and is so indicated by the replacement of $\{K_1 < \cdots < K_n\}$ with $\{t_1 < \cdots < t_n\}$.

The coefficients of $1/k^m$ in this expansion are exactly the Kontsevich integrals for the weight systems D_P. See Fig. 14.

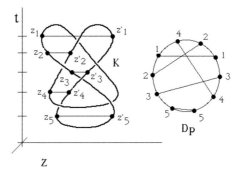

Fig. 14. Applying the Kontsevich integral.

It was Kontsevich's insight to see (by different means) that these integrals could be used to construct Vassiliev invariants from arbitrary weight systems satisfying the four-term relations. Here we have seen how these integrals arise naturally in the axial gauge fixing of the Witten functional integral.

Remark. The reader will note that we have not made a discussion of the role of the maxima and minima of the space curve of the knot with respect to the height direction (t). In fact one has to take these maxima and minima very carefully into account and to divide by the corresponding evaluated loop pattern (with these maxima and minima) to make the Kontsevich integral well-defined and actually invariant under ambient isotopy (with appropriate framing correction as well). The corresponding difficulty appears here in the fact that because of the gauge choice the Wilson lines are actually only defined in the complement of the maxima and minima and one needs to analyze a limiting procedure to take care of the inclusion of these points in the Wilson line.

Acknowledgments

We thank students and colleagues for many stimulating conversations on the themes of this paper, and we thank the organizers of the Conference on 60 Years of Yang–Mills Gauge Field Theories (25 to 28 May 2015) for the invitation and opportunity to speak about these ideas in Singapore.

References

1. S. Albevario and A. Sen-Gupta, *Commun. Math. Phys.* **186**, 563 (1997).
2. A. Ashtekar, C. Rovelli and L. Smolin, *Phys. Rev. Lett.* **69**, 237 (1992).
3. D. Altschuler and L. Freidel, *Commun. Math. Phys.* **187**, 261 (1997).
4. M. F. Atiyah, *The Geometry and Physics of Knots* (Cambridge University Press, 1990).
5. D. Bar-Natan, *Topology* **34**, 423 (1995).

6. D. Bar-Natan, Perturbative aspects of the Chern–Simons topological quantum field theory, Ph.D. thesis, Princeton University, June 1991.
7. J. Birman and X. S. Lin, *Invent. Math.* **111**, 225 (1993).
8. C. Dewitt-Morette, P. Cartier and A. Folacci (eds.), *Functional Integration: Basics and Applications*, NATO ASI Series B Physics, Vol. 361 (Plenum Press, 1997).
9. J. Fröhlich and C. King, *Commun. Math. Phys.* **126**, 167 (1989).
10. H. Kleinert, *Path Integrals in Quantum Mechanics, Statistics and Polymer Physics*, 2nd edn. (World Scientific, Singapore, 1995).
11. L. H. Kauffman, *Amer. Math. Mon.* **95**, 195 (1988).
12. L. H. Kauffman and P. Vogel, *J. Knot Theory Ramifications* **1**, 59 (1992).
13. L. H. Kauffman, *Knots and Physics* (World Scientific, 1991 and 1993).
14. L. H. Kauffman, *J. Math. Phys.* **36**, 2402 (1995).
15. L. H. Kauffman, *AIP Conf. Proc.* **453**, 368 (1998).
16. L. H. Kauffman, *Physica A* **281**, 173 (2000).
17. H. Kleinert, *Grand Treatise on Functional Integration* (World Scientific, 1999).
18. J. M. F. Labastida and E. Pérez, *J. Math. Phys.* **39**, 5183 (1998).
19. P. Cotta-Ramusino, E. Guadagnini, M. Martellini and M. Mintchev, *Nucl. Phys. B* **330**, 557 (1990).
20. L. Smolin, *Mod. Phys. Lett. A* **4**, 1091 (1989).
21. L. Smolin, Quantum gravity with a positive cosmological constant, arXiv:hep-th/0209079.
22. J. Magueijo and L. Bethke, New ground state for quantum gravity, arXiv:1207.0637.
23. W. Wieland, Complex Ashtekar variables, the Kodama state and spinfoam gravity, arXiv:1105.2330.
24. A. Randono, In search of quantum de Sitter space: Generalizing the Kodama state, arXiv:0709.2905.
25. H. Kodama, *Int. J. Mod. Phys. D* **1**, 439 (1992), arXiv:gr-qc/9211022.
26. C. Soo, *Class. Quantum Grav.* **19**, 1051 (2002), arXiv:gr-qc/0109046.
27. J. Pullin, *AIP Conf. Proc.* **317**, 141 (1994), arXiv:hep-th/9301028.
28. E. Witten, A note on the Chern–Simons and Kodama wave functions, arXiv:gr-qc/0306083.
29. E. Witten, *Commun. Math. Phys.* **121**, 351 (1989).

The Scattering Equations and Their Off-Shell Extension

York-Peng Yao

Department of Physics, University of Michigan, Ann Arbor, MI 48109, USA
yyao@umich.edu

1. Introduction

There are two beautiful theories, both of which can be formulated on gauge principles. One is general relativity,[1] which has passed all the classical tests, but has resisted consistency with quantum corrections. The other is non-Abelian[2] (Yang–Mills) theory and its extensions, which are the backbones of electroweak and strong interactions. As we also know, while the gauge principle[3] is elegant, calculations based on perturbation expansion in the past were cumbersome. The situation has been fundamentally changed in the last 15 years or so, because analyticity has been better understood and used, which results in recurrence relations,[4] gauge freedom has been better deployed in the language of spinors[5] and taken advantage of, and most surprisingly, many hidden symmetries in the amplitudes have been unveiled.[6,7] The last discovery leads to a much better understanding (and more manageable) of the connection between gauge and graviton amplitudes.[8] We shall review some of these progresses, which give a historical perspective of the scattering equation formulation,[9] which seems to give the possibility of a new interaction principle, perhaps beyond local field theories.

In the original formulation, CHY (Cachazo–He–Yuan)[9] is for on-shell tree amplitudes. We[10] shall show that by a simple modification, which is mostly dictated by Möbius covariance and off-shell energy–momentum conservation, CHY is also valid for tree Green's functions. This establishes CHY as a viable unitary theory.

2. A Brief Review

The arduous work shared by all practitioners in serious gauge field calculations based on Feynman diagrams is well documented. A further frustration is that there are tremendous amounts of cancellations in the sum, which definitely imply that a brighter path should exist. The proposal by BCFW[4] to analytically continue the amplitudes by introducing a complex variable z into some of the momenta is to splice higher point amplitudes into a sum of products of lower point amplitudes, or to establish recurrence relations. Somewhat later, it was found that gauge amplitudes actually possess a lot more symmetries than the explicit local one, which historically guides the construction of the Yang–Mills interaction. Thus, if we write the color-dressed n particle amplitude as

$$A^{cd} = \sum \frac{c_i n_i}{d_i}, \qquad (1)$$

in which c_i are products of color coefficients, d_i products of appropriate propagators, and n_i are appropriate kinematical numerators, then for certain subsets of c_i, they satisfy Jacobi identities

$$c_j + c_k + c_l = 0. \qquad (2)$$

It was pointed out that one can shuffle terms in the numerators, so that they satisfy

$$n_j + n_k + n_l = 0 \qquad (3)$$

without changing the amplitude. This is known as the color-kinematics duality.[6] It reduces the number of independent c_i, n_i and the color-stripped amplitudes A_i (coefficients to independent c_i in Eq. (1)) to $(n-2)!$. For these independent quantities, we can write Eq. (1) as

$$|A\rangle = M|N\rangle, \qquad A^{cd} = \langle C|M|N\rangle, \qquad (4)$$

where M is an $(n-2)! \times (n-2)!$ symmetric matrix with elements made of sums of products of propagators. It was soon realized that one can perform proper shifts on the numerators, without changing the color stripped amplitudes, known as generalized gauge transformations.[7] The underlying reason is that the matrix has $(n-3) \times (n-3)!$ number of independent eigenvectors $|\lambda_i^0\rangle$ with zero eigenvalue.[11] Consequently, one can shift

$$|N'\rangle = |N\rangle + \sum_{j=1}^{(n-3)\times(n-3)!} f_j|\lambda_j^0\rangle, \qquad |A\rangle = M|N'\rangle, \qquad (5)$$

where f_j are arbitrary functions. The number of relevant n_i and A_i is reduced to $(n-3)!$.

A fascinating result from string theories is the connection between gauge theories and gravitational theories.[8] It relates an n-particle graviton amplitude to a sum of products of two color stripped n-particle amplitudes, multiplied by some kinematical factors. With the color-kinematics duality, it is restated as the double copy postulate[7]: If one set of n_i satisfies Eq. (3) and another set $\tilde{n}_i$, which may or may not be from a different gauge theory and may or may not satisfy Eq. (3), are put together, then up to some coupling constants, a corresponding graviton amplitude will emerge as

$$A^{gr} = c\langle \tilde{N}|M|N\rangle. \qquad (6)$$

As we shall see, some of these major advances are incorporated in the CHY formula.

3. CHY Formulation[9]

A revolutionary formulation appeared about three years ago. As we mentioned earlier, there is a huge number of Feynman diagrams for the amplitude of even a moderate number of gauge particles or gravitons. It was also stated that despite the large difference between the Lagrangian for gauge theories, which has only three- and four-particle vertices and that for gravity, which has an infinite number of them, there is a simple and intimate relation between the amplitudes of these two theories. For a given n, CHY is able to put the sum of all diagrams in one seemingly simple formula. Furthermore, by doubling some part of the kernel, CHY is able to summarize the double copy postulate. We shall first explain the on-shell formula and then go on to show what modifications we need to make, so that we promote it to become a Green's function. It is important to remind the reader that Green's functions are more general in a theory, because with them we can fuse some of the off-shell lines together to form higher point Green's functions or amplitudes and/or loops.

3.1. Double color scalars

From our discussion, it is clear that an important ingredient in the construction of gauge and graviton tree amplitudes is the set of propagator matrices M^n, on which the dynamics are hung as numerators. Let us recall that these matrices are constructed by using the total antisymmetry of the structure constants at the three vertices and Jacobi identities in the four-particle sub-diagrams, and color-kinematics duality. Another way to realize these requirements is just to consider a set of scalars with a bi-color interaction

$$\frac{1}{3!} f_{abc} g_{ijk} \phi^{ai} \phi^{bj} \phi^{ck}, \tag{7}$$

where f_{abc} are the totally antisymmetric structure constants of one Lie algebra and g_{ijk} are from a different Lie algebra. The n scalar amplitude will be proportional to

$$A_{\text{double-color scalar}} \approx \langle F|M|G \rangle, \tag{8}$$

in which $\langle F|$ is a set of $(n-2)!$ independent coefficients formed with f's and $|G\rangle$ from g's.

It turns out that there is another interesting way to obtain such a matrix. We first write down a set of equations

$$\hat{f}_i = \sum_{j \neq i} \frac{s_{ij}}{\sigma_{ij}}, \quad i,j = 1,\ldots,n, \tag{9}$$

in which σ_i is a set of numbers with

$$\sigma_{ij} = \sigma_i - \sigma_j = -\sigma_{ji}, \quad s_{ij} = -(k_i + k_j)^2. \tag{10}$$

Under a Möbius transformation

$$\sigma_i \to \sigma_i' = \frac{\alpha \sigma_i + \beta}{\gamma \sigma_i + \delta}, \quad \alpha\delta - \beta\gamma = 1 \tag{11}$$

we have
$$\hat{f}_i \to (\gamma\sigma_i + \delta)^2 \hat{f}_i + \gamma(\gamma\sigma_i + \delta)s_{ii}. \tag{12}$$

We find from Eq. (9)
$$\sum_i \hat{f}_i = 0, \quad \sum_i \hat{f}_i\sigma_i = -\sum_i s_{ii}, \quad \sum_i \hat{f}_i(\sigma_i)^2 = -2\sum_i s_{ii}\sigma_i, \tag{13}$$

which implies that when all the particles are on-shell, there are only $n-3$ independent $\hat{f}_i$. The solutions of these scattering equations will give σ_i as functions of s_{ij}. As there are three free constants (α, β, γ, δ, $\alpha\delta - \beta\gamma = 1$) in the Möbius transformations, we can use them to assign three of the σ_i to some fixed values. These will be called constant lines σ_p, σ_q, σ_r. Likewise, we can leave out three of the n $\hat{f}_i$'s, which will be chosen to be $i = p, q, r$ also.

For n particles with color ordering
$$a = (1, i_2, \ldots, i_{n-1}, n), \quad b = (1, j_2, \ldots, j_{n-1}, n), \tag{14}$$

in which i and j run over the $(n-2)!$ permutations of the color indices, one has the matrix elements for M
$$M^{ab} = \left(\frac{-1}{2\pi i}\right)^{n-3} \oint \left(\Pi_{i \neq p,q,r} \frac{d\sigma_i}{\hat{f}_i}\right) \frac{(\sigma_{pqr})^2}{\tilde{\Sigma}(a)\tilde{\Sigma}(b)}, \tag{15}$$

in which the σ's are made complex and the contour is to enclose all the simultaneous poles of the $n-3$ $\hat{f}_i \to 0$. Clearly, this needs some collaboration from the dynamical factors $\tilde{\Sigma}(a)$ and $\tilde{\Sigma}(b)$. p, q, r are the three indices of i which have been specially picked. Also,
$$\sigma_{pqr} = \sigma_{pq}\sigma_{qr}\sigma_{rp}, \tag{16}$$
$$\tilde{\Sigma}(a) = \sigma_{1i_2}\sigma_{i_2i_3}\cdots\sigma_{i_{n-1}n}\sigma_{n1}, \quad \tilde{\Sigma}(b) = \sigma_{1j_2}\sigma_{j_2j_3}\cdots\sigma_{j_{n-1}n}\sigma_{n1}. \tag{17}$$

One can show that M^{ab} is Möbius invariant and that the result is independent of p, q and r.

If one strictly follows the instruction to find the zeros of $\hat{f}_i$, one will find that the resulting solutions are many and that they have all kinds of algebraic roots with radicals. Such radicals must cancel, because we are working only at the tree level. A more pedagogical and intuitive way to carry out the integrations is to recognize that there is a close correspondence between the propagators which appear in M^{ab} from double-color scalar field theory and the related inner structures of $\tilde{\Sigma}(a)$ and $\tilde{\Sigma}(b)$ in Eq. (15).

As an example, let us consider the amplitude
$$M^{12345\,12345} = \left(-\frac{1}{2\pi i}\right)^2 \oint \frac{\sigma_{(135)}^2 \, d\sigma_2 \, d\sigma_4}{\hat{f}_2 \hat{f}_4 \sigma_{(12345)} \sigma_{(12345)}}, \tag{18}$$

where

$$\hat{f}_2 = \frac{s_{21}}{\sigma_{21}} + \frac{s_{23}}{\sigma_{23}} + \frac{s_{24}}{\sigma_{24}} + \frac{s_{25}}{\sigma_{25}},$$
$$\hat{f}_4 = \frac{s_{41}}{\sigma_{41}} + \frac{s_{42}}{\sigma_{42}} + \frac{s_{43}}{\sigma_{43}} + \frac{s_{45}}{\sigma_{45}}$$
(19)

and

$$\sigma_{(135)} = \sigma_{13}\sigma_{35}\sigma_{51}, \qquad \sigma_{(12345)} = \sigma_{12}\sigma_{23}\sigma_{34}\sigma_{45}\sigma_{51}. \tag{20}$$

It is known to yield the result[9,11–14]

$$M^{12345\ 12345} = \frac{1}{s_{12}s_{45}} + \frac{1}{s_{15}s_{34}} + \frac{1}{s_{15}s_{23}} + \frac{1}{s_{12}s_{34}} + \frac{1}{s_{23}s_{45}}. \tag{21}$$

The dynamical factor $\sigma^2_{(12345)}$ contains factors σ^2_{12}, σ^2_{23}, σ^2_{34}, σ^2_{45}, σ^2_{51} in Eq. (18), whereas $M^{12345\ 12345}$ is made of propagators $\frac{1}{s_{12}}, \frac{1}{s_{23}}, \frac{1}{s_{34}}, \frac{1}{s_{45}}, \frac{1}{s_{51}}$ in Eq. (21). Note that the pairs of indices in each factor is of the form $\sigma_{i\,i+1}$ or $s_{j\,j+1}$, i.e. they are sequentially next to each other. We call $1, 3$ and 5 constant lines, because they are not integrated over in Eq. (1) and $2, 4$ variable lines, because they are.

The factors $\sigma_{i\,i+1}$ are all squared. When some pairs of them tend to zero, except for σ_{15}, poles will develop in both integrations in Eq. (17). Together with $\hat{f}_i$ around these neighborhoods, we obtain the corresponding terms in Eq. (20). (The $\frac{1}{s_{15}}$ propagator in this example is produced by considering the neighborhood of $\sigma_{23} \to 0$, $\sigma_{34} \to 0$ simultaneously. One may note that $s_{15} = s_{234}$.)

The general procedure[12] is to try to partition the lines in the two dynamical factors into subsets consecutively. Each subset in one has a corresponding subset in the other dynamical factor and they consist of the same group of external lines, which can be differently ordered. If such a partition can be done completely, one can carry out the integrations with a certain set of rules and identify such a partition with a Feynman diagram and thus extract the propagators.

3.2. Yang–Mills

The color-ordered amplitude is given by

$$M^n = \left(\frac{-1}{2\pi i}\right)^{n-3} \oint (\sigma_{pqr})^2 \Pi^n_{a \neq p,q,r} \frac{d\sigma_a}{\hat{f}_a} \frac{T_n}{v_{12\cdots n}}, \tag{22}$$

where

$$T_n = \frac{(-1)^{i+j}}{\sigma_{ij}} Pf\Psi^{ij}_{ij} \tag{23}$$

and $Pf\Psi^{ij}_{ij}$ is a reduced Pfaffian with i, j rows and i, j columns removed of a $2n \times 2n$ determinant

$$\Psi = \begin{pmatrix} A & -C^T \\ C & B \end{pmatrix}, \tag{24}$$

for $i \neq j$

$$A_{ij} = -A_{ji} = \frac{-\frac{1}{2}s_{ij}}{\sigma_{ij}}, \quad B_{ij} = -B_{ji} = \frac{\epsilon_i \cdot \epsilon_j}{\sigma_{ij}}, \quad C_{ij} = \frac{\epsilon_i \cdot k_j}{\sigma_{ij}} \quad (25)$$

and the diagonal elements for C, $C_{ii} = -\sum_{j \neq i}^n C_{ij}$. ϵ_i is the polarization of particle i. The rows and columns are color ordered according to $1 i_2 i_3 \cdots i_{n-1} n$. M^n is Möbius invariant, independent of i, j and p, q, r.

3.3. Amputated Green's functions (off-shell amplitudes[10])

The CHY formula shown above in its original form is a tree amplitude for massless particles. In order to implement unitarity, generalization to loop amplitudes is required. To facilitate such a generalization and to understand better its connection to quantum field theory, it is necessary to study the off-shell behavior of these scattering amplitudes. We have done so for scalar and gauge particles for general n, even for the massive ones.[10,13,15] The gravitational amplitudes, through the KLT and BCJ relations, will be just related to the square of the off-shell gauge amplitudes, as in the on-shell situation.

The hint for such an extension is to realize that the off-shell double-color scalar Green's functions are identical to the on-shell amplitudes (e.g. Eq. (21)). Our strategy to extend CHY to off-shell is therefore first to modify the scattering equations (9) in such a way that they will produce such a result. It turns out that the critical driving criterion is Möbius covariance of the scattering equations, imposed by off-shell kinematics. To give an example, let us look at $n = 5$. We generalize the catering equations to

$$\hat{f}_1 = \frac{s_{12}}{\sigma_{12}} + \frac{s_{13} + x_{13}}{\sigma_{13}} + \frac{s_{14} + x_{14}}{\sigma_{14}} + \frac{s_{15}}{\sigma_{15}},$$

$$\hat{f}_2 = \frac{s_{21}}{\sigma_{21}} + \frac{s_{23}}{\sigma_{23}} + \frac{s_{24} + x_{24}}{\sigma_{24}} + \frac{s_{25} + x_{25}}{\sigma_{25}},$$

$$\hat{f}_3 = \frac{s_{31} + x_{31}}{\sigma_{31}} + \frac{s_{32}}{\sigma_{32}} + \frac{s_{34}}{\sigma_{34}} + \frac{s_{35} + x_{35}}{\sigma_{35}}, \quad (26)$$

$$\hat{f}_4 = \frac{s_{41} + x_{41}}{\sigma_{41}} + \frac{s_{42} + x_{42}}{\sigma_{42}} + \frac{s_{43}}{\sigma_{43}} + \frac{s_{45}}{\sigma_{45}},$$

$$\hat{f}_5 = \frac{s_{51}}{\sigma_{51}} + \frac{s_{52} + x_{52}}{\sigma_{52}} + \frac{s_{53} + x_{53}}{\sigma_{53}} + \frac{s_{54}}{\sigma_{54}}.$$

It is easy to check that

$$\sum_{i=1}^{5} \hat{f}_i = \sum_{i=1}^{5} \sigma_i \hat{f}_i = \sum_{i=1}^{5} \sigma_i^2 \hat{f}_i = 0, \quad (27)$$

because of momentum conservation, and

$$\hat{f}_i(\sigma) = (\gamma \sigma_i + \delta)^2 \hat{f}_i(\sigma') \quad (28)$$

under Möbius transformation equation (11), when we take

$$x_{31} = (k_1^2 + k_2^2 + k_3^2)$$
$$x_{41} = (k_1^2 + k_4^2 + k_5^2), \quad x_{42} = (k_2^2 + k_3^2 + k_4^2), \quad (29)$$
$$x_{52} = (k_1^2 + k_2^2 + k_5^2), \quad x_{53} = (k_3^2 + k_4^2 + k_5^2).$$

Equation (27) implies that again only two of the $\hat{f}_i$'s are independent. One interesting comment is that these changes should be made to calculate $M^{12345,1i_2i_3i_45}$ only, where i_1, i_2, i_3 are permutations of $2, 3, 4$. For other color configurations, one should re-label the indices to make them fall into this form.

We have shown how to treat general n. The case for $n = 4$ is somewhat different, where we should have

$$x_{13} = x_{42} = \sum_{1}^{4} k_i^2, \quad (30)$$

and there are no shifts for the other s_{ij}.

One can show that the gauge amplitude given by Eq. (22) is Möbius invariant, if we make the same change as above, i.e. $s_{ij} \to s_{ij} + x_{ij}$ in the Pfaffian in Eq. (25). We have looked into the case of $n = 4$ ($1^+2^-3^+4^-$) and compared the Green's function so obtained to that from space-cone calculation[16,17] and found that there is a difference

$$(G^4_{\text{space-cone}} - G^4_{\text{gauge}})(1^+2^-3^+4^-)$$
$$= \frac{1}{s_{12}} \frac{1}{2(k_1 + k_2)} (k_4 K_1^2 - k_1 K_4^2 + k_3 K_2^2 - k_2 K_3^2)$$
$$+ \frac{1}{s_{14}} \frac{1}{2(k_1 + k_4)} (k_2 K_1^2 - k_1 K_2^2 + k_3 K_4^2 - k_4 K_3^2), \quad (31)$$

where lower case k_i's are certain components of the momenta in light-like coordinates, and upper case K_i's denote momentum four vectors. However, this difference is superfluous, because it can be resolved by making a field re-definition of the helicity fields in space-cone gauge

$$a_{b'}^- \to a_{b'}^{'-} = a_{b'}^- + \frac{1}{2} f_{bac} f_{b'a'c} \frac{1}{\hat{\partial}^2} \left[\left(\frac{1}{\hat{\partial}^2} \frac{1}{\hat{\partial}} (a_b^- \hat{\partial}^2 a_a^+) \right) \partial a_{a'}^- \right] \quad (32)$$

and

$$a_{b'}^+ \to a_{b'}^{'+} = a_{b'}^+ + \frac{1}{2} f_{bac} f_{b'a'c} \frac{1}{\hat{\partial}^2} \left[\left(\frac{1}{\hat{\partial}^2} \frac{1}{\hat{\partial}} (a_b^+ \hat{\partial}^2 a_a^-) \right) \partial a_{a'}^+ \right], \quad (33)$$

where $\hat{\partial}^2 = \partial^\mu \partial_\mu$. The equivalence theorem tells us that all physical consequences are not affected. It is also interesting to note that the difference in Eq. (31) is related to the violation of color-kinematic duality of $n = 4$, when the particles are off-shell.[18]

4. Conclusion

CHY is a fascinating formalism. A multitude of diagrams is summed into a single formula. The connection between gauge theories and gravity is exceedingly simple, by replacing $\frac{T_n}{\sigma_{123\cdots n}}$ in Eq. (22) with T_n^2. This resembles the double copy hypothesis. One connection some of us would like to find is why the Yang–Mills dynamics in Lagrangian form is simple, whereas that for gravity is seemingly complicated. Perhaps this disparity has to do with proper identifications of fields or equivalent in describing the two systems. We have extended CHY to off-shell, so that the transcription to field theoretics is made more readied, with which one can deal with non-perturbative issues as well. The challenge now is to break the inscribed codes between these two complementary dynamics.

Acknowledgments

I would like to thank the organizers of this conference for inviting me to this august occasion. D. Vaman and C.-S. Lam have helped me in understanding the nuances of the topics covered here.

References

1. A. Einstein, *Ann. Phys.* **49**, 769 (1916).
2. C.-N. Yang and R. L. Mills, *Phys. Rev.* **96**, 191 (1954).
3. R. Utiyama, *Phys. Rev.* **101**, 1597 (1956); T. W. Kibble, *J. Math. Phys.* **2**, 212 (1961).
4. R. Britto, F. Cachazo, B. Feng and E. Witten, *Phys. Rev. Lett.* **94**, 181602 (2005).
5. F. A. Berends, R. Kleiss, P. De Causmaecker, R. Gastmans and T. T. Wu, *Phys. Lett. B* **103**, 124 (1981); Z. Xu, D.-H. Zhang and L. Chang, *Nucl. Phys. B* **291**, 392 (1987).
6. Z. Bern, J. J. M. Carrasco and H. Johansson, *Phys. Rev. D* **78**, 085011 (2008).
7. Z. Bern, J. J. M. Carrasco and H. Johansson, *Phys. Rev. Lett.* **105**, 061602 (2010); Z. Bern, T. Dennen, Y.-T. Huang and M. Kiermaier, *Phys. Rev. D* **82**, 065003 (2010).
8. H. Kawai, D. C. Lewellen and S.-H. H. Tye, *Nucl. Phys. B* **1**, 269 (1986).
9. F. Cachazo, S. He and E. Y. Yuan, *Phys. Rev. D* **90**, 065001 (2014); *Phys. Rev. Lett.* **113**, 17161 (2014); *JHEP* **1407**, 033 (2014); *ibid* **1501**, 121 (2015); *ibid* **1507**, 149 (2015); *Phys. Rev. D* **92**, 065030 (2015).
10. C.-S. Lam and Y.-P. Yao, arXiv:1511.05050 (see here for more references).
11. D. Vaman and Y.-P. Yao, *JHEP* **1011**, 028 (2010).
12. C.-S. Lam and Y.-P. Yao, to be published.
13. L. Dolan and P. Goddard, *JHEP* **1407**, 029 (2014).
14. C. Baadsgaard, N. E. J. Bjerrum-Bohr, J. D. Bourjaily and P. H. Damgaard, *JHEP* **1509**, 136 (2015).

15. S. G. Naculich, *JHEP* **1409**, 029 (2014); *ibid.* **1505**, 050 (2015); *ibid.* **1509**, 141 (2015).
16. G. Chalmers and W. Siegel, *Phys. Rev. D* **59**, 045013 (1999).
17. D. Vaman and Y.-P. Yao, *JHEP* **0604**, 030 (2006).
18. D. Vaman and Y.-P. Yao, *JHEP* **1412**, 036 (2014).

Feynman Geometries

Sen Hu* and Andrey Losev†

*School of Mathematical Sciences, Wu Wen-Tsun Key Lab of Mathematics,
Chinese Academy of Sciences, University of Science and Technology of China,
96 Jinzhai Rd., Hefei, Anhui 230026, China
shu@ustc.edu.cn

†Department of Mathematics, Higher School of Economics,
20, Myasnitskaya St., Moscow, 101000, Russia
aslosev2@gmail.com

In this paper we introduce a notion of Feynman geometry on which quantum field theories could be properly defined. A strong Feynman geometry is a geometry when the vector space of A_∞ structures is finite dimensional. A weak Feynman geometry is a geometry when the vector space of A_∞ structures is infinite dimensional while the relevant operators are of trace-class. We construct families of Feynman geometries with "continuum" as their limit.

1. Introduction

Non-Abelian gauge symmetry, as discovered by C. N. Yang and R. Mills, is a basic concept in modern physics. There is a basic principle established by A. Einstein and C. N. Yang: symmetry dictates interactions, e.g. Lagrangians are determined by symmetries. Gauge invariance, conformal invariance and renormalizability of the theory determine the Lagrangian for the standard model. Quantum Yang–Mills theory is now the foundation of most of elementary particle theory, and its predictions have been tested at many experimental laboratories, but its mathematical foundation is still unclear.

Symmetries used by A. Einstein and C. N. Yang follow from differential geometry. In the case of A. Einstein theory of general relativity it is a diffeomorphism symmetry that is a fundamental equivalence between smooth manifolds. In the case of C. N. Yang theory the gauge symmetry comes from the geometry of principal fiber bundles, and may be understood as diffeomorphisms that preserve bundle structure.

Would these symmetries persist in the quantum world or should they be modified? The answer to this question depends on a more fundamental question: should we consider the geometry of the space–time in quantum field theory (QFT) the same classical differential geometry or it has to be modified? The following paper discusses this issue and concludes that most probably the geometry has to be modified to what we call Feynman geometry.

There are mainly two approaches in the construction of QFTs: Dirac–Segal's approach and Feynman's approach. In Dirac's approach one constructs quantum theories by abandoning classical physical concepts while keeping classical concept of space-time. This approach was pursued by many prominent physicists and mathematicians, namely by Atiyah–Segal–Witten and by Belavin–Polyakov–Zamalochikov, and many fruitful mathematical results were obtained by this approach.

In Feynman's approach it saves many classical concepts such as space-time, trajectories, actions, etc. It replaces the minimal action principle in classical mechanics by path integral to calculate various correlation functions. This approach is more familiar to physicists.

Here we propose a new approach to construct quantum gauge field theories by constructing quantum many-body systems over some spaces we call it de Rham type of differential graded algebra (DGA). As long as we have de Rham type of DGA we could construct all kinds of actions. We demonstrate this in Sec. 4.

The space of de Rham type of DGA are characterized by Feynman geometries. A strong Feynman geometry is a geometry when the vector space of A_∞ structures is finite dimensional. A weak Feynman geometry is a geometry when the vector space of A_∞ structures is infinite dimensional while the relevant operators are of trace-class. This notion unifies several important works, including lattice theory, fuzzy supersphere, Feynman geometry of Hodge type (momentum cutoff), Mnev's construction of A_∞ structure via Whitney forms, Costello's construction of A_∞ structures via homotopy associativity, and Zwiebach's string field theory. We construct families of Feynman geometries with "continuum" as their limit.

We propose to construct quantum field theories over an continuum as limit of a family of theories of Feynman geometries. The advantage of such a construction is its nonperturbative feature.

2. Dirac–Segal and Feynman approaches to QFT

The main question of theoretical physics is what is the definition of Quantum Field Theory (QFT). Despite of many attempts this question still does not have a completely satisfactory solution. These attempts may be separated into two different approaches that we may call Dirac–Segal approach and Feynman approach.

2.1. *Dirac–Segal approach and its classical limit*

In the Dirac–Segal approach the space-time is considered a very classical Riemann geometry — smooth manifolds with the boundary, but "physics" has nothing to do with classical physics — instead of fields and integrals over the space of fields, we have functors from the category of cobordisms (equipped with some geometrical structure) into category of vector spaces. Here we use Dirac's name before Segal's since Dirac was first to understand that strongly quantum mechanics, where cobordisms are just intervals, geometrical data — their lengths, the image of an interval is an evolution operator $I(t)$ acting on the abstract space of states V, and functoriality

is just a semigroup condition:
$$I(t_1 + t_2) = I(t_1)I(t_2), \quad I(t) \in \text{End}(V) \tag{1}$$

that may be solved in an almost standard form:
$$I(t) = \exp(tH). \tag{2}$$

It is very important to note that there is no Planck constant in Dirac–Segal approach (it is absolutely clear from the explicit solution — Dirac solution — in one-dimensional case). The "Planck constant" actually appears if we consider a family of theories Th_h, depending on h, with the following properties. For each h there is an algebra $\text{End}(V_h)$ generated by the set of operators $O(h)_a$, such that

$$[O(h)_a, O(h)_b] = hf(h)^c_{ab} O^h_c \tag{3}$$

and $f(h)$ has a finite limit when h tends to zero, and relations between generators have a finite limit at $h = 0$. The Hamiltonian has the form

$$H(h) = \frac{H^{cl}(h)}{h}, \quad H^{cl}(h) = y^a(h) O_a(h), \tag{4}$$

and the coefficients $y^a(h)$ are regular at $h = 0$. Let us call the family of observables $A(h)$ having a classical limit if it can be expanded through the generators with coefficients $z^a(h)$,

$$A(h) = z^a(h) O_a(h), \tag{5}$$

and $z^a(h)$ are regular at $h = 0$. Then, at $h = 0$ the symbols $O_a(0)$ form a commutative associative algebra, with its spectrum may be considered as a Poissonian manifold (may be with singularities), its Poisson structure is determined by

$$\{O_a(0), O_b(0)\}_f(0) = f^c_{ab}(0) O_c(0) \tag{6}$$

and the evolution is given by a classical mechanics equation

$$\frac{dA}{dt} = \{H^{cl}(0), A(0)\}. \tag{7}$$

Warning. Note that there is no h in the basic evolution equation. The denominator h appears only when evolution operator is written in terms of operators that have regular classical limit, i.e.

$$\exp(tH) = \exp\left(\frac{H^{cl}}{h}\right). \tag{8}$$

Example: Fuzzy sphere.

Consider $(N+1)$-dimensional representation of $su(2)$,

$$\{T_a, T_b\} = \epsilon_{abc}T_c, \quad \sum_{a=1}^{3} T_a^2 = N(N+1). \tag{9}$$

Consider a generating set

$$O_a = \frac{T_a}{N}, \tag{10}$$

then the generators O_a form an algebra

$$\{O_a, O_b\} = \frac{1}{N}\epsilon_{abc}O_c, \quad \sum_{a=1}^{3} O_a^2 = 1 + \frac{1}{N}. \tag{11}$$

If we consider $h = 1/N$, then it is a particular case of fuzzy sphere.

3. Feynman Approach and Feynman Geometry

The alternative to Dirac–Segal approach is the Feynman approach. This approach not just preserves but actively uses the basic notions of classical physics such as fields and Lagrangians. The only modification is the principal of minimal (extremal) action. The condition of extremality is replaced by functional integral.

However, such an integral in most cases does not exist as a mathematical object. In early days of functional integral, these difficulties were considered as temporary ones. It was believed that heuristic manipulation could produce correct results, and mathematical justification would come later. However, in the last 65 years, nobody provided such justification. Mathematicians do not even consider it as a problem and physicists do not want to think about it. We think that the main problem is that the integral has to be taken over infinite-dimensional space and such integrals rarely exist.

The problem can easily be pointed out in perturbative approach. Infinite dimensional space of integration in most cases leads to divergences.

The standard way out is to make regularization and then renormalization. Regularization basically replaces the space of functions on the classical space by another space, like space of function on the set of points (vertexes of auxillary lattice) or by space of functions with limited derivative (momentum cutoff). In the standard approach such replacement is often considered as a technical tool, a trick for computation of ideal infinite-dimensional integral (that does not exist). In dreaming about such ideal integral, it is mostly assumed that the property of the regularized integrals should be similar to properties of ideal integrals, or at least the difference in properties would somehow go away when regularized space of function would tend to original infinite-dimensional one.

In this way of thinking, the ideal action often has a geometrical meaning formulated in terms of conventional differential geometry, and has symmetries prescribed by differential geometry. Nobody have ever asked about what kind of geometry corresponds to the regularized space of function.

Actually, the regularized space of function does not even form an associative algebra. Consider as an example the space of function on a circle of length 2π. Let us make a momentum cutoff, restrict functions to be linear combinations of $\exp(kx)$ with $|k| < 3N$. Consider the three functions

$$f = \exp(2Nx), \quad g = \exp(2Nx), \quad h = \exp(-2Nx) \tag{12}$$

and let us simply restrict the multiplication on the reduced space. Then multiplying them in one way we get

$$(f \cdot g) \cdot h = 0 \cdot h = 0. \tag{13}$$

Since the product of f and g before momentum cutoff equals to $\exp(4Nx)$, it is outside the allowed momentum range and has to be considered as zero. At the same time multiplying in other order

$$f \cdot (g \cdot h) = f \cdot 1 = f. \tag{14}$$

So we have a violation of associativity. It was actually expected since the space of functions of momentum higher than $3N$ does not form an ideal (actually, it does not even form an algebra). Therefore, according to traditional setup the regularized space of function does not form an algebra and, therefore, there is no geometrical object that may be attached to it through conventional algebra-geometric correspondence. In particular, all properties and symmetries of the classical action are violated and we do not have any control of this violation. We may only hope that they will be recovered after cutoff of the renormalized theory would be taken to infinity. We think that such a complete loss of geometry in regularized theory means loosing control of symmetries and cannot be a part of proper definition of functional integral.

The situation of lattice regularization is more tricky but the basic phenomena persist. The algebra of functions on vertexes is an honest commutative and associative algebra, but this is not enough to construct the action. What is actually enough (as we will show in the following section) is the de Rham DGA (differential graded algebra). It is a supercommutative associative algebra. In the regularization this algebra should be replaced by an algebra of cochains. However, there is no supercommutative associative algebra structure on the space of cochains.

It is the moment of truth. In order to save the idea of functional integral, we have to find some kind of algebraic structure on the regularized space of differentials forms (it does not matter, with momentum or lattice cutoff). This structure should go to supercommutative DGA structure in the limit when we go to the ideal "continuum" limit. One may expect that algebro-geometric correspondence would somehow be extended to this structure, and we call the geometry that corresponds to such structure the Feynman geometry.

The good news is that such algebraic structure already exists in mathematics, and it is called A_∞ structure. The brief outline of this structure and its relation to BV (Batalin–Vilkovisky) language is presented in Appendix.

Let us call the A_∞ structure with operations belonging to the trace-class **Feynman geometry**. Here we may consider two types of Feynman geometry. The first type is **strong** Feynman geometry when vector space of A_∞ structure is finite dimensional. The second type is **weak** Feynman geometry when vector space is infinite dimensional but operations are of trace-class, so there are no divergences in Feynman diagrams computations.

Thus, our proposal is to develop Feynman approach to QFT by considering quantum field theories over Feynman geometry instead of continuum geometry.

The deviations of Feynman geometry from conventional geometry (as we will see below) happen in two directions. Feynman geometry can be nonsupercommutative but associative DGA, or Feynman geometry may include higher operations. General Feynman geometry deviates from conventional in both directions.

Therefore, our proposal is not to concentrate on each particular example of Feynman geometry but rather consider QFT over general Feynman geometry. Then, consider the variation of QFT when Feynman geometry is changing — a kind of connection in the bundle of theories over parameters of Feynman geometry and, finally, define a Feynman QFT over continuum geometry as a limit of QFT over Feynman geometry tending to continuum limit (kind of universal renormalization).

Before we come to examples of Feynman geometry, we would like to stress that our world is actually described by Feynman geometry. Experiments can only measure that our geometry is continuum to some accuracy, and they can only improve this accuracy. If QFT may be defined over Feynman geometry, we may only ask why in our world this geometry is so close to continuum, but this looks like the question that asks why cosmological constant is so small, and it cannot be answered with the present knowledge of physics.

4. All Actions of Traditional QFT Written in Terms of de Rham DGA

4.1. *Pure gravity*

The fundamental description of gravity is given by the following data — D-dimensional smooth manifold with a principal $\mathrm{Spin}(D)$ bundle S, equipped with the connection ∇ and a morphism e from the tangent bundle to the vector bundle L (Lorentzian bundle, that is a vector bundle associated with S by D-dimensional (vector) representation of $\mathrm{Spin}(D)$). In this way e may be considered as a L-valued 1-form, and connection ∇ may also be rewritten in terms of the 1-form ω with values in the second external power of L (in the case of trivialized bundle L)

$$\nabla = d + \omega. \tag{15}$$

Thus, the Einstein–Cartan action

$$S^{\text{EC}} = \int (e^{a_1} \cdots e^{a_{D-2}} (d\omega^{a_{D-1} a_D} + \omega^{a_{D-1} b} \cdot \omega^{b a_D}) \epsilon_{a_1 \ldots a_D} \quad (16)$$

may be written in the language of de Rham DGA.

4.2. *Chiral fields in 2D and self-dual Yang–Mills*

By chiral fields we mean $b - c$ or $\beta - \gamma$ systems. The easiest thing to modify is the spin 0 chiral system with the action

$$S^\chi = \int P \bar{\partial} X = \int P d X, \quad (17)$$

where P is a $(1, 0)$ form. We would like to express subdivision of differential forms into (p, q)-types using only e-field. To do this we propose to consider P field as a composite field and rewrite it as

$$P = e^+ p, \quad (18)$$

where e^+ is and eigenvalue $+i$ eigenvector of e^a, $a = 1, 2$ with respect to $SO(2)$ rotation, and p is a complex scalar field. Thus, the action written in terms of DGA (with an integral) reads

$$S^\chi_{\text{DGA}} = \int p e d X. \quad (19)$$

The action of self-dual Yang–Mills may be rewritten in a similar fashion. Namely, standard action of self-dual Yang–Mills has the form

$$S_{\text{SDYM}} = \int \text{Tr } P(dA + A^2), \quad (20)$$

where P is a self-dual 2-form with values in adjoint representation of the gauge group, A is the connection form in the gauge bundle (we assume that the gauge bundle is trivial and therefore connection may be considered as a 1-form with values in adjoint representation of the gauge algebra). Like in the chiral field case we will consider field P as a composite field, namely, we introduce the field p_{ab} with values in adjoint representation of the gauge group times the self-dual 2-tensors of L, thus

$$P = e^a e^b p_{ab}, \quad (21)$$

so the action takes the form that is written in the DGA language

$$S^{\text{SDYM}}_{\text{DGA}} = \int e^a e^b p_{ab} (dA + A^2). \quad (22)$$

Remark 1. If we will not be interested in dynamical gravity, i.e. only consider it as a fixed background, we can use just a 2-form E^{ab} with values in the antisymmetric self-dual tensors of L. Then, the action would take the form

$$S^{\text{SDYM}}_{\text{DGA,E}} = \int E^{ab} p_{ab} (dA + A^2). \quad (23)$$

It is interesting to note that the E field background is more general than e field since not all E are decomposable into a product of two e. It opens a previously unexplored possibility to study self-dual Yang–Mills and (as we will see later) ordinary Yang–Mills theory in a novel background.

Remark 2. An alternative approach would be to introduce the new structure in DGA, a polarization of middle degree elements, that is decomposition of this subspace into a sum of two spaces, invariant under multiplication over degree zero elements. Thus, we may consider the first-order actions as actions already written in the language of polarised DGA. Thus, actions (3) and (6) may be considered as written already in terms of polarized DGA. The disadvantage of this approach is that it is not clear how to write Einstein action on the space of polarizations.

4.3. Spinor fields in arbitrary dimensions

The very notion of spin connection comes from the desire to couple spinors to gravity. Therefore it is not surprising that there is such a coupling. What we want to stress is that this coupling is also written in terms of DGA. Namely, spinor field should be considered as a section of the spinor bundle associated with the Spin(D) bundle, that basically means that for the case of trivial bundles spinor fields ψ_α are zero forms with values in the spinor representation of Spin(D). The action for the Dirac spinor reads

$$S = \int \bar{\psi}_\alpha (d\psi_\beta + \omega_\beta^\delta \psi_\delta) \gamma_{a_1}^{\alpha\beta} e^{a_2} \cdots e^{a_D} \epsilon_{a_1 \ldots a_D} + m \bar{\psi}_\alpha \psi_\alpha vol \qquad (24)$$

where

$$vol = e^{a_1} \ldots e^{a_D} \epsilon_{a_1 \ldots a_D} \qquad (25)$$

and m is the mass of the fermion field. Note the similarity of this action to action for two-dimensional chiral field — really, this is how chiral fermions are described in $D = 2$. Similarly one can write the coupling of the spinor field to the gauge field.

4.4. Sigma models

There are two ways to write a free scalar field action in the language of DGA.

The first way works only in dimension 2 and utilizes the previously written action for a chiral field. We consider two copies of chiral action (actually, chiral and anti-chiral) and add a standard $P\bar{P}$ term that now looks as $p\bar{p}e^+e^-$ term. This procedure may be generalized to the general Riemannian metric and a B-field background using approach of Ref. 1.

The second way works in any geometry of the world-sheet and target, and is based on the first-order representation of the D-dimensional sigma model. Consider the field P_μ that is a $(D-1)$-form with values in the pullback of the cotangent bundle to the target manifold. Consider the action

$$S = \int (P_\mu dX^\mu + P_\mu * P_\nu G^{\mu\nu}(X)). \tag{26}$$

Integrating the field P out we get the standard sigma-model action. The first term looks perfect from the DGA language point of view, however, the second does not — it involves the Hodge star operation that cannot be formulated in the DGA language. To deal with this problem we have to go to composite fields like we did in the case of chiral field, namely, we use a field $p_{a_1...a_{D-1}}$ that is a section of the $D-1$ external power of $L*$, and make a $(D-1)$-form out of it using e-fields (taking a pullback to under the maps between L^* and T^*):

$$P_\mu = e^{a_1} \ldots e^{a_{D-1}} p_{a_1 \cdots a_{D-1}, \mu}. \tag{27}$$

Then action (12) becomes a legal expression in DGA language, i.e.

$$S = \int e^{a_1} \cdots e^{a_{D-1}} p_{a_1...a_{D-1},\mu} dX^\mu + p_{a_1...a_{D-1},\mu} p_{a_1...a_{D-1},\nu} G^{\mu\nu}(X)) vol(e), \tag{28}$$

where $vol(e)$ was defined in Eq (11). It is clear that the two ways of writing down sigma model in the first-order formalism are classically equivalent, but there is the question if they are really equivalent on the quantum level.

4.5. Yang–Mills theory

The Yang–Mills theory in arbitrary dimension can be written along the second way. Namely, we write down the first-order action

$$S = \text{Tr}(P(dA + A^2)) + g_0^2 \text{Tr}(P * P), \tag{29}$$

where P is the $(D-2)$-form with values in the adjoint representation. Similar to the sigma-model case, this form can be expressed through the section of the $(D-2)$-external power of $L*$,

$$P(p, e) = e^{a_1} \ldots e^{a_{D-2}} p_{a_1...a_{D-2}}, \tag{30}$$

and Yang–Mills action takes the form

$$S = \int \text{Tr}(P(p,e)(dA + A^2)) + \text{Tr}(p_{a_1...a_{D-2}} p_{a_1...a_{D-2}}) vol(e). \tag{31}$$

The case of $D = 4$ is distinguished since there is another representation of the Yang–Mills similar to the first way of describing the $D = 2$ sigma model. Namely, in this case we just add to the action of the self-dual Yang–Mills the $E^2 p^2$ term, namely

$$S^{\text{YM}} = \int E^{ab} \text{Tr}(p_{ab}(dA + A^2)) + g_0^2 \text{Tr}(p_{ab} E^{ab} p_{ce} E^{ce}). \tag{32}$$

In order to have standard interaction with gravity, we need to consider

$$E^{ab} = e^a e^b.$$

5. Noncommutative Feynman Geometry of the DGA Type

It seems impossible to have a family of finite-dimensional supercommutative DGA that have a de Rham DGA as a limit (may be is possible to have a rigorous proof of this statement but we do not know such a proof). Therefore, to have finite dimensionality we should sacrifice either supercommutativity or associativity (associativity would be replaced by homotopical associativity). In this section, we explain what could be done if we sacrifice supercommutativity condition. The case of homotopical associativity will be treated in the next section.

We know two apparently different ways to obtain nonsupercommutative finite-dimensional DGA that are close in some sense to de Rham DGA. The first example is the well-known DGA that people use to explain the ring structure on cohomology. We call it lattice DGA. The second one is related to noncommutative geometry of the manifold.

5.1. *Lattice type*

Consider simplicial complex. Physicists often consider it as a triangulation of a manifold. Differential forms on the manifold may be considered as cochains of simplicial complex since they can be integrated against simplexes.

Consider the set of simplicial complexes obtained by subsequent barycentric subdivisions. Given two smooth differential forms, we can always find such a subdivision that would produce different cochains. Therefore, we would like to equip cochains on simplicial complex with the structure of differential graded algebra. The question of taking a limit when the number of barycentric subdivision is going to infinity and its relation to de Rham DGA is subtle and we will discuss it elsewhere.

It is easier to discuss the chains rather than cochains, and to give a formula for comultiplication. The problem of finding of supercommutative coassociative comultiplication is known as a Kolmogorov problem and is known to have a negative answer.

It is enough to study the case of a complex associated with a single ordered n-simplex. The complex is a set of linear combinations of basic elements that are increasing sequences of integers between 1 and n. We will denote them as $(n_1, \ldots, n_k)$. The boundary operator ∂ acts as

$$\partial(n_1, \ldots, n_k) = -(n_2, \ldots, n_k)$$
$$+ \sum_{j=2}^{k} (-1)^l (n_1, \ldots, n_{j-1} n_{j+1} \ldots n_k) + (-1)^k (n_1, \ldots, n_{k-1}). \quad (33)$$

The associative comultiplication reads

$$\partial(n_1,\ldots,n_k) = \sum_{j=1}^{k}(n_1,\ldots,n_j) \otimes (n_j,\ldots,n_k). \qquad (34)$$

The associativity of this comultiplication can be explicitly checked, but it is definitely not the cocommutative multiplication.

In the simplest example of one-dimensional triangulation, namely, let us subdivide a circle into set of intervals. Let us take orientation on the circle and order end points belonging to each interval according to this orientation. Then we have comultiplication on chains that induces multiplication on cochains. In the continuum limit, this multiplication tends to the standard de Rham multiplication, however, for the small size of the interval (let as call it a), the nonsupercommutativity of multiplications is reflected in the commutator

$$f \cdot \omega - \omega \cdot f = \omega a \partial_x f. \qquad (35)$$

5.2. *Fuzzy supersphere*

Consider polynomials of two even and two odd variables $C[z_0, z_1, \theta_0, \theta_1]$, where z are even and θ -odd. Assign to all of these variables grading 1, and consider a subspace P_N of $C[z_0, z_1, \theta_0, \theta_1]$ of polynomials of degree N, i.e. define an Euler vector field E

$$E = \sum_{i=0}^{1} \theta_i \frac{\partial}{\partial \theta_i} + z_i \frac{\partial}{\partial z_i} \qquad (36)$$

then

$$(E - N)P_N = 0. \qquad (37)$$

Consider standard σ-matrixes σ_a orthonormal basis in traceless hermitian operators on C^2). Note, that the following operators commute with E and therefore act on P_N: Operators of noncommutative coordinates x

$$x_a = 1/N \sum_{i,j} \sigma_{a,ij} z_i \frac{\partial}{\partial z_j}. \qquad (38)$$

Noncommutative de Rham operator

$$D = \sum_{i=0}^{1} \theta_i \frac{\partial}{\partial \theta_i} \qquad (39)$$

and operators of supercoordinates

$$\psi_a = 1/N \sum_{i,j} \sigma_{a,ij} \theta_i \frac{\partial}{\partial z_j}. \qquad (40)$$

Note, that operators of supercoordinates are commutators of de Rham operator and operators of coordinates

$$\psi_a = [D, x_a]. \qquad (41)$$

It is clear that the algebra generated by x and ψ is a subalgebra of $End(P_N)$ and thus is finite dimensional. Moreover, like in the case of fuzzy sphere, in the limit $N \to +\infty$, this algebra tends to the de Rham algebra on the 2-dimensional sphere.

Note, not only x fails to commute just like in the example of ordinary fuzzy sphere (with the standard poisson bracket), but also x and Dx are not commuting:

$$[x_a, Dx_b] = 1/N \epsilon_{abc} Dx_c \qquad (42)$$

while supercoordinates supercommute

$$[Dx_a, Dx_b] = 0. \qquad (43)$$

It would be interesting to find proper generalization of the superfuzzy sphere construction — like fuzzy sphere can be generalized to finite-dimensional algebra of operators acting on sections of the very ample line bundle on a Kaeler manifold.

6. Strong Feynman Geometry of A-infinity type on a Subcomplex

6.1. *Reminder on induction of A-infinity structure on a subcomplex from DGA on the complex*

In this section, we consider the procedure that produces Feynman geometries from infinite-dimensional DGA's with finite-dimensional cohomology. This construction was developed in parallel in mathematical physics and in mathematics with different motivation and in different terms, but the construction is basically the same.

Mathematicians were mostly interested in inducing the algebraic structure on cohomology, and they found that, besides the ring structure that comes from restriction of multiplication to cohomology (that everybody knows), there are also higher multiplications (Massey operations). If we combine these higher operations with multiplications, we get the A_∞ structure on cohomology with zero differential. Later, mathematicians generalized these constructions and they started talking about A_∞ structures obtained by contraction of the acyclic subcomplex.

In mathematical physics, people studied solutions of BV master equations. In particular, they studied solutions of classical limit of BV equation. As it is well-known, BV action may be considered as a polyvector field. In the special case when BV action is a vector field, classical BV equations lead to either L_∞ or A_∞ structure. From this perspective the procedure of contraction of acyclic subcomplex get the following "physical" interpretation.

The main property of solutions of BV equation is the following. If we decompose BV variables into background variables and dynamical variables, and integrate out (against a Lagrangian submanifold) dynamical variables, we will get the effective action for background variables, and this effective action would also satisfy BV equations.

Therefore, if DGA contains contractible subcomplex as a direct summand (as a subcomplex), we may consider it as dynamical variables and treat the rest as a background. One can show that homotopy (that inverts differential on contractable subcomplex) determines the Lagrangian submanifold, and variables of this subcomplex may be integrated out. The effective action would be a sum of Feynman diagramms, and in defining induced A_∞ structure, we have to take only tree diagrams. The trivalent vertices in this diagrams correspond to multiplication in the algebra and propagator is given by homotopy.

Let us reformulate this in the mathematical language. Suppose that there is a DGA structure on the space V with differential $D : V \to V$ and multiplication $m_2 : V \otimes V \to V$. Suppose that V has a decomposition as a complex

$$V = V_{\text{IR}} \oplus V_{\text{UV}}, \tag{44}$$

with inclusion

$$\iota : V_{\text{IR}} \to V$$

and projection

$$\pi : V \to V_{\text{IR}}$$

and homotopy h

$$h : V \to V$$

with the properties

$$Dh + hD = 1 - \iota\pi \tag{45}$$

(note that the right-hand side is the projection on V_{UV})

$$h^2 = 0. \tag{46}$$

Then there are induced operations

$$M_k^{\text{IR}} : V_{\text{IR}}^{\otimes k} \to V_{\text{IR}}$$

given as a sum over rooted flat trees with k leaves and one root. For the exact formula, reader may consult Ref. 2, but we will give the simplest cases:

$$M_2^{\text{IR}}(u, v) = \pi m_2(\iota u, \iota v), \tag{47}$$

$$M_3^{\text{IR}}(u, v, w) = \pi m_2(h m_2(\iota u, \iota v), \iota w) + \pi m_2(\iota u, h m_2(\iota v, \iota w)). \tag{48}$$

6.2. A-infinity algebras of Hodge type as strong Feynman geometry

Suppose X is a compact manifold with a metric. Then for any positive number Λ there is a Feynman geometry constructed as follows. Let us start with de Rham DGA and decompose it into direct sum of subcomplexes Ω_E corresponding to Laplacian eigenvalues. Let us form V_{UV} from subcomplexes with eigenvalue larger than Λ and V_{IR} from subcomplexes with eigenvalues smaller than Λ,

$$\Omega^* = (\oplus_{E \leq \Lambda} \Omega_E) \oplus (\oplus_{E \geq \Lambda} \Omega_E). \tag{49}$$

The space

$$V_{\text{IR}} = \oplus_{E \leq \Lambda} \Omega_E$$

is definitely finite-dimensional and the space

$$V_{\text{UV}} = \oplus_{E \geq \Lambda} \Omega_E$$

is acyclic. In particular, we may take as a homotopy

$$h = d^*/E. \tag{50}$$

Thus we have Feynman geometry depending on Λ and we approach de Rham geometry as Λ goes to infinity.

The parameter Λ resembles a momentum cutoff, however, there is a fundamental difference between Feynman geometry approach and old fashioned cutoff approach. In terms of algebraic structures, "old cutoff" means that we are using the algebra with nonassociative multiplication and, therefore, we are breaking many symmetries like gauge symmetry (just recall that to protect gauge symmetry people invented such things as Pauli–Willars regularization). In the Feynman geometry approach, we should properly take into account higher multiplications and write action over a general Feynman geometry.

6.3. *Strict A-infinity algebras of lattice type as strong Feynman geometry*

In Subsec. 5.1 we studied noncommutative associative algebra of cochains associated to simplicial complex. However, it is possible to get supercommutative A_∞ structure based on decomposition

$$\Omega^* = \text{Ker} I \oplus \text{Cochains}. \tag{51}$$

In order to do this, we need to find an embedding of cochains into differential forms compatible with the differential. This problem was solved by Whitney and the corresponding forms are called Whitney forms. In particular, in the one-dimensional case the Whitney forms are

$$C_0 \to \omega_0(C_0) = C_0(A)(1-t) + C_0(B)t, \tag{52}$$

$$C_1 \to \omega_1(C_1) = C_1(BA)dt. \tag{53}$$

It is quite instructive to write $\mathrm{Ker} I$ explicitly. In degree 1, it consists of 1-forms $\omega(t)dt$ with zero integral:

$$\int_0^1 \omega(t)dt = 0 \tag{54}$$

and in degree 0, it consists of functions that may be represented as

$$f - \omega = \int_0^t \omega(u)du. \tag{55}$$

It is clear that f_ω takes zero values at 0 and at 1, thus belongs to Ker I, and actually spans it. The homotopy operator is just

$$h : \omega \to f_\omega \tag{56}$$

and zero otherwise.

These structures were studied in Ref. 2.

In higher dimensions there is no distinguished homotopy and therefore there are many A_∞ structures.

7. Weak Feynman Geometry

Weak Feynman geometry is an A_∞ structure that may not be finite dimensional as a vector space but whose operations are of trace class. Namely, the m_k should be such that

$$m_k(v_1, \ldots, v_{k-1}, \cdot)$$

considered for a given $v_1, \ldots, v_{k-1}$ as a linear operator from V to V should be of the trace class. Note, that the standard multiplication in the de Rham complex is not weakly Feynman, say, multiplication by 1 is an identity operator in the Hilbert space that is not of trace class.

7.1. Weak Feynman geometry of Costello type

The idea of Costello–Feynman geometry is to do the two-step procedure. First, modify multiplication to make it trace class. Then, most probably, the modified multiplication would not be associative. But it could and we will see actually it would be homotopically associative, thus providing A_∞ structure.

Let us consider definition of index of an operator as a prototype. Index was actually a supertrace of multiplication by 1 that did not exist. So, the result of multiplication was smoothed by

$$\exp(-\beta\Delta).$$

It gives the idea to replace m_2 of de Rham DGA by

$$m_2^\beta(v, u) = \exp(-\beta\Delta)m_2(v, u). \tag{57}$$

Then one can show that multiplication would not be associative. However, it could be completed to A_∞ structure by introducing the following higher operation.

$$m_3^\beta(u,v,w) = \exp\left(-\beta\Delta\right) m_2\left(\int_0^\beta \exp(-t\Delta) h dt m_2(u,v), w\right)$$
$$+ \pi m_2\left(u, \int_0^\beta \exp(-t\Delta) h dt m_2(v,w)\right) \quad (58)$$

This A_∞ structure was discovered by Costello[3] and for topological strings by Costello and Li.[4]

7.2. *String theory in form of Zwiebach as weak Feynman geometry*

The second example of the A_∞ structure is given by string theory in the works of Zwiebach.[5-7] Thus, string theory development is not orthogonal to our proposed program but rather it can be included as an important example in the program.

8. Instead of Conclusion: Open Questions

8.1. *Promotion of actions from de Rham DGA to A_∞ structures*

We have shown in Sec. 4 that all actions of classical field theory may be formulated in terms of de Rham DGA. However, it is not clear a priori how to promote them to general A_∞ structure, specifically taking into account possible noncommutativity of operations and existence of higher operations. Proper promotion should take into account symmetries of the theory, thus, rigorously the problem looks as promotion of BV action that includes gauge symmetries to general A_∞ structure. There is a well-known case where such promotion always exist, which is the case of so-called BF theories.

Originally, they were formulated as non-Abelian gauge D-dimensional theories coupled to a field B that is a $(D-2)$-form with values in adjoint representation of the gauge group with the action

$$S = \int \mathrm{Tr}(B(dA + A^2)). \quad (59)$$

In terms of A_∞ structures, this may be reformulated as follows. Consider two A_∞ structures. The first one would be de Rham DGA (considered as A_∞ structure) and the second one is the matrix algebra considered as A_∞ structure with zero differential. Take a tensor product of these two structures and form a new A_∞ structure. Then write a canonical action for this structure (see Appendix). The outcome would be exactly the BF theory. The only subtle moment here is the issue of integral and Tr in the action. Actually, in canonical action for A_∞ structure, we have pairing with a dual field. In the case of de Rham DGA, fields dual to differential forms may be represented also as differential forms with canonical pairing replaced by an integral. In the case of matrix algebra, the dual object can also be considered as a matrix with pairing given by a trace of multiplication.

This explains how to promote BF theory over general Feynman geometry — we should just replace de Rham DGA by a general Feynman geometry and write down canonical action.

Therefore, we can cover in this way 2D Yang–Mills with zero coupling constant and 3D gravity.

The most interesting question is whether it is possible to promote other actions from Sec. 3 to general Feynman geometry. If it is not possible in general, what is the obstruction? Could it be a new selection rule on classical actions?

If such obstruction may vanish on certain Feynman geometry, what is a subclass of Feynman geometry where such obstruction vanish?

These are quite new types of questions in QFT, that we are going to address in the following publications.

8.2. Dirac–Segal approach to QFT versus Feynman geometry approach and the dream of ultimate QFT

In this paper we studied two approaches to QFT: Dirac–Segal and Feynman approaches. One may think that they are orthogonal to each other. In Dirac–Segal approach, the geometry of the space–time is a standard continuum geometry, but there is no Planck constant, no classical fields and Lagrangians. In the Feynman approach, there are classical fields, Lagrangians and Planck constant, but the geometry of space-time is not classical anymore. It may look that these approaches exclude each other.

However, it is not the case. From our experience in dimensions 1 and 2, we know that there are theories that have description in both approaches, and the resulting theory is the same. The best known example is the WZWN that can be described and actually solved in Dirac–Segal approach through integrable representations of affine Lie algebras. At the same time this theory has a functional integral representation.

We may conjecture that the Ultimate Definition of QFT would combine these two paths in the following way — both geometry and physics would be nonclassical. Namely, we expect that it would be possible to generalize Dirac–Segal axioms to an arbitrary Feynman geometry and it would be the goal of axiomatic development of QFT. At the same time Feynman integrals over fields on Feynman geometry would serve as a constructive examples of solutions of these new axioms.

A. Appendix: BV master equation and A_∞ structures

For physicist the proper way to understand A_∞ structures is through the BV language. Consider the superspace (called BV space) of the form $\Pi T^* X$ with coordinates x^a on the base X and x_a^* on the cotangent fiber with opposite parity. Functions on BV superspace may be considered as polyvector fields on the space X. The main operator is the BV operator on functions on BV space

$$\Delta_{\text{BV}} = \sum_a \frac{\partial^2}{\partial x^a \partial x_a^*}. \tag{60}$$

The geometrical meaning of Δ_{BV} is the divergence of the polyvector field; actually, if we apply it to a vector field

$$v = x_a^* V^a(x), \tag{61}$$

we would get

$$\Delta_{\text{BV}} v = \sum_a \partial_a V^a. \tag{62}$$

The main object in BV language is a BV action $S(h)$ — even function on the BV space with values in formal power series in h. The BV master equation reads

$$\Delta_{\text{BV}} \exp S(h)/h = 0. \tag{63}$$

If the BV action represents a vector field, then the corresponding vector field is odd and the classical limit of BV equation states that it is a homological vector field:

$$Q = V^a(x) \frac{\partial}{\partial x^a} \quad Q^2 = 0. \tag{64}$$

Expanding the homological vector field in coordinates, we get a set of operations

$$V^a(x) = \sum_{k=1}^{\infty} \mu_{a_1 \ldots a_k}^a x^{a_1} \ldots x^{a_k}/k!, \tag{65}$$

where $\mu_{a_1 \ldots a_k}^a$ is considered as an operation

$$\mu_k : V^{\otimes k} \to V$$

written in a particular basis. The set of quadratic equations on operations μ_k is called set of conditions for L_∞ structure. The first condition states: that μ_1 is a differential, the second: that μ_1 together with μ_2 form a Leibnitz rule, the third: that Jacobi equation for μ_2 is true up to a commutator of μ_1 and μ_3 and so on.

Note, that L_∞ structure is a structure on supercommutative operations.

However, there is a possibility to encode nonsupercommutative operations in homological vector field. Namely, let us consider take X to be C^{MN^2} with coordinates $x_\alpha^{i,\beta}$, that we will denote simply as $\hat{x}^i$.

Consider the BV action that is still a vector field, but of a very special form

$$A = \sum_{k=1}^{\infty} m_{i_1 \ldots i_k}^j \text{Tr}(\hat{x}^*_j \hat{x}^{i_1} \ldots \hat{x}^{i_k})/k!. \tag{66}$$

If N is big enough then conditions that A solves classical master equation is exactly the set of quadratic equations for operations m that determine A_∞ structure.

They start with m_1 being a differential, Leibnitz rule for m_1 and m_2, associativity of m_2 up to a commutator of m_1 and m_3 and so on.

References

1. A. Losev, A. Marshakov and A. Zeitlin, On First Order Formalism in String Theory, *Phys. Lett. B* **633**, 375 (2006), arXiv:hep-th/0510065.
2. A. Alekseev and P. Mnev, One-dimensional Chern-Simons theory, *Commun. Math. Phys.* **307**, 185 (2011), arXiv:1005.2111.
3. K. Costello, Renormalization and effective field theory, AMS, 2011.
4. K. J. Costello and S. Li, Quantum BCOV theory on Calabi-Yau manifolds and the higher genus B-model, arXiv:1201.4501.
5. B. Zwiebach, Closed string field theory: Quantum action and the Batalin-Vilkovisky master equation, *Nucl. Phys. B* **390**, 33 (1993), arXiv:hep-th/9206084.
6. E. Witten, Non-commutative geometry and open string field theory, *Nucl. Phys. B* **268**, 253 (1986).
7. D. J. Gross and V. Periwal, String field theory, noncommutative Chern-Simons theory and Lie algebra cohomology, *JHEP* **0108**, 008 (2001), arXiv:hep-th/0106242.

Particle Accelerator Development: Selected Examples

Jie Wei

Facility for Rare Isotope Beams, Michigan State University,
East Lansing, MI 48824, USA
wei@frib.msu.edu

About 30 years ago, I was among several students mentored by Professor Yang at Stony Brook to enter the field of particle accelerator physics. Since then, I have been fortunate to work on several major accelerator projects in USA and in China, guided and at times directly supported by Professor Yang. The field of accelerator physics is flourishing worldwide both providing indispensable tools for fundamental physics research and covering an increasingly wide spectrum of applications beneficial to our society.

Keywords: Accelerator; particle; beam.

1. Introduction: Professor Yang as the Teacher, Mentor, and Master

During the years when Prof. Yang directed the Institute of Theoretical Physics and taught at the Department of Physics and Astronomy at Stony Brook (Fig. 1), he mentored several students to enter the field of accelerator science. Among them, I have been most grateful to be guided and at times directly supported by Prof. Yang to pursue a career that I feel most enjoyable and rewarding.

After obtaining a bachelor's degree from Tsinghua University, I traveled from Beijing to Stony Brook entering the graduate program. I took PHY515 Quantum Mechanics taught by Prof. Yang. At his office on the top floor of the Math Tower at Stony Brook, Prof. Yang patiently mentored many fellow students with his profound insight and wisdom.

Professor Yang believed that the future of high-energy physics was in accelerator physics. As the cost of constructing collider accelerators to reach high energy becomes prohibitively high, the future of high-energy physics would rely on breakthroughs in new acceleration principles and techniques. He also felt that important inventions like the strong-focusing principle by Dr. Courant were essential to the development of accelerator physics and thus high-energy physics.

In 1986, Prof. Yang introduced me to Dr. Courant at Brookhaven National Laboratory (BNL). In 1989, I obtained a Ph.D. in the field of accelerator physics under the supervision of Prof. Yang and Dr. Courant (Fig. 2).[1] Since then, I have been fortunate to work on several major accelerator projects at the frontier of both beam energy and beam intensity/power: the Relativistic Heavy Ion Collider (RHIC); the US Part of Large Hadron Collider (US-LHC); the Spallation Neutron Source

Fig. 1. Prof. Yang among the faculty, staff and students of the Department of Physics and Astronomy, Stony Brook University, New York, USA, 1986.

Fig. 2. Photograph taken after J. Wei's Ph.D. dissertation defense, Stony Brook, 1989.[1] From left to right: S. Y. Lee, J. Kirz, C. N. Yang, E. D. Courant and J. Wei (photograph taken by R. Ma).

(SNS); the China Spallation Neutron Source (CSNS); the Compact Pulsed Hadron Source (CPHS); and the Facility for Rare Isotope Beams (FRIB).[2]

Accelerators serve not only as tools to fundamental physics studies but also as platforms for applications beneficial to our society in general. Secondary-beam platforms widely used for accelerator applications include:

- Light sources and free-electron lasers (FELs) driven by electron beams through synchrotron or undulator radiation.
- Neutron sources driven by proton beams through target bombardment.
- Isotope harvesting facilities driven by heavy ion, proton, or electron beam through target bombardment.

Primary beams by themselves are also used for medical and industrial applications like the proton and carbon therapies.

Professor Yang has been an advocate for such applied accelerator projects, many of them at the accelerator power frontier and brilliance frontier instead of energy-frontier colliders for fundamental research. In 2005, Prof. Yang supported me as I led the design and R&D of the China Spallation Neutron Source Project. In 2009, he helped me to start the Compact Pulsed Hadron Source Project at Tsinghua University.[2] This educational project shares the technology that is important for applications like hadron therapy and material research as well as larger-scale facilities like accelerator-driven subcritical systems (ADS) for nuclear waste transmutation and power generation.

2. Frontiers of Particle Accelerators

In this section, we discuss examples of recent developments in the major frontiers of accelerator science: energy/luminosity, power/intensity, and brightness/coherence.

2.1. *Energy and luminosity frontier accelerators*

Energy and luminosity frontier accelerators are either hadron or lepton colliders for fundamental physics studies. Figure 3 shows the evolution of particle center-of-mass (c.m.) energy attained with such colliders.[3,4] Obviously, the present technology no longer supports the traditional exponential growth for the past decades in particle energy.

2.1.1. *Circular hadron and lepton colliders*

Circular proton–proton colliders are the main experimental tools available for exploring particle physics in the energy range of tens of TeV. Presently, the Large Hadron Collider at CERN operates at 14 TeV c.m. energy with counter-circulating protons in a tunnel of 26.7 km.[5] The proposed Future Circular Collider (FCC)[6] is designed to raise the collision energy 100 TeV, using bending magnets of 16 T using advanced superconductor in a tunnel of about 100 km circumference. The same

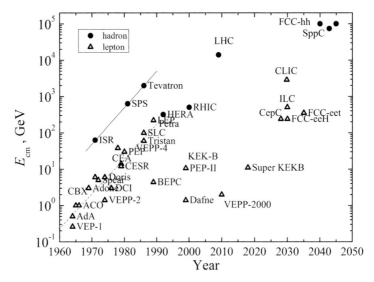

Fig. 3. Center-of-mass energy as a function of year that is either attained (before 2015) or planned with hadron and lepton colliders (courtesy of V. D. Shiltsev[3] and F. Zimmermann[4]).

tunnel infrastructure could accommodate a high-luminosity circular e^+e^- collider (FCC-ee), operating at 90–350 GeV (Fig. 3). In parallel, the Super Proton Proton Collider (SPPC) at 70 TeV c.m. and the "Higgs factory" Circular Electron Position Collider (CEPC) at 240 GeV c.m. are proposed to be placed in a tunnel of about 54 km circumference using similar technologies.[7]

2.1.2. *Linear lepton colliders*

Collision of fundamental particles like electrons and positrons are preferred over hadrons for precision experiments. However, the attainable energy of leptons in circular accelerators is much lower in comparison due to synchrotron radiation energy loss that is inversely proportional to the fourth power of the particle mass. Linear lepton colliders are thus a viable alternative.

The Stanford Linear Collider (SLC) reached c.m. energy of 100 GeV in a tunnel length of about 4 km using room-temperature radio-frequency accelerating structures.[8] The proposed International Linear Collider (ILC) is baselined to be a 200–500 GeV (c.m.) high luminosity linear electron–positron collider of about 31 km length, based on 1.3 GHz superconducting radio-frequency (SRF) accelerating technology.[9] The proposed Compact Linear Collider (CLIC) adopts a novel two-beam acceleration concept for efficient RF power transfer from a low-energy, high-intensity drive beam to the room-temperature accelerating structures for the main beam, accelerating the electrons and positrons to up to 3 TeV (c.m.) in a tunnel of about 50 km length[10] (Fig. 3).

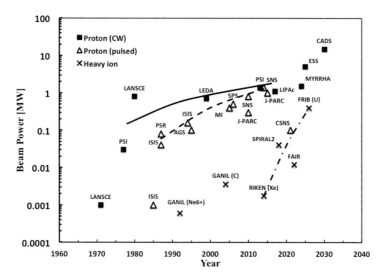

Fig. 4. Beam power on target as a function of year that is either attained (before 2015) or planned with cyclotrons and linacs for continuous-wave mode, and synchrotrons and accumulator rings for pulsed mode, for both proton and heavy ion primary beams.

2.1.3. *Advanced accelerator concepts for high gradient acceleration*

Based on so-called advanced accelerator concepts like laser-plasma acceleration and plasma wakefield acceleration, conceptual facility proposals exist. For example, the laser-plasma linear colliders aim at achieving 1 TeV c.m. energy in a distance of 100 m for each beam.[11] Presently, the highest energy of a high-quality electron beams achieved to date with a laser-plasma accelerator is about 1 GeV in a plasma structure of 3 cm long.[12]

2.2. *Power and intensity frontier accelerators*

Power and intensity frontier accelerators refer to moderate-energy, high-intensity hadron accelerators for both neutron source and isotope harvesting applications and fundamental high energy and nuclear physics studies.[13] The accelerated beams bombarding a fixed target generate secondary beams (neutrons, muons, Kaons, neutrinos, and rare isotopes). At a desired energy range, the yield of the secondary beam is proportional to the beam power of the primary beam. Figure 4 shows the evolution of average beam power for accelerators of proton and heavy ion beams. Key to reaching a high beam power is to limit the uncontrolled beam loss, thus facilitating hands-on maintenance of the facility.

2.2.1. *Continuous-wave proton accelerator*

Continuous-wave (CW) or long-pulse proton applications use cyclotrons or linacs where a high average beam power is desired while the beam time structure is not

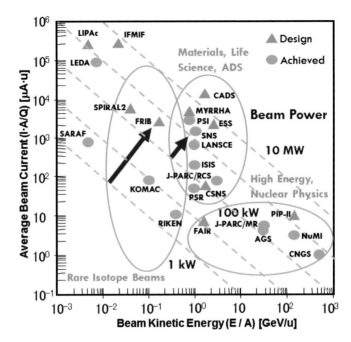

Fig. 5. Beam power on target as a function of the primary beam energy for some of the power-frontier hadron accelerators.[13]

important. In '80s, the LANSCE linac reached the power level of 1 MW for nuclear research and neutron applications.[14] The PSI cyclotron holds the CW power record of 1.4 MW as a neutron source.[15] The European Spallation Source (ESS) is under construction to generate 5 MW beam power at 2 GeV using largely superconducting radiofrequency accelerating cavities.[16] The proposed China Accelerator Driven Subcritical System project (CADS) aims at 15 MW to 30 MW beam power for nuclear waste transmutation and power generation.[17]

2.2.2. Pulsed proton accelerator

Synchrotrons and accumulator rings are typically used to stack the beam produced from linacs to reach high peak intensity and power for pulsed applications.[18] The Spallation Neutron Source (SNS) holds the pulsed power record of 1.4 MW with proton pulse length of 1 μs at 60 Hz repetition rate bombarding a liquid mercury target. A full-energy linac consisting of both room temperature and superconducting radiofrequency cavities accelerates H$^-$ beam to 1 GeV for injection into an accumulator ring for stacking.[19] The Japan Particle Accelerator Research Complex (J-PARC) uses a room-temperature linac to partially accelerate the H$^-$ beam to inject into a rapid cycling synchrotron reaching 1 MW beam power at 3 GeV to drive a spallation neutron source and a muon source for material and life science applications.[20] Part of the 3 GeV proton beam is also accelerated further in the main ring to 30 GeV to create heavy hadronic particles such as pions and kaons. The main ring beam is also used to create neutrino beams for analysis at the Kamioka detector, located approximately 300 km to the west.

2.2.3. *Heavy ion accelerator*

Acceleration becomes increasingly difficult with particles of heavier mass. The Facility for Rare Isotope Beams (FRIB) currently under construction at Michigan State University will be driven by a heavy ion linac designed to advance the heavy ion power frontier by more than two-order-of-magnitudes to 400 kW (Fig. 4).[21] All stable ion species from proton to uranium can be accelerated to above 200 MeV per nucleon bombarding a rotating graphite target to produce rare isotopes for both fundamental research and practical applications.

2.3. *Brightness and coherence frontier accelerators*

Brightness and coherence frontier light sources are application platforms driven by electron beams. Early generation facilities are based on storage rings where electrons passing a bend magnet emit light through synchrotron radiation. Later generation facilities use low emittance electrons and insertion devices like undulators for increased brightness. Latest generation facilities achieving even higher brightness, laser-like coherence, and short pulse duration include free-electron lasers (FELs) based on storage rings or linacs. The Linear Coherent Light Source-II (LCLS-II) is an X-ray free-electron laser designed to deliver photons between 200 eV and 5 keV at repetition rates up to 1 MHz using a superconducting radiofrequency linac.[22]

3. Technologies for Particle Accelerators

Cutting edge technologies continuously developed for accelerator systems have sustained the growth in beam energy/luminosity, power/intensity, and brightness/coherence.[23] Such technologies are also increasingly often used in industrial accelerator serving our society.

3.1. *Superconducting radiofrequency (SRF)*

Superconducting RF technology plays an increasingly important role in all frontiers of accelerators like the linear collider ILC,[9] the fourth generation light source LCLS-II,[22] the SNS high power proton linac,[19] and the FRIB heavy ion driver linac.[21] The high accelerating gradient is essential for lepton linear accelerators like the energy frontier ILC and brightness frontier FEL LCLS-II. The energy efficiency is essential for hadron linear accelerators like the power frontier ESS[16] and FRIB[21] of high beam duty factor.

Recent high-β (particle speed βc approaching the speed of light) accelerators like ILC, LCLS-II, and PIP-II[24] share 1.3 GHz frequency SRF technology.

For medium- and low-β accelerators, SRF technology is extensively used in the SNS linacs for the high energy efficiency, high accelerating gradient, and operational robustness.[19] For pulsed operations, resonance control by means of fast tuners and feedforward techniques is often required to counteract Lorentz force detuning, and the need of higher order mode damping is to be expected. FRIB as a heavy ion continuous-wave (CW) linac extends SRF to low energy of 500 keV/u. 330 low-β cavities are housed in 48 cryomodules. These cryomodules contain resonators (at 2 K temperature) and magnets (at 4.5 K) supported from the bottom to facilitate alignment and the cryogenic headers suspended from the top for vibration isolation. High performance subsystems including resonator, coupler, tuner, mechanical damper, solenoid and magnetic shielding are necessary.

3.2. Superconducting magnet

Energy frontier circular accelerators rely on the development of economical high-field magnets. The magnets of the present colliders like LHC and RHIC are made from Nb–Ti superconductor, which supports a maximum field of about 10 T. Nb_3Sn superconductor capable of reaching a practical magnetic field up to 16 T are to be used for dipole and quadrupole magnets for the high luminosity upgrade of LHC, representing an important milestone towards the FCC.[4]

FRIB prototyped high-temperature superconducting magnet with yttrium copper oxide (YBCO) material radiation resistant in the area of isotope production target and primary beam dump.

3.3. Large-scale cryogenics

For modern accelerator facilities using superconducting technology, large-scale cryogenics is an integral part of the SRF and magnet system. An integrated design of the cryogenic refrigeration, distribution, and cryomodule systems is key to efficient operations.

The FRIB refrigeration system adopts the floating pressure process — Ganni Cycle for efficient adaptation to the actual loads. Distribution lines are segmented and cryomodules are connected with the U-tubes to facilitate stage-wise commissioning and maintenance.

3.4. Machine protection, collimation, loss detection and control

Machine protection is crucial to the availability of frontier accelerators. At the LHC top energy the beam intensity is about three order-of-magnitudes above the destruction limit of the superconducting magnet coil and 11 orders of the fast loss quench limit. The protection consists of passive components including collimators, absorbers and masks and active components including beam loss monitors, beam interlock system and beam dumps.

FRIB adopts multi-time scale, multi-layer approaches: the fast protection system (FPS) is designed to prevent damage in less than 35 μs from acute beam loss by quickly activating the beam inhibit device; the run permit system (RPS) continuously queries the machine state and provides permission to operate with beam; the even slower but highly sensitive RPS prevent slow degradation of SRF system under small beam loss.

Challenges remain for intense low-energy heavy ion beams due to the low detection sensitivity and high power concentration/short range. Innovative techniques include the halo monitor ring and thermometry sensors for high-sensitivity loss detection and current monitoring modules for critical magnet power supply inhibition. ADS machines like MYRRHA demand mean-time-between-failure of trips exceeding 3 s to be longer than 250 h.[25]

3.5. Source

Among a wide range of ion sources meeting different hadron primary-beam requirements, ECR sources are essentially the only choice for high intensity (CW), high charge state beams. ECRs continue to move to higher RF frequency and magnetic field. High power ECR sources operate at frequencies up to 28 GHz and RF power of ~15 kW. The required SC sextupole and solenoid push the state-of-the-art in SC technology. Cesium-seeded, volume production sources are most promising for the demand on high current, long pulse, low emittance H$^-$ beams.

Brightness and coherence frontier light sources usually require electron sources of high brightness, high level of polarization, low angular divergence, low emittance, wide tenability in energy and wavelength, and short pulse duration. The LCLS-II free electron laser will use a normal-conducting photo-cathode electron gun providing a beam of 750 keV which is then bunched with a 1.3 GHz buncher cavity before being injected into a 1.3 GHz superconducting cryomodule where it is captured and accelerated.

3.6. Charge stripping

Intense heavy ions at low energies may cause severe damage on stripping material. Innovative stripping mechanisms are under development worldwide. RIKEN uses helium gas with differential pumping. Plasma windows are being tested to establish a high gas density. FRIB uses a liquid lithium film moving at ~50 m/s speed. Tests with a proton beam produced by the LEDA source demonstrated that power depositions similar to the FRIB uranium beams could be achieved without destroying the film.

Injection of intense H$^-$ beams into rings require sophisticated charge stripping designs. Innovative schemes like laser stripping are tested. Stripping can also be used to split H$^-$ beam to multiple beam lines.

3.7. *Target, radiation-resistant magnet and handling*

Target scenario is chosen based on secondary-beam requirements. High-power primary beams often demand non-stationary targets like circulating liquid or rotating solid targets. For pulsed neutron production at MW level, both SNS and J-PARC/RCS use liquid mercury. Target pitting issues are largely mitigated by vessel surface treatment, mercury flow and bubble controls. For lower-energy neutron production both SARAF[26] and IFMIF[27] use liquid lithium while SPIRAL2[28] prefers a rotating carbon wheel. MYRRHA's ADS target uses liquid Pb–Bi eutectic. For in-flight RIB production FRIB needs to focus 400 kW of heavy ion beam onto an area of 1 mm diameter ($\sim$60 MW/cm^3). A radiation-cooled multi-slice graphite target of 30 cm diameter rotates at 5000 revolutions-per-minute. While neutron targets are designed to absorb most beam power, FRIB's RIB target is designed to absorb $\sim$25% power; targets for high-energy physics (ν, μ, K) typically absorb <5% power.

Radiation resistance is important for magnets in the target region. Quadrupoles wound with mineral-insulated cables are built as an integral part of the shielding in front of the SNS target. Quick-disconnect vacuum flanges and remote water fittings allow easy access.

3.8. *Collider technology*

Pioneering heavy ion collider RHIC faced several design and construction challenges. As the first superconducting magnet accelerator crossing the transition energy leading to complications including chromatic nonlinear effects, electron cloud effects and collective instabilities; a "transition jump" is essential in alleviating this performance bottleneck during acceleration; Coulomb scatterings between high charge state ions (intra-beam scattering) lead to beam emittance growth, beam loss and luminosity degradation; stochastic cooling is most effective in countering this performance limiting mechanism during storage.[2]

Experience in RHIC design was also shared by subsequent collider projects like LHC. An example is the local correction system of the interaction region magnetic errors for low-β^* operations.

3.9. *Rapid cycling synchrotron technology*

High power pulsed hadron facilities that use synchrotron acceleration require rapid cycling.[18] At the AGS Booster, resonance corrections of magnetic nonlinearities are essential during high-intensity operations. Back-leg winding driven sextupole correctors allowed for metal vacuum chambers in rapid cycling synchrotrons avoiding ceramic chamber complications. ISIS uses dynamic tune variation to mitigate space charge, chromaticity, instability and coupling issues. Successfully hardware systems include collimators and ceramic vacuum chambers with supported internal stainless-steel wires, interrupted with ceramic-chip capacitors to allow the passage only of beam image charge at high frequency.

J-PARC advanced technologies pioneered by AGS and ISIS for rapid-cycling synchrotrons including introducing main magnets built with braided aluminum coil, high-gradient wideband RF cavity built with water-cooled magnetic alloy, and large aperture ceramic vacuum chamber with RF shielding. A large beam chamber aperture and accurate magnet tracking limit the uncontrolled beam loss below 1%.[20]

3.10. Accumulator ring technology

High power hadron accumulator rings typically accept H^- ions accelerated by full energy linear accelerators stacking them to form high peak intensity pulses. This alternative approach to rapid cycling synchrotron avoids many technical challenges discussed above. In fact, the SNS accumulator holds the world record of 1.4 MW pulsed power in a ring.

Accumulator rings face their own challenges in design and construction. The design must minimize premature stripping of the H^- ions at high energy due to Lorentz magnetic stripping, ionization stripping, black body stripping, and intra-beam stripping. To mitigate electron cloud effects that limited the performance of earlier proton accumulator PSR, SNS design exhaustively adopted preventive measures suppressing electron generation and enhancing Landau damping.[18]

3.11. Medical and industrial accelerator technology

More than 1000 medical industrial accelerators are produced each year covering many applications including medical diagnostics and treatment, ion implantation, electron beam material processing, electron beam irradiator, radioisotope production, ion beam analysis, high energy X-ray inspection, neutron generator, and compact synchrotron radiation source.[29]

An example of accelerator technological application is hadron therapy facilities. Hadron (usually proton or carbon) therapy is an effective and non-invasive way to treat tumors deep within the body because the ions deliver most of their dose just before stopping in the Bragg peak.[30,31] To facilitate treatment flexibility, device reliability, compactness and robustness, spot scanning capability, breathing synchronized, and radiation safety, modern acceleration, delivery, and controls technologies are used or proposed based on room temperature and superconducting cyclotrons, slow and rapid cycling synchrotrons, and gantries with fixed-field alternating gradient (FFAG) design.

4. Discussion and Summary

For more than 80 years since the invention of cyclotron, particle accelerators have evolved from a simple tool for physics research to its own field of science and engineering essential to not only high-energy physics but almost all disciplines of fundamental research and practical applications.

Existing accelerators in the energy frontier are based upon several major breakthroughs including the invention of synchrotron, the strong focusing principle, the collider, the stochastic cooling, and the superconducting technology. To reduce the high construction cost of energy-frontier accelerators, advanced acceleration concepts have been pursued. However, such concepts are yet to be matured for facility construction.

On the other hand, accelerators in the power and brightness frontiers are flourishing worldwide providing application platforms of neutron sources, light sources, rare isotope facilities, nuclear waste treatment, and power generation. Technologies developed for frontier accelerators have been benefiting the development of medical and industrial applications directly serving our society.

To me, the accelerator profession is so uniquely rewarding that physical ideas can be turned into reality through the execution of a construction project. Throughout its completion one experiences endless learning in physics, technology, teamwork, and creating friendship. I am profoundly indebted to Prof. Yang for his mentoring, guidance, and support since 30 years ago.[2]

Acknowledgment

This work is supported by the U.S. Department of Energy Office of Science under Cooperative Agreement DE-SC0000661 and the National Science Foundation under Cooperative Agreement PHY-1102511.

References

1. J. Wei, Longitudinal dynamics of the non-adiabatic regime on alternating-gradient synchrotrons, Ph.D. thesis, State University of New York at Stony Brook (USA, 1989).
2. J. Wei, in *Proc. 2010 Int. Part. Accel. Conf. (IPAC'10)*, Kyoto, Japan, 2010, p. 3658.
3. V. D. Shiltsev, *Phys. Usp.* **55**, 965 (2012).
4. M. Benedikt, F. Zimmermann, D. Schulte and J. Wenninger, in *Proc. 2014 Int. Part. Accel. Conf. (IPAC'14)*, Dresden, Germany, 2014, p. 1.
5. L. R. Evans, in *Proc. 2007 Asia Part. Accel. Conf. (APAC'07)*, Indore, India, 2007, p. 46; http://home.cern/topics/large-hadron-collider.
6. http://cern.ch/fcc.
7. Y. F. Wang, in *Proc. 2015 Int. Part. Accel. Conf. (IPAC'15)*, Richmond, USA, 2015, RYGB2; http://cepc.ihep.ac.cn/preCDR/main_preCDR.pdf.
8. J. T. Seeman, in *Proc. 1990 Linear Accel. Conf. (LINAC'90)*, Albuquerque, USA, 1990, p. 3.
9. B. C. Barish, in *Proc. Int. Part. Accel. Conf. (IPAC'13)*, Shanghai, China, 2013, p. 1101; https://www.linearcollider.org/ILC.

10. S. Stapnes, in *Proc. 2012 Int. Part. Accel. Conf. (IPAC'12)*, New Orleans, USA, 2012, p. 2076; http://clic-study.web.cern.ch/.
11. C. B. Schroeder, E. Esarey, C. G. R. Geddes, C. Benedetti and W. P. Leemans, *Phys. Rev. ST-AB.* **13**, 101301 (2010).
12. W. P. Leemans, B. Nagler, A. J. Gonsalves, C. Toth, K. Nakamura, C. G. R. Geddes, E. Esarey, C. B. Schroeder and S. M. Hooker, *Nat. Phys.* **2**, 696 (2006).
13. J. Wei, in *Proc. 2014 Int. Part. Accel. Conf. (IPAC'14)*, Dresden, Germany, 2014, p. 17.
14. D. E. Nagle, in *Proc. 1972 Linear Accel. Conf. (LINAC'72)*, Los Alamos, USA, 1972, p. 4.
15. M. Seidel, S. Adam, A. Adelmann, C. Baumgarten, Y. J. Bi, R. Doelling, H. Fitze, A. Fuchs, M. Humbel, J. Grillenberger, D. Kiselev, A. Mezger, D. Reggiani, M. Schneider, J. J. Yang, H. Zhang and T. J. Zhang, in *Proc. 2010 Int. Part. Accel. Conf. (IPAC'10)*, Kyoto, Japan, 2010, p. 1309.
16. M. Lindroos, S. Molloy, D. McGinnis, C. Darve and H. Danared, in *Proc. 2012 Linear Accel. Conf. (LINAC'12)*, Tel-Aviv, Israel, 2012, p. 768.
17. W. Zhan, in *Proc. 2013 Int. Part. Accel. Conf. (IPAC'13)*, Shanghai, China, 2013, MOXAB101.
18. J. Wei, *Rev. Mod. Phys.* **75**, 1383 (2003).
19. S. Henderson, in *Proc. 2010 Linear Accel. Conf. (LINAC'10)*, Tsukuba, Japan, 2010, p. 11.
20. Y. Yamazaki, in *Proc. 2009 Part. Accel. Conf. (PAC'09)*, Vancouver, Canada, 2009, p. 18.
21. J. Wei *et al.*, in *Proc. 2015 Heavy Ion Accel. Conf. (HIAT'15)*, Yokohama, Japan, 2015, MOM1I02; http://www.frib.msu.edu/.
22. J. N. Galayda, in *Proc. 2014 Linear Accel. Conf. (LINAC'14)*, Geneva, Switzerland, 2014, p. 404; https://portal.slac.stanford.edu/sites/lcls_public/lcls_ii/Pages/default.aspx.
23. Eds. A. W. Chao, K. H. Mess, M. Tigner and F. Zimmermann, *Handbook of Accelerator Physics and Engineering*, 2nd edn. (World Scientific, 2003).
24. S. Holmes, P. Derwent, V. Lebedev, S. Mishra, D. Mitchell and V. P. Yakovlev, in *Proc. 2015 Int. Part. Accel. Conf. (IPAC'15)*, Richmond, USA, 2015, p. 3982.
25. D. Vandeplassche, J.-L. Biarrotte, H. Klein and H Podlech, in *Proc. 2011 Int. Part. Accel. Conf. (IPAC'11)*, San Sebastián, Spain, 2011, p. 2718.
26. D. Berkovits *et al.*, in *Proc. 2012 Linear Accel. Conf. (LINAC'12)*, Tel-Aviv, Israel, 2012, p. 100.
27. J. Knaster, P. Cara, A. Mosnier, S. Chel, A. Facco, J. Molla and H. Suzuki, in *Proc. 2013 Int. Part. Accel. Conf. (IPAC'13)*, Shanghai, China, 2013, p. 1090.
28. R. Ferdinand, P. Bertrand, X. Hulin, M. Jacquemet and E. Petit, in *Proc. 2013 Int. Part. Accel. Conf. (IPAC'13)*, Shanghai, China, 2013, p. 3755.

29. R. W. Hamm and M. E. Hamm, in *Proc. 2013 Int. Part. Accel. Conf. (IPAC'13)*, Shanghai, China, 2013, p. 2100.
30. K. Noda, in *Proc. 2011 Int. Part. Accel. Conf. (IPAC'11)*, San Sebastián, Spain, 2011, p. 3784.
31. M. Schillo, in *Proc. 2014 Int. Part. Accel. Conf. (IPAC'14)*, Dresden, Germany, 2014, p. 1912.

A New Storage-Ring Light Source

Alex Chao

SLAC National Accelerator Laboratory, Stanford, California, USA
achao@slac.stanford.edu

A recently proposed technique in storage ring accelerators is applied to provide potential high-power sources of photon radiation. The technique is based on the steady-state microbunching (SSMB) mechanism. As examples of this application, one may consider a high-power DUV photon source for research in atomic and molecular physics or a high-power EUV radiation source for industrial lithography. A less challenging proof-of-principle test to produce IR radiation using an existing storage ring is also considered.

Keywords: Coherent radiation source; storage ring; steady-state microbunching; lithography.

1. Coherent E&M Radiation Field

This conference is about Yang–Mills field. Here I wish to talk about the coherent radiation of the electromagnetic field, the very first known Yang–Mills field (although Abelian). Let me start with a short remind of how the technology of coherent E&M radiation has evolved over time. Generation of coherent E&M radiation first started with the technology of vacuum tubes of the early 1900s. The uses included radios, antenna, television, and then sped up by the war efforts, the radars. The invention of klystron in 1937 pushed this line of technology to high sophistication. We also have microwave ovens and wi-fi.

But the need of coherent radiation continued to grow, particularly towards higher frequencies. The approach based on vacuum tubes and klystron technology ran into a severe limit when the radiation frequency is pushed up towards 100 GHz (10^{11} Hz). As the desired frequency increases, the electron-beam/vacuum-tube system gets smaller due to the necessary boundary conditions of the radiation fields, and the required technology becomes increasingly difficult. The difficulty originates from the dominance of the boundary conditions for the coherent radiation field imposed by its being confined to the interior of the devices.

[Courtesy: Stanford News Service]

A breakthrough occurred in 1955 when laser was invented. Boundary condition imposed by the coherent field is no longer a limitation, and lasers completely took over as the ideal source of coherent radiation at high frequencies.

But lasers are still bounded. Although the radiation field is in free space, the radiating electrons are bound by the atoms.[a] The energy levels available to lasers are limited by the available discrete energy levels in atoms and molecules, and as

[Courtesy: AP Photo/File]

---

[a] Alternatively one might say that the radiation still has to satisfy the boundary condition imposed by the quantum mechanical wave equation within the confinement of the atoms, in contrast to the earlier boundary condition imposed by the Maxwell wave equation.

[Courtesy: Chuck Painter/Stanford News Service]

a result, the radiation frequency is limited to 10^{15} Hz, or deep ultraviolet frequencies. For a long while, there seemed no hope for coherent radiation with higher frequencies, such as soft or hard X-ray (approaching 10^{18} Hz) simply because there were no available discrete energy levels at those frequencies.

Then the history is such that the electron beam technology returned to the stage, this time without boundary conditions.[2] With the invention of the free electron laser in 1971, not only the electromagnetic fields have no boundary conditions, but also neither do the radiating electrons — thus the term "free" electron laser. Coherent X-rays became possible this way. FEL is a tremendous game changer. The trick here is to somehow make the free electron beam to "microbunch."

Microbunching is a wonderful thing. The microbunched electron beam not only allowed the radiation frequency to be raised to X-rays, but also due to the coherence of the radiation process, the power of radiation is increased by a factor of N, where N, the number of electrons in the microbunch, is a very large number. The peak power of the FEL radiation is therefore extremely high. When the LCLS FEL radiation was commissioned in 2009,[3] the peak X-ray brilliance was raised by 10 orders of magnitude overnight, mostly due to the microbunching factor of N. Not only the power source has changed, all the experimental instrumentation, detection, and methodology all had to change.

So the fact that these E&M radiations are coherent serves *two* purposes, both are important. First and more obvious, when coherent, the radiation is a laser and comes with all its analyzing and signal carrying power. Second and perhaps less noted, the coherence also makes the raw radiation wattage extremely high due to the additional factor of N.

In passing, I should mention that the pathway to free up the bound electrons as a means to reach higher frequencies applied to both the electron beam and the laser technologies. In the laser technology, high harmonic generation (HHG) technique was invented,[4] which is another idea trying to free up the bound electrons in atoms. Perhaps we summarize the landscape with the following table.

	Bound systems	$\rightarrow$	Free systems
Electron beam technology	vacuum tubes klystrons (10^{11} Hz)	$\rightarrow$	free electron lasers (FEL) (10^{18} Hz)
Laser technology	molecular lasers atomic lasers (10^{15} Hz)	$\rightarrow$	high harmonic generation lasers (HHG) (10^{18} Hz)

As mentioned, the peak power has been raised to extreme high levels by the FEL. However, application demand never ceases. FELs have the drawback that although its peak power is extremely high, its average power is low. This is because FELs use linear accelerators, and linear accelerators have notoriously low repetition rates.

The issue of repetition rate brings us to the consideration of storage rings. The traditional synchrotron radiation from a storage ring produces high power mainly because the beam circulates with high repetition rates — the beam is reused every revolution. There are of course many electrons in the beam and they all radiate, and that gives a high radiation power, but the electrons all radiate individually, and that is no comparison with the high peak power in the FELs when electrons radiate coherently.

It seems apparent that to make a next step in the development of high power coherent radiation sources, one must try to keep the extremely high peak power, while somehow maintain a high repetition rate. Repetition rate is readily available from storage rings. The high peak power requests the beam to be microbunched. This leads to the introduction of a "steady state microbunching" (SSMB) technique to be applied to electron storage rings as powerful radiation sources.[1] If developed, the very large number of coherent electrons will also likely outcompete the HHG.

2. Lithography and Moore's Law

Powerful radiation sources have applications in many areas, from research tools to industrial applications. Perhaps let me mention one of the possible industrial applications, namely to lithography, as one prominent example.

Semiconductor industry is important contributor in our economy. Its 2012 worldwide business:

$$10-20B\$ \quad \text{wafer fabrication equipment}$$

$$250B\$ \quad \text{semiconductor devices}$$

$$\sim 2T\$ \quad \text{consumer electronics}$$

And the industry is still growing rapidly. Its growth is best represented by the Moore's law (Gordon Moore, 1965):

> The number of transistors that could be fit on a chip of a given size at an acceptable cost doubles every two years.

(Incidentally, in accelerator field, there is a famous "Livingston chart," which says the equivalent accelerator beam energy doubles every two years — same rate as Moore's law.)

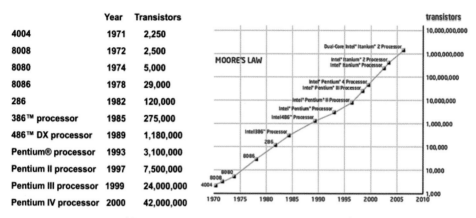

	Year	Transistors
4004	1971	2,250
8008	1972	2,500
8080	1974	5,000
8086	1978	29,000
286	1982	120,000
386™ processor	1985	275,000
486™ DX processor	1989	1,180,000
Pentium® processor	1993	3,100,000
Pentium II processor	1997	7,500,000
Pentium III processor	1999	24,000,000
Pentium IV processor	2000	42,000,000

[Courtesy: 林合俊 陳虹宇 周展弘 陳玟儒 吳萱郁, 2014]

Note that Moore's law is not just a passive curiosity. It is a result of market demand (2 T$, and growing). The semiconductor industry runs on the Moore model and is compelled to follow it. Breaking from Moore's law has severe economic consequences.

Moore's law is maintained by continually miniaturizing the transistors. As the resolution of lithography to produce the chips becomes finer, the required wavelength of the lithography light becomes shorter. To fulfill the industrial needs, the power of the radiation also needs to be very high. Present day lithography uses DUV (deep ultraviolet, 365, 248, 193 nm), but DUV is falling behind. We are presently dangerously near the point of departure from Moore's law.

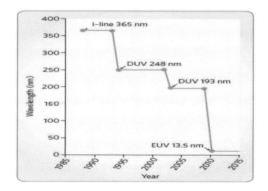

[Courtesy: 林合俊 陳虹宇 周展弘 陳玟儒 吳萱郁, 2014]

We need a radiation source of high power with shorter wavelengths. The next advance proposes to use EUV (extreme ultraviolet, 13.5 nm) light. EUV is presently industry's best hope for keeping up with Moore's law. Choice of 13.5 nm is due to a window of high reflectivity of multilayered mirrors. Bandwidth is $\sim \pm 2\%$. If made available, EUV light can sustain Moore's law for $\sim$10 more years. Soft X-rays or electron beams may be considered after EUV.

But EUV photons are difficult to generate. Existing techniques fall short on the required power. ASML, the Netherlands, the world's leading provider of lithography systems, after much R&D efforts and spending on EUV, is pushing the technology of laser-produced plasma (LPP) devices. Their record power so far is $\sim$75 W per tool. The industry aims for $\gtrsim 1$ kW per tool. Since too much is at stake, there has been a need of ideas to provide a high power EUV light source.

3. The SSMB Approach

One presently conceived approach for producing EUV light of kW level is to use accelerators. There are two types of accelerators: circular ones (storage rings) and linear ones (linacs).

- Linacs (FELs) produce EUV radiation that has high peak power, but low average power because of their low repetition rates.
- Storage rings have high repetition rates but low peak power.

To achieve kW power, we need both high peak power and high repetition rate.

The low repetition rate of FELs can be partially solved by invoking superconducting linac technology, raising repetition rate from 60 or 120 Hz to $\sim$1 MHz. However, the electron beam is thrown away after usage, so it requires too much power to run, unless one invokes another technology called energy recovery linacs to recover and reuse the electron beam energy. Superconducting, energy recovery FEL EUV source is a feasible approach. It is also expensive.

An alternative approach, adopted by SSMB, is to combine the strengths of linacs and storage rings.[1] With a conventional room-temperature electron storage ring, the SSMB utilizes only off-the-shelf hardware and its cost can be a small fraction of the SC/ERL linac FEL alternative. However, being a new proposed technique untested in existing storage rings, it is still in an early R&D stage. R&D efforts and a proof-of-principle test are required before it can be considered a feasible alternative for actual applications.

The basic idea of SSMB is to manipulate the beam's dynamics in a storage ring so that its distribution is not the conventional Gaussian with a typical bunch length $\sim$ a few millimeters, but microbunched with each microbunch having a length $<$ a few μm, and in the case of EUV application, $<$13.5 nm so that all electrons in each microbunch and its neighboring microbunches radiate coherently. The two cases are sketched below (note the very different length scales):

Conventional storage ring

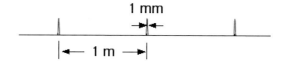

SSMB storage ring

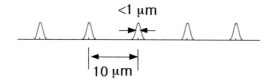

The functioning of SSMB lies in finding a way to make the beam microbunched in the first place, and then in addition make them stay microbunched in the turn-by-turn environment of a storage ring.

- The beam is <u>microbunched</u> and strongly focused, so it readily radiates at the desired short wavelength (13.5 nm for lithography) at an appropriate radiator, yielding high peak power $\propto N^2$ instead of $\propto N$.
- The beam is microbunched in a <u>steady state</u> in a storage ring, so it radiates every turn with a high repetition rate. With a bunch spacing of $10\,\mu$m, the repetition rate is 300 GHz.

These two features, microbunching and steady-state, lead to the term "steady state microbunching."

With the high repetition rate, radiation per bunch passage is far weaker than what is demanded by a superconducting linac FEL (by 5–6 orders of magnitude). The electron beams are not disrupted by each passage through the radiator. This device is not an FEL.

SSMB uses a natural equilibrium state of the electron beam. The microbunch structure can not be produced by chopping the beam with fast electronics or laser techniques using present technologies. Instead, it is reached because the electron beam chooses it as its steady state when the storage ring environment is so provided.

The required hardwares are simple: Take an electron storage ring of $\sim 1\,$GeV, and insert three appropriate undulator magnets, each about 3–4 m long, in its circumference. To two of the undulators, add an IR laser (called "seed laser"). The storage ring is conventional, the undulators are routine components in accelerator applications. The seed laser requires high power, comparable to what is needed for the superconducting linac option.

SSMB can be scaled to a range of frequencies from IR to EUV. As mentioned, kW EUV sources are useful for the industry. Multi-kW IR or DUV radiation are potentially research tools for atomic or molecular physics.

A schematic of the layout of the facility looks like the following figure. It shows a facility with two SSMB radiation tools arranged back to back in a single storage ring.

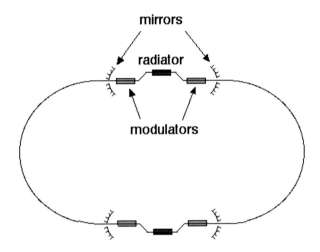

Each tool consists of a radiator sandwiched between two modulators. A radiator is an undulator magnet that resonates at the desired radiation wavelength λ (13.5 nm for EUV, for example). This is where coherent radiation is emitted and delivered to the users. A modulator is also an undulator magnet, resonant at a longer wavelength $M\lambda$ where M is an integer, typically 15–20. The two modulators around each radiator are "seeded" with a laser with wavelength $M\lambda$. By the action of this seed laser, the electron beam will find its microbunched steady state and stay there turn by turn if other storage ring parameters are chosen properly.

The seed laser's another important function is to strongly focus the microbunches to allow the harmonic generation factor M. The SSMB storage ring is therefore "strong focusing" in its longitudinal beam dynamics, a new regime of storage ring operation that provides tight control of the beam's longitudinal emittance and consequently promises further possible applications beyond SSMB.

Around the seed laser, we have two mirrors to form a laser cavity. One technical requirement is that the mirrors must have high reflectivity (higher than ~ 0.999) at the desired wavelength $M\lambda$.

Table below shows some example applications of SSMB as IR, DUV and EUV sources. (As proposed in Ref. 1, SSMB can also be applied to *lower* the coherent radiation frequency, e.g. to THz, by beating two modulations with nearby IR frequencies.) The advertised radiation power for the three cases are in the range of multiples of kW, readily orders of magnitude higher than other approaches, again due to the high repetition rate of a storage ring and the microbunched coherence nature of the electron beam. In case large radiation power is demanded from the SSMB source, the storage ring can be provided a conventional RF system, an induction linac section, or a barrier RF bucket system to refurbish the lost beam energy. However, there are caveats.

		IR SPEAR3	DUV SPEAR3	EUV dedicated ring	
E_0	beam energy	900	900	580	MeV
C	ring circumference	234	234	100	m
α_C	ring mom. comp. factor	1.9	0.57	0.16	10^{-6}
I_0	average beam current	8.5	4.7	1.12	A
L_m	modulator length	3.7	3.2	3.4	m
λ_m	seed laser wavelength	13.2	3.5	0.37	μm
P_{seed}	seed laser power	15	15.7	11	kW
L_r	radiator length	3.5	3.3	3.54	m
λ_r	SSMB rad. wavelength	0.94	0.205	0.0133	μm
P_r	SSMB rad. power	85	41	4.06	kW

Two caveats:

(1) The first is that the required seed laser has high power and the mirrors will have to stand its radiation heating while maintaining their high reflectivity. In case the seed laser power must be reduced, the intended radiation power will be reduced by the same factor. For example, in the IR case, if the seed laser is limited to 1 kW, the IR radiation per radiator is reduced to 5.6 kW, which is still a high level. One approach to reduce the needed seed laser power is to divide the modulators into smaller pieces. With reduced Rayleigh length, the seed laser power is reduced. Another approach, much more intriguing and exciting, is to self-seed the modulators, sparing the seed laser altogether, although mirrors will still be needed. On the other hand, if high power seed laser is available, one can consider yet another option of single shot seeding without mirrors.

(2) The SSMB mechanism has only existed on paper and on computer simulation studies. It has not been scrutinized with experimental tests. Feasibility is not established. One of the main next activities is to try a proof-of-principle test on an existing storage ring.

4. Proof-of-Principle Test Proposal

We considered two operation cases using the existing storage ring SPEAR3.

(1) Without a seed laser: Setup consists of only a single modulator with mirrors, without a seed laser and without a radiator. The microbunching originates from the self-modulation due to the accumulated spontaneous radiation from the undulator. The microbunching is to be built up from the white noise of the beam distribution. This intriguing possibility is one of the ideas to be explored in the PoP test. This is work in progress.

(2) With a seed laser: Augment the single modulator with a seed laser. Fill 1% of the ring with electrons. The beam is expected to be fully microbunched at the modulation wavelength.

	Without seeding	With seeding	
Modulation wavelength (λ)	10	10	μm
Cavity Q	500	500	
Undulator strength parameter K	30	30	
Undulator period	20	20	cm
# of undulator periods	10	10	
Electron beam energy	1	1	GeV
Momentum compaction factor (α_C)	1.3	0.3	10^{-5}
Beam current	500	54	mA
Seed laser power	—	190	W

5. Conclusion

It is time to conclude. Instead of talking about how great SSMB is, I wish to conclude with a recollection of memory 41 years ago of what led to my standing here today talking about the SSMB.

That should begin with my first involvement in the field of accelerator physics. It all started when Professor Yang in early 1973 advised me to take an accelerator physics course by Professor Ernest Courant. It was all very innocent at the beginning. Following his advice, I happily took the course and wholly enjoyed it. It was real fun! In 1973 later, my last year of PhD study at Stony Brook, Professor Yang arranged me to study half time under Professor Courant at Brookhaven Lab, which I also enjoyed.

Then came the time of my graduation and decision was to be made of which field I should focus after graduation. My thesis was on high energy theory. Naturally I had initially considered high energy physics as my career choice. Professor Yang talked to me one day. He told me that I could find a job in high energy theory and that I could do well. But he advised me to instead consider accelerator physics as my career choice, arguing that HEP field is more crowded and that it is better and more rewarding for a young researcher to choose a field that is less crowded like is the case of accelerator physics. He must have lectured for 10 minutes, or at least it felt so. His lecture I summarize as follows.

Yang's lecture 1973 & 1974 on career choices:

Choose 僧少粥多 "Few monks, much congee"
(accelerator physics)

Don't choose 粥少僧多 "Many monks, little congee"
(high energy physics)

But now this was a much more serious matter for the then-young me. His argument was convincing and I did enjoy my studies on accelerator physics at the time. But hesitating I did.

I hesitated, if I recall, for a few weeks. After all, I enjoyed both high energy physics and accelerator physics. While still hesitating, Professor Yang asked me again about my decision. As I hardly started to explain my reasons of hesitation that he became impatient with me — for the one and only time that I could recall. He said in a raised voice, and I quote, "I strongly dispute your reasons," and continued to give his lecture again. The time was about April 1974.

To make the story short, I did not actually disagree with him, but his persuasion pushed me into a final decision. For 41 years now, I have thoroughly enjoyed this field. It is so much fun and indeed few other "monks" came sharing my "congee." I wish to thank Professor Yang for his strong and timely advice at a time when I most needed it.

After me, Professor Yang also advised several talented physicists to accelerator physics. Examples abound, but comes to mind quickly are Juinn-Ming Wang, Sam Krinsky, Ron Ruth, Bill Weng, Jie Wei, Lihua Yu, Shyh-Yuan Lee, and Steve Tepekian. They all did very well. The field of accelerator physics owes much to professor Yang's early vision.

Without Yang's advice at a critical time, I would not be here today talking about SSMB.

Acknowledgments

- Reason I am here: Professor C. N. Yang
- Collaborators: Daniel Ratner and Xiaobiao Huang
- Many thanks: Claudio Pellegrini, Kwang-Je Kim, Gennady Stupakov, Juhao Wu, Zhirong Huang, Ron Ruth, Kai Tien, Bob Hettel, Marc Levenson, and Xiaozhe Shen.

This work was supported by U.S. DOE Contract No. DE-AC02-76SF00515.

References

1. D. F. Ratner and A. W. Chao, *Phys. Rev. Lett.* **105**, 154801 (2010).
2. J. Madey, *Rev. Accel. Sci. & Tech.* **3**, 1 (2010).
3. P. Emma *et al.*, *Nature Photonics* **4**(9), 641 (2010).
4. P. A. Franken, A. E. Hill, C. W. Peters and G. Weinreich, *Phys. Rev. Lett.* **7**, 118 (1961).

New Contributions to Physics by Prof. C. N. Yang: 2009–2011

Zhong-Qi Ma

Institute of High Energy Physics, Beijing 100049, P. R. China
mazq@ihep.ac.cn

In a seminal paper of 1967, Professor Chen Ning Yang found the full solution of the one-dimensional Fermi gas with a repulsive delta function interaction by using the Bethe ansatz and group theory. This work with a brilliant discovery of the Yang–Baxter equation has been inspiring new developments in mathematical physics, statistical physics, and many-body physics. Based on experimental developments in simulating many-body physics of one-dimensional systems of ultracold atoms, during a period from 2009 to 2011, Prof. Yang published seven papers on the exact properties of the ground state of bosonic and fermionic atoms with the repulsive delta function interaction and a confined potential to one dimension. Here I would like to share my experience in doing research work fortunately under the direct supervision of Prof. Yang in that period.

Keywords: Systems of bosonic and fermionic atoms; repulsive delta function interaction; confined potential to one dimension.

1. A Seminal Paper of 1967

In a seminal paper of 1967,[1] Prof. Chen Ning Yang found the full solution of the one-dimensional Fermi gas with a repulsive delta function interaction by using the Bethe ansatz and group theory. This work has been continuously inspiring new developments in physics and mathematics.

The brilliant discovery of the Yang–Baxter equation in this paper[1] has been inspiring the developments of the theory on quantum group.[2] This paper[1] has been inspiring the Yang–Yang equilibrium statistical mechanics for integrable models of interacting fermions, bosons and spins.[3] Recent experiments on ultracold bosonic and fermionic atoms confined to one dimension have provided a better understanding of the quantum statistical and dynamical effects in quantum many-body systems.[4,5]

In Yang's paper,[1] the Hamiltonian of the one-dimensional Fermi gas only with a repulsive delta function interaction:

$$H = -\sum_{i=1}^{N} \frac{\partial^2}{\partial x_i^2} + 2c \sum_{i>j} \delta(x_i - x_j). \tag{1}$$

In experiment,[4,5] the fermion has to be confined to quasi-one dimension by a trap $V(x_i)$, say, by a harmonic trap $V(x_i) = x_i^2/2$:

$$H = \sum_{i=1}^{N} \left[-\frac{\partial^2}{\partial x_i^2} + V(x_i) \right] + g \sum_{i>j} \delta(x_i - x_j). \tag{2}$$

During 2009–2011 Yang decided to study the exact properties of the ground state of N particles, bosonic and fermionic atoms, with the repulsive delta function interaction and a confined potential to one dimension as N goes to infinity, and published seven papers in Chinese Physics Letters.[6–12] It was new contribution to physics by Prof. C. N. Yang.

In a talk Prof. Yang summarized his research experiences by 3P: Perception, Persistence and Power. Here I would like to share my experience in doing research work fortunately under the direct supervision of Prof. Yang during that period, and especially to share my feeling on the 3P.

2. The Ground State Energy

Professor Yang first studied the fundamental trend of the ground state energy of fermions versus the coupling coefficient g in (2) by group theory.[6] This is the first step in a series of research works, showing his *perception*.

Assume that the system of N fermions with spin-$\frac{1}{2}$ has definite total spin $J = J_z = N/2 - M$ where the number of up-spin fermions is $N - M$ and that of down-spin is M. From group theory the space wave function of the system has the permutation symmetry denoted by Young pattern $Y = [2^M, 1^{N-2M}]$. There are two exact solutions of (2) as g goes to zero and goes to infinity[13]:

$$E = \begin{cases} N^2/4 + J^2, & g \to 0, \\ N^2/2, & g \to \infty. \end{cases} \tag{3}$$

Define the subgroup $H = S_{N-M} \otimes S_M$ of the permutation group S_N of N particles, where S_{N-M} is the permutation group of the first $N - M$ particles and S_M is that of the last M particles. Then, the whole space Ω is divided by the subgroup H:

$$\Omega = \bigcup_{R \in H} R\Omega_Y,$$

$$\Omega_Y = \left\{ \begin{array}{l} -\infty < x_1 < x_2 < \cdots < x_{N-M} < \infty, \\ -\infty < x_{N-M+1} < x_{N-M+2} < \cdots < x_N < \infty \end{array} \right\}. \tag{4}$$

If we apply the Young operator $\mathcal{Y} = \mathcal{QP}$ where $\mathcal{P}$ is the symmetrizer and $\mathcal{Q}$ is the antisymmetrizer, the space wave function with the permutation symmetry Y has to be zero at the boundary of Ω_Y. Thus Prof. Yang established two theorems by group theory.[6]

Theorem 2.1. *For any value of g, consider two different eigenvalue problems:*
  (A) *Hamiltonian H in (2) with symmetry Y in full space* ∞^N;
  (B) *Hamiltonian H in (2) in region* Ω_Y *with the boundary condition that the wave function vanishes on its surface.*
  The eigenvalues of the two problems are identical, and the corresponding unnormalized wave function are proportional in region Ω_Y.

Theorem 2.2. *For any value of g, the ground state wave function for problem B has no zeros in the interior of* Ω_Y, *and is not degenerate.*

From those two theorems and the well-known Sturm–Liouville theorem, one comes to the conclusion that the eigenvalue of ground state energy increases monotonically as g increases from zero to infinity:

$$\frac{N^2}{4} + J^2 \to \frac{N^2}{2}. \tag{5}$$

Due to the exact solution of one pair of fermions with total spin-0 for large negative g (say, see Ref. 14), where the total energy is $-g^2/4$, it is reasonable to estimate the ground state energy to be $-Mg^2/4$ for large N as $g \to -\infty$.[7] Noting (3) and $M/N = 1/2 - J/N$, we have

$$\frac{E}{N^2} = \begin{cases} \dfrac{1}{2}, & g \to \infty, \\ \dfrac{1}{4} + \left(\dfrac{J}{N}\right)^2, & g \to 0, \\ -\dfrac{1}{4}\left(\dfrac{1}{2} - \dfrac{J}{N}\right)\left(\dfrac{g}{\sqrt{N}}\right)^2, & g \to -\infty. \end{cases} \tag{6}$$

Then, the behaviors of physical quantities as N goes to infinity are as follows:

$$E \sim N^2, \quad J \sim N, \quad g \sim \sqrt{N}, \quad \text{as } N \to \infty. \tag{7}$$

Thus, in terms of group theory, Prof. Yang obtained the curves of the ground state energy E/N^2 versus $g/\sqrt{N}$ for different J/N even though the curves in detail have to be studied further. Those curves with different J/N are never intersection due to the Lieb–Mattis Theorem[15]: For the Hamiltonian (2) the ground state energy E increases as the spin J increases.

3. The Thomas–Fermi Method

From his *perception*, Prof. Yang proved that for a given J/N the ground state energy E/N^2 increases monotonically from $1/4 + (J/N)^2$ to $1/2$ as $g/\sqrt{N}$ increases from zero to infinity, and those curves with different J/N are never intersection. The remaining problem is to calculate the energy E/N^2 for any coupling coefficient

$g/\sqrt{N}$ with given J/N. It is a difficult task. Prof. Yang and I have made the *consistent* effort for solving this task. There were a great deal of calculations by Yang's own hand. At last, we made a conjecture: The Thomas–Fermi method gives *the correct limit of energy* as $N \to \infty$.

By the Thomas–Fermi method, in a given small interval dx there is no traping potential and the total energy E for the ground state of H in (2) is denoted by

$$E = \int e_1\, dx + \int \rho(x) V(x) dx\,, \qquad e_1 = \rho g^2 \zeta\left(\frac{\rho}{g}\right), \qquad (8)$$

where the result for the trapless fermions given in 1967[1] can be used at the interval dx such that the length of interval $L \to dx$, the energy $\mathcal{E} \to e_1(x) dx$, and the particle number $\mathcal{N} \to \rho(x) dx$. The approximate analytic form $e_1(x)$ is simulated numerically from the Fredholm equation in Ref. 1 by Dr. Y. Z. You.[16] Varying $\rho(x)$ to obtain the minimum for the total energy E, under the condition that $\int \rho\, dx$ is fixed, we[10] are able to calculate E/N^2 for $g/\sqrt{N}$ with a given J/N. Given enough computer time, all calculation errors can be made smaller than any pre-assigned value. In this sense we have solved the problem of calculating the exact limit of E/N^2 as $N \to \infty$ at fixed $g/\sqrt{N}$ and J/N. However, is this conjecture true?

Fortunately, there is an exact solution of the fermion system (2) in a ferromagnetic state ($M = 0$ and $J = N/2$) for arbitrary g and N:

$$\Psi = \psi_A(x_1, x_2, \ldots, x_N) \chi_1(1) \cdots \chi_1(N)\,,$$
$$\psi_A(x_1, x_2, \ldots, x_N) = \det[u_0(x_1) u_1(x_2) \cdots u_{N-1}(x_N)]\,, \qquad (9)$$
$$J = N/2\,, \qquad E = N^2/2\,, \qquad \rho_N = \sum_{n=0}^{N-1} |u_n(x)|^2\,,$$

where χ_1 is the up-spinor, $u_n(x)$ denotes the nth excited state in the harmonic oscillator potential, and $\rho(x)$ is the density function. Comparing the density function calculated by the Thomas–Fermi method as N goes to infinity with that of the exact solution shows[7] that the Thomas–Fermi method does give the correct limit of energy for the ferromagnetic state as $N \to \infty$. In the comparison a trick is needed to avoid infinity.[7] Thus, it is convinced that the conjecture holds for the case with any total spin J.

In terms of the Thomas–Fermi method, Prof. Yang and I calculated the exact limit of the ground state energy E/N^2 of the fermion system in one dimension as $N \to \infty$ at fixed $g/\sqrt{N}$ and found that there are kinks in the density distribution $\rho(x)$ where the total spin J/N is set the values between 0 and 0.5 (see Fig. 3(c) in Ref. 10). Perhaps these is possible to achieve experimentally with magneto-optical technologies.

4. w-Component Bosons in 1D

Discuss w-component fermions and bosons with repulsive delta function interaction (1) in 1D. Lieb and Liniger[17] solved the problem for spinless bosons in 1963. Yang[1] solved the problem for two-component fermions in 1967. Sutherland[18] generalized it to three-component fermions in 1968 and the case for w-component fermions is straightforward. How to solve the problem for w-component bosons?

For the balance cases of 1D w-component fermions and bosons with repulsive delta function interaction (1), the space wave function of the system has the permutation symmetry denoted by Young patterns $Y = [w^{N/w}]$ and $[(N/w)^w]$ (see Fig. 1).

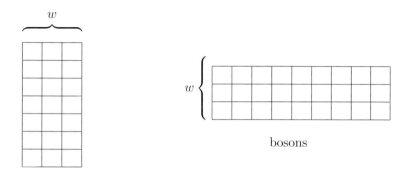

Fig. 1. The Young patterns $Y = [w^{N/w}]$ and $[(N/w)^w]$.

Prof. Yang and Dr. You[12] define

$$Y_{Fw} = \left(\frac{3L^2}{\pi^2 N^3}\right) E_{Fw}, \qquad Z = \left(\frac{N}{L}\right) c, \qquad (10)$$

for the system (1) of N fermions in the balanced case where N is large. Similarly, define Y_{Bw} for the system (1) of N bosons.

The second Lieb–Mattis theorem[15] shows that if Hamiltonian has no spin- or velocity-dependent forces, the ground state energies E and E' respectively correspond to Young diagrams Y and Y' with the same boxes satisfy

$$E \leq E', \quad \text{if } \sum_{i=1}^{a} n_i \geq \sum_{i=1}^{a} n'_i, \quad a = 1, 2, 3, \ldots, \qquad (11)$$

where n_i and n'_i are the number of the ith column of Y and Y', respectively. Thus, Prof. Yang and Dr. You[12] obtain

$$Y_{F1}(Z) \geq Y_{F2}(Z) \geq Y_{F3}(Z) \geq \cdots \geq Y_{B3}(Z) \geq Y_{B2}(Z) \geq Y_{B1}(Z). \qquad (12)$$

It is proved from the Fredholm equation[12] that

$$\lim_{w \to \infty} Y_{Fw}(Z) = Y_{F\infty}(Z) = Y_{B1}. \qquad (13)$$

Therefore,
$$Y_{F\infty}(Z) = Y_{Bw}(Z) = Y_{B1}. \qquad (14)$$

5. Conclusion

Based on experimental developments in simulating many-body physics of one-dimensional systems of ultracold atoms, Prof. Yang made new contribution to physics by publishing seven papers during a period from 2009 to 2011. From my experience in doing research work under the direct supervision of Prof. Yang in that period, I realized the 3P in Prof. Yang's research. We also see that the method of group theory is a very good method: Simple and clear in physical meaning.

Acknowledgments

The author would like to thank Prof. K. K. Phua and the Organizing Committee for the invitation to the Conference on 60 Years of Yang–Mills Gauge Field Theories and the financial support during the conference.

References

1. C. N. Yang, *Phys. Rev. Lett.* **19**, 1312 (1967).
2. L. D. Faddeev, N. Y. Reshetikhin and L. A. Takhtajan, Quantization of Lie groups and Lie algebras, in *Algebraic Analysis* (Academic Press, 1988), p. 129.
3. C. N. Yang and C. P. Yang, *J. Math. Phys. (N. Y.)* **10**, 1115 (1969).
4. V. A. Yurovsky, M. Olshanii and D. S. Weiss, *Adv. At. Mol. Opt. Phys.* **55**, 61 (2008).
5. Y. Liao, A. S. C. Rittner, T. Paprotta, W. Li, G. B. Partridge, R. G. Hulet, S. K. Baur and E. J. Mueller, Nature*Nature* **467**, 567 (2010).
6. C. N. Yang, *Chin. Phys. Lett.* **26**, 120504 (2009).
7. Z. Q. Ma and C. N. Yang, *Chin. Phys. Lett.* **26**, 120505 (2009).
8. Z. Q. Ma and C. N. Yang, *Chin. Phys. Lett.* **26**, 120506 (2009).
9. Z. Q. Ma and C. N. Yang, *Chin. Phys. Lett.* **27**, 020506 (2010).
10. Z. Q. Ma and C. N. Yang, *Chin. Phys. Lett.* **27**, 080501 (2010).
11. Z. Q. Ma and C. N. Yang, *Chin. Phys. Lett.* **27**, 090505 (2010).
12. C. N. Yang and Y. Z. You, *Chin. Phys. Lett.* **28**, 020503 (2011).
13. L. Guan, S. Chen, Y. Wang and Z. Q. Ma, *Phys. Rev. Lett.* **102**, 160402 (2009).
14. J. Y. Zeng, *Quantum Mechanics*, Vol. 1, §3.5.2, 4th edn. (Science Press, 2007) (in Chinese).
15. E. Lieb and D. Mattis, *Phys. Rev.* **125**, 164 (1962).
16. Y. Z. You, *Chin. Phys. Lett.* **27**, 080305 (2010).
17. E. Lieb and W. Liniger, *Phys. Rev.* **130**, 1605 (1963).
18. B. Sutherland, *Phys. Rev. Lett.* **20**, 98 (1968).

Brief Overview of C. N. Yang's 13 Important Contributions to Physics

Yu Shi

Department of Physics, Fudan University, Shanghai 200433, China
yushi@fudan.edu.cn

We give a brief overview of Professor Chen Ning Yang's 13 important contributions to physics, especially his contributions to gauge theory. The great impact of his relevant papers is analyzed. Commentary is made on Yang's distinctive style and trailblazing role in the history of physics.

1. Introduction

Professor Chen Ning Yang published two volumes of *Selected Papers with Commentaries* in 1983 and 2014 respectively.[1,2] In the preface of the first volume, Yang quoted the following two lines from Tu Fu (712–770):

> *A piece of literature*
> *Is meant for the millennium*
> *But its ups and downs are known*
> *Already in the author's heart*

On this volume of Yang, Freeman Dyson commented[3]:

> *Some of the chosen papers are important and others are unimportant. Some are technical and others are popular. Every one of them is a gem. Frank was not trying to cram as much hard science and possible into five hundred pages. He was trying to show us in five hundred pages the spirit of a great scientist, and he magnificantly succeeded.*

A question arises then: among all these papers, which are most important? This question was answered by a selection of 13 contributions carved on a black cube, which was a gift for Yang's 90th birthday in 2012 (Fig. 1). On the top of the cube, the above quote of Tu Fu is engraved. On the four vertical surfaces are listed Yang's 13 important contributions to four areas of physics: statistical mechanics, condensed matter physics, particle physics and field theory.

Fig. 1. The black cube engraved with C. N. Yang's 13 contributions in four areas of physics. Upper picture: front and right sides. Lower picture: back and left sides.

Among the 13 contributions, those on statistical mechanics and condensed matter physics are,

- 1952 Phase Transition,
- 1957 Bosons,
- 1961 Flux Quantization,
- 1962 Off-diagonal Long-range Order,

- 1967 Yang–Baxter Equation,
- 1969 Finite Temperature,

the contributions to particle physics phenomenology are

- 1956 Parity Nonconservation,
- 1957 T, C and P,
- 1960 Neutrino Experiment,
- 1964 CP Nonconservation,

and the contributions to field theory are

- 1954 Gauge Theory,
- 1974 Integral Formalism,
- 1975 Fiber Bundle.

A commentary on these 13 contributions was recently made in considering their historic contexts in physics, their scientific beauty, and the taste and style of Yang.[4] Here we make some analyses on their impacts on the development of the relevant subjects. We make use of the citation database in *Web of Science*. Note that all the citation data are as of May 21, 2015.

2. Citation Analysis

Yang's 1952 papers on phase transitions include one on magnetization of the two-dimensional Ising model,[5] as well as two on the lattice gas model with T. D. Lee.[6,7] The total number of citations to these three papers are 698, 1181 and 1337, respectively. The number of citations to these three papers in total each year is shown in Fig. 2. One can observe that there was a jump in 1960s, and that it has

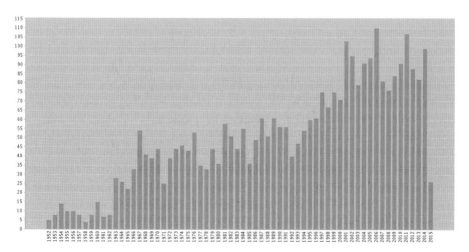

Fig. 2. The number of citations each year to Yang's three papers on phase transition of 1952. The sum is 3216.

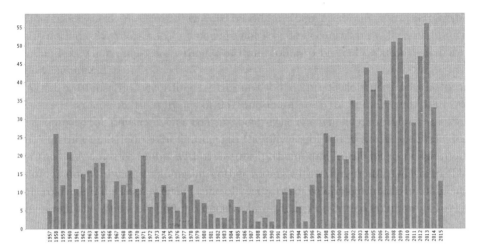

Fig. 3. The number of citations each year to Yang's two 1957 papers on dilute Bose gases. The sum is 1033.

almost kept increasing up to now. This is an "anomaly" in comparison with the usual trend that the number of citations each year eventually decreases with time.

Yang's 1957 papers on Bosons with his collaborators Kerson Huang and T. D. Lee[8,9] were among a series of papers on the dilute hard sphere Bose gas, which they chose as a mathematically well defined model yet motivated by the superfluidity of liquid helium. The total number of citations to these two papers is 643 and 390, respectively. The number of citations to these two papers in total each year is shown in Fig. 3. One can observe that it has kept increasing since the 1990s, apparently because of experimental breakthroughs brought about by modern techniques of cooling and trapping atoms.

Yang's 1961 paper with Byers on quantization of the magnetic flux trapped in a superconducting ring was a result of his interaction with solid state experimentalists at Stanford University.[10] The total number of citations to this paper is 702. The number of citations to this paper each year is shown in Fig. 4. One can see that there was a peak immediately following the publication, and there was a much more pronounced peak in 1990s and a plateau ever since, indicating great advance many years after the publication.

In 1962, Yang published his paper on off-diagonal long-range order as a unified framework of superfluidity and superconductivity.[11] The total number of citations to this paper is 932. The number of citations to this paper each year is shown in Fig. 5. It can be seen that there was a peak around 1968, moreover, there has been a plateau since the late 1980s, indicating much activity on quantum condensation phenomena in more recent years.

In 1967, after moving from Institute for Advanced Study at Princeton to Stony Brook, Yang published his paper on what has later been called Yang–Baxter equation.[12] As a theory of one-dimensional Fermions, it also provided a basis for

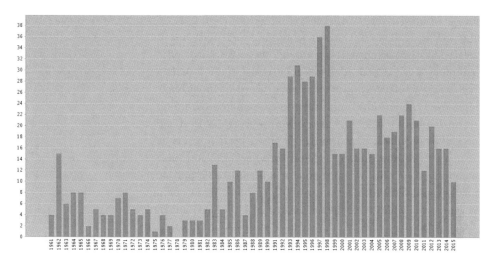

Fig. 4. The number of citations each year to Yang's 1961 paper on flux quantization of a superconducting ring. The sum is 702.

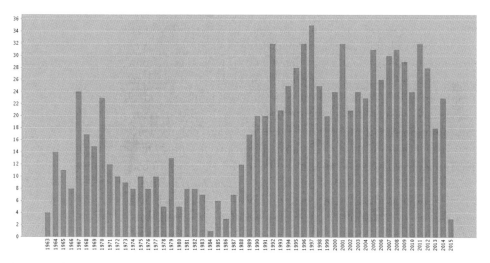

Fig. 5. The number of citations each year to Yang's 1964 paper on off-diagonal long-range order. The sum is 932.

analyzing many one-dimensional cold atom experiments of recent years. The total number of citations to this paper is 1393. The number of citations to this paper each year is shown in Fig. 6. It can be seen that there was a peak in 1980s and a much more pronounced peak in 1990s, and it has steadily increased since 2000.

In 1969, Yang published the exact solution of Bosons in one-dimensional repulsive potential at finite temperatures, with his brother C. P. Yang.[13] The model and its solution have also been realized and confirmed in cold atom experiments in recent years. The total number of citations to this paper is 817. The number of

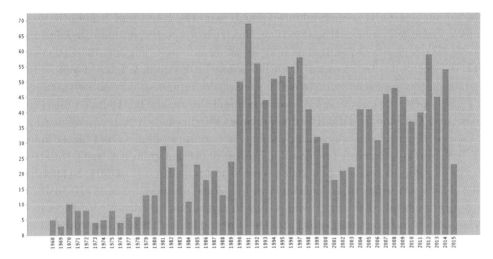

Fig. 6. The number of citations each year to Yang's 1967 paper on what has later been called Yang-Baxter equation. The sum is 1393.

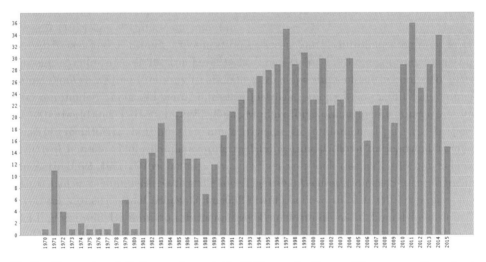

Fig. 7. The number of citations each year to the 1969 paper by Yang and Yang on one-dimensional Bosons at finte temperatures. The sum is 817.

citations to this paper each year is shown in Fig. 7. One can observe peaks in 1971, in 1980s, in 1990s, and in recent years. The peaks have become more and more pronounced.

Now we turn to Yang's papers on particle physics phenomenology. Yang's 1956 paper with T. D. Lee famously questioning whether parity is conserved in weak interactions earned them Nobel Prize in physics the very next year.[14] The speed of recognition is the highest one in all history of Nobel Prizes. The total number of citations to this paper is 1358. The number of citations each year is shown in Fig. 8.

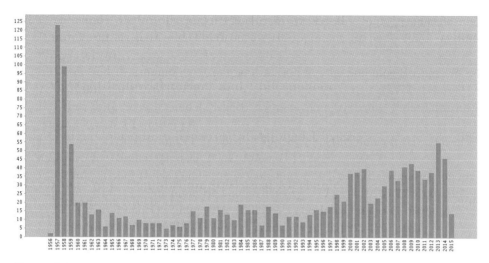

Fig. 8. The number of citations each year to the 1956 paper by Yang and Lee on parity violation in weak interactions. The sum is 1358.

The extremely high peak in 1957 was obviously because of the parity revolution. Interestingly, the citation each year has started increasing again since the 1990s.

In 1957, with Lee and R. Oehme, Yang published the paper on the relationship among violations of the discrete symmetries of time reversal (T), charge conjugation (C) and parity (P).[15] This paper had decisive impact on all theoretical analyses on CP violation later discovered in 1964. The total number of citations to this paper is 346. The number of citations to this paper each year is shown in Fig. 9. One can observe peaks following the parity revolution in 1957, and following the 1964 CP violation discovery, as well as in the 2000s.

In 1960, following the experimentalist M. Schwartz's idea that weak interactions at high energies can be studied by using neutrino beams, Lee and Yang explored theoretically the importance of such experiments,[16] which led to future development in neutrino physics. The total number of citations to this paper is 218. The number of citations to this paper in each year is shown in Fig. 10. There was a pronounced high peak immediately following the publication of this paper.

In 1964, Yang and T. T. Wu published their phenomenological analysis of the violation of CP invariance following its the experimental discovery that year.[17] The number of citations to this paper is 278. The number of citations to this paper each year is shown in Fig. 11. There was a pronounced high peak immediately following the publication of this paper.

Now we come to Yang's greatest contribution to physics, that is, Yang–Mills gauge theory or nonabelian gauge theory, which was founded in two short papers of 1954.[18,19] The first paper, which is less well known, was an abstract of Yang's talk at the April meeting of the American Physical Society. The total numbers of citations to these two papers are 19 and 1740 respectively. The number of citations to these

512 Y. Shi

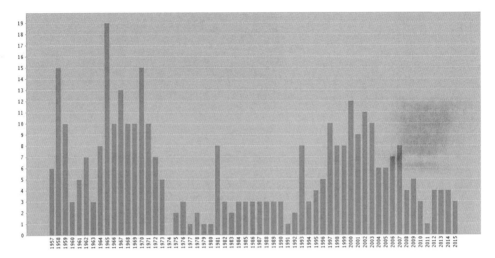

Fig. 9. The number of citations each year to the 1957 paper by Yang with Lee and Oehme on relationship among violations of T, C, and P in weak interactions. The sum is 346.

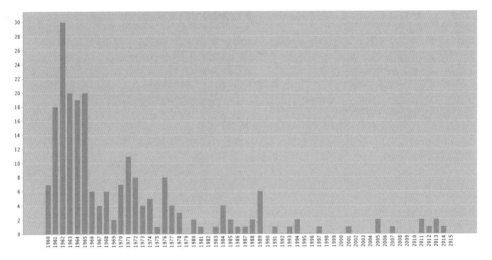

Fig. 10. The number of citations each year to the 1960 paper by Lee and Yang on high energy neutrinos. The sum is 218.

two papers in total each year is shown in Fig. 12. One can observe a huge peak in the 1970s, and it has kept increasing since 1990s. However, one should be cautioned that the original papers are often not cited in referring to Yang–Mills theory, just like that Einstein's 1905 paper is usually not cited in referring to special relativity. In *Web of Science*, for example, on May 21 2015, searching "Yang Mills" in titles returned 5101 papers while searching "Yang Mills" as a topic returned 14967 papers.

In 1974, Yang published his paper on the integral formalism of gauge theory.[20] The total number of citations to this paper is 396. The number of citations to this

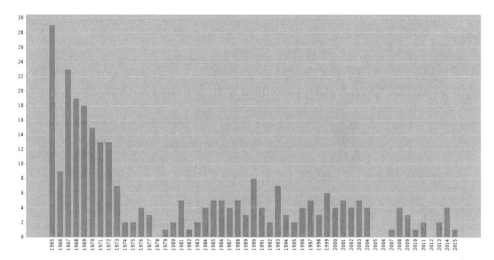

Fig. 11. The number of citations each year to the 1964 paper by Yang and Wu on CP violation. The sum is 278.

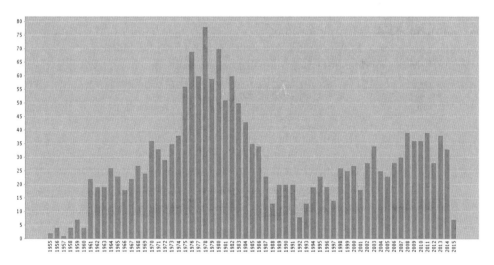

Fig. 12. The number of citations each year to the original 1954 papers by Yang and Mills on nonabelian gauge theory. The sum is 1759.

paper each year is shown in Fig. 13. There was a pronounced peak following the publication of this paper.

In 1975, Yang and T. T. Wu published the paper on the correspondence between gauge theory and fiber bundle theory,[21] which inspired the subsequent reunion between mathematics and theoretical physics in last forty years. The number of citations to this paper is 993. The number of citations to this paper each year is shown in Fig. 14. There were two broad peaks in 1980s and in 2000s, without significant decay in between.

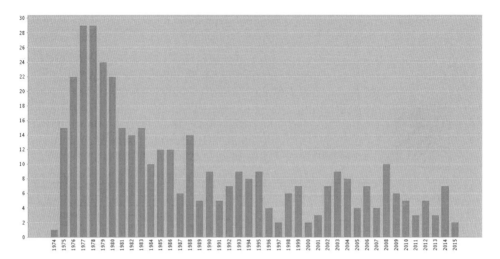

Fig. 13. The number of citations each year to the 1974 paper by Yang on integral formalism of gauge theory. The sum is 396.

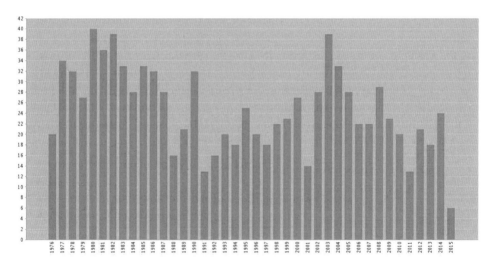

Fig. 14. The number of citations each year to the 1975 paper by Yang and Wu on the correspondence between gauge theory and fiber bundle theory. The sum is 993.

3. Two Types of Impact

We can classify the impact of these 13 contributions into two types.

Some of Yang's contributions were quickly confirmed by experiments or followed by others works, as indicated by immediate citation peaks. Moreover, citations may revive later because of new developments. Among the 13 important contributions, the following belong to this type:

- 1956 Parity nonconservation.
- 1957 T, C and P.
- 1960 Neutrino experiments.
- 1964 CP nonconservation.
- 1974 Integral formalism.
- 1975 Fiber bundle.

For the other nine important contributions, the importance may or may not have been recognized immediately following the publication, but they all received much more recognition after many years' further development. These contributions are:

- 1952 phase transition.
- 1957 Bosons.
- 1961 Flux quantization.
- 1962 ODLRO.
- 1967 Yang–Baxter equation.
- 1969 Finite temperature.
- 1954 Yang–Mills gauge theory.

Amongst these nine contributions, there are two extreme cases. For the 1952 phase transition, the number of citations each year has almost been increasing monotonically. For 1954 Yang–Mills gauge theory, many authors do not cite the original papers in referring to Yang–Mills theory.

Finally, we note that up to May 21, 2015, the total number of citations to the 16 papers representing the 13 contributions is 13867. The total number of citations to all the papers of Yang is 27124, with h-index 68. Moreover, at least 46 papers have been cited in 2015, including 15 among the 16 papers representing the 13 contributions. This indicates that the choice of the 13 contributions is indeed a good one.

The total number of citations each year to 13 contributions is shown in Fig. 15, while the total number of citations each year to all the papers by Yang is shown in Fig. 16. The two diagrams are similar in trend, indicating that the 13 contributions are indeed representative. In each of these two diagrams, there was a peak in 1957 because of parity revolution. Note that the number of citations each year is really the rate of the change of the total citations in all years up to that year. Thus its increase indicates the acceleration of the total number of citations up to that year. What is the "energy" driving this acceleration?

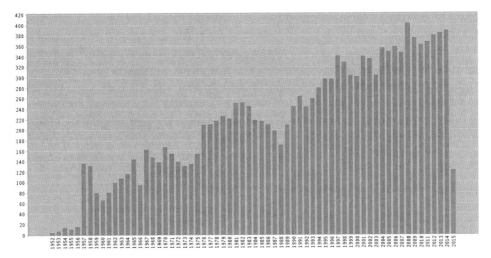

Fig. 15. The number of citations each year to the 16 papers representing the 13 contributions of by Yang. The sum is 13867.

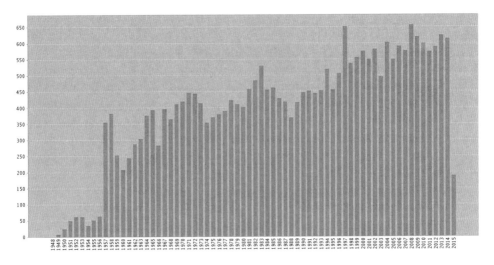

Fig. 16. The number of citations each year to all the papers of Yang. The sum is 27124.

4. Characteristic of Yang's Work

This leads us to the characteristics of Yang's work. Yang's work is distinguished by distinctive style and taste, which can be summarized as: *originality, elegance, power and physical relevance*. Yang has always paid great attention to experimental facts, basic structure of physics and beauty of theoretical form. He has always kept close to his own thoughts, ideas and intuition, rather than following fashions or jumping on bandwagons.

In the preface to his recently published collection of articles,[22] Dyson elaborates that his 1999 paper quoted above[3]

> *is a banquet speech, celebrating my friend Yang Chen Ning on the occasion of his retirement from Stony Brook in 1999. I called attention to three outstanding qualities of Yang that are rarely combined. First, a marvellous mathematical skill, enabling him to solve technical problems. Second, a deep understanding of nature, enabling him to ask important questions. Third, a community spirit, enabling him to play a major role in the rebirth of Chinese civilization. Together, these three qualities make him what he is, a conservative revolutionary who values the past and leads the way to the future.*

5. Yang–Mills Theory in Nobel Lectures

Yang–Mills theory, together with the ideas of spontaneous symmetry breaking and asymptotic freedom developed later by many authors, led to the Standard Model of particle physics, which has dominated research in fundamental physics in all subsequent years. The two 1954 papers by Yang and Mills opened the door of the principle *"Symmetry Dictates Interaction"*, which has been the spirit of the main conceptual advance of theoretical physics in the last half century, and will continue to be general guidance for future development.

The building of the Standard Model has been recognized by many Nobel Prizes in physics. In the following, we make some quotations from the Nobel Lectures of these prizes.[23]

In his 1979 Nobel Lecture, Glashow said:

> *Today we have what has been called a "standard theory" of elementary particle physics in which strong, weak, and electromagnetic interactions all arise from a local symmetry principle.... Not only is electromagnetism mediated by photons, but it arises from the requirement of local gauge invariance. This concept was generalized in 1954 to apply to non-Abelian local symmetry groups.*

In his 1979 Nobel Lecture, Weinberg said:

> *The extension to more complicated groups was made by Yang and Mills in 1954 in a seminal paper in which they showed how to construct an SU(2) gauge theory of strong interactions.... To a remarkable degree, our present detailed theories of elementary particle interactions can be understood deductively, as consequences of symmetry principles and of a principle of renormalizability which is invoked to deal with the infinities.*

In his 1999 Nobel Lecture, Veltman recalled:

I therefore concluded that Yang–Mills theories are probably the best one can have with respect to renormalizability.... Starting then on the study of diagrams in a Yang–Mills theory I established the vanishing of many divergencies, provided the external legs of the diagrams were on the mass shell.

In his 1999 Nobel Lecture, 't Hooft recalled:

Back in 1971, I carried out my own calculations of the scaling properties of field theories, and the first theory I tried was Yang–Mills theory... Quantum chromodynamics, a Yang–Mills theory with gauge group $SU(3)$, could therefore serve as a theory for the strong interactions.

't Hooft's prize-winning paper is entitled "Renormalizable Lagrangians for massive Yang–Mills fields".

In his 1999 Nobel Lecture, Gross recalled:

Gerard 't Hooft's spectacular work on the renormalizability of Yang–Mills theory reintroduced non-Abelian gauge theories to the community ... We decided that we would calculate the β function for Yang–Mills theory ... Politzer went ahead with his own calculation of the function for Yang–Mills theory ... Our abstract reads: It is shown that a wide class of non-Abelian gauge theories have, up to calculable logarithmic corrections, free-field asymptotic behavior ... His abstract reads: An explicit calculation shows perturbation theory to be arbitrarily good for the deep Euclidean Greens functions of any Yang–Mills theory and of many Yang–Mills theories with fermions...

In his 2008 Nobel Lecture, Nambu commented:

Thus the beautiful properties of electromagnetism was extended to the $SU(2)$ non-Abelian gauge field.

In his 2008 Nobel Lecture, Kobayashi commented:

In the framework of the gauge theory, flavor mixing arises from mismatch between gauge symmetry and particle states.

In his 2008 Nobel Lecture, Maskawa recalled:

We started to study the CP violation based on the quartet quark model in the electroweak unified gauge theory.

In his 2013 Nobel Lecture, Englert said:

As a valid theory of short range interactions clearly required quantum consistency, we were naturally driven to take, as a model of the corresponding long range interactions, the generalization of quantum electrodynamics, known as Yang–Mills theory... To transmute long range interactions into short range ones in the context of Yang–Mills theory it would suffice to give these generalized photons a mass, a feature that, as we just indicated, is apparently forbidden by the local symmetries... The disappearance of the NG boson is thus an immediate consequence of local symmetry... Therefore the coupling of the would be NG boson to the gauge field must render the latter massive. This is the essence of the BEH mechanism.

In his 2013 Nobel Lecture, Higgs recalled:

Anderson remarked "The Goldstone zero-mass difficulty is not a serious one, because we can probably cancel it off against an equal Yang–Mills zero-mass problem."... Schwinger had, as recently as 1962, written papers in which he demolished the folklore that it is gauge invariance alone that requires photons to be massless. He had provided examples of some properties of a gauge theory containing massive "photons"... During the weekend 18–19 July it occurred to me that Schwingers way of formulating gauge theories undermined the axioms which had been used to prove the Goldstone theorem. So gauge theories might save Nambu's programme.

6. Conclusion

In 2009, Dyson observed[24]:

Yang took Weyl's place as the leading bird among my generation of physicists... With non-Abelian gauge fields generating nontrivial Lie algebras, the possible forms of interaction between fields become unique, so that symmetry dictates interaction. This idea is Yang's greatest contribution to physics. It is a contribution of a bird flying high over the rain forest of little problems in which most of us spend our lives.

In 1999, Dyson assessed[3]:

The discovery of non-Abelian gauge fields was a laying of foundations for new intellectual structures that have taken 30 years to build. The nature of matter as described in modern theories and confirmed by modern experiments is a soup of non-abelian gauge fields, held together by the mathematical symmetries that Yang first conjectured 45 years ago... Professor Yang is, after Einstein and Dirac, the preeminent stylist of the 20th century physics... He is a conservative revolutionary.

Let me conclude by adapting the two lines of Tu Fu:

Yang–Mills Theory
Is meant for the millennium
But its ups and downs are known
Already in the authors' heart

Acknowledgment

I thank Professor Chen Ning Yang for enlightening discussions.

References

1. C. N. Yang, *Selected Papers 1945–1980 with Commentary* (W. H. Freeman and Company, 1983); 2005 Edition (World Scientific, Singapore, 2005).
2. C. N. Yang, *Selected Papers II with Commentary* (World Scientific, Singapore, 2013).
3. F. Dyson, A conservative revolutionary, *Int. J. Mod. Phys. A* **14**, 1455 (1999); also in A. Goldhaber, R. Shrock, J. Smith, G. Sterman, P. van Niuwenhuizen and W. Weisberger (eds.), *Symmetry and Modern Physics* (World Scientific, Singapore, 2003).
4. Y. Shi, *Int. J. Mod. Phys. A* **29**, 1475001 (2014). The author's talk at the Conference on 60 Years of Yang–Mills Gauge Field Theories was based partly on this article and partly on the present one.
5. C. N. Yang, The spontaneous magnetization of a two-dimensional Ising model, *Phys. Rev.* **85**, 808 (1952).
6. C. N. Yang and T. D. Lee, Statistical theory of equations of state and phase transitions. I. Theory of condensation, *Phys. Rev.* **87**, 404 (1952).
7. T. D. Lee and C. N. Yang, Statistical theory of equations of state and phase transitions. II. Lattice gas and Ising model, *Phys. Rev.* **87**, 410 (1952).
8. T. D. Lee and C. N. Yang, Many-body problem in quantum mechanics and quantum statistical mechanics, *Phys. Rev.* **105**, 1119 (1957).
9. T. D. Lee, K. Huang, and C. N. Yang, Eigenvalues and eigenfunctions of a Bose system of hard spheres and its low-temperature properties, *Phys. Rev.* **106**, 1135 (1957).
10. N. Byers and C. N. Yang, Theoretical considerations concerning quantized magnetic flux in superconducting cylinders, *Phys. Rev. Lett.* **7**, 46 (1961).
11. C. N. Yang, Concept of off-diagonal long-range order and the quantum phases of liquid He and of superconductors, *Rev. Mod. Phys.* **34**, 694 (1962).
12. C. N. Yang, Some exact results for the many-body problem in one dimension with repulsive delta-function interaction, *Phys. Rev. Lett.* **19**, 1312 (1967).
13. C. N. Yang and C. P. Yang, Thermodynamics of a one-dimensional system of bosons with repulsive delta-function interaction, *J. Math. Phys.* **10**, 1115 (1969).

14. T. D. Lee and C. N. Yang, Question of parity conservation in weak interactions, *Phys. Rev.* **104**, 254 (1956).
15. T. D. Lee, R. Oehme, and C. N. Yang, Remarks on possible noninvariance under time reversal and charge conjugation, *Phys. Rev.* **106**, 340 (1957).
16. T. D. Lee and C. N. Yang, Theoretical discussions on possible high-energy neutrino experiments, *Phys. Rev. Lett.* **4**, 307 (1960).
17. T. T. Wu and C. N. Yang, Phenomenological analysis of violation of CP invariance in decay of K^0 and $\bar{K}^0$, *Phys. Rev. Lett.* **13**, 380 (1964).
18. C. N. Yang and R. Mills, Isotopic spin conservation and a generalized gauge invariance, *Phys. Rev.* **95**, 631 (1954).
19. C. N. Yang and R. L. Mills, Conservation of isotopic spin and isotopic gauge invariance, *Phys. Rev.* **96**, 191 (1954).
20. C. N. Yang, Integral formalism for gauge fields, *Phys. Rev. Lett.* **33**, 445 (1974).
21. T. T. Wu and C. N. Yang, Concept of nonintegrable phase factors and global formulation of gauge fields, *Phys. Rev. D* **12**, 3845 (1975).
22. F. J. Dyson, *Birds and Frogs (Selected Papers, 1990–2014)* (World Scientific, Singapore, 2015).
23. http://nobelprize.org
24. F. Dyson, Birds and frogs, *Notices of AMS* **56**(2), 212 (2009).